R.G. Behrens 1988

Mass Transport Phenomena

CHRISTIE J. GEANKOPLIS, The Ohio State University

Mass Transport Phenomena

Holt, Rinehart and Winston, Inc.
New York Chicago San Francisco Atlanta Dallas Montreal Toronto London Sydney

Copyright © 1972 by Holt, Rinehart and Winston, Inc.
All rights reserved
Library of Congress Catalog Card Number: 79-154348
ISBN: 0-03-085233-1
Printed in the United States of America
2 3 4 5   038   9 8 7 6 5 4 3 2 1

*To my father and mother
for their concern and encouragement
And also to my brother, Deno*

Preface

During recent years the field of mass transport has broadened and spread to many engineering and applied science disciplines. Once concerned mainly with gases and liquids, it now includes solids and combinations of fluids and solids.

This text provides a unified and broad treatment in some depth of mass transport phenomena in gases, liquids, and solids, and it relates this field, for better understanding, to momentum and heat transport. It delves into the scientific principles of mass transport, emphasizing the mathematical-analytical approach. It is not intended as a design text, but numerous examples are treated to emphasize the use of the theory.

Recently there has been a marked increase in sophistication in the mathematical treatment of mass transport. This text uses modern analytical and machine computational tools, including vectors, Laplace transforms, Fourier series, numerical methods, difference equations, analog computation, and digital computation.

The text assumes only an elementary knowledge of digital machine programming. It includes actual Fortran programs for typical problems but requires no previous knowledge of analog computer programming. An introduction to analog computational methods is given, and many sample problems are solved. Elementary analog computation should help the reader to better understand the basic physical and computational phenomena.

The machine and analog computational methods are integrated in the text, as are all of the mathematical-analytical techniques used, so that the reader need not consult other sources to obtain a background. This should be quite helpful to the practicing engineer or scientist who is using this text to advance his knowledge of the field.

As to instructional level, this book is designed for use of last-semester juniors or seniors and for beginning graduate students in engineering and applied science. The student will find it helpful but not necessary to have had an introduction to transport phenomena and/or a background in momentum and heat transport. Parts of Chapters 1 and 2 do review the fundamentals of transport phenomena for those with no background. This text is also suited for use in the last third of the undergraduate transport-phenomena sequence of momentum, heat, and mass usually taught to chemical engineers. The entire book, however, is too lengthy to be covered completely in a one-quarter or one-semester course. At the option of the reader or instructor various sections can be omitted without markedly harming the continuity.

Chapter 1 integrates momentum, heat, and mass transport and gives the basic background for mass transport. Sections 1.1 to 1.6 introduce the elementary molar flux equation; then successive sections introduce the concepts of a molar flux relative to a moving fluid and of other types of fluxes. Chapter 2 gives the basic equations of change, which may be a review for chemical engineering students and a partial review for some students who have already had some momentum and heat transport. Different instructors may select different chapters, since some may be omitted without loss of continuity as indicated below.

The following chapter sequences are suggested to the instructor. For chemical engineering, use Chapter 1, Sections 2.1 and 2.5 to 2.9 in Chapter 2, and Chapters 3, 4, 6, 7, and 8, with Chapters 5, 9, and 10 being optional. For certain other fields such as mechanical engineering, other engineering, and applied science, one might use Chapters 1, 2, 4, 5, and 6, with Chapter 10 being optional, depending on whether one wants to cover analog computation. The selected problems on digital machine computation and analog computation in the various chapters are so designed that their omission at the discretion of the instructor or reader will not affect the continuity of the chapters.

In some cases the text briefly introduces new and advanced topics without going into great detail — for example, in the section on irreversible thermodynamics. The purpose is to make the reader aware of new areas that are being developed. References for further study also are included.

I am grateful to my associates here at The Ohio State University for their many discussions with me and especially to J. H. Koffolt and Aldrich Syverson for their help and encouragement. Also, I am indebted to A. N. Hixson, University of Pennsylvania, for his teaching me mass transport as a graduate student.

Columbus, Ohio
September 1971

Christie J. Geankoplis

Contents

1   Molecular Mass Transport Phenomena in Fluids   1

- **1.1** INTRODUCTION TO TRANSPORT PHENOMENA   1
  A. Purpose   B. General Transport Equation
- **1.2** TRANSPORT PHENOMENA FOR MOMENTUM, HEAT, AND MASS TRANSPORT   4
  A. Introduction to Molecular Transport   B. Molecular Momentum Transport   C. Molecular Heat Transport   D. Molecular Mass Transport
- **1.3** INTRODUCTION TO MOLECULAR MASS TRANSPORT   11
- **1.4** MOLECULAR TRANSPORT EQUATIONS   13
  A. Fick's Law   B. Equimolar Counterdiffusion
- **1.5** COMPARISON OF MOLECULAR TRANSPORT FOR MOMENTUM, HEAT, AND MASS TRANSFER   18
- **1.6** MOLECULAR DIFFUSION COEFFICIENTS FOR BINARY GASES   20
  A. Experimental Determination of Diffusion Coefficients   B. Experimental Data   C. Prediction of Diffusion Coefficients
- **1.7** DERIVATION OF EQUATION FOR MOLECULAR TRANSPORT PLUS CONVECTIVE FLOW USING MOLAR FLUXES   35
  A. Derivation Using the Sum of Diffusion Plus Convective Velocities—Method I   B. Derivation Using the Stefan-Maxwell Equations—Method II   C. Integration of the General Equation for Diffusion Plus Convection at Steady-State
- **1.8** DIFFUSION AND CHEMICAL REACTION   45
  A. Heterogeneous Reaction on a Solid Surface   B. Homogeneous Reaction in a Phase
- **1.9** SOLUTION OF MASS TRANSPORT DIFFERENTIAL EQUATIONS BY NUMERICAL METHODS AND DIGITAL COMPUTER   50
  A. Introduction   B. Initial-Value Problems   C. Boundary-Value Problems

1.10 STEADY-STATE MULTICOMPONENT DIFFUSION OF GASES   59
A. Component A Diffusing through a Stagnant Multicomponent Mixture   B. Components A and B Diffusing through Stagnant C C. Equimolar Diffusion of Three Components   D. General Case for Diffusion of Two or More Components

2   Transport Phenomena and the Basic Equations of Change   78

2.1 INTRODUCTION TO EQUATIONS OF CHANGE   78
A. Purpose and Use of Equations of Change   B. Physical and Chemical Properties and Boundary Conditions   C. Discussion of Types of Derivatives with Respect to Time   D. Introduction to Scalars, Vectors, and Differential Operators

2.2 EQUATION OF CONTINUITY FOR A PURE FLUID   84

2.3 EQUATION OF MOTION FOR A PURE FLUID   86
A. Equation for All Fluids   B. Equation for Newtonian Fluids   C. Use of the Equations of Continuity and Motion

2.4 EQUATION OF ENERGY CHANGE FOR A PURE FLUID   91
A. The General Equation   B. Special Equations   C. Use of Equation of Energy Change

2.5 RELATION BETWEEN DIFFERENT TYPES OF FLUXES AND FICK'S LAW IN BINARY SYSTEMS   95

2.6 EQUATIONS OF CONTINUITY FOR A BINARY MIXTURE   99
A. The General Equation   B. Special Equations   C. Use of Equations of Continuity for a Binary Mixture

2.7 EQUATIONS OF ENERGY CHANGE AND CONTINUITY FOR A BINARY MIXTURE   104

2.8 MASS FLUXES FOR MULTICOMPONENT SYSTEMS IN TERMS OF TRANSPORT PROPERTIES   107

2.9 DIMENSIONAL ANALYSIS OF EQUATIONS OF CHANGE   108

3   Molecular Mass Transport Phenomena in Liquids   119

3.1 MOLECULAR TRANSPORT EQUATIONS FOR LIQUIDS   119
A. General Equation   B. Diffusion of A through Stagnant B C. Equimolar Counterdiffusion of A and B

3.2 MOLECULAR DIFFUSION COEFFICIENTS FOR LIQUIDS   124
A. Experimental Determination of Diffusion Coefficients   B. Experimental Data   C. Prediction of Diffusivities for Binary Solutions   D. Prediction of Diffusivities for Multicomponent Solutions

3.3 MOLECULAR TRANSPORT AND CHEMICAL REACTION IN LIQUIDS   131
A. Steady-State Reaction and Diffusion   B. Unsteady-State Reaction and Diffusion in a Semi-Infinite Medium

3.4 INTRODUCTION TO THERMODYNAMICS OF IRREVERSIBLE PROCESSES AND ONSAGER'S RECIPROCAL RELATIONS   136

4   Mass Transport Phenomena in Solids   143

4.1 STEADY-STATE MASS TRANSPORT EQUATIONS FOR SOLIDS   143
A. Introduction and General Types of Diffusion in Solids   B. Diffusion That Does Not Depend on Structure

4.2 DIFFUSION IN POROUS SOLIDS THAT DEPENDS ON STRUCTURE   149
   A. Introduction and Diffusion of Liquids   B. Diffusion of Gases in Pores
   C. Flux Ratios for Diffusion of Gases   D. Multicomponent Diffusion in
   the Transition Region   E. Forced Flow of Gases in Pores due to Total
   Pressure Differences   F. Diffusion and Reaction in Porous Solids
4.3 DIFFUSION AND REACTION IN SOLIDS   163
   A. Diffusion and Reaction in a Single Phase   B. Diffusion and Reaction
   Where a Second Phase Forms
4.4 STEADY-STATE DIFFUSION IN TWO DIMENSIONS   166
   A. Derivation of Basic Equation and Analytical Solution   B. Numerical
   Methods for Solution of Steady-State Diffusion
4.5 DETERMINATION OF DIFFUSIVITIES IN SOLIDS   180
   A. Steady-State Methods   B. Unsteady-State Methods

5   Unsteady-State Diffusion   193

5.1 DERIVATION OF BASIC EQUATIONS FOR UNSTEADY-STATE
   DIFFUSION   193
5.2 SOLUTION OF EQUATIONS BY ANALYTICAL METHODS USING
   THE METHOD OF SEPARATION OF VARIABLES   195
   A. Basic Derivation   B. Formalized Method Using Euler Formulas
   C. Unsteady-State Charts
5.3 SOLUTION OF EQUATIONS BY ANALYTICAL METHODS USING
   THE LAPLACE TRANSFORM   210
   A. Basic Theory Of Laplace Transform   B. Solution Of Unsteady-State
   Diffusion By Laplace Transform
5.4 INTRODUCTION TO NUMERICAL METHODS USING THE EXPLICIT
   METHOD WITH THE DIGITAL COMPUTER   216
5.5 NUMERICAL METHODS USING THE IMPLICIT METHOD WITH
   THE DIGITAL COMPUTER   224
5.6 NUMERICAL METHODS FOR OTHER PHYSICAL GEOMETRIES   229
   A. Flat Slabs in Series   B. Resistance between Flat Slabs in Series
5.7 COMBINED NUMERICAL-ANALOG COMPUTER METHOD   238
5.8 NUMERICAL METHODS FOR HEAT OR MASS TRANSFER   239
5.9 UNSTEADY-STATE DIFFUSION IN COMPOSITE MEDIA   239
   A. Two Slabs in Series, No Interface Resistance   B. Two Slabs in Series,
   Interface Resistance Present   C. Diffusion and Phase Change

6   Mass Transfer Coefficients in Laminar and Turbulent Flow   250

6.1 INTRODUCTION TO TURBULENT TRANSPORT   250
6.2 DERIVATION OF GENERAL EQUATION FOR MASS TRANSFER
   COEFFICIENTS   252
6.3 DEFINITION OF VARIOUS MASS TRANSFER COEFFICIENTS   254
   A. Introduction to Theory   B. Equimolar Counterdiffusion   C. Diffusion
   of A through Stagnant B D. Multicomponent Turbulent Transport
6.4 MASS TRANSFER COEFFICIENTS IN LAMINAR FLOW   261
   A. Introduction   B. Mass Transport from a Tube Wall in Laminar Flow
   C. Mass Transport from a Gas into a Falling Liquid Film in Laminar Flow

xii   Contents

- **6.5** MASS TRANSFER COEFFICIENTS IN TURBULENT FLOW   268
  - A. Introduction and Dimensionless Numbers Used to Correlate Data
  - B. Analogies between Mass, Heat, and Momentum Transport
  - C. Dimensional Analysis and Buckingham Pi Theorem
- **6.6** CONCENTRATION DRIVING FORCES TO USE WITH MASS TRANSFER COEFFICIENTS   277
  - A. Ln-Mean Driving Force   B. Graphical Integration   C. Analog Computer Solution for Integration   D. Numerical Integration and Simpson's Rule   E. Digital Computer Solution for Numerical Integration
- **6.7** EXPERIMENTAL MASS TRANSFER COEFFICIENTS   286
  - A. Experimental Methods to Determine Mass Transfer Coefficients
  - B. Correlations of Mass Transfer Coefficients
- **6.8** MASS TRANSFER COEFFICIENTS AND TURBULENCE THEORIES AND MODELS   299
  - A. Film Theory   B. Penetration Theory   C. Eddy Diffusivity Theory   D. J Factor Empirical Model   E. Boundary Layer Theory

7   Interphase Mass Transport   313

- **7.1** INTRODUCTION TO INTERPHASE MASS TRANSPORT   313
- **7.2** EQUILIBRIUM RELATIONS BETWEEN PHASES   314
  - A. Experimental Determination   B. Use of Phase Rule and Equilibrium Relations
- **7.3** MASS TRANSPORT BETWEEN TWO PHASES   318
  - A. Concentration Profiles   B. Interface Compositions and Film Mass Transfer Coefficients   C. Overall Mass Transfer Coefficients and Driving Forces
- **7.4** MASS TRANSPORT AND CHEMICAL REACTION IN TWO PHASES   332
  - A. Introduction to Two-Phase Reactions   B. Slow First-Order Irreversible Reaction Using the Film Theory   C. Very Fast Second-Order Irreversible Reaction and Film Theory   D. Very Fast Second-Order Irreversible Reaction and Unsteady-State or Penetration Theory   E. Slow Second-Order Reaction

8   Continuous Two-Phase Mass Transport Processes   348

- **8.1** INTRODUCTION TO CONTINUOUS TWO-PHASE MASS TRANSPORT PROCESSES   348
  - A. Classification of Two-Phase Continuous Processes   B. Introduction to the Theory of Two-Phase Continuous Processes
- **8.2** MATERIAL BALANCES AND OPERATING LINES   350
  - A. Countercurrent Processes   B. Cocurrent Processes
- **8.3** DESIGN METHODS FOR CONTINUOUS COUNTERCURRENT PROCESSES   369
  - A. Method Using Film Mass Transfer Coefficients   B. Method Using Overall Mass Transfer Coefficients   C. Method Using Transfer Units   D. Simplified Design Methods for Dilute Gases and A Diffusing through Stagnant B
- **8.4** ESTIMATION OF MASS TRANSFER COEFFICIENTS IN PACKED TOWERS   394

A. Introduction and Experimental Methods   B. Correlations for Film Coefficients
8.5   HEAT EFFECTS IN ABSORPTION COLUMNS   401

9   Mass Transport in Stage Processes   408

9.1   INTRODUCTION TO STAGE PROCESSES   408
9.2   PHASE EQUILIBRIUM RELATIONS   409
A. Vapor-Liquid Equilibrium   B. Liquid-Liquid Equilibrium   C. Gas-Liquid Equilibrium
9.3   THEORY FOR ANALYTICAL AND GRAPHICAL CALCULATIONS IN A SINGLE STAGE   421
A. Derivation of Equations   B. Single Equilibrium Stage Process
9.4   MULTIPLE-STAGE PROCESSES   424
9.5   COUNTERCURRENT MULTISTAGE PROCESSES   426
A. Countercurrent Multistage Process and Overall Balances   B. Stage-to-Stage Calculations in Countercurrent Processes   C. Simplified Procedures for Countercurrent Stage Processes

10   Analog Computer Methods   440

10.1   INTRODUCTION TO ANALOG COMPUTATION   440
10.2   BASIC ANALOG COMPUTER ELEMENTS OR FUNCTIONS   441
A. Introduction   B. Summer or Summation Element   C. Integrator Element   D. Potentiometer Element   E. Initial Conditions
10.3   USE OF ANALOG COMPUTER IN SOLVING DIFFERENTIAL EQUATIONS   447
A. Solving Single Differential Equations   B. Solving Simultaneous Differential Equations
10.4   SCALING OF VARIABLES FOR COMPUTER   450
A. Introduction and Output Devices   B. Time Scaling   C. Amplitude Scaling
10.5   NONLINEAR AND OTHER FUNCTION ELEMENTS   454

A.1   Fundamental Constants   460

A.1-1   STANDARD CONSTANTS   460
A.1-2   TRANSPORT PROPERTY CONSTANTS   462

A.2   Physical Properties   464

A.2-1   VAPOR PRESSURE OF LIQUID WATER FROM 0 TO 100°C   464
A.2-2   DENSITY AND VOLUME OF WATER, −10 TO +250°C   464
A.2-3   VISCOSITIES OF GASES AT 1 ATM   465
A.2-4   VISCOSITIES OF LIQUIDS   467
A.2-5   VISCOSITY OF WATER   470

A.3 Unsteady-State Charts 471

- **A.3-1** CHART FOR DETERMINING CONCENTRATION HISTORY AT SURFACE OF FLAT PLATE ($n = 1.0$) 471
- **A.3-2** CHART FOR DETERMINING CONCENTRATION HISTORY AT CENTER ($n = 0$) AND MIDPLANE ($n = 0.5$) OF FLAT PLATE 472
- **A.3-3** CHART FOR DETERMINING CONCENTRATION HISTORY AT CENTER OF FLAT PLATE 473
- **A.3-4** CHART FOR DETERMINING CONCENTRATION HISTORY AT SURFACE OF INFINITELY LONG CYLINDER 474
- **A.3-5** CHART FOR DETERMINING CONCENTRATION HISTORY AT CENTER ($n = 0$) AND HALF-RADIUS ($n = 0.5$) OF INFINITELY LONG CYLINDER 475
- **A.3-6** CHART FOR DETERMINING CONCENTRATION HISTORY AT CENTER OF INFINITELY LONG CYLINDER 476
- **A.3-7** POSITION CORRECTION FACTORS FOR DIMENSIONLESS CONCENTRATION RATIOS FOR FLAT PLATE ($X > 2$) 477
- **A.3-8** POSITION CORRECTION FACTORS FOR DIMENSIONLESS CONCENTRATION RATIOS FOR INFINITELY LONG CYLINDER ($X > 2$) 477

A.4 Equilibrium Data 478

- **A.4-1** HENRY'S LAW CONSTANTS FOR GASES IN WATER 478
- **A.4-2** EQUILIBRIUM DATA FOR AMMONIA-WATER SYSTEM 478
- **A.4-3** EQUILIBRIUM DATA FOR SO_2-WATER SYSTEM 479
- **A.4-4** EQUILIBRIUM DATA FOR METHANOL-WATER SYSTEM 479
- **A.4-5** EQUILIBRIUM DATA FOR ACETONE-WATER SYSTEM AT 68°F 479
- **A.4-6** ACETIC ACID-WATER-ISOPROPYL ETHER SYSTEM, LIQUID-LIQUID EQUILIBRIA AT 20°C 480
- **A.4-7** STYRENE-ETHYL BENZENE-DIETHYLENE GLYCOL SYSTEM, LIQUID-LIQUID EQUILIBRIA 480
- **A.4-8** EQUILIBRIUM DATA FOR ETHANOL-WATER SYSTEM AT 1 ATM 481

Index 483

Mass Transport Phenomena

1 Molecular Mass Transport Phenomena in Fluids

1.1 INTRODUCTION TO TRANSPORT PHENOMENA

1.1A Purpose

This text will present a unified treatment in depth of mass transport phenomena, relating them—when illuminating—to the phenomena of momentum and heat transport. Since its purpose is not the study of applications per se, the reader is referred to the many excellent design texts available, such as those by Treybal (T3), Perry (P2), McCabe and Smith (M4), and others. This text attempts to delve into the scientific principles of mass transport, and it emphasizes the mathematical, analytical approach, although numerous applications will be presented to help the reader understand the theory.

Many modern computational tools used by today's engineer-scientist are briefly reviewed for the reader, and numerous applications are included. These tools include vectors, Laplace transforms, numerical methods, analog computation, and digital computation. These methods will be applied and learned best by the reader when he encounters them as an integral part of the text.

To begin with, the simple analogies between momentum, heat, and mass transport phenomena are covered to enable the reader to understand their elementary interrelations. Then the text covers a wide variety of topics that are important in mass transport. The reader will note that the analogies soon break down, since topics such as diffusion in porous solids, multicomponent diffusion, and the like have no counterparts in the other transport phenomena.

Since analog computer solutions will also be given for some examples, it is suggested that the reader study Sections 10.1 through 10.3A in this chapter before proceeding further. Eventually he should master all of Chapter 10 on analog computers before proceeding past Chapter 1.

It is assumed that the reader has had simple Fortran programming as a background for understanding the sample machine computational problems given in this text. The mathematical tools needed are covered in the text just before these tools are used in a problem. This saves the reader the necessity of resorting to mathematical texts for a review of the theory.

If desired by the reader, the digital and/or analog computer problems can be omitted without affecting the continuity of the text.

1.1B General Transport Equation

All of the three transport phenomena of momentum, heat, and mass are characterized in the elementary sense by the same general type of equation. Transport of electrical current can also be included in this category. We start by noting that

$$\text{rate of a transport process} = \frac{\text{driving force}}{\text{resistance}} \tag{1.1-1}$$

This equation states what many know intuitively: that we need a driving force to overcome a resistance in order to transport a property. This is similar to Ohm's law in physics, where the flow of electrical current is proportional to the voltage drop (driving force) and inversely proportional to the resistance.

We can formalize our discussion above by writing a mathematical equation for Eq. (1.1-1):

$$\Psi_x = -\delta \frac{d\psi}{dx} \tag{1.1-2}$$

where Ψ_x is defined as the flux of a property whose units are amount of property being transferred per unit time per unit cross-sectional area perpendicular to the x direction of flow, δ is a proportionality constant, ψ is a concentration or amount of the property per unit volume, and x is the distance in the direction of the flux.

If the process is steady-state, the flux Ψ_x is constant. Rearranging Eq. (1.1-2) and integrating,

$$\Psi_x \int_{x_1}^{x_2} dx = -\delta \int_{\psi_1}^{\psi_2} d\psi \tag{1.1-3}$$

$$\Psi_x = \frac{\delta(\psi_1 - \psi_2)}{x_2 - x_1} \tag{1.1-4}$$

1.1 Introduction to Transport Phenomena

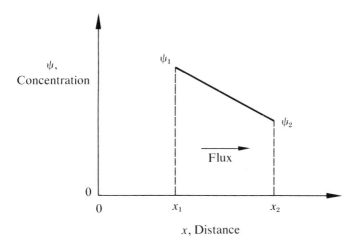

FIGURE 1.1-1 Plot of concentration versus distance for a transport process

Closer inspection of Eq. (1.1-4) reveals that a plot of the concentration ψ versus x is a straight line in Fig. 1.1-1. Also, the flux is in the direction of decreasing concentration, which accounts for the negative sign in Eq. (1.1-2).

If we are at steady-state, then of course we can write for no generation of the property the equation that rate of input = rate of output or

$$\Psi_x|_x = \Psi_x|_{x+\Delta x}$$

Hence, Ψ_x is a constant in this case.

For the very simplified case of unsteady-state diffusion in the x direction with no convection or generation of the property,

$$\text{rate of input} = \text{rate of output} + \text{rate of accumulation} \qquad (1.1\text{-}5)$$

In Fig. 1.1-2 a small element Δx thick is shown with unit cross-sectional area perpendicular to the x direction of flow. The input and output fluxes at x and $x+\Delta x$ are shown. For the accumulation term we can

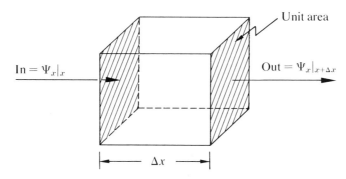

FIGURE 1.1-2 Unsteady-state general property balance

proceed as follows. The amount of the property in the volume $(1)(\Delta x)$ cm³ at time t is $\psi(1)(\Delta x)$. To obtain the accumulation we write the time rate of change of the property in the volume element as

$$\text{rate of accumulation} = \frac{\partial \psi}{\partial t}[(1)(\Delta x)] \qquad (1.1\text{-}6)$$

Substituting the input of $\Psi_x|_x$, output of $\Psi_x|_{x+\Delta x}$, and the rate of accumulation of $(\partial \psi/\partial t)\Delta x$ into Eq. (1.1-5),

$$\Psi_x|_x = \Psi_x|_{x+\Delta x} + \frac{\partial \psi}{\partial t}\Delta x \qquad (1.1\text{-}7)$$

Then, dividing by Δx and taking the limit as Δx goes to zero,

$$-\frac{\partial \Psi_x}{\partial x} = \frac{\partial \psi}{\partial t} \qquad (1.1\text{-}8)$$

Since Eq. (1.1-2) is an expression for Ψ_x, we can differentiate it with respect to x with δ and t constant and substitute it into Eq. (1.1-8) to obtain

$$\frac{\partial \psi}{\partial t} = \delta \frac{\partial^2 \psi}{\partial x^2} \qquad (1.1\text{-}9)$$

This final equation relates the concentration of the property ψ to position x and time, and it will be used in unsteady-steady problems discussed later in this text.

Equations (1.1-2) and (1.1-9) are the real bases for the analogies between momentum, heat, and mass transport phenomena; they will be considered more thoroughly later in this chapter.

An alternate method often used to obtain Eq. (1.1-8) is to write the output as $\Psi_x|_x + (\partial \Psi_x/\partial x)\Delta x$ obtained by using the first two terms of a Taylor series expansion. This is then substituted into Eq. (1.1-7), which gives Eq. (1.1-8) directly.

In summary, Eqs. (1.1-2) and (1.1-9) are the elementary diffusion equations and do not include terms for transport by convection or for generation (chemical reaction). Later in this chapter, in Section 1.7, the term for convection will be included. In Chapter 2, Section 2.6, the final general conservation equation which includes diffusion, convection, and generation will be derived.

1.2 TRANSPORT PHENOMENA FOR MOMENTUM, HEAT, AND MASS TRANSPORT

1.2A Introduction to Molecular Transport

Molecular transport of a property such as momentum, heat, or mass occurs in a gas because of the random movements of individual molecules. Each individual molecule containing the property being transferred moves randomly in all directions, and there are fluxes of the property in all directions. Hence, if we have a system in which there is a concentration gradient of the property, there will be a net flux of the property from the high to the low concentration. This occurs because equal numbers of molecules diffuse in each direction between the high-concentration and low-concentration regions.

If the property is heat, then the hotter molecules migrate to the area of the cooler molecules. Similarly, molecules with a greater directed momentum diffuse to a region where molecules have a less concentration of directed momentum. For mass transfer the same phenomena hold.

1.2B   Molecular Momentum Transport

The important physical property that differentiates fluids from solids is fluid viscosity. When a stress is applied to a solid, it deforms and in some cases may actually break apart. When such a stress is applied to a fluid in the usual sense it will flow, and its velocity will increase as the stress is increased. When a fluid flows, then we are actually inducing momentum transfer in the fluid.

It is common to explain the action of viscosity by considering two very long and very wide parallel plates as in Fig. 1.2-1. We hold the top plate stationary and pull the lower plate with a constant force in the x direction so that the velocity is a constant V in cm/sec. The force per unit area is called the shear stress τ_{yx} where

$$g_c \tau_{yx} = -\mu \frac{dv_x}{dy} \qquad (1.2\text{-}1)$$

where μ is the viscosity and v_x the velocity in the x direction. The velocity v_x is a function of y, and $v_x(0) = V$. One is struck by the similarity of Eq. (1.2-1), which is called Newton's law for fluids, and Eq. (1.1-2). The plot of velocity versus distance in Fig. 1.2-1 is linear just like that in Fig. 1.1-1. Hence we can say we have an "analogy" between the two equations.

To see how momentum is being transferred in the y direction upward in Fig. 1.2-1, let us start at the lower plate. The fluid directly adjacent to this bottom plate has a velocity in the x direction, or x-directed momentum. A short distance above, the fluid has a smaller velocity in the x direction or a smaller x-momentum component, and a velocity gradient

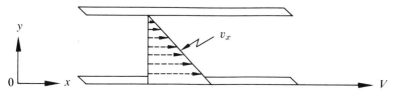

FIGURE 1.2-1   Fluid flow between two parallel plates at steady-state

6 Molecular Mass Transport Phenomena in Fluids

exists. By random diffusion of molecules there is an exchange of molecules, an equal number moving in each direction between the faster-moving layer and the adjacent layer. Hence, the x-directed momentum in the bottom layer has been transferred to the adjacent layer in the y direction. Thus we can look at τ_{yx} as the flux of x-directed momentum in the y direction.

Brief mention should be made of the engineering system of units often employed. These are

$$\tau_{yx} = \text{g force/cm}^2 \text{ or lb force/ft}^2$$
$$\mu = \text{g mass/cm-sec or lb mass/ft-sec}$$
$$v_x = \text{cm/sec or ft/sec}$$
$$y = \text{cm or ft}$$
$$g_c = 980 \text{ g mass-cm/g force-sec}^2 \text{ or } 32.17 \text{ lb mass-ft/lb force-sec}^2$$
$$1 \text{ poise (viscosity)} = 100 \text{ centipoises} = 1 \text{ g mass/cm-sec}$$

Often another set of units is used, where the g_c is omitted in Eq. (1.2-1) and τ_{yx} is then expressed as dynes/cm^2 or poundals/ft^2.

We can write the term $\tau_{yx}g_c$ with units of (g mass-cm/sec)/(sec-cm^2), which is momentum transferred per sec per unit area. Then the kinematic viscosity is expressed as μ/ρ in cm^2/sec and the concentration as $(v_x\rho)$ or

Table 1.2-1 Viscosities of Typical Fluids

Substance	Temp., °C	Viscosity, cp	Ref.
Gases (1 atm)			
Air	20	0.01813	(N1)
CH$_4$	100	0.02173	(N1)
	20	0.01089	(R1)
CO$_2$	0	0.01370	(R1)
	100	0.01828	(R1)
Hexane	35	0.00709	(R1)
SO$_2$	40	0.01350	(R1)
Liquids (1 atm)			
Water	20	1.787	(S4)
	100	0.2821	(S4)
Benzene	5	0.826	(R1)
	40	0.492	(R1)
Ethyl alcohol	0	1.770	(R1)
Glycerol	20	1069	(L2)
Hg	20	1.55	(R2)
	100	1.21	(R2)
Pb	441	2.116	(R2)
Na	250	0.381	(R2)

momentum/cm^3. Then Eq. (1.2-1) becomes for constant ρ,

$$g_c \tau_{yx} = -\frac{\mu}{\rho} \frac{d(v_x \rho)}{dy} \qquad (1.2\text{-}2)$$

which is in the same form as Eq. (1.1-2).

Brief mention should be made of the transport property viscosity for various kinds of materials. Table 1.2-1 gives a brief survey of some viscosities for comparison purposes. The viscosities for the gases as expected are the lowest and do not differ much from gas to gas. The viscosities increase with increasing temperature in gases. For liquids the viscosities are much higher and decrease markedly as temperature increases. Also, the range of viscosities for liquids is quite large, being 1069 centipoises for glycerol.

Example 1.2-1 Flux of Momentum with Water by Analytical Method and Analog Computer

In Fig. 1.2-1 the distance between the two plates is 0.10 cm and the velocity of the lower plate is 10 cm/sec. The fluid is water at 20°C. (a) Calculate the steady-state momentum flux. (b) Solve this equation using the analog computer.

Solution

For part (a), integrating Eq. (1.2-1) analytically,

$$\tau_{yx} g_c \int_0^{0.10} dy = -\mu \int_{10}^0 dv_x$$

$$\tau_{yx} = \frac{\mu}{g_c} \frac{(10-0)}{(0.10-0)} = \frac{(1.787/100)(10-0)}{980(0.10-0)}$$

$$= 1.83 \times 10^{-3} \text{ g force/cm}^2$$

For part (b), Eq. (1.2-1) is rearranged in solving for the derivative. (The reader should have studied Sections 10.1 to 10.3A in Chapter 10 in order to follow this simple example.)

$$\frac{dv_x}{dy} = -\frac{g_c \tau_{yx}}{\mu} \qquad (1.2\text{-}1a)$$

Calling v_x the voltage e, and y the time τ on the computer, we can write Eq. (1.2-1a) as follows, where the right-hand side is an unknown constant:

$$\frac{de}{d\tau} = -\frac{g_c \tau_{yx}}{\mu} \qquad (1.2\text{-}1b)$$

The computer diagram is just a simple integrator in Fig. 1.2-2. The initial condition IC is the value of v_x at y (or time τ) $= 0$, which is $(v_x)_{y=0} = V$.

8 Molecular Mass Transport Phenomena in Fluids

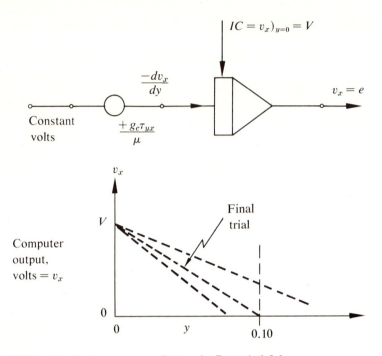

FIGURE 1.2-2 Analog computer diagram for Example 1.2-1

The unknown is τ_{yx}. To make a run, τ_{yx} is guessed, then the potentiometer is set using this "trial" value. A run is made and a plot of v_x (or e) is made versus y (or τ). At $y = 0.10$ cm the plot should give $v_x = 0$. If not, a new trial is made.[1]

1.2C Molecular Heat Transport

Fourier's law of molecular heat conduction in fluids or solids states that the flux of heat is directly proportional to the temperature gradient. This empirical law can be written for a pure substance as

$$q_y = -k \frac{dT}{dy} \qquad (1.2\text{-}3)$$

where q_y is the energy or heat flux in the y direction in cal/sec-cm², T is temperature °K, and k is the thermal conductivity of the material in

---

[1]This analog-computer example was given to help the reader learn the computer techniques. Normally the computer is not used to solve cases where simple analytical solutions exist, as they do for the example above.

1.2 Phenomena for Momentum, Heat, and Mass Transport

cal-cm/(sec-cm^2-°K). Equation (1.2-3) can be written in a form very similar to Eq. (1.1-2) for constant ρ and c_p as

$$q_y = -\alpha \frac{d(T\rho c_p)}{dy} \qquad (1.2\text{-}4)$$

where $\alpha = k/\rho c_p$ and is called the thermal diffusivity in cm^2/sec, c_p is the heat capacity in cal/g mass-°K, and ρ is the density in g mass/cm^3.

When there is a temperature gradient, equal numbers of gas molecules diffuse in each direction between the hot region and the colder region. In this way energy in gases is transferred in the y direction.

Table 1.2-2 gives a short list of values of the transport property thermal conductivity for various types of materials. There is a very wide range in the values of k in this table. Gases have quite low values, solid metals quite large values, and liquids intermediate values.

Table 1.2-2 Thermal Conductivities of Materials

Substance	Temp., °C	Thermal conduc., cal-cm/(sec-cm^2-°K)	Ref.
Gases (1 atm)			
Air	0	5.572 × 10^{-5}	(L2)
	100	7.197 × 10^{-5}	(L2)
H$_2$	0	39.6 × 10^{-5}	(L2)
	100	49.94 × 10^{-5}	(L2)
Benzene	0	2.094 × 10^{-5}	(L2)
CO$_2$	0	3.393 × 10^{-5}	(L2)
Liquids (1 atm)			
Water	12	1.36 × 10^{-3}	(L2)
	40.8	1.555 × 10^{-3}	(L2)
Benzene	12	0.333 × 10^{-3}	(L2)
CCl$_4$	12	0.252 × 10^{-3}	(L2)
Glycerol	12	0.670 × 10^{-3}	(L2)
Hg	0	19.6 × 10^{-3}	(R2)
K	200	107.3 × 10^{-3}	(R2)
Pb	330	39 × 10^{-3}	(R2)
Solids			
Cu	0	0.920	(L2)
Pb	18	0.0827	(L2)
Concrete	20	2.2 × 10^{-3}	(L2)
Cork	30	0.128 × 10^{-3}	(L2)

10 Molecular Mass Transport Phenomena in Fluids

Example 1.2-2 Flux of Heat by Conduction in a Fluid

A truly stagnant fluid is situated between two parallel flat plates in a conduction experiment. The plates are 0.15 cm apart, and the top plate is held constant at 80.20°C and the lower plate at 79.60°C. The thermal conductivity is 4.3×10^{-4} cal-cm/(sec-cm^2-°K) and can be considered constant over this range. Predict the heat flux at steady-state.

Solution

Equation (1.2-3) must be integrated.

$$q_y \int_0^{0.15} dy = -k \int_{80.20}^{79.60} dT$$

$$q_y = \frac{k(80.20 - 79.60)}{(0.15 - 0)} = 4.3 \times 10^{-4} \frac{(0.60)}{(0.15)}$$

$$= 0.00172 \text{ cal/sec-cm}^2$$

1.2D Molecular Mass Transport

Fick's law of molecular mass transport tells us that the flux of component A is directly proportional to the concentration gradient, where the flux is relative to the molar average velocity of the mixture (defined in Section 1.7A):

$$J_{Ay}^* = -cD_{AB}\frac{dx_A}{dy} \tag{1.2-5}$$

Table 1.2-3 Typical Values of Molecular Diffusivity*

System	Temp., °C	Diffusivity, cm^2/sec
Gases (1 atm)		
N_2-NH_3	25	0.230
N_2-He	25	0.687
CH_4-He	25	0.675
Liquids (dilute)		
CO_2 in water	25	2.00×10^{-5}
Tin in Hg	20	1.63×10^{-5}
Solids		
H_2 in Ni	85	1.16×10^{-8}
Cd in Cu	20	2.7×10^{-15}

*For a more complete listing see Tables 1.6-1, 3.2-1, 4.4-1, and 6.7-2.

where J_{Ay}^* is the molar flux of component A in g mole A/sec-cm^2, D_{AB} is the molecular diffusivity of the pair A-B in cm^2/sec, c is the total concentration of A and B in g mole/cm^3, and x_A is the mole fraction of A in the mixture of A and B. Again this equation is very similar to the general transport equation (1.1-2). In the rest of this text the flux J_{Ay}^* will be written as J_A^* when the direction of diffusion is obvious. A boldface symbol $\mathbf{J}_A^*$ indicates a vector with x, y, and z components.

Table 1.2-3 gives a few values of experimental diffusivities. It is clearly evident that the highest diffusion coefficients are for gases. This is to be expected, since the gas molecules are far apart and have large kinetic energies. The values for liquids and liquid metals are smaller than gases by a factor of 10^{-4} to 10^{-5}. Solids generally have the smallest values. More extensive tables of diffusivity data for gases, liquids, and solids will be given in this and succeeding chapters.

Example 1.2-3 Molar Diffusion Flux of Methane

A mixture of CH_4 and He gas is contained in a vessel at 25°C and 1 atm total pressure. At one point the partial pressure p_{A1} of methane is 0.60 atm, and at a distance 2.0 cm away $p_{A2} = 0.20$ atm. The total pressure is constant throughout. Calculate the flux of methane at steady-state.

Solution

The diffusivity D_{AB} is 0.675 cm^2 from Table 1.2-3. Since P is constant, we can substitute P/RT for c in Eq. (1.2-5). Then, since $Px_A = p_A$, we obtain

$$J_{Ay}^* = -\frac{D_{AB}}{RT}\frac{dp_A}{dy} \qquad (1.2\text{-}6)$$

Integrating at steady-state,

$$J_{Ay}^* = \frac{D_{AB}(p_{A1}-p_{A2})}{RT(y_2-y_1)} \qquad (1.2\text{-}7)$$

Substituting the known values into Eq. (1.2-7) and solving,

$$J_{Ay}^* = \frac{0.675(0.60-0.20)}{82.06(298)(2.0-0)} = 5.53 \times 10^{-6} \text{ g mole } A/\text{sec-cm}^2$$

1.3 INTRODUCTION TO MOLECULAR MASS TRANSPORT

The kinetic theory of gases gives us a good physical interpretation of the motion of molecules in fluids. Because of their kinetic energy we visualize the molecules in rapid movement, often colliding with each other, and sometimes colliding with the walls of the container if the fluid is enclosed.

12 Molecular Mass Transport Phenomena in Fluids

We can define molecular diffusion or molecular transport as the transfer or movement of individual molecules through a truly stagnant fluid by means of the random, individual movements of the molecules. If the fluid is in laminar flow, molecular diffusion can occur in all directions in the fluid. The molecules can be imagined as traveling in a blind manner only in straight lines and changing direction by bouncing off other molecules after collisions. Since the molecules travel in a random path, molecular diffusion is often called a random-walk process.

Figure 1.3-1 shows this molecular diffusion process schematically for A molecules diffusing through B molecules. The dotted lines show the random path that an A molecule might take in diffusing from point (1) to (2). If at the start there are more A molecules near point (1) than at point (2), then it is obvious that more A molecules will diffuse from (1) to (2) than from (2) to (1). This means we have a concentration that is higher at (1). The net diffusion will be from the region of high to that of low concentration.

As another example of molecular diffusion, suppose a drop of blue ink dye is placed in a beaker containing water that is truly stagnant and not mixed by convection or stirring. A high concentration difference of blue dye will exist between the drop of dye added and the adjacent water. The dye molecules will diffuse slowly by molecular diffusion to all parts of the water. Eventually the fluid will be almost completely mixed, but this process will take many days. If instead the contents are rapidly stirred with a stirring rod — that is, turbulence or bulk motion of the fluid

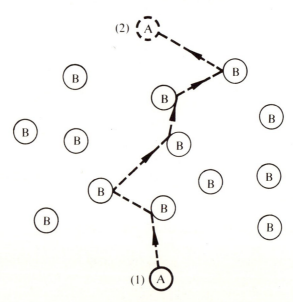

FIGURE 1.3-1 Diagram of molecular diffusion or random-walk diffusion process

is introduced—then the solution will be completely mixed in a few seconds. Hence, one can see that mass transport by turbulent motion as well as molecular diffusion is very rapid compared to molecular diffusion alone.

It should be evident that the rate of molecular diffusion in liquids will be many times slower than in gases. Referring again to Fig. 1.3-1, if B is a liquid, the molecules will be very close together compared to B as a gas. Hence, the molecules of A will collide with molecules of liquid B more often and diffuse more slowly. Also, mean free paths of gaseous molecules are much greater than liquid molecules. In general, the diffusivity coefficient in a gas will be of the order of magnitude of 10^5 times larger than in a liquid. However, the flux in a gas is not that much greater, being about 100 times faster, since concentrations in liquids are quite high compared to those in gases.

1.4 MOLECULAR TRANSPORT EQUATIONS

1.4A Fick's Law

As discussed in Section 1.2, the rate of molecular diffusion or flux of A in gram moles of A diffusing per square centimeter of cross-sectional area perpendicular to the flux per second is proportional to the concentration gradient. This means Eq. (1.2-5) can be written as follows. (Note that the subscript indicating the direction of the flux has been dropped since the direction is obvious.)

$$J_A^* = \frac{\bar{J}_A^*}{A} = -cD_{AB}\frac{dx_A}{dz} \qquad (1.4\text{-}1)$$

where $\bar{J}_A^*$ is the flux in g mole A/sec, x_A is the mole fraction of A in the mixture, A is cross-sectional area perpendicular to the direction of diffusion in cm^2, z is the distance in the direction of diffusion in cm, and D_{AB} is the proportionality constant called diffusivity in cm^2/sec. Often the cross-sectional area A is combined with the $\bar{J}_A^*$ in one term J_A^*, which is $\bar{J}_A^*/A$. This equation is analogous to the Fourier equation for heat transfer, where the heat flux is proportional to the temperature gradient with the thermal conductivity k as the proportionality constant. Fick's law was first obtained empirically. It can also be derived from the kinetic theory of gases.

The molar diffusion flux J_A^* (Fick's law) is due to a concentration gradient and is the flux of A relative to the molar average velocity v^* of the whole fluid defined as $v^* = x_A v_A + x_B v_B$, where v_A is the velocity of A in cm/sec relative to stationary coordinates and v_B the velocity of B.

Another molar flux often used is N_A, which is the total molar flux of A relative to stationary coordinates and is equal to the sum of the molar diffusion flux of A plus the convective flux of A:

N_A (total molar flux) $= J_A^*$ (diffusion flux) $+ c_A v^*$ (convective flux)

where c_A is the concentration of A in g mole/cm³.

These molar fluxes will be discussed in greater detail in Section 1.7A. In the present section we will be concerned with the flux J_A^* and with a fluid that is not flowing.

1.4B Equimolar Counterdiffusion

Figure 1.4-1 shows a diagram of two chambers with gases A and B at constant total pressure P connected by a tube where molecular diffusion is occurring. The concentration in the left chamber is kept uniform by mixing and it is also uniform in the right. The partial pressure $p_{A1} > p_{A2}$ and $p_{B2} > p_{B1}$. If one observes at a given point in the tube, he will be able to see the number of A molecules diffusing per unit time to the right because of the concentration gradient and the rate of diffusion of the B molecules to the left. Since the total pressure P is constant throughout,

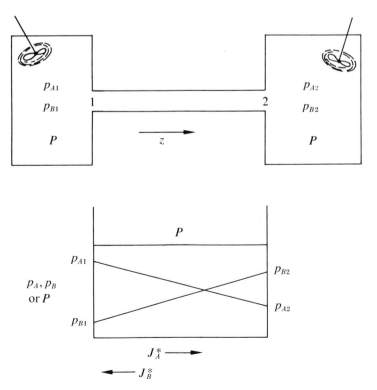

FIGURE 1.4-1 Physical model for equimolar counterdiffusion of gases A and B

1.4 Molecular Transport Equations

the moles of A to the right must be equal to the moles of B to the left; otherwise the total pressure will not remain constant throughout. Or,

$$\frac{\bar{J}_A^*}{A} = -\frac{\bar{J}_B^*}{A} \tag{1.4-2}$$

The fluxes $\bar{J}_A^*/A$ and $\bar{J}_B^*/A$ are in the z direction. When they are in the x, y, and z directions, they are written as vectors.

Another result can be obtained from Eq. (1.4-2) by first setting

$$P = p_A + p_B \tag{1.4-3}$$

Also, one can write Fick's law for B, where $c_B = cx_B$.

$$\frac{\bar{J}_B^*}{A} = -D_{BA}\frac{dc_B}{dz} \tag{1.4-4}$$

Differentiating Eq. (1.4-3) with respect to distance z,

$$\frac{dp_A}{dz} = -\frac{dp_B}{dz} \tag{1.4-5}$$

Using the perfect gas law,

$$p_A V = n_A RT \tag{1.4-6}$$

where V is the total volume in cm³ and n_A the g moles of A. Also,

$$\frac{p_A}{RT} = \frac{n_A}{V} = c_A \tag{1.4-7}$$

Hence, Eq. (1.4-7) relates concentration c_A and partial pressure p_A. Equation (1.4-2) can be written as follows by substituting Eqs. (1.4-4) and (1.4-1) into (1.4-2):

$$-D_{AB}\frac{dc_A}{dz} = +D_{BA}\frac{dc_B}{dz} \tag{1.4-8}$$

Substituting Eq. (1.4-7) for A into (1.4-5) and a similar one for B,

$$\frac{dc_A}{dz} = -\frac{dc_B}{dz} \tag{1.4-9}$$

Combining Eqs. (1.4-9) and (1.4-8),

$$D_{AB} = D_{BA} \tag{1.4-10}$$

This shows that for a binary gas mixture the diffusivity coefficient for A diffusing in B is the same as for B in A. Equation (1.4-1) can be rewritten in terms of partial pressures:

$$\frac{\bar{J}_A^*}{A} = -\frac{D_{AB}}{RT}\frac{dp_A}{dz} \tag{1.4-11}$$

16 Molecular Mass Transport Phenomena in Fluids

Returning to Fig. 1.4-1, we can integrate the equation for steady-state. This means the boundary conditions at points 1 and 2 and the fluxes remain constant. Rearranging Eq. (1.4-1) for a constant cross-sectional area A,

$$\frac{\bar{J}_A^*}{A}\int_{z_1}^{z_2} dz = -D_{AB}\int_{c_{A1}}^{c_{A2}} dc_A \qquad (1.4\text{-}12)$$

Integrating,

$$\frac{\bar{J}_A^*}{A} = D_{AB}\frac{c_{A1}-c_{A2}}{z_2-z_1} = \frac{D_{AB}}{RT}\frac{p_{A1}-p_{A2}}{z_2-z_1} \qquad (1.4\text{-}13)$$

Hence, as shown in Fig. 1.4-1, plots of p_A and p_B versus z should be a straight line.

Example 1.4-1 Equimolar Counterdiffusion

Ammonia (A) is diffusing through a straight uniform tube 8 cm long containing N_2 (B) at 25°C and 1.0 atm total pressure. The partial pressure of ammonia at one end of the tube is constant at 85 mm Hg and at the other end is 15 mm Hg. Both ends of the tube open into large mixed chambers at 1.0 atm. The diffusivity of ammonia in N_2 is 0.230 cm²/sec at 25°C and 1 atm. (a) Calculate the flux of A at steady-state. (b) Calculate the flux of B.

Solution

Since the tube is uniform, the cross-sectional area A is constant and Eq. (1.4-13) applies. $D_{AB} = 0.230$ cm²/sec, $P = 1.0$ atm, $z_2 - z_1 = 8.0$ cm, $R = 82.06$ cm³-atm/g mole-°K, $T = 298$°K, $p_{A1} = 85/760 = 0.112$ atm, $p_{A2} = 15/760 = 0.0198$ atm, $p_{B1} = 1 - 0.112 = 0.888$ atm, $p_{B2} = 1 - 0.0198 = 0.9802$ atm. Substituting into Eq. (1.4-13),

$$J_A^* = \frac{\bar{J}_A^*}{A} = \frac{D_{AB}}{RT}\frac{p_{A1}-p_{A2}}{z_2-z_1} = \frac{0.230(0.112-0.0198)}{82.06(298)(8.0)}$$

$$= 1.08 \times 10^{-7} \text{ g mole/sec-cm}^2 \qquad \text{[Ans. to (a)]}$$

Rewriting Eq. (1.4-13) for component B,

$$\bar{J}_B^* = \frac{J_B^*}{A} = \frac{D_{AB}}{RT}\frac{p_{B1}-p_{B2}}{z_2-z_1} = \frac{0.230(0.888-0.9802)}{82.06(298)(8.0)}$$

$$= -1.08 \times 10^{-7} \text{ g mole/sec-cm}^2 \qquad \text{[Ans. to (b)]}$$

1.4 Molecular Transport Equations

Example 1.4-2 Equimolar Diffusion with a Varying Cross-Sectional Area

Benzene gas (A) is diffusing at steady-state through N_2 (B) in a pipe at 25°C and a total pressure of 1 atm. The partial pressure of benzene at one end of the pipe is p_{A1} and is 0.120 atm, and the cross-sectional area (circular) is 1.0 ft². At the other end of the pipe (a distance of 3.0 ft) the p_{A2} of benzene is 0.025 atm and the cross-sectional area (circular) is 1.5 ft². The pipe is tapered uniformly like a cone. The diffusivity of benzene in N_2 at 25°C and 1 atm is 0.088 cm²/sec. (a) Calculate the molar flux of benzene at steady-state in lb moles/hr. (b) Make a plot of p_A vs. distance from the small end. Do this by first calculating p_A at $z = 1.5$ ft. Plot the three values of p_A vs. z.

Solution

Equation (1.4-12) does not apply here, since the cross-sectional area A is not constant. Hence, we must start with the basic differential equation (1.4-11). $D_{AB} = 0.088$ cm²/sec or 0.342 ft²/hr, $P = 1.0$ atm, $z = 3.0$ ft, $R = 0.730$ ft³-atm/lb mole-°R, $T = 537$°R, $p_{A1} = 0.120$ atm, $p_{A2} = 0.025$ atm. Figure 1.4-2 gives the diagram. For a cross-sectional area of 1.0 ft² the radius $r_1 = 0.564$ ft, and $r_2 = 0.690$ ft at the other end. A relation

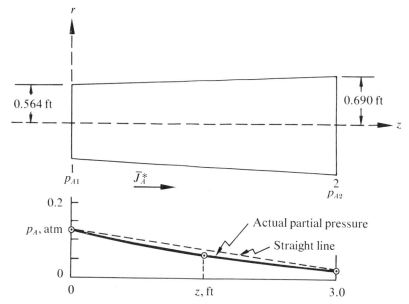

FIGURE 1.4-2 Process diagram for Example 1.4-2

18 Molecular Mass Transport Phenomena in Fluids

between the radius r and the distance z must be found so that Eq. (1.4-11) can be integrated.

The equation of the straight line relating the radius r to z is $r = Kz + b$, where K and b are constants. With two unknowns K and b and $r = r_1$ at $z = 0$ and $r = r_2$ at $z = 3.0$ ft,

$$r = \frac{r_2 - r_1}{3}z + r_1 = 0.0420z + 0.564$$

The cross-sectional area A is

$$A = \pi r^2 = \pi(0.0420z + 0.564)^2$$

Substituting the value for A into Eq. (1.4-11), rearranging, and integrating,

$$\frac{\bar{J}_A^*}{\pi}\int_{z_1=0}^{z_2=3.0} \frac{dz}{(0.0420z + 0.564)^2} = \frac{-0.342}{0.730(537)}\int_{p_{A1}=0.120}^{p_{A2}=0.025} dp_A$$

Integrating and substituting numerical values,

$$-\frac{\bar{J}_A^*}{\pi}\left(\frac{1}{0.042(0.042 \times 3 + 0.564)} - \frac{1}{0.042(0 + 0.564)}\right) = \frac{-0.342(0.025 - 0.120)}{0.730(537)}$$

$$\bar{J}_A^* = 3.36 \times 10^{-5} \text{ lb mole benzene/hr} \quad \text{[Ans. to (a)]}$$

For calculating the p_A at 1.50 ft, the limits on the integration will be between $z_1 = 0$, $p_{A1} = 0.120$ atm, and $z_2 = 1.50$ ft, $p_{A2} = p_{A2}$. Substituting these values,

$$p_{A2}(1.5 \text{ ft}) = 0.068 \text{ atm} \quad \text{[Ans. to (b)]}$$

The three values of p_A are plotted in Fig. 1.4-2. As expected, the plot of partial pressure versus z is not linear. If it were linear, the p_A at $z = 1.5$ ft would be $(0.120 + 0.025)/2$ or 0.0725 atm.

1.5 COMPARISON OF MOLECULAR TRANSPORT FOR MOMENTUM, HEAT, AND MASS TRANSFER

As pointed out earlier, the equations for molecular diffusion for mass, heat, and momentum given below are very similar.

$$J_{Az}^* = -D_{AB}\frac{dc_A}{dz} \quad \text{(Fick's law, constant } c\text{)} \quad (1.5\text{-}1)$$

$$q_z = -\alpha\frac{d(\rho c_p T)}{dz} \quad \text{(Fourier's law, constant } \rho, c_p\text{)} \quad (1.5\text{-}2)$$

$$\tau_{zx} = -\nu\frac{d(v_x\rho)}{dz} \quad \text{(Newton's law, constant } \rho\text{)} \quad (1.5\text{-}3)$$

1.5 Comparison of Molecular Transport Phenomena

All of the fluxes on the left-hand sides of Eqs. (1.5-1), (1.5-2), and (1.5-3) have as units transfer of a quantity of mass, heat, or momentum per unit time per unit area. The transport properties D_{AB}, α, and ν all have units of cm²/sec. The concentrations are represented as g mole/cm³, calories/cm³, or momentum/cm³, respectively.

In correlating data for heat, momentum, and mass transfer certain dimensionless numbers are useful.

$$N_{Pr} = \frac{\text{shear component of diffusivity for momentum}}{\text{diffusivity for heat}} = \frac{\mu/\rho}{k/\rho c_p} = \frac{c_p \mu}{k} \tag{1.5-4}$$

$$N_{Sc} = \frac{\text{shear component of diffusivity for momentum}}{\text{diffusivity for mass}} = \frac{\mu/\rho}{D} = \frac{\mu}{\rho D} \tag{1.5-5}$$

$$N_{Le} = \frac{\text{diffusivity for heat}}{\text{diffusivity for mass}} = \frac{k/\rho c_p}{D} = \frac{k}{\rho c_p D} \tag{1.5-6}$$

If simultaneous heat transfer and momentum transfer are occurring, such as for heating a fluid flowing, the Prandtl number is important in correlating and predicting the data. For mass transfer of flowing fluids the Schmidt number is used, and for combined heat and mass transfer the Lewis number.

The magnitude of each of the dimensionless numbers is important, since the ratios give the relative rates of molecular transport of each process. In the Schmidt number the ratio of the momentum diffusivity to the mass diffusivity is about 1 for gases. This occurs because the transport is mainly by random motions. In liquids the molecules are close together, and momentum can be transmitted by an additional mechanism as well as by random motion. In this case in liquids, the principal mechanism for momentum transfer is by actual collisions of the molecules. For mass transport it is very difficult for a molecule to diffuse through this close-packed group.

Table 1.5-1 gives Schmidt numbers of a few typical gases and liquids.

Table 1.5-1 Some Typical Schmidt Numbers

Material	Temp., °C	N_{Sc}
Gases		
Benzene–air	0	1.7
Water–air	0	0.6
CO_2–air	0	0.9
Liquids (Dilute Solutions)		
CO_2 in water	20	570
Acetic acid in water	20	1140
CO_2 in ethanol	20	445

1.6 MOLECULAR DIFFUSION COEFFICIENTS FOR BINARY GASES

1.6A Experimental Determination of Diffusion Coefficients

There are a number of different experimental methods to determine the molecular diffusivity for binary gases. The most important methods will be briefly summarized below. It has been shown that there is only a very small effect of composition on the diffusivity for a binary mixture (H1).

1. *Evaporation of a Pure Liquid in a Narrow Tube.* One of the two components must be a liquid at the temperature used. A second gas is passed over the top of a narrow tube. The liquid in the tube evaporates and the fall of the liquid level is noted. When the liquid evaporates, it diffuses by molecular diffusion through the tube.

2. *Unsteady-State Diffusion of Two Gases.* Smith (S1) describes the method where two pure gases are placed in separate sections of a tube separated by a partition. The partition is carefully removed and the diffusion proceeds. After a measured time the partition is reinserted and the gas in each section analyzed. These results are then compared with the integrated equation.

3. *Evaporation of Drops.* Diffusion coefficients of naphthalene, iodine, benzene, and many other materials have been obtained by measuring the time for spheres of solid or liquid to evaporate completely. Perry (P1) derives the final equation as follows for the sphere of A with radius r_p evaporating into a pure gas B.

$$\theta_0 = \frac{\rho_A r_p^2 RT}{2 M_A D_{AB} P \ln [P/(P-P_A)]} \qquad (1.6\text{-}1)$$

4. *Two-Bulb Method.* Ney and Armistead (N2) discuss a relatively recent method to experimentally determine the diffusivities of two gases. The apparatus consists of two glass bulbs with volumes V_1 and V_2 cm^3, connected by a capillary of known size, of cross-sectional area A and length L, whose volume is small compared to V_1 and V_2 as shown in Fig. 1.6-1. Pure gas A is put in V_1 and B in V_2 at the same pressures.

The valve is opened between the two chambers, and diffusion proceeds for a predetermined time. Then, after mixing has taken place in a bulb, the contents of this bulb are sampled. The equations can be derived by neglecting the capillary volume and assuming each bulb is always of a uniform concentration. Assuming quasi-steady-state in the capillary,

$$J_A^* = -D_{AB}\frac{dc}{dz} = -\frac{D_{AB}(c_2 - c_1)}{L} \qquad (1.6\text{-}2)$$

1.6 Molecular Diffusion Coefficients for Binary Gases

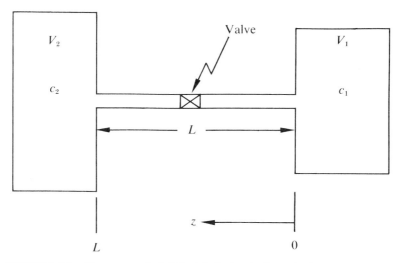

FIGURE 1.6-1 Measurement of diffusivity by two-bulb method

where c_2 is the concentration in V_2 at t and c_1 in V_1. It is assumed that the volume of the capillary is negligible. The flux in the capillary going to bulb V_2 is equal to the accumulation in V_2.

$$AJ_A^* = \frac{-D_{AB}A(c_2-c_1)}{L} = V_2\frac{dc_2}{dt} \qquad (1.6\text{-}3)$$

The average value c_{ave} at equilibrium can be calculated from the starting compositions c_2^0 and c_1^0 and volumes.

$$(V_1+V_2)c_{\text{ave}} = V_1c_1^0 + V_2c_2^0 \qquad (1.6\text{-}4)$$

At the time t a similar balance gives

$$(V_1+V_2)c_{\text{ave}} = V_1c_1 + V_2c_2 \qquad (1.6\text{-}5)$$

Substituting the value of c_1 from Eq. (1.6-5) into (1.6-3) and rearranging,

$$\frac{dc_2}{dt} = -\beta c_2 + \beta c_{\text{ave}} \qquad (1.6\text{-}6)$$

where β is a constant.

$$\beta = \frac{D_{AB}(V_1+V_2)}{(L/A)V_2V_1} \qquad (1.6\text{-}7)$$

Integrating for $c_2 = c_2^0$ at $t = 0$ and c_2 at t,

$$\frac{c_{\text{ave}}-c_2}{c_{\text{ave}}-c_2^0} = e^{-\beta t} \qquad (1.6\text{-}8)$$

Hence, if an experimental value c_2 is obtained at t, the diffusivity can be calculated. More details are given elsewhere (N2).

Table 1.6-1 Molecular Diffusivities of Gases at 1 Atm Abs Pressure

System	Temp., °C	Diffusivity, cm²/sec	Ref.
Air–NH_3	0	0.198	(W3)
Air–H_2O	0	0.220	(I1)
	42	0.288	(M1)
Air–CO_2	3	0.142	(H2)
	44	0.177	(H2)
Air–ethanol	42	0.145	(M1)
	25	0.135	(M1)
Air–acetic Acid	0	0.106	(I1)
Air–n-hexane	21	0.080	(C1)
Air–toluene	25.9	0.086	(G1)
	59.0	0.104	(G1)
Air–hydrogen	0	0.611	(I1)
Air–n-butanol	0	0.0703	(I1)
	25.9	0.087	(G1)
	59.0	0.104	(G1)
Air–n-pentane	21	0.071	(C1)
H_2–Ar	22.4	0.83	(W4)
	175	1.76	(W4)
	796	8.10	(W4)
He–Ar	25	0.729	(S3)
	225	1.728	(S3)
CH_4–Ar	25	0.202	(C2)
CH_4–He	25	0.675	(C2)
N_2–He	25	0.687	(S3)
CH_4–H_2	0	0.625	(C3)
N_2–NH_3	25	0.230	(M1)
	85	0.328	(M1)
H_2–NH_3	25	0.783	(M1)
	85	1.093	(M1)
H_2–N_2	25	0.784	(M1)
	85	1.052	(M1)
H_2O–N_2	34.4	0.256	(S5)
	55.4	0.303	(S5)
H_2O–CO_2	34.3	0.202	(S5)
	55.4	0.211	(S5)
SO_2–CO_2	343°K	0.108	(S6)
C_2H_5OH–CO_2	67	0.106	(T4)
$(C_2H_5)_2O$–air	19.9	0.0896	(T4)
$(CH_3)_2O$–SO_2	30	0.0672	(C4)
$(C_2H_5)_2O$–NH_3	26.5	0.1078	(S7)

1.6B Experimental Data

Some of the typical data are listed in Table 1.6-1. A longer more complete list is given in Perry (P1, P2) and Reid and Sherwood (R1). The relation between the diffusivity in cm^2/sec and ft^2/hr is

$$D \text{ (ft}^2/\text{hr)} = 3.87 \text{ (cm}^2/\text{sec)}$$

1.6C Prediction of Diffusion Coefficients

1. *Elementary Kinetic Theory Prediction.* To give the reader some insight into the classical kinetic theory of gases, a simplified derivation will be given for predicting the diffusivity of a binary gas mixture in the dilute gas region—that is, at low pressures near atmospheric. This is similar to that of Bird *et al.* (B1).

The gas is assumed to consist of rigid spherical particles that are completely elastic on collision with another molecule, which implies momentum is conserved. We will assume that there are no attractive or repulsive forces between the molecules. We will assume also that the molecules A and B are isotopes with similar molecular weights of m_A and m_B, where $m_A = m = m_B$. The concentration of the molecules is n molecules per unit volume.

To simplify matters we assume an average velocity $\bar{u}$ of each molecule relative to the molar average velocity $\mathbf{v}^*$ of the gas stream defined as $\mathbf{v}^* = x_A \mathbf{v}_a + x_B \mathbf{v}_B$, where $\mathbf{v}_A$ and $\mathbf{v}_B$ are the velocities of A and B relative to stationary coordinates. Of course, the velocities are completely random in any direction, and the average is

$$\bar{u} = \sqrt{\frac{8kT}{m\pi}} \tag{1.6-9}$$

where T is in °K and k the Boltzmann constant. We can write an equation for S, the frequency with which molecules strike one side of a stationary plane exposed to the gas (Fig. 1.6-2).

$$S = \frac{n\bar{u}}{4} \tag{1.6-10}$$

We start the derivation for the fluid moving in the z direction only with velocity v_z^* and assume a linear concentration profile so dx_A/dz remains constant where x_A is mole fraction of A. The convective or bulk flow flux of A at z is $(x_A v_z^*)_z n/N$, where N is Avogadro's number. The diffusion flux in the z direction is the number of moles of A going from $z-\delta$ to z minus those going from z to $z+\delta$ or

$$\frac{1}{N}(Sx_A)_{z-\delta} - \frac{1}{N}(Sx_A)_{z+\delta} = \frac{1}{N}\frac{n}{4}(\bar{u}x_A)_{z-\delta} - \frac{1}{N}\frac{n}{4}(\bar{u}x_A)_{z+\delta} \tag{1.6-11}$$

24 Molecular Mass Transport Phenomena in Fluids

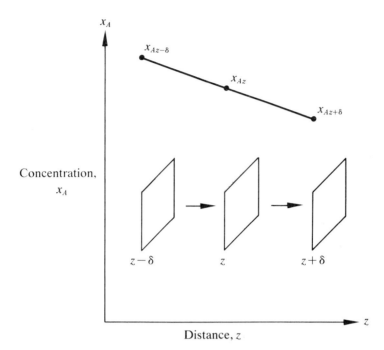

FIGURE 1.6-2 Diagram for diffusion of gas molecules

The total flux N_{Az} is made up of the sum of the convective flux plus the diffusion flux, or

$$N_{Az} = \frac{n}{N}(x_A v_z^*)_z + \frac{n}{4N}(\bar{u}x_A)_{z-\delta} - \frac{n}{4N}(\bar{u}x_A)_{z+\delta} \quad (1.6\text{-}12)$$

Next we can write the following, since the concentration profile over this short distance is linear.

$$(x_A)_{z-\delta} = x_A - \frac{dx_A}{dz}\delta$$

$$(x_A)_{z+\delta} = x_A + \frac{dx_A}{dz}\delta \quad (1.6\text{-}13)$$

Using $\bar{u} = \bar{u}_{z+\delta} = \bar{u}_{z-\delta}$, since we used an average $\bar{u}$, noting $c = n/N$, and $N_{Az} + N_{Bz} = cv_z^*$, we can make these substitutions into Eq. (1.6-12) and obtain

$$N_{Az} = \frac{c\bar{u}}{4}\left(-2\delta\frac{dx_A}{dz}\right) + x_A(N_{Az} + N_{Bz}) \quad (1.6\text{-}14)$$

The mean free path λ is defined as the average distance a molecule has traveled between collisions, where

$$\lambda = \frac{1}{\sqrt{2}\,\pi n d_A^2} \quad (1.6\text{-}15)$$

1.6 Molecular Diffusion Coefficients for Binary Gases

where d_A is the molecule diameter. Also, we will select a value of δ where

$$\delta = \tfrac{2}{3}\lambda \tag{1.6-16}$$

This means 50 percent of the molecules reaching a plane had their last collision at this distance δ from the plane.

In Eq. (1.6-14) the diffusion flux relative to the stream is equal to J_{Az}^*.

$$\frac{c\bar{u}}{4}\left(-2\delta\frac{dx_A}{dz}\right) = J_{Az}^* = -cD_{AB}\frac{dx_A}{dz} \tag{1.6-17}$$

Combining Eqs. (1.6-16) and (1.6-17) and solving for D_{AB},

$$D_{AB} = \tfrac{1}{3}\bar{u}\lambda. \tag{1.6-18}$$

Substituting Eqs. (1.6-9) and (1.6-15) into (1.6-18) and using $P = nkT$,

$$D_{AB} = \frac{2}{3}\left(\frac{k^3}{\pi^3}\right)^{1/2}\frac{T^{3/2}}{m_A^{1/2}Pd_A^2} \tag{1.6-19}$$

This final equation represents the diffusivity of two species A and B that have the same mass and diameter. This is an approximate equation, but the pressure effect is correct and the temperature effect qualitatively so.

2. *Chapman-Enskog Theory Prediction.* A more accurate and rigorous treatment must, of course, consider the intermolecular forces between molecules and the different sizes of molecules A and B. Chapman and Enskog (H1) solved the Boltzmann equation, which does not use the mean free path but a distribution function.

However, to solve this equation the potential energy of interaction $\varphi(r)$ between a given pair of molecules in the gas phase must be known. For nonpolar molecules a reasonable approximation to the forces between molecules is the Lennard-Jones function:

$$\varphi_{AB}(r) = 4\epsilon_{AB}\left[\left(\frac{\sigma_{AB}}{r}\right)^{12} - \left(\frac{\sigma_{AB}}{r}\right)^{6}\right] \tag{1.6-20}$$

where ϵ_{AB} and σ_{AB} are empirical parameters (see Fig. 1.6-3). The r^{-12} term accounts for the repulsion forces and r^{-6} for attraction. If one experimentally measures the viscosity of a pure gas, the parameters can be obtained for the pure gas, and various combining rules can be used to give the parameters for a mixture.

The resultant equation for predicting the diffusivity of a binary gas pair A-B at low density using the Chapman-Enskog theory and the Lennard-Jones potential energy function is

$$D_{AB} = 0.0018583\left(\frac{1}{M_A} + \frac{1}{M_B}\right)^{1/2}\frac{T^{3/2}f_D}{P\sigma_{AB}^2\Omega_{D,AB}} \tag{1.6-21}$$

26   Molecular Mass Transport Phenomena in Fluids

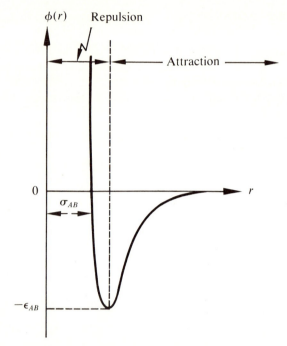

FIGURE 1.6-3   Lennard-Jones potential-energy function $\phi(r)$ as a function of the distance r between molecules

where D_{AB} = diffusivity, cm²/sec
   T = °K
   P = total pressure, atm
   σ_{AB} = "collision diameter," Å
   = $(\sigma_A + \sigma_B)/2$ (see Table 1.6-2 for σ_A and σ_B)
   ϵ_{AB} = energy of molecular interaction, ergs

$\epsilon_{AB}/k = \sqrt{\dfrac{\epsilon_A}{k}\dfrac{\epsilon_B}{k}}$, °K (see Table 1.6-2 for ϵ_A/k and ϵ_B/k)

   k = Boltzmann constant
   $\Omega_{D,AB}$ = collision integral based on the Lennard-Jones potential, Table 1.6-3 (Ω_D symbol is often used)
   f_D = a small second-order correction factor, which is close to 1.0 and is dropped

The collision integral $\Omega_{D,AB}$ is really a ratio giving the deviation of a gas with Lennard-Jones type interactions compared to a gas composed of rigid, elastic spheres. The value of this integral would be 1.0 for a gas with no interactions. Potential energy functions other than the Lennard-Jones can be used (R1).

Table 1.6-2 Force Constants and Collision Diameters as Determined from Viscosity Data* (R1)

Molecule	Compound	σ, Å	ϵ_0/k, °K
A	Argon	3.542	93.3
He	Helium	2.551†	10.22
Kr	Krypton	3.655	178.9
Ne	Neon	2.820	32.8
Xe	Xenon	4.047	231.0
Air	Air	3.711	78.6
AsH_3	Arsine	4.145	259.8
BCl_3	Boron chloride	5.127	337.7
BF_3	Boron fluoride	4.198	186.3
$B(OCH_3)_3$	Methyl borate	5.503	396.7
Br_2	Bromine	4.296	507.9
CCl_4	Carbon tetrachloride	5.947	322.7
CF_4	Carbon tetrafluoride	4.662	134.0
$CHCl_3$	Chloroform	5.389	340.2
CH_2Cl_2	Methylene chloride	4.898	356.3
CH_3Br	Methyl bromide	4.118	449.2
CH_3Cl	Methyl chloride	4.182	350
CH_3OH	Methanol	3.626	481.8
CH_4	Methane	3.758	148.6
CO	Carbon monoxide	3.690	91.7
COS	Carbonyl sulfide	4.130	336.0
CO_2	Carbon dioxide	3.941	195.2
CS_2	Carbon disulfide	4.483	467
C_2H_2	Acetylene	4.033	231.8
C_2H_4	Ethylene	4.163	224.7
C_2H_6	Ethane	4.443	215.7
C_2H_5Cl	Ethyl chloride	4.898	300
C_2H_5OH	Ethanol	4.530	362.6
C_2N_2	Cyanogen	4.361	348.6
CH_3OCH_3	Methyl ether	4.307	395.0
CH_2CHCH_3	Propylene	4.678	298.9
CH_3CCH	Methylacetylene	4.761	251.8
C_3H_6	Cyclopropane	4.807	248.9
C_3H_8	Propane	5.118	237.1
$n\text{-}C_3H_7OH$	n-Propyl alcohol	4.549	576.7
CH_3COCH_3	Acetone	4.600	560.2
CH_3COOCH_3	Methyl acetate	4.936	469.8
$n\text{-}C_4H_{10}$	n-Butane	4.687	531.4
$iso\text{-}C_4H_{10}$	Isobutane	5.278	330.1
$C_2H_5OC_2H_5$	Ethyl ether	5.678	313.8
$CH_3COOC_2H_5$	Ethyl acetate	5.205	521.3
$n\text{-}C_5H_{12}$	n-Pentane	5.784	341.1
$C(CH_3)_4$	2,2-Dimethylpropane	6.464	193.4
C_6H_6	Benzene	5.349	412.3

Table 1.6-2 (continued)

Molecule	Compound	σ, Å	ϵ_0/k, °K
C_6H_{12}	Cyclohexane	6.182	297.1
$n\text{-}C_6H_{14}$	n-Hexane	5.949	399.3
Cl_2	Chlorine	4.217	316.0
F_2	Fluorine	3.357	112.6
HBr	Hydrogen bromide	3.353	449
HCN	Hydrogen cyanide	3.630	569.1
HCl	Hydrogen chloride	3.339	344.7
HF	Hydrogen fluoride	3.148	330
HI	Hydrogen iodide	4.211	288.7
H_2	Hydrogen	2.827	59.7
H_2O	Water	2.641	809.1
H_2O_2	Hydrogen peroxide	4.196	289.3
H_2S	Hydrogen sulfide	3.623	301.1
Hg	Mercury	2.969	750
$HgBr_2$	Mercuric bromide	5.080	686.2
$HgCl_2$	Mercuric chloride	4.550	750
HgI_2	Mercuric iodide	5.625	695.6
I_2	Iodine	5.160	474.2
NH_3	Ammonia	2.900	558.3
NO	Nitric oxide	3.492	116.7
NOCl	Nitrosyl chloride	4.112	395.3
N_2	Nitrogen	3.798	71.4
N_2O	Nitrous oxide	3.828	232.4
O_2	Oxygen	3.467	106.7
PH_3	Phosphine	3.981	251.5
SF_6	Sulfur hexafluoride	5.128	222.1
SO_2	Sulfur dioxide	4.112	335.4
SiF_4	Silicon tetrafluoride	4.880	171.9
SiH_4	Silicon hydride	4.084	207.6
$SnBr_4$	Stannic bromide	6.388	563.7
UF_6	Uranium hexafluoride	5.967	236.8

*R. A. Svehla, *NASA Tech. Rept.* R-132, Lewis Research Center, Cleveland, Ohio (1962).
†The potential σ was determined by quantum-mechanical formulas.
SOURCE: R. C. Reid and T. K. Sherwood, *The Properties of Gases and Liquids*, 2d ed. (New York: McGraw-Hill, Inc., 1966), pp. 632–633. With permission.

In cases where the force constants are not available, they may be estimated from

$$\sigma_A = 1.18 V_b^{1/3} \quad (1.6\text{-}22)$$

or

$$\epsilon_0/k = 1.21 T_b \quad (1.6\text{-}23)$$

$$\epsilon_0/k = 0.75 T_c \quad (1.6\text{-}24)$$

Table 1.6-3 Values of the Collision Integral Ω_D Based on the Lennard-Jones Potential (H1)

kT/ϵ_0†	Ω_D†	kT/ϵ_0	Ω_D	kT/ϵ_0	Ω_D
0.30	2.662	1.65	1.153	4.0	0.8836
0.35	2.476	1.70	1.140	4.1	0.8788
0.40	2.318	1.75	1.128	4.2	0.8740
0.45	2.184	1.80	1.116	4.3	0.8694
0.50	2.066	1.85	1.105	4.4	0.8652
0.55	1.966	1.90	1.094	4.5	0.8610
0.60	1.877	1.95	1.084	4.6	0.8568
0.65	1.798	2.00	1.075	4.7	0.8530
0.70	1.729	2.1	1.057	4.8	0.8492
0.75	1.667	2.2	1.041	4.9	0.8456
0.80	1.612	2.3	1.026	5.0	0.8422
0.85	1.562	2.4	1.012	6	0.8124
0.90	1.517	2.5	0.9996	7	0.7896
0.95	1.476	2.6	0.9878	8	0.7712
1.00	1.439	2.7	0.9770	9	0.7556
1.05	1.406	2.8	0.9672	10	0.7424
1.10	1.375	2.9	0.9576	20	0.6640
1.15	1.346	3.0	0.9490	30	0.6232
1.20	1.320	3.1	0.9406	40	0.5960
1.25	1.296	3.2	0.9328	50	0.5756
1.30	1.273	3.3	0.9256	60	0.5596
1.35	1.253	3.4	0.9186	70	0.5464
1.40	1.233	3.5	0.9120	80	0.5352
1.45	1.215	3.6	0.9058	90	0.5256
1.50	1.198	3.7	0.8998	100	0.5130
1.55	1.182	3.8	0.8942	200	0.4644
1.60	1.167	3.9	0.8888	400	0.4170

†Hirschfelder uses the symbols T^* for kT/ϵ_0 and $\Omega^{(1,1)*}$ in place of Ω_D.

SOURCE: J. O. Hirschfelder, C. F. Curtiss, and R. B. Bird, *Molecular Theory of Gases and Liquids*, (New York: John Wiley & Sons, Inc., 1954). With permission.

where V_b = molar volume at the normal boiling point, cm³/g mole (estimate this from Table 1.6-4)
T_b = normal boiling point, °K
T_c = critical temperature, °K

Table 1.6-4 Atomic and Molar Volumes at the Normal Boiling Point (L1, P2, T2)

Material	Atomic Volume, cm³/g mole
Carbon	14.8
Hydrogen	3.7
Oxygen (except as noted below)	7.4
Doubly bound as carbonyl	7.4
Coupled to two other elements	
In aldehydes and ketones	7.4
In methyl esters	9.1
In methyl ethers	9.9
In ethyl esters and ethers	9.9
In higher esters and ethers	11.0
In acids (−OH)	12.0
Joined to S, P, N	8.3
Nitrogen	
Doubly bonded	15.6
In primary amines	10.5
In secondary amines	12.0
Bromine	27.0
Chlorine as in RCHClR′	24.6
Chlorine as in RCl (terminal)	21.6
Fluorine	8.7
Iodine	37.0
Sulfur	25.6
Phosphorus	27.0
Mercury	19.0
Silicon	32.0
Chromium	27.4
Tin	42.3
Titanium	35.7
Lead	46.5–50.1
Zinc	20.4
Ring, three-membered as in ethylene oxide	−6
Four-membered	−8.5
Five-membered	−11.5
Six-membered	−15
Naphthalene ring	−30
Anthracene ring	−47.5

Table 1.6-4 (continued)

Material	Molecular Volume, cm^3/g mole
Air	29.9
O_2	25.6
N_2	31.2
Br_2	53.2
Cl_2	48.4
CO	30.7
CO_2	34.0
COS	51.5
H_2	14.3
H_2O	18.8
H_2S	32.9
I_2	71.5
NH_3	25.8
NO	23.6
N_2O	36.4
SO_2	44.8

This method can be used with reasonably good results to predict values with an average deviation of about 8 percent up to about 1000°K (R1).

For mixtures of a polar gas and nonpolar gas Stiel and Thodos (S2) give methods to calculate the force constants and collision diameters for the polar molecule. Monchick and Mason (M1, M2) found that if the correct force constant is used for the polar gas—that is, the constant determined by a viscosity experiment or estimated by the special method above (S2)—that Eq. (1.6-21) can be used with about the same accuracy as for nonpolar pairs.

For systems that contain water as the one polar compound, the predicted values are consistently low by about 10 percent (R1). Hence, it is recommended that one multiply the predicted value by 1.09 when using Eq. (1.6-21).

For polar–polar gas pairs the intermolecular forces differ because of dipole–dipole interactions. The potential function commonly used is the Stockmayer potential. Monchick and Mason (M2) give details and a table of the collision integral using the Stockmayer potentials.

The effect of composition of the binary mixture A-B is included in the factor f_D in Eq. (1.6-21), which was assumed as 1.0 and is close to that value. In solving the Boltzmann equation for the diffusivity, Chapman and Cowling (C3) used a Sonine polynomial expansion. This was expressed as an infinite series, and as an approximation only the first term was used, since the series converges very rapidly.

32   Molecular Mass Transport Phenomena in Fluids

In the use of this expansion in this manner, called the first approximation, only the interactions between molecules A and B were considered, and the mole fractions x_A and x_B did not enter in. In carrying the equation's solution further, the second approximation which is f_D includes the concentrations and also interactions between A and A and B and B. This equation for f_D is a complicated function of the mole fractions and other physical properties.

For rigid spheres with no interactions the maximum effect of concentration was predicted as 13.2 percent (C3). However, for real gases with interactions this is reduced to about 4 percent, and at room temperatures the maximum effect is about 3 percent. For equimolar mixtures this effect is only about 1.5 percent. Hence, in conclusion, the concentration effect of f_D is generally neglected. This has been discussed by others (W1).

3. *Empirical Method of Fuller, Schettler, and Giddings.* This empirical method of Fuller *et al.* (F1) was obtained by correlating many recent data. The equation uses atomic volumes from Table 1.6-5, which are summed for each gas molecule. These numbers differ numerically from the Le Bas (L1) molar volumes in Table 1.6-4. The equation is

$$D_{AB} = \frac{0.00100 T^{1.75}\sqrt{\frac{1}{M_A}+\frac{1}{M_B}}}{P[(\Sigma v_A)^{1/3}+(\Sigma v_B)^{1/3}]^2} \qquad (1.6\text{-}25)$$

where Σv_A = sum of structural volume increments, Table 1.6-5. This method is simple to use for moderate temperature ranges. No additional correction factor for mixtures containing water is needed. The method can be used for mixtures of nonpolar gases or for a polar and a nonpolar mixture.

4. *Effect of Temperature on Diffusivity.* Using Eq. (1.6-21), we can see that

$$D_{AB} \propto \frac{T^{3/2}}{\Omega_D} \qquad (1.6\text{-}26)$$

This correction factor can also be used to correct experimental data from one temperature to another up to 1000°K, but the force constants must be estimated for the Ω_D. A simpler method to use over temperature ranges up to about 1000°K is Eq. (1.6-25), which says that D_{AB} is proportional to $T^{1.75}$. To explain the effects on diffusion at higher temperatures, the kinetic energy and velocity of the molecules increase; hence, they will diffuse more rapidly.

The Schmidt number varies very little with temperature and can be assumed independent of temperature over moderate ranges. The Schmidt number is calculated for a dilute gas A in B by using the viscosity of B, density of B, and D_{AB}.

1.6 Molecular Diffusion Coefficients for Binary Gases

Table 1.6-5 Atomic Diffusion Volumes for Use with the Fuller, Schettler, and Giddings Method (F1)

A. Atomic and Structural Diffusion Volume Increments, v

C	16.5	(Cl)	19.5
H	1.98	(S)	17.0
O	5.48	Aromatic ring	−20.2
(N)	5.69	Heterocyclic ring	−20.2

B. Diffusion Volumes for Simple Molecules, Σv

H_2	7.07	CO	18.9
D_2	6.70	CO_2	26.9
He	2.88	N_2O	35.9
N_2	17.9	NH_3	14.9
O_2	16.6	H_2O	12.7
Air	20.1	(CCl_2F_2)	114.8
Ar	16.1	(SF_6)	69.7
Kr	22.8	(Cl_2)	37.7
(Xe)	37.9	(Br_2)	67.2
Ne	5.59	(SO_2)	41.1

Note: Parentheses indicate that the value listed is based on only a few data points.

SOURCE: From E. N. Fuller, P. D. Schettler, and J. C. Giddings, *Ind. Eng. Chem.* **58**, 19 (1966). With permission.

5. *Effect of Pressure on Diffusivity.* The diffusivity is shown to be inversely proportional to pressure. This relation is good up to about 100 atm and will not introduce additional errors due to the pressure effect of more than about 10 percent (R1). The Schmidt number is essentially independent of pressure up to moderate pressures of about 10 atm. To explain the effect of pressure on the D_{AB}, it is apparent that for a higher pressure, the molecules will be closer together. The diffusing molecule will meet more resistance and, hence, diffuse more slowly.

If an experimental value of D_{AB} is available at a given T and P and it is desired to have a value of D_{AB} at another T and P, one should correct the experimental value to the new conditions by using the $T^{1.75}$ and $1/P$ corrections. This is generally preferred compared to predicting a new value completely.

Example 1.6-1 Estimation of Diffusivities of Gases

Butanol is diffusing through air at 25.9°C and 1 atm. (a) Estimate the diffusivity at 25.9°C and 59.0°C using the Chapman-Enskog method. (b) Estimate it using the Fuller *et al.* method. (c) Compare these with the experimental data.

Solution

For part (a), $P = 1.0$ atm, $T = 298.9°K$, $M_A = 74.1$, $M_B = 29$. From Table 1.6-2 for air, $\epsilon_B/k = 78.6°K$, $\sigma_B = 3.711$ Å. For butanol Eq. (1.6-22) and Eq. (1.6-23) must be used. From Table 1.6-4 for butanol (C_2H_5-CH_2CH_2OH), $V_A = 4(14.8) + 10(3.7) + 7.4 = 103.6$. Hence, $\sigma_A = 1.18(103.6)^{1/3} = 5.55$. The normal boiling point is $T_b = 273 + 117 = 390°K$. Then, $\epsilon_A/k = 1.21(390) = 472$.

$$\sigma_{AB} = \frac{3.711 + 5.55}{2} = 4.630$$

$$\frac{\epsilon_{AB}}{k} = \sqrt{78.6(472)} = 193.5$$

$$\frac{kT}{\epsilon_{AB}} = \frac{298.9}{193.5} = 1.545$$

From Table 1.6-3 $\Omega_D = 1.183$

$$\sqrt{\frac{1}{M_A} + \frac{1}{M_B}} = \sqrt{\frac{1}{74.1} + \frac{1}{29}} = 0.219$$

From Eq. (1.6-21),

$$D_{AB} = \frac{(0.0018583)298.9^{3/2}(0.219)}{1.0(4.630)^2(1.183)}$$

$$= 0.0835 \text{ cm}^2/\text{sec at } 25.9°C \quad \text{[Ans. to (a)]}$$

For 59.0°C, $T = 332$, $kT/\epsilon_{AB} = 332/193.5 = 1.715$, $\Omega_D = 1.136$

$$D_{AB} = 0.101 \text{ cm}^2/\text{sec at } 59.0°C \quad \text{[Ans. to (a)]}$$

From Table 1.6-5 for part (b), $\Sigma v_A = 4(16.5) + 10(1.98) + 1(5.48) = 91.28$, $\Sigma v_B = 20.1$ for air (B). From Eq. (1.6-25),

$$D_{AB} = \frac{0.00100(298.9)^{1.75}(0.219)}{1.0[(91.28)^{1/3} + (20.1)^{1/3}]^2}$$

$$= 0.0905 \text{ cm}^2/\text{sec at } 25.9°C \quad \text{[Ans. to (b)]}$$

$$D_{AB} = 0.109 \text{ cm}^2/\text{sec at } 59.0°C \quad \text{[Ans. to (b)]}$$

The experimental values from Table 1.6-1 are 0.087 cm²/sec at 25.9°C and 0.104 cm²/sec at 59.0°C. The predicted values deviate a maximum of 4.8 percent [Ans. to (c)].

1.7 DERIVATION OF EQUATION FOR MOLECULAR TRANSPORT PLUS CONVECTIVE FLOW USING MOLAR FLUXES

1.7A Derivation Using the Sum of Diffusion Plus Convective Velocities — Method I

In this section we will discuss in detail the two most important fluxes, J_A^* and N_A, which are molar fluxes and are related to the molar average velocity v^* of the stream. These fluxes are the most convenient to use. In Chapter 2, Section 2.5, other types such as mass fluxes will be considered. The relationship of these mass fluxes to the molar fluxes and to the mass and/or molar average velocity of the stream will also be discussed in Section 2.5.

In Section 1.4A we used Fick's law in a nonflow system of no bulk or convective flow of the whole fluid and for molecular diffusion relative to a stagnant fluid. Since the fluid was stagnant, the flux J_A^* relative to the stagnant fluid is also equal in this case to the flux relative to a fixed or stationary point. However, often the fluid is not stagnant but flowing. We will define the diffusion flux J_A^* as the flux of A relative to the average molar velocity, which for a gas is the same as the average volumetric velocity of all constituents in the stream. In many diffusional processes there is a net flow of the entire phase in the direction of diffusion of component A. This flow of the entire phase will also move component A from one point to another completely independently of the diffusion flux of A itself. If component B is diffusing in the reverse direction, the movement of the entire phase in the opposite direction will slow the net flow of B when viewed from a stationary position.

As a result of the discussion above, we define the flux $\overline{N}_A/A$ or N_A, which is the net flux of A in g mole/sec-cm² relative to stationary coordinates. This flux, a vector with direction and magnitude, is the sum of the diffusion flux vector J_A^* and the convective flux vector.

An analogy, for example, is that of a fish swimming in a flowing river in the direction of the current. The fish swims at a flux of J_A^* relative to the flowing river. The convective flux can be looked at as the flux of the fish due to the bulk motion of the river relative to the shore, which is a stationary point. The total flux N_A relative to the shore can be considered as the sum of the flux of the fish relative to the river plus the flux of the fish resulting from the convective motion of the river.

To derive the equation for N_A for diffusion and convection, consider Fig. 1.7-1, which is a small element of a binary fluid A and B. Since $c_{A1} > c_{A2}$, then A diffuses to the right in the z direction and B diffuses to the left. Consider a thin slice of thickness dz located a distance of z from point 1, and consider the process containing the binary fluid A and B. The velocity of movement of A relative to stationary coordinates is v_A cm/sec.

36 Molecular Mass Transport Phenomena in Fluids

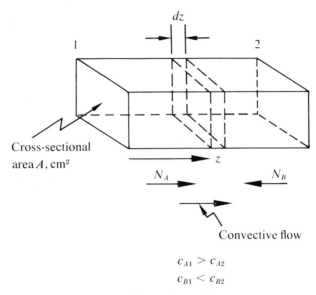

FIGURE 1.7-1   Diffusion of A and B in a binary mixture

Hence, $N_A = v_A c_A$ and $N_B = v_B c_B$. Table 1.7-1 lists this nomenclature. We will derive the equation for all directions so that $\mathbf{v}_A$, $\mathbf{N}_A$, and so on are vectors. Now the bulk flow or convective flux of the entire phase is $\mathbf{N}_A + \mathbf{N}_B$ g mole of A plus B per sec per cm². Hence, the molar average velocity of the entire phase is $\mathbf{v}^*$ cm/sec, and $\mathbf{v}^* c$ is also the flux of the entire phase.

$$\mathbf{v}^* = \frac{\mathbf{v}_A c_A + \mathbf{v}_B c_B}{c} = \frac{\mathbf{N}_A + \mathbf{N}_B}{c} \tag{1.7-1}$$

The total flux, $\mathbf{N}_A$, of component A relative to a stationary point will be equal to the sum of the diffusion flux of A relative to the molar average velocity of the stream, $\mathbf{J}_A^*$, plus the convective flux of A relative to stationary coordinates, $\mathbf{v}^* c_A$. Or

$$\begin{array}{ccc} \mathbf{N}_A & = & \mathbf{J}_A^* & + & \mathbf{v}^* c_A. \end{array} \tag{1.7-2}$$
(total flux of A relative to a fixed point)   (flux of A relative to molar average velocity — that is, diffusion flux)   (flux of A due to convective flow relative to a fixed point)

Substituting Eq. (1.7-1) into (1.7-2),

$$\mathbf{N}_A = \mathbf{J}_A^* + \frac{c_A}{c}(\mathbf{N}_A + \mathbf{N}_B) = \mathbf{J}_A^* + x_A(\mathbf{N}_A + \mathbf{N}_B) \tag{1.7-3}$$

Also for $\mathbf{N}_B$,

$$\mathbf{N}_B = \mathbf{J}_B^* + x_B(\mathbf{N}_A + \mathbf{N}_B) \tag{1.7-4}$$

1.7 Equation for Molecular Transport plus Convection

Table 1.7-1 Notation for Binary Systems

Velocities
v_A = velocity of A relative to stationary coordinates, cm/sec.
v_B = velocity of B relative to stationary coordinates, cm/sec.
$\mathbf{v} = w_A \mathbf{v}_A + w_B \mathbf{v}_B = (1/\rho)(\rho_A \mathbf{v}_A + \rho_B \mathbf{v}_B)$ = mass average velocity of bulk stream, cm/sec.
$\mathbf{v}_A - \mathbf{v}^*$ = velocity of A relative to the molar average velocity of the bulk stream—that is, the diffusion velocity, cm/sec.
$\mathbf{v}_B - \mathbf{v}^*$ = velocity of B relative to the molar average velocity of the bulk stream, cm/sec.
$\mathbf{v}^* = \dfrac{c_A \mathbf{v}_A + c_B \mathbf{v}_B}{c} = x_A \mathbf{v}_A + x_B \mathbf{v}_B$ = molar average velocity of bulk stream, cm/sec.

Fluxes†
$\mathbf{J}_A^* = \bar{\mathbf{J}}_A^*/A = c_A(\mathbf{v}_A - \mathbf{v}^*)$ = diffusion flux of A relative to the molar average velocity of the bulk stream, g mole A/sec-cm².
$\mathbf{J}_B^* = \bar{\mathbf{J}}_B^*/A = c_B(\mathbf{v}_B - \mathbf{v}^*)$ = diffusion flux of B relative to the molar average velocity of the bulk stream, g mole B/sec-cm².
$\mathbf{N}_A = c_A \mathbf{v}_A$ = total flux of A relative to stationary coordinates, g mole A/sec-cm².
$\mathbf{N}_B = c_B \mathbf{v}_B$ = total flux of B relative to stationary coordinates, g mole B/sec-cm².

Concentrations
$c_A = \rho_A/M_A$ = concentration of A, g mole A/cm³ solution.
ρ_A = concentration of A, g mass A/cm³.
M_A = g mass A/g mole A.
x_A = mole fraction of A.
$c = c_A + c_B$ = total concentration, g mole $A + B$/cm³.
$\rho = \rho_A + \rho_B$ = total concentration, g mass $A + B$/cm³.
$w_A = x_A M_A/(x_A M_A + x_B M_B) = \rho_A/\rho$ = mass fraction of A.
$w_A + w_B = 1.0$.
$M = x_A M_A + x_B M_B = \rho/c$ = g mass $A + B$/g mole $A + B$.

†For additional definitions of fluxes and relations between fluxes see Tables 2.5-1 and 2.5-2 in Chapter 2.

Equations (1.7-3) and (1.7-4) are the general equations to use for diffusion whether there is convective flow or not. They can be used for molecular diffusion of fluids or for turbulent diffusion. The only change that occurs between molecular or turbulent diffusion is in the form of equation to use in $\mathbf{J}_A^*$.

For molecular diffusion of components A and/or B the following general equations can be obtained by substituting Fick's law in Eqs. (1.7-3) and (1.7-4) for the z direction only.

$$N_A = -cD_{AB}\frac{dx_A}{dz} + \frac{c_A}{c}(N_A + N_B) \quad (1.7\text{-}5)$$

$$N_B = -cD_{BA}\frac{dx_B}{dz} + \frac{c_B}{c}(N_A + N_B) \quad (1.7\text{-}6)$$

For a gas phase with constant pressure P (equivalent to constant c) and constant T, adding Eqs. (1.7-5) and (1.7-6),

$$N_A + N_B = -cD_{AB}\frac{dx_A}{dz} - cD_{BA}\frac{dx_B}{dz} + \frac{c_A + c_B}{c}(N_A + N_B)$$

Hence,

$$-cD_{AB}\frac{dx_A}{dz} = cD_{BA}\frac{dx_B}{dz}$$

Now, since $\Sigma x = 1.0 = x_A + x_B$, then $dx_A/dz = -dx_B/dz$ and

$$D_{AB} = D_{BA}$$

Equation (1.7-5) was written for diffusion and flow in only the z direction. Actually, diffusion and flow can occur in all three directions, so Eq. (1.7-5) can be written as

$$\mathbf{N}_A = -cD_{AB}\left(\frac{\partial x_A}{\partial x} + \frac{\partial x_A}{\partial y} + \frac{\partial x_A}{\partial z}\right) + x_A(\mathbf{N}_A + \mathbf{N}_B)$$

$$= -cD_{AB}\nabla x_A + x_A(\mathbf{N}_A + \mathbf{N}_B). \quad (1.7\text{-}7)$$

The reader should note that for $\mathbf{J}_A^*$, $-D_{AB}\nabla c_A$ is used in place of $-cD_{AB}\nabla x_A$ only when c is a constant as in many gas systems.

1.7B Derivation Using the Stefan-Maxwell Equations — Method II

An alternate method that is sometimes used to derive Eq. (1.7-5) is to use the method of Stefan and Maxwell (W2), who used the kinetic theory of gases. The basic assumptions are as follows for a binary gas A and B.

1. The resistance to diffusion of gas A is proportional to the number of molecules of diffusing gas A in a given volume, or c_A.
2. The resistance is proportional to the number of molecules of gas B through which A diffuses, or c_B.
3. The resistance is proportional to the difference between the velocity v_A of gas A in the direction of net diffusion and the velocity v_B, or $v_A - v_B$.
4. The resistance is proportional to the length of path of diffusion, or dz.
5. The driving force for diffusion of A is the change in partial pressure in the direction of diffusion, or $-dp_A$. Combining these terms into an equation at constant pressure P,

$$-dp_A = \alpha_{AB} c_A c_B (v_A - v_B)\, dz \quad (1.7\text{-}8)$$

1.7 Equation for Molecular Transport plus Convection

where α_{AB} is a proportionality constant for diffusion of A through B. Multiplying out Eq. (1.7-8),

$$-dp_A = \alpha_{AB}(c_B c_A v_A - c_A c_B v_B)\, dz \tag{1.7-9}$$

But $N_A = c_A v_A$ and $N_B = c_B v_B$, and $c_A = p_A/RT$. So

$$-RT\, dc_A = \alpha_{AB}(c_B N_A - c_A N_B)\, dz \tag{1.7-10}$$

Define D_{AB} as

$$D_{AB} = \frac{R^2 T^2}{\alpha_{AB} P} \tag{1.7-11}$$

Substituting Eq. (1.7-11) into (1.7-10) and $c_B = c - c_A$,

$$-dc_A = \frac{RT}{D_{AB} P}(cN_A - c_A N_A - c_A N_B)\, dz \tag{1.7-12}$$

Rearranging,

$$cN_A - c_A N_A - c_A N_B = -D_{AB}\frac{P}{RT}\frac{dc_A}{dz} \tag{1.7-13}$$

Solving for N_A,

$$N_A = \frac{-D_{AB}}{c}\frac{P}{RT}\frac{dc_A}{dz} + \frac{c_A}{c}(N_A + N_B) \tag{1.7-14}$$

Now, substituting $c = P/RT$ into Eq. (1.7-14), we get a result identical to Eq. (1.7-5).

$$N_A = -cD_{AB}\frac{dx_A}{dz} + \frac{c_A}{c}(N_A + N_B) \tag{1.7-5}$$

Equation (1.7-5) can be rewritten for gases in terms of partial pressures.

$$N_A = \frac{-D_{AB}}{RT}\frac{dp_A}{dz} + \frac{p_A}{P}(N_A + N_B) \tag{1.7-15}$$

In Chapter 2, Sections 2.5 and 2.6, further discussions of fluxes and Fick's law are given and the general equation of continuity is derived.

1.7C Integration of the General Equation for Diffusion Plus Convection at Steady-State

Equation (1.7-5) derived above is completely general for two-component systems and accounts for any combination of the individual fluxes N_A and N_B. The equation can be integrated for fluids for steady-state and different combination of the fluxes. If the cross-sectional area A is assumed constant, the equation can be integrated over a finite thickness or distance. The D_{AB} will be considered constant. This, of course, is justified for gases.

40   Molecular Mass Transport Phenomena in Fluids

1. *General Integrated Equation.* Equation (1.7-5) can be rearranged and the variables separated so that for c, N_A, and N_B constant, $c_A = c x_A$ and

$$\int_{c_{A1}}^{c_{A2}} \frac{-dc_A}{\left(\dfrac{cN_A}{N_A+N_B} - c_A\right)} = \frac{N_A+N_B}{cD_{AB}} \int_{z_1}^{z_2} dz \tag{1.7-16}$$

Integrating, rearranging, and multiplying both sides by N_A,

$$N_A = \frac{N_A}{N_A+N_B} \frac{D_{AB}c}{z_2-z_1} \ln\left[\frac{c_{A2}/c - N_A/(N_A+N_B)}{c_{A1}/c - N_A/(N_A+N_B)}\right] \tag{1.7-17}$$

In terms of partial pressures Eq. (1.7-17) becomes as follows, if we let $z = z_2 - z_1$:

$$N_A = \frac{N_A}{N_A+N_B} \frac{D_{AB}P}{RTz} \ln\left[\frac{p_{A2}/P - N_A/(N_A+N_B)}{p_{A1}/P - N_A/(N_A+N_B)}\right] \tag{1.7-18}$$

Several important special cases of Eqs. (1.7-17) and (1.7-18) will now be considered for steady-state.

2. *Equimolar Counterdiffusion of A and B.* This situation of equimolar counterdiffusion at steady-state can arise in several ways in diffusional processes. In gases this occurs where the total pressure P is constant and the two ends of the diffusion path are permeable to both A and B. This is shown in Fig. 1.4-1. The flux N_A must equal $-N_B$ for the total pressure to remain constant. For this case $N_A/(N_A+N_B)$ is indeterminate, so we will start with Eq. (1.7-15) and set $N_A = -N_B$.

$$N_A = \frac{-D_{AB}}{RT}\frac{dp_A}{dz} + \frac{p_A}{P}(N_A + -N_A) \tag{1.7-19}$$

$$N_A \int_{z_1}^{z_2} dz = \frac{-D_{AB}}{RT} \int_{p_{A1}}^{p_{A2}} dp_A \tag{1.7-20}$$

$$N_A = \frac{D_{AB}}{RTz}(p_{A1} - p_{A2}) \tag{1.7-21}$$

This is the same as Eq. (1.4-13). This physical situation can also arise in distillation, where equimolar diffusion often occurs. In the case of equimolar counterdiffusion it is seen that $N_A = J_A^*$, since there is no convective flow.

3. *Diffusion of A through Stagnant or Nondiffusing B.* This case of diffusion of A through stagnant B at steady-state often occurs in absorption. In this case one boundary of the system is impermeable to component B so it cannot pass through. One example would be in evaporation of a pure liquid such as benzene at the bottom of a narrow

1.7 Equation for Molecular Transport plus Convection

tube where air is passed over the top open end of the tube. The benzene vapor (A) diffuses through the air (B) in the tube. The boundary at the surface of the liquid is impermeable to the air, since the air is insoluble in the benzene liquid. Another example is in the absorption of ammonia (A) vapor from an air-ammonia mixture by water. The water interface is impermeable to the air (B), since the air is only very slightly soluble in the water. Thus $N_B = 0$ or

$$\frac{N_A}{N_A + N_B} = \frac{N_A}{N_A + 0} = 1 \qquad (1.7\text{-}22)$$

Hence, Eq. (1.7-18) becomes

$$N_A = \frac{D_{AB}P}{RTz} \ln\left(\frac{P - p_{A2}}{P - p_{A1}}\right) \qquad (1.7\text{-}23)$$

Now, since $P = \text{constant} = p_{A1} + p_{B1} = p_{A2} + p_{B2}$, $P - p_{A2} = p_{B2}$, $P - p_{A1} = p_{B1}$, and $p_{B2} - p_{B1} = p_{A1} - p_{A2}$,

$$N_A = \frac{D_{AB}P}{RTz} \ln\left(\frac{p_{B2}}{p_{B1}}\right) \qquad (1.7\text{-}24)$$

Defining a ln mean value of the inert B as

$$p_{BM} = \frac{p_{B2} - p_{B1}}{\ln(p_{B2}/p_{B1})} = \frac{p_{A1} - p_{A2}}{\ln(p_{B2}/p_{B1})} \qquad (1.7\text{-}25)$$

and substituting Eq. (1.7-25) into (1.7-24),

$$N_A = \frac{D_{AB}P}{RTz\,p_{BM}}(p_{A1} - p_{A2}) \qquad (1.7\text{-}26)$$

Equation (1.7-26) could have been easily derived from Eq. (1.7-15) by letting $N_A = 0$ and integrating:

$$N_A = \frac{-D_{AB}}{RT}\frac{dp_A}{dz} + \frac{p_A}{P}(N_A + 0) \qquad (1.7\text{-}27)$$

It should be noted that the total flux of A is equal to the diffusion flux plus the convective flux of A.

Since B has a partial pressure gradient, it must be diffusing also at a rate relative to the bulk flow of

$$J_B^* = \frac{-D_{AB}}{RT}\frac{dp_B}{dz}$$

However, the bulk or convective flow carries along B at a velocity equal to the negative of the diffusion velocity, so that the net $N_B = 0$. This is similar to a fish swimming upstream at the same velocity as the river that flows downstream. The net velocity relative to a fixed point is zero.

Example 1.7-1 Analytical and Analog Computer Solution for Diffusion of A through Stagnant B

A thin layer of water 0.0100 ft thick in a flat pan is held at a constant temperature of 68°F (20°C) and total pressure of dry air above the pan of 1.01 atm and at 68°F. Assuming the evaporation to take place by molecular diffusion through a gas film of constant thickness of 0.0500 ft, calculate the time in hours for the water to evaporate completely. Also write the analog computer diagram solution.

Solution

This is the case of water (A) diffusing through stagnant air (B). From Table 1.6-1, $D_{AB} = 0.220$ cm²/sec at 0°C and 1.0 atm. Correcting this for temperature and pressure,

$$D_{AB} = 0.220 \left(\frac{293}{273}\right)^{1.75} \left(\frac{1.00}{1.01}\right) = 0.248 \text{ cm}^2/\text{sec}$$

or $0.248(3.87) = 0.960$ ft²/hr. Calling the surface of the water at the water-air interface point 1 and the other end of the path in the air point 2, then p_{A1} = vapor pressure of water at 20°C = 0.0231 atm, $p_{B1} = 1.0100 - 0.0231 = 0.9869$ atm, $p_{A2} = 0$, $p_{B2} = 1.010$ atm. Using Eq. (1.7-25),

$$p_{BM} = \frac{p_{B2} - p_{B1}}{\ln(p_{B2}/p_{B1})} = \frac{1.010 - 0.9869}{\ln(1.010/0.9869)} = 0.998 \text{ atm}$$

Since p_{B1} is very close to p_{B2}, the linear mean can be used and will be very close to the ln mean. Substituting into Eq. (1.7-26),

$$N_A = \frac{D_{AB}P}{RTzp_{BM}}(p_{A1} - p_{A2}) = \frac{0.960(1.010)(0.0231 - 0)}{0.730(528)(0.0500)(0.998)}$$

$$= 1.16 \times 10^{-3} \frac{\text{lb mole } A}{\text{hr-ft}^2}$$

Assuming 1.0 ft² of pan area, the

lb mole to evaporate = $0.0100(1)(62.4)(1/18) = 0.0346$ lb mole;

time to evaporate = $(0.0346)/(1.16 \times 10^{-3}) = 29.8$ hr.

For the analog computer solution we must go back to the original differential equation. (The reader should have read Sections 10.1 through 10.3 in Chapter 10 to understand this problem clearly.) Using Eq. (1.7-27) and rearranging, calling p_A/P as x_A and P/RT as c,

$$N_A = -cD_{AB}\frac{dx_A}{dz} + x_A N_A \tag{1.7-27a}$$

1.7 Equation for Molecular Transport plus Convection

Solving for the derivative,

$$\frac{dx_A}{dz} = -\left(\frac{N_A}{cD_{AB}}\right) + \left(\frac{N_A}{cD_{AB}}\right)x_A \quad (1.7\text{-}27b)$$

The analog computer diagram is shown in Fig. 1.7-2. To generate dx_A/dz, a constant voltage is added directly to the integrator, and another constant voltage times x_A is also put into integrator number 1. The IC is set at the value of $-x_A$ at $z = 0$, which is $x_{A1} = p_{A1}/P$. Since N_A is unknown, a value is guessed, the potentiometers are set accordingly and the run started. At $z = 0.0500$ (in terms of time τ), the value of x_A should be $x_{A2} = p_{A2}/P$. If not, a new guess is made. The reader should note that the computer is not usually used to solve cases that have analytical solutions. However, the purpose here is to acquaint the reader with use of the analog computer.

4. *Diffusion of A and B Controlled by Stoichiometry or Other Factors.* Often in catalytic chemical reactions where A and B are diffusing to and from a catalyst surface, the relation between the fluxes of A and B at steady-state is controlled by stoichiometry. An example is gas A diffusing to a catalyst surface and reacting instantaneously, with gas B diffusing back.

$$2A \rightarrow B$$

Here, $N_B = -\tfrac{1}{2}N_A$ and

$$\frac{N_A}{N_A + N_B} = \frac{N_A}{N_A - \tfrac{1}{2}N_A} = 2$$

Then, substituting into Eq. (1.7-18) or into (1.7-15) and integrating,

$$N_A = \frac{2D_{AB}P}{RTz} \ln\left(\frac{p_{A2}/P - 2}{p_{A1}/P - 2}\right) \quad (1.7\text{-}28)$$

In many cases where components A and B are being separated by distillation, the relation between the fluxes N_A and N_B is determined by

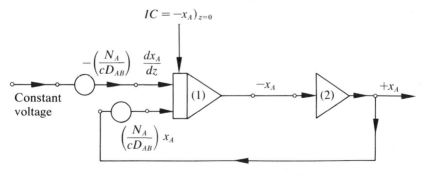

FIGURE 1.7-2 Analog computer diagram for Example 1.7-1

44 Molecular Mass Transport Phenomena in Fluids

a heat balance. For example, if the latent heats of A and B are equal, and sensible heat effects are neglected, then condensation of 1.0 mole of A will release enough heat to vaporize 1.0 mole of B – or, $N_A = -N_B$.

Example 1.7-2 Effects of Flux Ratios on Flux of A

Gas components A and B are diffusing under steady-state conditions at 1.0 atm total pressure and 0°C. The partial pressure p_{A1} is 100 mm Hg and p_{A2} is 50 mm at a distance of 10 cm apart. The diffusivity for the mixture is 0.185 cm²/sec. Calculate the total flux N_A of component A and N_B of B, if (a) the boundary at point 1 is impermeable to B, (b) both boundaries are permeable to A and B, (c) species A diffuses to the surface of a catalyst with solid C so the stoichiometry is

$$\tfrac{1}{2}A + C(\text{solid}) \to B.$$

(d) Tabulate N_A and N_B.

Solution

Part (a) is the case of A diffusing through stagnant B. $p_{A1} = 100/760 = 0.1317$ atm, $p_{A2} = 50/760 = 0.0658$ atm, $p_{B1} = 1.0 - 0.1317 = 0.8683$, $p_{B2} = 1 - 0.0658 = 0.9342$ atm. Then

$$p_{BM} = \frac{0.9342 - 0.8683}{\ln(0.9342/0.8683)} = 0.901 \text{ atm}$$

To calculate the total flux, use Eq. (1.7-26).

$$N_A = \frac{D_{AB}P}{RTzp_{BM}}(p_{A1} - p_{A2}) = \frac{0.185(1)(0.1317 - 0.0658)}{82.06(273)(10)(0.901)}$$

$N_A = 6.03 \times 10^{-8}$ g mole A/sec-cm², $\quad N_B = 0 \quad$ [Ans. to (a)]

For case (b), this is just Fick's law for J_A^*, Eq. (1.4-13) or (1.7-21).

$$N_A = J_A^* = \frac{D_{AB}}{RTz}(p_{A1} - p_{A2}) = \frac{0.185(0.1317 - 0.0658)}{82.06(273)(10)}$$

$$= 5.46 \times 10^{-8} \text{ g mole A/sec-cm}^2$$

$N_B = -N_A = -5.46 \times 10^{-8}$ g mole B/sec-cm² $\quad$ [Ans. to (b)]

For case (c), $N_B = -2N_A$. Hence, Eq. (1.7-18) is used. Here,

$$\frac{N_A}{N_A + N_B} = \frac{N_A}{N_A - 2N_A} = -1$$

$$N_A = (-1)\frac{(0.185)(1)}{82.06(273)(10)} \ln\left(\frac{0.0658 - (-1)}{0.1317 - (-1)}\right)$$

$$= 4.96 \times 10^{-8} \text{ g mole A/sec-cm}^2$$

$$N_B = -2N_A = -2(4.96 \times 10^{-8})$$
$$= -9.92 \times 10^{-8} \text{ g mole } B/\text{sec-cm}^2 \qquad [\text{Ans. to (c)}]$$

Tabulation for part (d):

Case	N_A	N_B
(a)	6.03×10^{-8}	0
(b)	5.46×10^{-8}	-5.46×10^{-8}
(c)	4.96×10^{-8}	-9.92×10^{-8}

Hence, as the flux of B is increased, the flux of A changes, owing to convection effects.

1.8 DIFFUSION AND CHEMICAL REACTION

1.8A Heterogeneous Reaction on a Solid Surface

Often catalytic solid particles are used to speed up chemical reactions. The gaseous component diffuses from the bulk gas to the catalyst surface, where it reacts, and the products diffuse back to the bulk gas stream. Two general cases will be considered.

1. *Instantaneous Reaction.* Consider the reaction where $A \rightarrow 2B$, where A is decomposed at the catalyst. In Fig. 1.8-1, A diffuses from point 1 in the gas a distance of δ to point 2 on the surface. It reacts irreversibly and instantaneously, and B diffuses back.

At steady-state 1 mole of A diffuses to the catalyst for every 2 moles B diffusing away, or $N_B = -2N_A$. Using Eq. (1.7-5) in terms of mole fractions,

$$N_A = -cD_{AB}\frac{dx_A}{dz} + x_A(N_A + N_B) \qquad (1.8\text{-}1)$$

Substituting $N_B = -2N_A$,

$$N_A = -cD_{AB}\frac{dx_A}{dz} - x_A(N_A) \qquad (1.8\text{-}2)$$

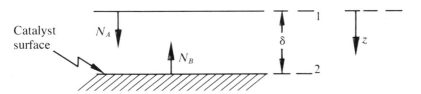

FIGURE 1.8-1 Diffusion and reaction at a catalyst surface

46 Molecular Mass Transport Phenomena in Fluids

Rearranging and integrating with c constant, since P is constant,

$$N_A \int_{z_1}^{z_2} dz = -cD_{AB} \int_{x_{A1}}^{x_{A2}=0} \frac{dx_A}{1+x_A} \qquad (1.8\text{-}3)$$

$$N_A = \frac{cD_{AB}}{\delta} \ln\left(\frac{1+x_{A1}}{1+0}\right) \qquad (1.8\text{-}4)$$

If the reaction is instantaneous, x_{A2} will be zero, since no A can exist next to the catalyst surface. Equation (1.8-4) describes the rate of the overall process of diffusion plus chemical reaction. This is called a diffusion-controlled process, since the diffusion process is the slow one and limits the overall rate.

2. *Slow Reaction at the Surface.* Since the reaction is slow at the surface for the reaction $A \rightarrow 2B$,

$$N_{Az=\delta} = k'_1 c_A = k'_1 c x_A \qquad (1.8\text{-}5)$$

where k'_1 is the first-order reaction velocity constant. We proceed exactly as before for steady-state, except the boundary condition at $z = \delta$ (point 2) is Eq. (1.8-5) and we solve for x_A.

$$x_{A2} = \frac{N_{Az=\delta}}{k'_1 c} = \frac{N_A}{k'_1 c} \qquad (1.8\text{-}6)$$

Since the process is steady-state, $N_{Az=\delta} = N_A$. Substituting Eq. (1.8-6) for the integration limit at x_{A2} into Eq. (1.8-3) and integrating,

$$N_A = \frac{cD_{AB}}{\delta} \ln\left(\frac{1+x_{A1}}{1+N_A/(k'_1 c)}\right) \qquad (1.8\text{-}7)$$

Once again, Eq. (1.8-7) describes the overall rate of the process. The rate is smaller than in Eq. (1.8-4), since the ln term in Eq. (1.8-7) is smaller.

1.8B Homogeneous Reaction in a Phase

1. *First-Order Reaction.* In some cases the gaseous component A undergoes chemical reaction in a gas phase to form another gas. For example, for the reaction $A \rightarrow B$ in the gaseous phase at constant P and steady-state when only A and B are present, molecular diffusion of A occurs. If x_A, the mole fraction of A in the gas phase, is small—that is, the gas is dilute in A—then Eq. (1.7-5) becomes

$$N_{Az} = -D_{AB}\frac{dc_A}{dz} + 0 \qquad (1.8\text{-}8)$$

The irreversible rate of generation is equal to $-k'c_A$, where $k' = \text{sec}^{-1}$. The gas mixture is in a tubular reactor of L cm and radius r cm. The differential equation will be derived for steady-state, where the inlet concentra-

1.8 Diffusion and Chemical Reaction

tion at $z = 0$, $c_A = c_{A1}$, and at $z = L$, $c_A = c_{A2}$. The c_{A1} and c_{A2} are constants. Component A diffuses in the reactor by molecular diffusion. This is shown in Fig. 1.8-2. Writing, for the Δz element, rate of input + rate of generation = rate of output,

$$AN_{Az}|_z - k'c_A A\, \Delta z = AN_{Az}|_{z+\Delta z} \tag{1.8-9}$$

Next we divide through by $A\, \Delta z$ and let Δz approach zero, which gives us the derivative $-dN_{Az}/dz$. Substituting Eq. (1.8-8) for N_{Az},

$$\frac{d^2 c_A}{dz^2} = \frac{k'}{D_{AB}} c_A \tag{1.8-10}$$

This is a second-order linear differential equation with constant coefficients and can be easily solved.

$$c_A = \frac{c_{A2} \sinh\left(\sqrt{\frac{k'}{D_{AB}}} z\right) + c_{A1} \sinh\left(\sqrt{\frac{k'}{D_{AB}}} (L - z)\right)}{\sinh\left(\sqrt{\frac{k'}{D_{AB}}} L\right)} \tag{1.8-11}$$

Hence, the equation above can be used to calculate the concentration of c_A at any z. The equations derived in this Section 1.8B can be used for diffusion and reaction in gases or liquids when the solute A is dilute.

Example 1.8-1 Analog Computer Solution for Diffusion and Reaction

The analog computer will also be used to solve Eq. (1.8-10). (The reader should have read Sections 10.1–10.4.) First the equation will be scaled. The range of the computer used is e of 0 to 10 volts. The machine or computer time is τ in seconds. Since the data will be plotted by a recorder that is slow, the run should be completed in not less than 10 sec. First do time scaling and scale the distance z in terms of machine τ. Let $\tau = \alpha z$ and do the amplitude scaling by letting $e = \beta c_A$.

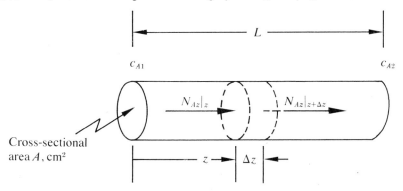

FIGURE 1.8-2 Diffusion and chemical reaction

Solution

The first derivative is $d\tau = \alpha\, dz$, and $de = \beta\, dc_A$. Hence,

$$\frac{dc_A}{dz} = \frac{de/\beta}{d\tau/\alpha} = \frac{\alpha}{\beta}\frac{de}{d\tau} \tag{1.8-12}$$

The second derivative is

$$\frac{d^2c_A}{dz^2} = \frac{d(dc_A/dz)}{dz} = \frac{d\left(\frac{\alpha}{\beta}\frac{de}{d\tau}\right)}{d\tau/\alpha} = \frac{\alpha^2}{\beta}\frac{d^2e}{d\tau^2} \tag{1.8-13}$$

Substituting e/β for c_A and Eq. (1.8-13) into Eq. (1.8-10), the scaled equation is

$$\frac{d^2e}{d\tau^2} = \left(\frac{k'}{\alpha^2 D_{AB}}\right)e \tag{1.8-14}$$

The variables are machine time τ and volts e. If $c_{A1} = 1 \times 10^{-5}$ g mole/cm³, $c_{A2} = 0.3 \times 10^{-5}$ g mole/cm³, $k' = 0.010$ sec⁻¹, $L = 10$ cm, $D_{AB} = 0.70$ cm²/sec, calculate the scale factors α and β as follows. For $\tau = \alpha L$, 10 sec = α (10 cm), and $\alpha = 1.0$ sec/cm. For $e = \beta c_A$, 10.0 volts = $\beta(1 \times 10^{-5})$ g mole/cm³ and $\beta = 10^6$ volts/(g mole/cm³). The maximum value that c_A can reach is 1.0×10^{-5}, so it was used to scale the volts.

Substituting the scaling factors into Eq. (1.8-14):

$$\frac{d^2e}{d\tau^2} = \frac{(0.010)}{(1)^2(0.70)}e = 0.0143e \tag{1.8-15}$$

For the boundary conditions, when $c_{A1} = 1 \times 10^{-5}$, $e_1 = 10^6(1 \times 10^{-5}) = 10$ volts at $\tau = 0$. This is an initial condition shown on integrator 5 in Fig. 1.8-3. At $\tau = 10.0$ sec or $z = L$, $e = 10^6(0.3 \times 10^{-5}) = 3.0$ volts. To draw the analog computer diagram, it is assumed that the $d^2e/d\tau^2$ is available. It is integrated twice to generate e, and then the e is multiplied by $k'/(\alpha^2 D_{AB})$ to generate $d^2e/d\tau^2$ in Fig. 1.8-3.

The potentiometer is set at a value of 0.0143 for $k'/(\alpha^2 D_{AB})$. Before the run is started, the initial condition on integrator 5 is set at 10.0 volts. This is the value that e has for $\tau = 0$—that is, $z = 0$. On the first integrator the initial condition is the negative slope $-de/d\tau$. This is unknown and is trial and error. A value is guessed, then the run is made. If the voltage at $\tau = 10$ sec is 3.0 volts, the guess was correct. Figure 1.8-3 shows the curves obtained for $k' = 0.010, 0.100$, and also for 0 (no chemical reaction), which is a straight line, since it is Fick's law. Other parametric curves are easily and rapidly obtained. Again, since an analytical solution is available, an analog computer solution is not usually obtained. The purpose here is to illustrate scaling of an equation.

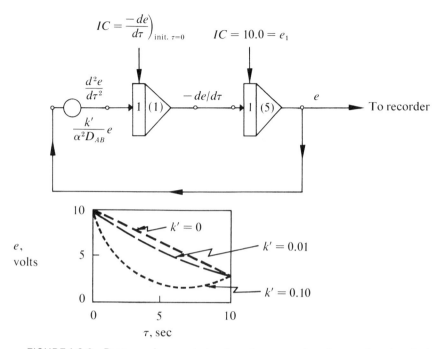

FIGURE 1.8-3 Programming symbols of analog computer diagram for chemical reaction

2. *Empirical Reaction Rate.* Often experimental reaction rate data are obtained where the reaction rate constant varies. Over a narrow concentration range an empirical straight-line equation can often be used. For the reaction $A \rightarrow B$ as before, the irreversible reaction rate for a narrow concentration range is $-(k'_3 c_A + k'_4)$. Using the same assumptions as the previous example the material balance as in Eq. (1.8-9) can be written, and the final equation is

$$\frac{d^2 c_A}{dz^2} = \frac{k'_3}{D_{AB}} c_A + \frac{k'_4}{D_{AB}} \tag{1.8-16}$$

This is a linear second-order differential equation, which can easily be solved by direct integration.

3. *First-order Reaction with a Variable Diffusivity.* In liquids in a few cases the diffusivity may not be constant but can be approximated as a linear function of the concentration c_A or $D_{AB} = K_1 + K_2 c_A$. If, as before, $A \rightarrow B$ and the generation rate is $-k' c_A$, then Eq. (1.8-9) holds, but

$$\frac{dN_A}{dz} = -\frac{d[(K_1 + K_2 c_A) dc_A/dz]}{dz} \tag{1.8-17}$$

50 Molecular Mass Transport Phenomena in Fluids

Substituting Eq. (1.8-17) into the resultant of Eq. (1.8-9), and rearranging,

$$(K_1 + K_2 c_A)\frac{d^2 c_A}{dz^2} + K_2 \left(\frac{dc_A}{dz}\right)^2 = k' c_A \qquad (1.8\text{-}18)$$

Since the diffusivity in gases is constant, this case does not occur in gases.

Example 1.8-2 Reaction with Variable Diffusivity and Analog Computer Solution

Solve Eq. (1.8-18) by the analog computer.

Solution

Solving Eq. (1.8-18) for the highest derivative,

$$\frac{d^2 c_A}{dz^2} = \left(\frac{k'}{K_1 + K_2 c_A}\right) c_A - \frac{K_2}{K_1 + K_2 c_A}\left(\frac{dc_A}{dz}\right)^2 \qquad (1.8\text{-}19)$$

Since Eq. (1.8-19) is difficult to solve on the digital computer, the analog computer will be used. In Fig. 1.8-4 the analog computer diagram is drawn without the equation's being scaled. First it is assumed that $d^2 c_A/dz^2$ is available, and it is integrated in integrator number 1 to give $-dc_A/dz$. This is then integrated to give c_A. The function $(K_1 + K_2 c_A)$ is generated in summer number 7 by multiplying c_A by K_2 in the potentiometer and adding a constant K_1. The functions c_A and $(K_1 + K_2 c_A)$ are then brought to a function divider to give $10 c_A/(K_1 + K_2 c_A)$, which is fed back to integrator number 1. To generate $(dc_A/dz)^2$, the output of integrator number 1 is brought to a function multiplier, where (dc_A/dz) is multiplied by (dc_A/dz). This function $(dc_A/dz)^2$ is then brought to a divider, where it is divided by $(K_1 + K_2 c_A)$ to give the proper result, which is fed into integrator number 1. The reader is cautioned to refer to Section 10.5 on function multipliers and dividers to insure that he understands the use of these elements. If K_2 is negative then the voltage $K_2 c_A$ must be put through a sign changer before entering integrator number 7.

1.9 SOLUTION OF MASS TRANSPORT DIFFERENTIAL EQUATIONS BY NUMERICAL METHODS AND DIGITAL COMPUTER

1.9A Introduction

Often in mass transport phenomena the differential equations representing these phenomena are incapable of solution by analytical or exact methods. As a result, various types of numerical methods are available to approximate the solutions to these equations. Often these equations can be solved by both analog computer methods and numerical methods.

1.9 Numerical Solutions of Mass Transport Equations

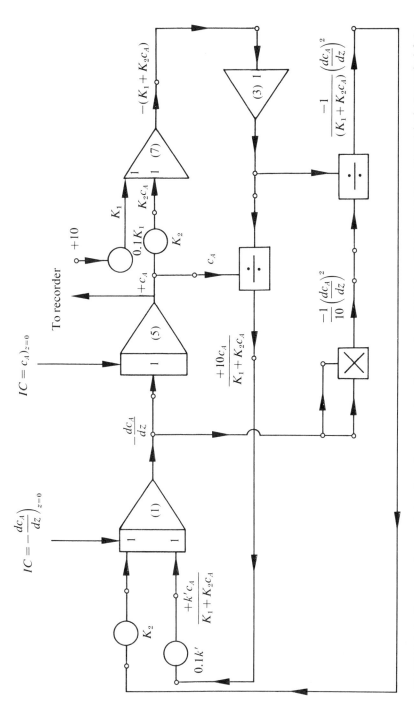

FIGURE 1.8-4 Programming symbols of analog computer diagram for chemical reaction with variable diffusivity in Example 1.8-2

For analog methods the reader is referred to the chapter on these methods. The present section will give a brief description or outline of numerical methods, followed by examples.

Methods for solution of ordinary differential equations fall into two broad classes: initial-value and boundary-value problems.

1.9B Initial-Value Problems

Assume we are given an equation of first order that is an initial-value problem:

$$\frac{dy}{dx} = f(x, y) \qquad (1.9\text{-}1)$$

where its initial conditions are $y = y_0$ at $x = x_0$. This equation can be rearranged and integrated as follows between x_0 and x:

$$\int_{y_0}^{y} dy = \int_{x_0}^{x} f(x, y)\, dx \qquad (1.9\text{-}2)$$

$$y = y_0 + \int_{x_0}^{x} f(x, y)\, dx = y_0 + \int_{x_0}^{x} \left(\frac{dy}{dx}\right) dx \qquad (1.9\text{-}3)$$

Or, integrating between x_n and x_{n+1},

$$y_{n+1} = y_n + \int_{x_n}^{x_{n+1}} \left(\frac{dy}{dx}\right) dx = y_n + \left(\frac{dy}{dx}\right)_n h \qquad (1.9\text{-}4)$$

where the constant $h = x_{n+1} - x_n$ and $n = 0, 1, 2, 3, \ldots$.

A simple Euler method to solve this numerically is as follows. A step increment h is selected and kept constant. Then for the first step $n = 0$, and

$$y_1 = y_0 + \left(\frac{dy}{dx}\right)_0 (x_1 - x_0) \qquad (1.9\text{-}5)$$

The value of $(dy/dx)_0$ is calculated by substituting y_0 and x_0 into Eq. (1.9-1) and assuming that this slope at the beginning of the interval holds throughout this interval. Now, using the new y_1 and x_1, n is set as 1 and y_2 calculated from Eq. (1.9-4). This is repeated and yields y as a function of x over the desired range.

Obviously, if the increment is large, the slope may change greatly over the interval $x_{n+1} - x_n$. Many excellent methods are available that improve on the average value of the slope to be used over the interval — the modified Euler, Milne, Runge-Kutta, predictor-corrector, and Adams methods, among others. A good discussion of these methods is given in Lapidus (L3, pp. 84–108), Perry (P2, pp. 2–61 to 2–62), Conte (C5, pp. 204–258), and Mickley et al. (M3, pp. 187–200), and the reader is referred there for more details. Conte gives actual Fortran programs for the most important methods.

Since applications in mass transport involving boundary-value differential equations are quite numerous, this area will be covered in some detail in the next section.

1.9C   Boundary-Value Problems

Differential equations of second order, linear, and of the following type:

$$\frac{d^2y}{dx^2} + g(x)y = p(x) \tag{1.9-6}$$

where $y = y_0$, $x = x_0$, and $y = y_{N+1}$, $x = x_{N+1}$ at the two boundaries, can often be solved by finite difference methods. These are called boundary-value problems. If a first derivative appears, it may be removed by a change of variable and take the form of Eq. (1.9-6).

One general method of approach is to use a finite difference for the derivatives as follows. For the first derivatives,

$$\frac{dy}{dx} = \frac{y_{n+1} - y_n}{h} \tag{1.9-7}$$

where the interval x_0 to x_{N+1} is divided into $N+1$ equally spaced intervals of width h. This means

$$N + 1 = \frac{x_{N+1} - x_0}{h} \tag{1.9-8}$$

where $h = x_{n+1} - x_n$. The second derivative is

$$\frac{d^2y}{dx^2} = \frac{d(dy/dx)}{dx}$$

$$= \frac{\frac{y_{n+1} - y_n}{h} - \frac{y_n - y_{n-1}}{h}}{h}$$

$$= \frac{y_{n+1} - 2y_n + y_{n-1}}{h^2} \tag{1.9-9}$$

Substituting Eq. (1.9-9) into Eq. (1.9-6) and rearranging (L3, pp. 108–113),

$$y_{n+1} - 2\left(1 - \frac{h^2}{2}g_n\right)y_n + y_{n-1} = h^2 p_n \tag{1.9-10}$$

The interval x_0 to x_{N+1} is divided into $N+1$ intervals each of width h, so that $x_1, x_2, \ldots, x_N$ are the interior points between the boundary values of x_0 and x_{N+1}, which are known values. Writing an equation for each value of n from 1 to N,

$$-2\left(1-\frac{h^2}{2}g_1\right)y_1 + y_2 = h^2 p_1 - y_0 \qquad (n=1)$$

$$y_1 - 2\left(1-\frac{h^2}{2}g_2\right)y_2 + y_3 = h^2 p_2 \qquad (n=2)$$

$$y_2 - 2\left(1-\frac{h^2}{2}g_3\right)y_3 + y_4 = h^2 p_3 \qquad (n=3)$$

$$\vdots \qquad (1.9\text{-}11)$$

$$y_{N-2} - 2\left(1-\frac{h^2}{2}g_{N-1}\right)y_{N-1} + y_N = h^2 p_{N-1} \qquad (n=N-1)$$

$$y_{N-1} - 2\left(1-\frac{h^2}{2}g_N\right)y_N = h^2 p_N - y_{N+1} \qquad (n=N)$$

All values on the right-hand side of the equal sign are constants.

Hence, there are N simultaneous linear algebraic equations and N unknowns in y from y_1 to y_N. The main problem that sometimes occurs in solving these simultaneous equations for y is in the large storage required on the digital computer. In boundary-value problems of this type there are no questions of stability; that is, errors do not build up and give meaningless results.

To solve n simultaneous equations in n unknowns of $x_1, x_2, x_3, \ldots, x_n$ (where x has been substituted for y), we can write them as follows (L3, p. 197; C5, p. 152; P2, p. 2–52):

$$\begin{aligned} a_{11}x_1 + a_{12}x_2 + \cdots + a_{1n}x_n &= c_1 \\ a_{21}x_1 + a_{22}x_2 + \cdots + a_{2n}x_n &= c_2 \\ &\vdots \\ a_{n1}x_1 + a_{n2}x_2 + \cdots + a_{nn}x_n &= c_n \end{aligned} \qquad (1.9\text{-}12)$$

where a_{ij} and c_i are constants with i the row index and j the column index. Equations (1.9-12) can be written in matrix form as

$$\mathbf{Ax} = \mathbf{c} \qquad (1.9\text{-}13)$$

where $\mathbf{A}$ is an $n \times n$ matrix, $\mathbf{x}$ an $n \times 1$ vector, $\mathbf{c}$ an $n \times 1$ vector, and

$$\mathbf{A} = \begin{bmatrix} a_{11} & a_{12} & \cdots & a_{1n} \\ a_{21} & a_{22} & \cdots & a_{2n} \\ \vdots & & & \\ a_{n1} & a_{n2} & \cdots & a_{nn} \end{bmatrix}, \quad \mathbf{x} = \begin{bmatrix} x_1 \\ x_2 \\ \vdots \\ x_n \end{bmatrix}, \quad \mathbf{c} = \begin{bmatrix} c_1 \\ c_2 \\ \vdots \\ c_n \end{bmatrix}$$

1.9 Numerical Solutions of Mass Transport Equations

The basic procedure for solving directly for the values of $x_1, x_2, \ldots, x_n$ is the Gauss elimination method (L3, pp. 243–245; C5, pp. 156–163; P2, p. 2–52). First, the first equation in (1.9-12) is divided by a_{11}; then this result used to eliminate x_1 from all succeeding equations. (If $a_{11} = 0$, reorder or renumber all equations and use the result to eliminate x_1 as before.) Then the modified second equation is divided by a'_{22} and x_2 eliminated (where a'_{22} is the modified a_{22}) from the succeeding equations. This is done n times, and this gives x_n directly. Then the solution to the rest of the x values is obtained by working backward from the final equation. A better modification is the Gauss-Jordon method (L3, pp. 246–247), which eliminates x_2 from the succeeding equations and also from the first equation as well, which shortens the number of operations required.

Generally when the solution of such simultaneous linear algebraic equations is required on a digital computer, it is preferable to use a standard subroutine. Most computers have one or several standard subroutines that can solve such systems. Hence, it is not necessary that the reader know the details, but only that he be able to use the subroutines in numerical solution of boundary-value problems.

Another general method used to solve a boundary-value linear differential equation is to convert it into an initial-value problem and solve it by the methods discussed in Section 1.9B (C5, L3, M3).

Example 1.9-1 Solution of Empirical Reaction Rate by Digital Computer

It is desired to solve Eq. (1.8-16) using a digital computer:

$$\frac{d^2 c_A}{dz^2} = \frac{k'_3}{D_{AB}} c_A + \frac{k'_4}{D_{AB}} \qquad (1.8\text{-}16)$$

In this equation the numerical values to be used are $c_{A1} = 1 \times 10^{-5}$ g mole/cm^3, $c_{A2} = 0.3 \times 10^{-5}$ g mole/cm^3, $L = 10$ cm (z varies between 0 and 10 cm), $D_{AB} = 0.7$ cm^2/sec, $k'_3 = 0.01$ sec^{-1}, $k'_4 = 0.2 \times 10^{-7}$ g mole/sec-cm^3. (a) Obtain the numerical values for this boundary-value problem by writing a Fortran IV program, and make a table of c_A vs. z from $z = 0$ to $z = L = 10$ cm. (b) Repeat part (a) but set $k'_4 = 0$. This then will be the same as Example 1.8-1, which solved Eq. (1.8-10).

Solution

Equation (1.8-16) can be put in the same form as Eq. (1.9-6):

$$\frac{d^2 y}{dz^2} - \frac{k'_3}{D_{AB}} y = \frac{k'_4}{D_{AB}} \qquad (1.9\text{-}14)$$

56 Molecular Mass Transport Phenomena in Fluids

Writing Eq. (1.9-14) in finite difference form similar to Eq. (1.9-9),

$$\frac{d^2y}{dz^2} = \frac{y_{i+1} - 2y_i - y_{i-1}}{h^2} \qquad (1.9\text{-}15)$$

where h is the spacing in the independent variable z and is $z_{i+1} - z_i$, and $y_i = c_A$. The overall number of spaces between z values is

$$N_0 = \frac{L-0}{h} \qquad (1.9\text{-}16)$$

Substituting Eq. (1.9-15) into (1.9-14),

$$\frac{1}{h^2}y_{i+1} - \frac{2}{h^2}y_i - \frac{1}{h^2}y_{i-1} - \frac{k_3'}{D_{AB}}y_i = \frac{k_4'}{D_{AB}} \qquad (1.9\text{-}17)$$

Rearranging,

$$\left(\frac{1}{h^2}\right)y_{i-1} + \left(-\frac{k_3'}{D_{AB}} - \frac{2}{h^2}\right)y_i + \left(\frac{1}{h^2}\right)y_{i+1} = \frac{k_4'}{D_{AB}} \qquad (1.9\text{-}18)$$

where $i = 1, 2, 3, \ldots, N_0$.

Since $L - 0 = 10 - 0$, a value of $h = 0.10$ is selected. Hence, by Eq. (1.9-16), $N_0 = 100$. The equations to solve are

$$\left(\frac{-k_3'}{D_{AB}} - \frac{2}{h^2}\right)y_1 + \left(\frac{1}{h^2}\right)y_2 = \frac{k_4'}{D_{AB}} - \left(\frac{1}{h^2}\right)y_0 \qquad (i=1),$$

$$\left(\frac{1}{h^2}\right)y_1 + \left(\frac{-k_3'}{D_{AB}} - \frac{2}{h^2}\right)y_2 + \left(\frac{1}{h^2}\right)y_3 = \frac{k_4'}{D_{AB}} \qquad (i=2),$$

$$\vdots$$

$$\left(\frac{1}{h^2}\right)y_{N-3} + \left(\frac{-k_3'}{D_{AB}} - \frac{2}{h^2}\right)y_{N-2} + \left(\frac{1}{h^2}\right)y_{N-1} = \frac{k_4'}{D_{AB}} \qquad (i = N-2),$$

$$\left(\frac{1}{h^2}\right)y_{N-2} + \left(\frac{-k_3'}{D_{AB}} - \frac{2}{h^2}\right)y_{N-1} = \frac{k_4'}{D_{AB}} - \left(\frac{1}{h^2}\right)y_N \qquad (i = NY-1),$$

where y_0 and y_N are the known boundary conditions. This system represents 99 equations and 99 unknowns, in which the unknowns in y can be solved by a standard subroutine for solving simultaneous linear algebraic equations.

The Fortran IV program to solve this problem for parts (a) and (b) is given in Table 1.9-1; the results are summarized in Table 1.9-2. The standard subroutine used is SIMQ (A,D,N,KS), which is available in the System 360 Scientific Subroutine Package (360A-CM-03X) Version III of IBM. It is usually available in this form or a modified form in the computer library of most computers. In this subroutine,

1.9 Numerical Solutions of Mass Transport Equations 57

A is the coefficient matrix stored columnwise in the computer memory.
D is the vector of known values initially. It is replaced by the solution vector at the end.
N is the number of equations, 99.
KS is an indicator that tells whether the coefficient matrix is singular. (Not used in this program.)

All programs in this text have been run using the IBM System 360 Fortran IV Level G Release 20 Compiler at The Ohio State University Instruction and Research Computer Center. The reader should note that individual machines of the same and different manufacturers sometimes differ in minor details in the implementation of Fortran.

Table 1.9-1 Fortran Program, Example 1.9-1

```
C         EXAMPLE PROBLEM - NUMERICAL SOLUTION OF BOUNDARY VALUE
C            DIFFERENTIAL EQUATION
C
C         EQUATION IS D2C/DZ2 = (K3/DAB)*C + K4/DAB
C
C         THE BOUNDARY CONDITIONS ARE AS FOLLOWS
C            C=CA1,  Z=0
C            C=CA2,  Z=L
C
          DIMENSION A(9801),D(99),Y(101)
          REAL K3,K4
    2     FORMAT(77H1EXAMPLE PROBLEM - NUMERICAL SOLUTION OF BOUNDARY VALUE
         1DIFFERENTIAL EQUATION)
    3     FORMAT(11H1END OF JOB)
    4     FORMAT(18H0THE INTERVAL H IS,F10.7/40H0THE NUMBER OF EQUATIONS TO
         1BE SOLVED IS,I5/19H0THE DIFFUSIVITY IS,F10.7/33H0THE REACTION RATE
         2 CONSTANT K3 IS,E10.2/33H0THE REACTION RATE CONSTANT K4 IS,E10.2)
    5     FORMAT(1H0/1H0,11X,1HZ,16X,2HCA)
    6     FORMAT(F18.8,E18.8)
          IO=6
          ZI=0.
          ZL=10.
          H=.1
          CA1=1.0E-05
          CA2=0.3E-05
          K3=0.01
          K4=0.2E-07
          DAB=0.7
          DO 100 IJK=1,2
          WRITE(IO,2)
          N=((ZL-ZI)/H-1.0)*1.000001
          M=N**2
C         ZERO MATRIX COEFFICIENTS
          DO 20 I=1,M
   20     A(I)=0
          WRITE(IO,4)H,N,DAB,K3,K4
          WRITE(IO,5)
          NY=N+1
          NYY=N+2
          Y(1)=CA1
          Y(NYY)=CA2
C         SET UP NONZERO MATRIX COEFFICIENTS
          RR=-K3/DAB-2./(H*H)
          AACC=1./(H*H)
          DD=K4/DAB
          NA=N-1
```

Table 1.9-1 (continued)

```
        A(1)=RR
        A(N+1)=AACC
        D(1)=DD-AACC*Y(1)
        J=0
        K=J+N
        L=K+N
        DO 30 I=2,NA
        A(J+I)=AACC
        A(K+I)=RR
        A(L+I)=AACC
        J=J+N
        K=K+N
        L=L+N
30      D(I)=DD
        A(M-N)=AACC
        A(M)=RR
        D(N)=DD-Y(NYY)*AACC
        CALL SIMQ (A,D,N,KS)
        X=ZI-H
        DO 33 I=2,NY
33      Y(I)=D(I-1)
        DO 40 I=1,NYY
        X=X+H
40      WRITE(IO,6)X,Y(I)
        K4=0.0
100     CONTINUE
        WRITE(IO,3)
        STOP
        END
```

Table 1.9-2 Results for Numerical Solution of Differential Equation for Example 1.9-1

Distance z, cm	Case (a) $k_4' = 2.0 \times 10^{-8}$	Case (b) $k_4' = 0$
	$c_A \times 10^5$	$c_A \times 10^5$
0	1.0000	1.0000
1	0.8757	0.8871
2	0.7668	0.7869
3	0.6717	0.6980
4	0.5892	0.6190
5	0.5179	0.5490
6	0.4569	0.4867
7	0.4052	0.4315
8	0.3623	0.3823
9	0.3273	0.3388
10	0.3000	0.3000

Frequently the number of simultaneous equations and the number of unknowns to be solved require excessive storage capacity in the computer. In Example 1.9-1 only 99 equations and 99 unknowns were used. Most computers can handle up to several times as large a system of equations. However, the limit is often reached when small increments are required.

Certain compact and more efficient schemes are available. If the matrix **A** is a tridiagonal type, then methods are available to solve this very efficiently (in terms of the number of operations required). Such a matrix consists of the main diagonal and the two adjacent diagonals with zeros elsewhere. (Example 1.9-1 is actually such a matrix.) These methods find wide use in the solution of ordinary and partial differential equations and are discussed in detail elsewhere in various texts (L3, pp. 224–230; C5, pp. 182–187). Conte (C5) gives an actual Fortran subroutine TRID for solving a tridiagonal matrix and shows the solution of an example.

1.10 STEADY-STATE MULTICOMPONENT DIFFUSION OF GASES

1.10A Component A Diffusing through a Stagnant Multicomponent Mixture

In this case component A is diffusing through stagnant components B, C, D, and so on. The derivation at steady-state is started using the Stefan-Maxwell method used in Section 1.7B. The same assumptions are used as in Eq. (1.7-8), and terms are also included for A with C, A with D, and so on as follows.

$$\frac{-dp_A}{dz} = \alpha_{AB} c_A c_B (v_A - v_B) + \alpha_{AC} c_A c_C (v_A - v_C) + \alpha_{AD} c_A c_D (v_A - v_D) + \cdots \quad (1.10\text{-}1)$$

Now v_B, v_C, and v_D are zero, since the components B, C, and D are stagnant. Writing Eq. (1.7-11) for the binary pairs A-B, A-C, A-D, and so on, converting to mole fractions, and substituting into Eq. (1.10-1),

$$\frac{-dp_A}{dz} = N_A RT \left(\frac{x_B}{D_{AB}} + \frac{x_C}{D_{AC}} + \frac{x_D}{D_{AD}} + \cdots \right) \quad (1.10\text{-}2)$$

Now the partial pressure of the inerts B, C, D, ... is $p_i = P - p_A = P(1 - x_A)$. Equation (1.10-2) is solved for N_A and the right-hand side is multiplied by $P(1 - x_A)/p_i$.

$$N_A = \frac{-(1-x_A)}{\dfrac{x_B}{D_{AB}} + \dfrac{x_C}{D_{AC}} + \dfrac{x_D}{D_{AD}} + \cdots} \frac{P}{RTp_i} \frac{dp_A}{dz} = -\frac{D_{Am} P}{RTp_i} \frac{dp_A}{dz} \quad (1.10\text{-}2a)$$

Hence,

$$D_{Am} = \frac{1 - x_A}{\dfrac{x_B}{D_{AB}} + \dfrac{x_C}{D_{AC}} + \dfrac{x_D}{D_{AD}} + \cdots} \quad (1.10\text{-}3)$$

Dividing top and bottom of Eq. (1.10-3) by $(1-x_A)$,

$$D_{Am} = \frac{1}{\dfrac{x'_B}{D_{AB}} + \dfrac{x'_C}{D_{AC}} + \dfrac{x'_D}{D_{AD}} + \cdots} \qquad (1.10\text{-}4)$$

where x'_B = moles B/mole inerts = $x_B/(1-x_A)$, x'_C = moles C/mole inerts, and so on. Since the mole ratios of inerts, x'_B, x'_C, x'_D, ... are constant, then D_{Am} is constant, even though composition varies in the diffusion path. Since $p_i = P - p_A$, Eq. (1.10-2a) can now be integrated to give

$$N_A = \frac{D_{Am}P}{RTzp_{iM}}(p_{A1} - p_{A2}) \qquad (1.10\text{-}5)$$

where p_{iM} is the ln mean of p_{i2} and p_{i1}.

This analytical equation has been verified experimentally by Wilke (W2).

Example 1.10-1 Diffusion of A through a Stagnant Multicomponent Mixture

Methane (A) is diffusing at steady-state, 25°C, and 1.0 atm total pressure through a mixture of argon (B) and helium (C), which are inert and nondiffusing. The partial pressures in atmospheres are $p_{A1} = 0.4$, $p_{B1} = 0.4$, $p_{C1} = 0.2$, $p_{A2} = 0.1$, $p_{B2} = 0.6$, $p_{C2} = 0.3$. The diffusivities from Table 1.6-1 are $D_{AB} = 0.202$ cm²/sec, $D_{AC} = 0.675$ cm²/sec, and $D_{BC} = 0.729$ cm²/sec. Points 1 and 2 are 0.10 cm apart. Calculate N_A.

Solution

x'_B = moles B/mole inerts = $x_B/(1-x_A) = 0.4/(1-0.4) = 0.667$ and $x'_C = 0.2/(1-0.4) = 0.333$. These values are constant through the path of diffusion.

Substituting into Eq. (1.10-4),

$$D_{Am} = \frac{1}{\dfrac{x'_B}{D_{AB}} + \dfrac{x'_C}{D_{AC}}} = \frac{1}{\dfrac{0.667}{0.202} + \dfrac{0.333}{0.675}} = 0.264 \text{ cm}^2/\text{sec}$$

To calculate p_{iM}, $p_{i1} = P - p_{A1} = 1.0 - 0.4 = 0.60$, $p_{i2} = P - p_{A2} = 1.0 - 0.1 = 0.90$. Then

$$p_{iM} = \frac{0.90 - 0.60}{\ln(0.9/0.6)} = 0.743 \text{ atm}$$

1.10 Steady-State Multicomponent Diffusion of Gases

Substituting into Eq. (1.10-5),

$$N_A = \frac{D_{Am}P}{RT z p_{iM}}(p_{A1} - p_{A2}) = \frac{0.264(1.0)(0.4-0.1)}{82.06(0.10)(298)(0.743)}$$

$$= 4.38 \times 10^{-5} \text{ g mole } A/\text{sec-cm}^2$$

1.10B Components A and B Diffusing Through Stagnant C

Gilliland (W2) derived two equations for the simultaneous diffusion of A and B through stagnant C at steady-state. He wrote the Stefan-Maxwell equations for component A as in Eq. (1.10-1) and a similar one for component B. Setting $v_C = 0$ and $p_C = P - p_A - p_B$, he derived the following two simultaneous equations:

$$\frac{N_A}{D_{AC}} + \frac{N_B}{D_{BC}} = \frac{P}{RT(z_2 - z_1)} \ln \frac{p_{C2}}{p_{C1}} \quad (1.10\text{-}6)$$

$$N_A + N_B = \frac{D_{AB}P}{RT(z_2 - z_1)}$$

$$\times \ln \left[\frac{\left(\frac{1}{D_{AB}} - \frac{1}{D_{AC}}\right)\frac{N_A+N_B}{N_B} p_{B2} - \frac{N_A+N_B}{N_B} p_{A2} + \left(\frac{1}{D_{AC}} - \frac{1}{D_{BC}}\right)P}{\frac{1}{D_{AB}} - \frac{1}{D_{BC}}}}{\frac{\left(\frac{1}{D_{AB}} - \frac{1}{D_{AC}}\right)\frac{N_A+N_B}{N_B} p_{B1} - \frac{N_A+N_B}{N_B} p_{A1} + \left(\frac{1}{D_{AC}} - \frac{1}{D_{BC}}\right)P}{\frac{1}{D_{AB}} - \frac{1}{D_{BC}}}} \right]$$

$$(1.10\text{-}7)$$

The two unknowns are N_A and N_B and the solution is trial and error. First a value of N_A is estimated using a binary diffusion equation; then N_B is solved for in Eq. (1.10-6). Then these values are substituted into both sides of Eq. (1.10-7). If the left side does not equal the right side, a new value of N_A is tried and the process repeated. The two equations above are not valid, as stated by Toor (T1), when the problem approaches one of equimolar diffusion where N_A becomes close to $-N_B$.

Example 1.10-2 Molecular Transport in a Multicomponent System

It is desired to determine the effects of simultaneous diffusion of two gases A and B through stagnant gas C. Gas A is diffusing through a mixture of gases C and B at a total pressure P of 0.200 atm and at 55°C. The diffusion takes place by molecular diffusion through a distance of

62 Molecular Mass Transport Phenomena in Fluids

$(z_2 - z_1) = 0.100$ cm. Component B is counterdiffusing into C and A. The data are as follows.

$P = 0.200$ atm, $T = 328°K$, $(z_2 - z_1) = 0.100$ cm

Point 1	Point 2	Diffusivity, cm^2/sec (at 0.20 atm)
(A) $p_{A1} = 0.006$ atm	$p_{A2} = 0$	$D_{AC} = 1.075$
(B) $p_{B1} = 0$	$p_{B2} = 0.0727$	$D_{BC} = 1.245$
(C) $p_{C1} = 0.194$	$p_{C2} = 0.1273$	$D_{AB} = 1.47$
0.200	0.2000	

(a) Calculate N_A using a binary diffusion equation assuming the other gases are inert C (that is, neglecting counterdiffusion of B). Calculate N_B assuming the other gases are inert C. Note that N_A diffuses from point 1 to 2 and is positive. Hence, N_B will be negative. (b) Calculate N_A and N_B using the method of Gilliland.

Solution

To calculate N_A assuming the other gases are inert we shall write Eq. (1.7-26) with p_{iM} for p_{BM}.

$$N_A = \frac{D_{AC}P}{RTzp_{iM}}(p_{A1} - p_{A2}) \quad (1.7\text{-}26a)$$

To calculate p_{iM}, $p_{i1} = p_{B1} + p_{C1} = 0 + 0.194 = 0.194$ atm. Also, $p_{i2} = p_{B2} + p_{C2} = 0.0727 + 0.1273 = 0.200$. Since p_{i1} and p_{i2} are so close together, $p_{iM} = (0.194 + 0.200)/2 = 0.197$ atm. Substituting into Eq. (1.7-26a),

$$N_A = \frac{1.075(0.200)(0.006 - 0)}{(82.06)(328)(0.100)(0.197)} = +0.244 \times 10^{-5} \text{ g mole } A/\text{sec-cm}^2$$

Similarly for N_B,

$$N_B = \frac{D_{BC}P(p_{B1} - p_{B2})}{RTzp_{iM}} \quad (1.7\text{-}26b)$$

For p_{iM}, $p_{i1} = p_{A1} + p_{C1} = 0.006 + 0.194 = 0.200$ atm. $p_{i2} = p_{A2} + p_{C2} = 0 + 0.1273 = 0.1273$. $p_{iM} = (0.200 - 0.1273)/\ln(0.200/0.1273) = 0.1609$ atm. Then, substituting into Eq. (1.7-26b),

$$N_B = \frac{1.245(0.200)(0 - 0.0727)}{(82.06)(328)(0.100)(0.1609)} = -4.19 \times 10^{-5} \text{ g mole } B/\text{sec-cm}^2$$

Here we have component B counterdiffusing in the opposite direction to A [Ans. to (a)].

For part (b) the data are first substituted into Eq. (1.10-6).

$$\frac{N_A}{1.075} + \frac{N_B}{1.245} = \frac{(0.200)}{(82.06)(328)(0.100)} \ln\left(\frac{0.1273}{0.194}\right)$$

Solving,

$$N_B = -3.89 \times 10^{-5} - 1.157 N_A$$

Substituting known values into Eq. (1.10-7),

$$N_A + N_B = \frac{(1.47)(0.200)}{(82.06)(328)(0.100)}$$

$$\times \ln \left[\frac{\dfrac{\dfrac{1}{1.47} - \dfrac{1}{1.075}}{\dfrac{1}{1.47} - \dfrac{1}{1.245}} \dfrac{N_A + N_B}{N_B}(0.0727) - \dfrac{N_A + N_B}{N_B}(0) + \dfrac{\dfrac{1}{1.075} - \dfrac{1}{1.245}}{\dfrac{1}{1.47} - \dfrac{1}{1.245}}(0.200)}{\dfrac{\dfrac{1}{1.47} - \dfrac{1}{1.075}}{\dfrac{1}{1.47} - \dfrac{1}{1.245}} \dfrac{N_A + N_B}{N_B}(0) - \dfrac{N_A + N_B}{N_B}(0.0060) + \dfrac{\dfrac{1}{1.075} - \dfrac{1}{1.245}}{\dfrac{1}{1.47} - \dfrac{1}{1.245}}(0.200)} \right]$$

We now have two equations and two unknowns N_A and N_B, which are solved by trial and error as follows. A value of N_A is assumed and substituted into the first equation to obtain N_B. As a first trial $N_A = 0.244 \times 10^{-5}$ is used. Next these two values are substituted into the second equation. If the right-hand side is equal to the left, the solution is correct; if not, a new N_A value is assumed.

Final values are $N_A = +0.211 \times 10^{-5}$ g mole A/sec-cm² and $N_B = -4.13 \times 10^{-5}$ g mole B/sec-cm². The true value of N_A of 0.211×10^{-5} is substantially less than the approximate value of 0.244×10^{-5}. This indicates that one should use "binary type" equations with care when approximating multicomponent diffusion.

1.10C Equimolar Diffusion of Three Components

Toor (T1) derived analytical solutions for equimolar diffusion of components A, B, and C in a ternary gas system at steady-state and constant total pressure, where

$$\sum_{i=1}^{3} N_i = N_A + N_B + N_C = 0 \qquad (1.10\text{-}8)$$

To obtain numerical values of the fluxes the solutions are trial and error. Toor shows in great detail that many interactions can occur with a three-component system as compared to a two-component one. Under certain conditions a component can undergo reverse diffusion, whereby it actually diffuses against its concentration gradient. In another case he shows how a component can be nondiffusing even though there is a concentration gradient present and no solubility barrier exists. Hence, approximate equations for multicomponent systems that employ effective diffusivity must be used with caution. He also gives a linearized approximate solution to his equations to avoid the trial-and-error procedures.

1.10D General Case for Diffusion of Two or More Components

The case of diffusion of three components in a ternary mixture where the fluxes are not equimolar can, of course, be solved analytically by using the Stefan-Maxwell equation (1.10-1), which, when generalized, becomes

$$\nabla x_i = \sum_{j=1}^{n} \frac{1}{cD_{ij}} (x_i \mathbf{N}_j - x_j \mathbf{N}_i) \tag{1.10-9}$$

We often desire to express the diffusion of a multicomponent mixture with a "binary-type" equation similar to

$$\mathbf{N}_A = -cD_{AB} \nabla x_A + x_A (\mathbf{N}_A + \mathbf{N}_B) \tag{1.10-10}$$

Hence, we write in a similar fashion

$$\mathbf{N}_i = -cD_{im} \nabla x_i + x_i \sum_{j=1}^{n} \mathbf{N}_j \tag{1.10-11}$$

where D_{iM} is an "effective binary diffusivity" of i in the mixture. Substituting Eq. (1.10-9) into (1.10-11), we obtain

$$\frac{1}{cD_{im}} = \frac{\sum_{j=1}^{n} \frac{1}{cD_{ij}} (x_j \mathbf{N}_i - x_i \mathbf{N}_j)}{\mathbf{N}_i - x_i \sum_{j=1}^{n} \mathbf{N}_j} = \frac{\sum_{j=1}^{n} \frac{1}{cD_{ij}} \left(x_j - x_i \frac{\mathbf{N}_j}{\mathbf{N}_i} \right)}{1 - x_i \left(\sum_{j=1}^{n} \mathbf{N}_j \right) / \mathbf{N}_i} \tag{1.10-12}$$

The main difficulty with Eq. (1.10-12) is that D_{im} is not a constant and depends on composition. Hence, Eq. (1.10-11) cannot be integrated easily and made convenient to use unless some sort of "average value" of D_{im} is used. Hsu and Bird (H3) have systematically done this for several cases.

For the case where gases A and B diffuse through a gas film from point 0 to δ at a catalyst surface and react instantaneously in the presence of an inert gas I, the stoichiometry is (H3).

$$aA + bB \rightleftharpoons pP + qQ \tag{1.10-13}$$

Hence,

$$N_B = \frac{b}{a} N_A, \qquad N_P = -\frac{p}{a} N_A, \qquad N_Q = -\frac{q}{a} N_A \tag{1.10-14}$$

1.10 Steady-State Multicomponent Diffusion of Gases

Writing Eq. (1.10-9) for the above case and in the z direction,

$$\frac{dx_A}{dz} = \frac{1}{cD_{AB}}(x_A N_B - x_B N_A) + \frac{1}{cD_{AI}}(0 - x_I N_A)$$

$$+ \frac{1}{cD_{AP}}(x_A N_P - x_P N_A) + \frac{1}{cD_{AQ}}(x_A N_Q - x_Q N_A) \quad (1.10\text{-}15)$$

Substituting Eq. (1.10-14) into (1.10-15),

$$-\frac{dx_A}{dz} = \frac{N_A}{c}\left[\frac{x_B}{D_{AB}} + \frac{x_P}{D_{AP}} + \frac{x_Q}{D_{AQ}} + \frac{x_I}{D_{AI}} - \frac{x_A}{a}\left(\frac{b}{D_{AB}} - \frac{p}{D_{AP}} - \frac{q}{D_{AQ}}\right)\right] \quad (1.10\text{-}16)$$

Solving for N_A,

$$N_A = -\frac{1}{\left[\frac{x_B}{D_{AB}} + \frac{x_P}{D_{AP}} + \frac{x_Q}{D_{AQ}} + \frac{x_I}{D_{AI}} - \frac{x_A}{a}\left(\frac{b}{D_{AB}} - \frac{p}{D_{AP}} - \frac{q}{D_{AQ}}\right)\right]} \frac{c\,dx_A}{dz} \quad (1.10\text{-}17)$$

Next we define

$$E_A = \frac{\sum_{j=1}^{n} N_j}{N_i} = \frac{N_A + N_B + N_P + N_Q}{N_A} = \frac{a+b-p-q}{a} \quad (1.10\text{-}18)$$

Multiplying numerator and denominator of Eq. (1.10-17) by $(1 - E_A x_A)$ we get

$$N_A = \frac{-(1 - E_A x_A)}{\left[\frac{x_B}{D_{AB}} + \frac{x_P}{D_{AP}} + \frac{x_Q}{D_{AQ}} + \frac{x_I}{D_{AI}} - \frac{x_A}{a}\left(\frac{b}{D_{AB}} - \frac{p}{D_{AP}} - \frac{q}{D_{AQ}}\right)\right]} \frac{c}{1 - E_A x_A} \frac{dx_A}{dz} \quad (1.10\text{-}19)$$

Comparison of Eqs. (1.10-12) and (1.10-19) shows that

$$D_{im} = D_{Am} = \frac{1 - E_A x_A}{\left[\frac{x_B}{D_{AB}} + \frac{x_P}{D_{AP}} + \frac{x_Q}{D_{AQ}} + \frac{x_I}{D_{AI}} - \frac{x_A}{a}\left(\frac{b}{D_{AB}} - \frac{p}{D_{AP}} - \frac{q}{D_{AQ}}\right)\right]} \quad (1.10\text{-}20)$$

and

$$N_A = -cD_{Am}\frac{dx_A}{(1 - E_A x_A)dz} \quad (1.10\text{-}21)$$

Integrating between $x_A = x_{A0}$ at $z = 0$ and $x_A = x_{A\delta}$ at $z = \delta$ and assuming D_{Am} is constant,

$$N_A = \frac{cD_{Am}}{\delta E_A}\ln\left(\frac{1 - E_A x_{A\delta}}{1 - E_A x_{A0}}\right) \quad (1.10\text{-}22)$$

The final equations to use are (1.10-20) and (1.10-22). Hsu and Bird (H3) recommend calculating D_{Am} by using the linear average of $x_A = (x_{A0} + x_{A\delta})/2$, and also a linear average of the other mole fractions. Using this method, they found reasonably close checks to the exact analytical expressions, which are cumbersome to use. They also recommended different ways to calculate D_{Am} that are more accurate.

To calculate N_B, N_P, and N_Q, Eq. (1.10-14) is used. In some cases it may be necessary to calculate D_{Bm}, D_{Pm}, D_{Qm}, or D_{Im} and the fluxes separately, particularly if $x_{i\delta}$ is unknown. These calculations are as follows:

$$E_B = \frac{a+b-p-q}{b} \tag{1.10-23}$$

$$E_P = \frac{a+b-p-q}{-p} \tag{1.10-24}$$

$$E_Q = \frac{a+b-p-q}{-q} \tag{1.10-25}$$

$$D_{Bm} = \frac{1-E_B x_B}{\left[\dfrac{x_A}{D_{BA}} + \dfrac{x_P}{D_{BP}} + \dfrac{x_Q}{D_{BQ}} + \dfrac{x_I}{D_{BI}} - \dfrac{x_B}{b}\left(\dfrac{a}{D_{BA}} - \dfrac{p}{D_{BP}} - \dfrac{q}{D_{BQ}}\right)\right]} \tag{1.10-26}$$

$$D_{Pm} = \frac{1-E_P x_P}{\left[\dfrac{x_A}{D_{PA}} + \dfrac{x_B}{D_{PB}} + \dfrac{x_I}{D_{PI}} + \dfrac{x_Q}{D_{PQ}} - \dfrac{x_P}{p}\left(\dfrac{-a}{D_{PA}} - \dfrac{b}{D_{PB}} + \dfrac{q}{D_{PQ}}\right)\right]} \tag{1.10-27}$$

$$D_{Qm} = \frac{1-E_Q x_Q}{\left[\dfrac{x_A}{D_{QA}} + \dfrac{x_B}{D_{QB}} + \dfrac{x_I}{D_{QI}} + \dfrac{x_P}{D_{QP}} - \dfrac{x_Q}{q}\left(\dfrac{-a}{D_{QA}} - \dfrac{b}{D_{QB}} + \dfrac{p}{D_{QP}}\right)\right]} \tag{1.10-28}$$

$$D_{Im} = \frac{a+b-p-q}{\left(\dfrac{a}{D_{IA}} + \dfrac{b}{D_{IB}} - \dfrac{p}{D_{IP}} - \dfrac{q}{D_{IQ}}\right)} \tag{1.10-29}$$

$$N_B = \frac{cD_{Bm}}{\delta E_B} \ln\left(\frac{1-E_B x_{B\delta}}{1-E_B x_{B0}}\right) \tag{1.10-30}$$

$$N_P = \frac{cD_{Pm}}{\delta E_P} \ln\left(\frac{1-E_P x_{P\delta}}{1-E_P x_{P0}}\right) \tag{1.10-31}$$

$$N_Q = \frac{cD_{Qm}}{\delta E_Q} \ln\left(\frac{1-E_Q x_{Q\delta}}{1-E_Q x_{Q0}}\right) \tag{1.10-32}$$

Example 1.10-3 Simultaneous Diffusion of Three Components

Derive the equations similar to Eqs. (1.10-20) and (1.10-22) for the following case:

$$aA \to pP + qQ \tag{1.10-33}$$

where $a = 1.0$, $p = 2$, $q = 1$. Also $D_{PQ}/D_{AP} = 1.0$, $D_{PQ}/D_{AQ} = 3.0$. $x_{A0} = 0.10$, $x_{P0} = 0.7$, $x_{Q0} = 0.2$, $x_{A\delta} = 0$, $x_{P\delta} = 0.778$, $x_{Q\delta} = 0.222$. Also solve the equation using numerical values.

1.10 Steady-State Multicomponent Diffusion of Gases 67

Solution

Using Eq. (1.10-20) for neither I nor B present, we get

$$D_{im} = D_{Am} = \frac{1 - E_A x_A}{\left[0 + \dfrac{x_P}{D_{AP}} + \dfrac{x_Q}{D_{AQ}} + 0 - \dfrac{x_A}{a}\left(0 - \dfrac{p}{D_{AP}} - \dfrac{q}{D_{AQ}}\right)\right]}$$

$$= \frac{(1 - E_A x_A) D_{PQ}}{(x_P + p x_A)\dfrac{D_{PQ}}{D_{AP}} + (x_Q + q x_A)\dfrac{D_{PQ}}{D_{AQ}}} \qquad (1.10\text{-}34)$$

Next, we calculate average values to use for D_{Am}. $x_A = (0.10 + 0)/2 = 0.05$, $x_P = (0.70 + 0.778)/2 = 0.739$, $x_Q = (0.20 + 0.222)/2 = 0.211$. Substituting into Eq. (1.10-18),

$$E_A = \frac{1 + 0 - 2 - 1}{1} = -2$$

Substituting into Eq. (1.10-34),

$$D_{Am} = \frac{(1 - (-2)(0.05)) D_{PQ}}{(0.739 + 2 \times 0.05)1.0 + (0.211 + 1 \times 0.05)3.0} = 0.679 D_{PQ}$$

Finally, substituting into Eq. (1.10-22),

$$\frac{N_A \delta}{c D_{PQ}} = \frac{0.679}{(-2)} \ln\left(\frac{1 - (-2)(0)}{1 - (-2)(0.1)}\right) = 0.0618$$

Hsu and Bird (H3) obtain an exact value of 0.0597, which is reasonably close to this approximate method. To calculate N_P and N_Q directly, Eq. (1.10-14) can be used, where $N_P = -(p/a)N_A$ and $N_Q = -(q/a)N_A$.

Example 1.10-4 Solution of Example 1.10-2 by the Approximate Method

Solve Example 1.10-2 again, using the approximate method.

Solution

In this case we have no relationship between the fluxes of A and of B; hence it is trial and error. For the first trial we will use the values for N_A and N_B estimated using the "binary" equations. Renaming these components, call A as A, B as P, and C as I. Then, we can write an assumed stoichiometric relation as

$$aA \rightarrow pP, \qquad N_P = -\frac{p}{a}N_A = -\frac{4.19 \times 10^{-5}}{0.244 \times 10^{-5}}N_A = -17.2 N_A$$

for our first trial. Calling $a = 1.0$ and $p = 17.2$ for use in Eq. (1.10-34),

68 Molecular Mass Transport Phenomena in Fluids

we must first convert to mole fractions. The diffusivities are $D_{AI} = 1.075$, $D_{PI} = 1.245$, $D_{AP} = 1.47$. At point 0,

$$x_A = \frac{0.006}{0.200} = 0.030, \qquad x_P = 0, \qquad x_I = 0.970$$

At point δ,

$$x_A = 0, \qquad x_P = 0.3635, \qquad x_I = 0.6365$$

Average values are $x_A = 0.0150$, $x_P = 0.1818$, $x_I = 0.8032$.
For E_A in Eq. (1.10-18), $E_A = (1 + 0 - 17.2 - 0)/1 = -16.2$. Substituting into Eq. (1.10-20),

$$D_{Am} = \frac{1 - (-16.2)(0.0150)}{\left[0 + \dfrac{0.1818}{1.47} + 0 + \dfrac{0.8032}{1.075} - \dfrac{0.015}{1.0}\left(0 - \dfrac{17.2}{1.47} - 0\right)\right]} = 1.187$$

Finally, using P/RT for c, we substitute into Eq. (1.10-22):

$$N_A = \frac{(0.200)(1.187)}{(82.06)(328)(0.100)(-16.2)} \ln\left(\frac{1 - (-16.2)(0)}{1 - (-16.2)(0.030)}\right)$$

$$= 0.215 \times 10^{-5} \text{ g mole } A/\text{sec-cm}^2$$

The value for N_B or N_P will not change appreciably on recalculation, since it has such a large concentration compared to x_A, so it will not be calculated. For a second trial, the value of 0.215×10^{-5} is used as the beginning estimate, and a final value of 0.212×10^{-5} is obtained. This is quite close to the true value of 0.211×10^{-5}. To check N_B, the value of p/a used in the final trial can be used to calculate N_B from N_A. Also, N_B can be calculated from Eqs. (1.10-26) and (1.10-30).

PROBLEMS

1-1 Momentum Flux in a Liquid Metal

An experiment is being performed to determine the momentum flux in liquid Hg that is situated between two infinitely wide and long parallel plates. The bottom plate is fixed, and Hg is situated between this plate and the upper plate, which is drawn at a constant velocity of 0.85 cm/sec at 20°C.

(a) Calculate the momentum flux at steady-state when the plates are 0.25 cm apart. Also calculate the velocity midway between the two plates.
(b) Does the fact that the density of Hg is so high affect the momentum flux?
(c) Repeat part (a) for air situated between the two plates at 20°C.

1-2 Energy Flux in a Liquid and Analog Computer Solution at Steady-State

Glycerol at 12°C is situated between two flat and parallel plates that are 0.11 cm apart; the top plate is held at 12.20°C and the lower plate at 11.80°C.

(a) Predict the energy flux by conduction through the glycerol.

(b) Do the same for solid Cu at 0°C situated between these plates with a temperature difference of 0.40°C also.
(c) Repeat for cork at 30°C and a temperature difference of 0.40°C.
(d) Solve part (a) using the analog computer. Draw the computer diagram, being sure to include the initial condition. State how to make the run and sketch a graph of the computer output.

1-3 Equimolar Counterdiffusion at Steady-State

Ammonia is diffusing through a straight uniform tube 2.0 ft long with a diameter of 0.080 ft containing N_2 at 25°C and 1.0 atm total pressure. Both ends of the tube are connected to large mixed chambers at 1.0 atm. The partial pressure of ammonia at one end is constant at 150 mm Hg and 50 mm at the other end.

(a) Calculate the flux of $NH_3(A)$ in lb mole/hr.
(b) Calculate the flux of $N_2(B)$.
(c) Calculate the partial pressures at a point 1.0 ft in the tube. Make a plot of p_A, p_B, and P vs. z.

1-4 Diffusion in a Nonuniform Area

Ammonia (A) is diffusing at steady-state through N_2 (B) in a metal conduit 4.0 ft long at 25°C and a total pressure of 1.0 atm. The partial pressure of ammonia at the left end is 0.25 atm and at the other end 0.05 atm. The cross section of the conduit is in the shape of an equilateral triangle, the length of each side of the triangle being 0.20 ft at the left end and tapering uniformly to 0.10 ft at the right end.

(a) Calculate for the molar flux of ammonia for equimolar counterdiffusion.
(b) Calculate the partial pressure of ammonia at a distance of 2.00 ft and plot the three values of p_A vs. z.

1-5 Prediction of Molecular Diffusivity

For a mixture of methane (A) and ethanol (B) predict the diffusivity as follows.

(a) At 1.0 atm and 25 and 100°C using the method of Chapman-Enskog.
(b) The same as (a) but using the method of Fuller *et al.*
(c) At 2.0 atm and 25°C using the method of Fuller *et al.*

1-6 Molecular Diffusion of Nitrogen Gas and Effect of Temperature and Pressure

Calculate the mass flux of nitrogen gas through a uniform tube of CO 11.0 cm long at 1.0 atm total pressure. The partial pressure of nitrogen is 80 mm Hg at the left face and 10 mm Hg at the distance 11.0 cm to the right. These partial pressures at the two points remain constant. Both ends of the tube are connected to large chambers completely mixed. Predict diffusivities by the Fuller *et al.* method.

(a) Calculate the flux in g mole/cm²-sec at 25°C for nitrogen.
(b) Do this at 200°C also. Does the flux increase with temperature?
(c) What is the flux of CO at 25°C? How is this related to the flux of N_2 in (a)?
(d) Suppose the total pressure is raised to 3.0 atm and the partial pressure of nitrogen remains at 80 and 10 mm as before. Calculate the new flux of nitrogen at 25°C. Does it increase or decrease, and why? Assume steady-state.

1-7 Time to Evaporate a Drop

A drop of liquid toluene at 25.9°C is suspended in pure air by a fine wire and is allowed to evaporate completely. The initial radius is 0.20 cm. Predict the time for the drop to evaporate completely. The vapor pressure of toluene at 25.9°C is 28.8 mm.

1-8 Mass Transfer and Chemical Reaction of a Sphere

A sphere of fixed radius r_1 is composed of a pure, nonporous solid component B with no vapor pressure. It is located in a stagnant gas of infinite volume (that is, a small sphere in a large volume). The gas is composed only of pure component A. At steady-state and constant total pressure, gas A diffuses from the medium to the surface of the sphere, where it undergoes a rapid chemical reaction with component B at the surface as follows:

$$A \text{ (gas)} + B \text{ (solid)} \rightarrow 2D \text{ (gas)}.$$

Component or product D at the surface is a gas and immediately vaporizes and diffuses to the surrounding medium. [*Note*: A physical situation somewhat similar to this is the combustion of a sphere of pure carbon in an oxygen stream.]

(a) Derive an equation assuming pseudo-steady-state and r_1 constant for N_A and the rate of disappearance of B. Integrate the equation to solve for the fluxes. [*Hint*: Start with the following equation for diffusion of two components E and F, where x_E is mole fraction of E in the gas. (How many components are diffusing in the gas?)

$$N_E = -cD_{EF}\frac{dx_E}{dz} + x_E(N_E + N_F).$$

Call point 1 the position at the surface and point 2 that in the bulk gas stream. Call N_A the flux at the surface of the sphere.]

(b) Derive the integrated equation for the time for the solid sphere to disappear completely.

$$\text{Ans.: (a) } N_A = N_B = \frac{D_{AD}P}{RTr_1}\ln\left(\frac{P+p_{A1}}{P+p_{A2}}\right)$$

$$\text{(b) } \theta_0 = \frac{\rho_B r_1^2}{2M_B cD_{AD}\ln\left(\frac{P+p_{A1}}{P+p_{A2}}\right)}$$

1-9 Heterogeneous Chemical Reaction and Analog Computer Solution

Consider the case where A at point 1 diffuses from a gas stream a distance δ to a catalyst surface at point 2, reacts to form a molecule B by $2A \rightarrow B$, and B diffuses from the catalyst back to the main gas stream. The process is steady-state.

(a) Derive the integrated equation for N_A where the reaction is instantaneous.
(b) Draw the analog computer diagram for the differential equation of part (a). [*Hints*: First solve for dx_A/dz. Then integrate to get $-x_A$. Add a constant 1.0 to $-\frac{1}{2}x_A$ in a summer.]

(c) Do the same as part (a) for a slow first-order reaction.

$$\text{Ans.: (a) } N_A = \frac{2cD_{AB}}{\delta} \ln\left(\frac{1-0}{1-\frac{1}{2}x_{A1}}\right)$$

$$\text{(c) } N_A = \frac{2cD_{AB}}{\delta} \ln\left(\frac{1-\frac{1}{2}N_A/ck_1'}{1-\frac{1}{2}x_{A1}}\right)$$

1-10 Evaporation from a Tank

A large rectangular tank 4 ft by 8 ft contains liquid n-hexane at 15.8°C, and its vapor pressure is 100 mm Hg at this temperature. At certain times the tank is open to the atmosphere so that a stagnant air layer 1.0 cm thick is above it. The air above this stagnant layer can be assumed to contain no hexane. Calculate the loss of hexane in lb mass/day from this tank when it is uncovered, assuming steady-state.

1-11 Evaporation Losses of Water

Water in an irrigation ditch is flowing at 25°C for a distance of 500 ft. The ditch is below ground and is covered to reduce evaporation losses. Every 50 ft there is a vent line 1.0 in. ID to the outside air at 25°C. The length of the vent line is 3.5 ft. Calculate the total evaporation loss of water in lb mass/day for the 10 vents, assuming the outside air is dry and the partial pressure of water vapor above the surface of the water is the vapor pressure, 23.76 mm Hg, at 25°C.

1-12 Diffusivity by Two-Bulb Method

Rederive Eq. (1.6-8), only do not neglect the volume AL in the capillary. Assume half of this volume belongs to V_2 and half to V_1.

$$\text{Ans.: } \frac{c_{\text{ave}}-c_2}{c_{\text{ave}}-c_2^0} = e^{-\delta t}, \text{ where } \delta = \frac{D_{AB}(V_1+V_2+AL)}{(L/A)(V_2+AL/2)(V_1+AL/2)}$$

1-13 Second-Order Reaction and Analog Computer Solution

Gaseous component A in the gas phase undergoes a steady-state second-order reaction of $2A \rightarrow B$ at constant total pressure P. The mole fraction of A is small in the gas phase. The chemical reaction rate is $-k_2'c_A^2$ which is second-order. The gas mixture is in a tabular reactor L cm long. At $z = 0$, $c_A = c_{A1}$ and at $z = L$, $c_A = c_{A2}$, which are constants. Component A diffuses by molecular diffusion.

(a) Derive the final differential equation.
(b) Write and scale the equation for the analog computer with $\tau = \alpha z$ and $e = \beta c_A$. Draw the analog computer diagram using programming symbols. Note that a function multiplier will be needed.
(c) If $L = 100$ cm, $c_{A1} = 1 \times 10^{-4}$ g mole/cm³, $c_{A2} = 5 \times 10^{-5}$ g mole/cm³, $D_{AB} = 0.100$ cm²/sec, and k_2' has values of 0.100, 0.010, and 0. Scale the differential equation, solving for β, τ, and potentiometer settings on the analog

computer. The recorder used is such that a run will be finished in 10 sec, and 10 volts can be used on the computer.

Ans.: (a) $\dfrac{d^2 c_A}{dz^2} = \dfrac{k_2'}{D_{AB}}(c_A)^2$

(b) $\dfrac{d^2 e}{d\tau^2} = \dfrac{k_2'}{\alpha^2 \beta D_{AB}} e^2$

1-14 Molecular Diffusion, Variable Diffusivity, and Analog Computer Solution

Liquids A and B are diffusing in a tube L cm long of uniform cross-sectional area at constant total concentration c. The diffusivity depends on concentration as follows: $D_{AB} = K_1 + K_2 c_A$, where K_1 and K_2 are constants. The concentrations at the two ends are c_{A1} and c_{A2} and are constant. It is desired to do as follows for steady-state.

(a) For equimolar counterdiffusion derive the final differential and integrated equations.
(b) Draw the analog computer diagram to solve the differential equation of part (a).
(c) For diffusion of A through stagnant B derive the final differential equation.

Ans.: (a) $N_A = \dfrac{K_1}{L}(c_{A1} - c_{A2}) + \dfrac{K_2}{2L}(c_{A1}^2 - c_{A2}^2)$

(c) $N_A = -\dfrac{(K_1 + K_2 c_A)}{(1 - c_A/c)} \dfrac{dc_A}{dz}$

1-15 Numerical Solution of Boundary-Value Problem Using Digital Computer

Repeat Example 1.9-1 with the following numerical values. $c_{A1} = 2 \times 10^{-5}$ g mole/cm³, $c_{A2} = 0.1 \times 10^{-5}$, $L = 5$ cm, $D_{AB} = 1.5$ cm²/sec, $k_3' = 0.02$ sec⁻¹, $k_4' = 0.3 \times 10^{-7}$ g mole/sec-cm³. Use a spacing $h = 0.04$ cm.

1-16 Molecular Multicomponent Diffusion

Ammonia (A) is diffusing at steady-state and 85°C and a total pressure of 2.0 atm through an inert gas mixture of nitrogen (B) and hydrogen (C), and the length of path is 0.20 cm. The mole fractions at the two ends of the path are $x_{A1} = 0.8$, $x_{B1} = 0.15$, $x_{C1} = 0.05$, $x_{A2} = 0.2$, $x_{B2} = 0.6$, $x_{C2} = 0.2$.
(a) Calculate the flux of ammonia.
(b) If the total pressure is dropped to 0.20 atm, calculate the flux of ammonia.

1-17 Multicomponent Diffusion and Reaction

Hydrogen is being oxidized on a solid catalyst surface at steady-state,

$$2H_2 + O_2 \rightarrow 2H_2O$$

At a given point in the reactor at 200°C and 1 atm abs total pressure the composition of the bulk gas phase (point 0) is for $H_2(A)$, $x_{A0} = 0.40$; for $O_2(B)$, $x_{B0} = 0.20$; and for $H_2O(C)$, $x_{C0} = 0.40$. Assume that the gas is in laminar flow, and the

H$_2$ and O$_2$ diffuse through a gas film 0.10 cm thick to the catalyst surface, react instantly, and the H$_2$O diffuses back through the same film. Estimate the maximum rate of reaction in g mole H$_2$/sec-cm^2 surface of catalyst that can be obtained if the overall reaction rate is diffusion controlled. The diffusivities are $D_{AB} = 0.80$ cm^2/sec at 25°C, $D_{AC} = 1.12$ cm^2/sec at 55.5°C, and $D_{BC} = 0.35$ cm^2/sec at 79.3°C. Correct the diffusivities to 200°C by using $D_{AB} \propto T^{1.75}$.
[*Hints*: This is a case of diffusion of three or more components and Eq. (1.10-22) can be used. First relate the fluxes N_A, N_B, and N_C by the stoichiometry of the reaction. This will be $N_A = 2N_B = -N_C$.]

1-18 Diffusion of Three Components

Repeat Example 1.10-3, if the ratios of the diffusivities are different and $D_{PQ}/D_{AP} = 6$ and $D_{PQ}/D_{AP} = 3$.

$$\text{Ans.:} \quad \frac{N_A \delta}{c D_{PQ}} = 0.0246 \ (\text{H3})$$

1-19 Simultaneous Diffusion of Two Components through a Third

A gas A is diffusing through a mixture of C and B at a total pressure of 0.200 atm and 55°C and is steady-state. The distance of diffusion is 0.200 cm. Component B is also diffusing in the mixture. The partial pressures are as follows:

	Point 1	Point 2
p_A	0.020	0.00
p_B	0	0.100
p_C	0.180	0.100
	0.200	0.200

Use physical property data from Example 1.10-2.
(a) Calculate N_A assuming all other gases are inert C. Calculate N_B assuming all other gases are inert C.
(b) Calculate N_A and N_B using the method of Gilliland.
(c) Calculate N_A and N_B using the approximate method of Hsu and Bird in Section 1.10.

NOTATION

(Boldface symbols are vectors.)

a constant
A cross-sectional area perpendicular to flux, cm^2 or ft^2
$\mathbf{A}$ matrix
b constant
c total concentration, g mole/cm^3 or lb mole/ft^3
$\mathbf{c}$ vector
c_A concentration of A, g mole A/cm^3 or lb mole A/ft^3
c_p heat capacity at constant pressure, cal/g mass-°C or Btu/lb mass-°F

d_A diameter of a molecule, cm
D_{AB} diffusivity, cm²/sec or ft²/hr
D_{Am} mean diffusivity given in Eq. (1.10-20)
e dependent variable on analog computer, volts
E_A flux ratio in Eq. (1.10-18)
f_D correction factor in Eq. (1.6-21)
g_c conversion constant, 980 g mass-cm/(g force-sec²) or 32.17 lb mass-ft/(lb force-sec²)
h interval $x_{n+1} - x_n$
$\bar{J}_A^*$ flux of A relative to the molar average velocity, g mole A/sec or lb mole A/hr
J_A^* flux of A relative to the molar average velocity, g mole A/sec-cm² or lb mole A/hr-ft²
k Boltzmann constant
k thermal conductivity, cal-cm/sec-cm²-°C or Btu-ft/hr-ft²-°F
k' homogeneous first-order reaction velocity constant, sec⁻¹
k'_1 heterogeneous first-order reaction velocity constant, cm/sec
k'_3, k'_4 empirical rate constants
K constant
K_1, K_2 empirical constants for $D_{AB} = K_1 + K_2 c_A$
L distance, cm
m_A molecular weight of A, g mass/molecule
M_A molecular weight of A, g mass A/g mole or lb mass A/lb mole
N Avogadro's number
N number
n_A g mole A or lb mole A
n concentration, molecules per unit volume
$\bar{N}_A$ flux of A relative to a fixed point, g mole A/sec or lb mole A/hr
N_A flux of A relative to a fixed point, g mole A/sec-cm² or lb mole A/hr-ft²
N_{Le} Lewis number $= k/(\rho c_p D)$, dimensionless
N_{Pr} Prandtl number $= c_p \mu / k$, dimensionless
N_{Sc} Schmidt number $= \mu / \rho D$, dimensionless
P total pressure, atm abs.
P_A vapor pressure of pure A, atm
p constant
p_A partial pressure of A, atm
p_{BM} ln mean of p_{B1} and p_{B2}, atm
p_i partial pressure of inerts, atm
p_{iM} ln mean of inerts p_{i1} and p_{i2}, atm
q constant
q flux of heat, cal/sec-cm² or Btu/hr-ft²
r radius, cm or ft
r distance of separation of two molecules defined by Eq. (1.6-20)
r_p initial radius of sphere, cm or ft
R universal gas law constant, 82.06 cm³-atm/gm mole-°K or 0.730 ft³-atm/lb mole-°R

Notation 75

S frequency of collision given by Eq. (1.6-10)
T absolute temperature, °K or °R
t time
T_b normal boiling point, °K
T_c critical temperature, °K
$\bar{u}$ mean velocity given by Eq. (1.6-9)
$\mathbf{v}$ mass average velocity relative to stationary coordinates, cm/sec or ft/hr
$\mathbf{v}_A$ velocity of A relative to stationary coordinates, cm/sec or ft/hr
$\mathbf{v}^*$ molar average velocity relative to stationary coordinates, cm/sec or ft/hr
v_A structural volume increment of A (see Table 1.6-5)
V volume, cm³ or ft³
V velocity, cm/sec
V_b molar volume of liquid at normal boiling point, cm³/g mole (see Table 1.6-4)
w_A mass fraction of A
x_A mole fraction of A
x'_B moles B/mole inerts
x variable
$\mathbf{x}$ vector
x distance in x direction, cm or ft
y distance in y direction, cm or ft
y variable
z distance in z direction, cm or ft

Greek Letters

α thermal diffusivity, $k/\rho c_p$, cm²/sec or ft²/hr
α analog computer time-scale factor in $\tau = \alpha t$
α_{AB} proportionality constant for diffusion of A and B
β analog computer amplitude-scale factor
β constant defined by Eq. (1.6-7)
δ distance of diffusion path, cm or ft
δ generalized transport property, length² per unit time
ϵ_{AB} energy of molecular interaction, ergs
θ_0 total time, sec or hr
λ mean free path given by Eq. (1.6-15)
μ viscosity, g mass/cm-sec or lb mass/ft-hr
ν momentum diffusivity = μ/ρ, cm²/sec or ft²/hr
ρ density, g mass/cm³ or lb mass/ft³
σ_A collision diameter of A, Å
σ_{AB} collision diameter of pair A-B, Å
τ_{yx} shear stress, g force/cm² or lb force/ft²
$\tau_{yx}g_c$ flux of momentum, g mass/cm-sec² or lb mass/ft-hr²
τ analog computer time, sec
φ_{AB} Lennard-Jones function defined by Eq. (1.6-20)

Ψ_x flux of property, amount per unit time per unit cross-sectional area
ψ concentration of property, amount per unit volume
Ω_D collision integral based on the Lennard-Jones potential

Subscripts

A component A
ave average
B component B
C component C
D component D
I component I
i component i
j component j
m effective
M molar average
n last of n components
n number
N number
P component P
Q component Q
x, y, z x, y, and z directions
δ position
1   beginning of diffusion path
2   end of diffusion path

REFERENCES

(B1) R. B. Bird, W. E. Stewart, and E. N. Lightfoot, *Transport Phenomena*. New York: John Wiley & Sons, Inc., 1960.

(C1) L. T. Carmichael, B. H. Sage, and W. N. Lacey, *AIChEJ* **1**, 385 (1955).

(C2) A. J. Carswell and J. C. Stryland, *Can. J. Phys.* **41**, 708 (1963).

(C3) S. Chapman and T. G. Cowling, *Mathematical Theory of Non-uniform Gases*, p. 252. New York: Cambridge University Press, 1939.

(C4) P. K. Chakraborti and P. Gray, *Trans. Farad. Soc.* **62**, 333 (1966).

(C5) S. D. Conte, *Elementary Numerical Analysis*, New York: McGraw-Hill, Inc., 1965.

(F1) E. N. Fuller, P. D. Schettler, and J. C. Giddings, *Ind. Eng. Chem.* **58**, 19 (1966).

(G1) E. R. Gilliland, *Ind. Eng. Chem.* **26**, 681 (1934).

(H1) J. O. Hirschfelder, C. F. Curtiss, and R. B. Bird, *Molecular Theory of Gases and Liquids*. New York: John Wiley & Sons, Inc., 1954.

(H2) J. N. Holsen and M. R. Strunk, *Ind. Eng. Chem. Fund.* **3**, 143 (1964).

(H3) H. W. Hsu and R. B. Bird, *AIChEJ* **6**, 516 (1960).

(I1) National Research Council, *International Critical Tables*, vol. V. New York: McGraw-Hill, Inc., 1929.

References

(L1) G. Le Bas, *The Molecular Volumes of Liquid Chemical Compounds.* New York: David McKay Company, Inc., 1915.
(L2) N. A. Lange, *Handbook of Chemistry*, 10th ed. New York: McGraw-Hill, Inc., 1967.
(L3) L. Lapidus, *Digital Computation for Chemical Engineers.* New York: McGraw-Hill, Inc., 1962.
(M1) E. A. Mason and L. Monchick, *J. Chem. Phys.* **36**, 2746 (1962).
(M2) L. Monchick and E. A. Mason, *J. Chem. Phys.* **35**, 1676 (1961).
(M3) H. S. Mickley, T. K. Sherwood, and C. E. Reed, *Applied Mathmatics in Chemical Engineering*, 2d ed. New York: McGraw-Hill, Inc., 1957.
(M4) W. L. McCabe and J. C. Smith, *Unit Operations of Chemical Engineering*, 2d ed. New York: McGraw-Hill, Inc., 1967.
(N1) National Bureau of Standards, *Tables of Thermal Properties of Gases*, circ. 464 (1955).
(N2) E. P. Ney and F. C. Armistead, *Phys. Rev.* **71**, 14 (1947).
(P1) J. H. Perry, *Chemical Engineers' Handbook*, 3d ed. New York: McGraw-Hill, Inc., 1950.
(P2) *Ibid.*, 4th ed., 1963.
(R1) R. C. Reid and T. K. Sherwood, *The Properties of Gases and Liquids*, 2d ed. New York: McGraw-Hill, Inc., 1966.
(R2) *Reactor Handbook*, vol. 2, Atomic Energy Commission, AECD-3646, Washington, D.C., May 1955.
(S1) A. S. Smith, *Ind. Eng. Chem.* **26**, 1167 (1934).
(S2) L. I. Stiel and G. Thodos, *AIChEJ* **10**, 266 (1964).
(S3) S. L. Seager, L. R. Geertson, and J. C. Giddings, *J. Chem. Eng. Data* **8**, 168 (1963).
(S4) J. F. Swindells, J. R. Coe Jr., and T. B. Godfrey, *J. Res. Nat. Bur. Standards* **48**, 1 (1952).
(S5) F. A. Schwertz and J. E. Brow, *J. Chem. Phys.* **19**, 640 (1951).
(S6) K. L. Schafer, *Z. Electrochem.* **63**, 111 (1959).
(S7) B. N. Srivastava and I. B. Srivastava, *J. Chem. Phys.* **38**, 1183 (1963).
(S8) T. K. Sherwood, *Absorption and Extraction*, 1st ed. New York: McGraw-Hill, Inc., 1937.
(T1) H. L. Toor, *AIChEJ* **3**, 198 (1957).
(T2) R. E. Treybal, *Mass-Transfer Operations*, 1st ed. New York: McGraw-Hill, Inc., 1955.
(T3) *Ibid.*, 2d ed., 1968.
(T4) M. Trautz and W. Muller, *Ann. Physik* **22**, 333 (1935).
(W1) C. R. Wilke and C. Y. Lee, *Ind. Eng. Chem.* **47**, 1253 (1955).
(W2) C. R. Wilke, *Chem. Eng. Progr.* **46**, 95 (1950).
(W3) V. E. Wintergerst, *Ann. Physik* **4**, 323 (1930).
(W4) A. A. Westenberg and G. Frazier, *J. Chem. Phys.* **36**, 3499 (1962).

2 Transport Phenomena and the Basic Equations of Change

2.1 INTRODUCTION TO EQUATIONS OF CHANGE

2.1A Purpose and Use of Equations of Change

In Chapter 1 the basic molecular transport equations for momentum, heat, and mass transport were introduced. Also the equation for mass transport plus convective flow was derived. These relations were used to calculate the fluxes and/or the concentration profiles for physical systems whose geometries were well defined. The problems were solved in most instances by setting up mass balances over a differential section.

The principle of the conservation of mass was used in the solution of these problems. For each new case considered, a new differential balance was set up. However, it is really not necessary to set up a new mass balance for each new mass transport problem. In this chapter we will derive general equations of change for the conservation of mass, momentum, and energy. Then, for each new problem, the unneeded terms will be simply discarded and the simplified equations used to obtain a solution.

These equations of change, together with the flux equations, are all that is necessary to solve problems in molecular transport of heat, mass, or momentum. The flux equations are those of Fick, Fourier, or Newton. If more than one chemical species is involved, then a separate conservation-of-mass equation is needed for each species.

In the first sections of this chapter we will derive the conservation-of-mass equation for a pure fluid when convective flow is occurring. Then the equation of motion will be described briefly, since it has some application in the solving of convective mass transport problems. The equation of energy change will also be reviewed briefly, since it has use in problems where combined heat and mass transport occur. Finally, the equations of change will be derived for a binary mixture.

Before starting the actual derivations we shall review certain basic physical phenomena and mathematical principles that will be used later.

2.1B Physical and Chemical Properties and Boundary Conditions

1. *Steady- and Unsteady-State.* We can define a transport process as being at steady-state when the concentrations, fluxes, temperatures, and all other physical and chemical properties are constant with time at any point in the process. In unsteady-state these properties are changing with time. Of course, the unsteady-state processes are more difficult to solve because of the additional variable of time.

2. *Chemical Reaction Rates and Generation.* Often in systems where two or more individual species occur, or even in a pure fluid, one species may react alone or with another species and a new species is formed. In a homogeneous chemical reaction we will use a simple reaction rate expression when A is being formed:

$$R_A = k_1' c_A \qquad (2.1\text{-}1)$$

where R_A is the generation rate in g mole A/sec-cm^3 of volume, c_A is the concentration g mole A/cm^3 of volume, and k' is a first-order reaction velocity constant, sec^{-1}. If A disappears, R_A is equal to $-k' c_A$.

3. *Physical Conditions at the Boundaries.* In many mass transport problems the molar ratio N_A/N_B is specified by the physical conditions occurring at the boundaries. For example, one boundary of the diffusion path may be impermeable to species B because B is insoluble in the adjacent phase at the boundary. Diffusion of ammonia (A) and air (B) through gas to a water phase at the boundary is such a case. The air is essentially insoluble in the water.

The flux ratios could also be set by stoichiometry of a reaction at a boundary—for example, for the reaction at the boundary of

$$aA \rightarrow pP \qquad (2.1\text{-}2)$$

Then the fluxes are related by

$$N_P = -\frac{p}{a} N_A \qquad (2.1\text{-}3)$$

Heat balance considerations also could affect or set the flux ratios.

The boundary conditions at the two ends of the diffusion path x_{A1} and x_{A2} can often be specified. For example, at one end of the path a large volume of the fluid phase can be flowing by with a fixed composition of x_{A1}. In some cases the concentration x_{A1} may be set by an equilibrium condition, whereby x_{A1} is in equilibrium with some fixed composition.

A chemical reaction can occur on a catalyst surface at point 2, which is the end of the diffusion path. Then the flux at point 2 is related to the reaction rate k_1' in cm/sec on the solid surface by

$$N_{A2} = k_1' c_A \quad (2.1\text{-}4)$$

In some cases the flux N_{A1} at point 1 is fixed by the rate at which species A can be supplied from an adjacent flowing fluid by

$$N_{A1} = k_c (c_{AL} - c_{A1}) \quad (2.1\text{-}5)$$

where k_c is a mass transfer convective coefficient in cm/sec and c_{AL} is the bulk fluid concentration in g mole A/cm^3. This coefficient k_c is an empirical coefficient and is a function of the system geometry of the surface, of fluid and flow properties, and to a small extent of the concentration difference. A more detailed explanation is given in Sections 6.2 and 6.3.

2.1C Discussion of Types of Derivatives with Respect to Time

In deriving the various equations of change in this chapter, we use different types of time derivatives in defining velocities and unsteady-state conditions. A brief review will be given of these time derivatives.

1. *The Partial Time Derivative.* In order to indicate the use of the various types of time derivatives we shall consider the classical example of fish moving in a flowing river. The fish are swimming and moving relative to the flowing river itself. We can think of the fish as molecules of species A, which diffuse or move relative to the main river stream. We are interested in the concentration c of the fish (or solute molecules A) in the river as a function of position x, y, z and also of time t.

The most common type of derivative used is the partial time derivative $\partial c / \partial t$. The partial means that the other variables x, y, and z were held constant—that is, at a fixed point—while we observed the change in c with respect to t.

2. *The Total Derivative with Respect to Time.* An observer can be noting the change in the concentration, dc, of the fish in the river while he himself is moving about, in, say a motorboat. This means that the concentration c is now a function not only of t but also of x, y, and z—that is,

$$\frac{dc}{dt} = \frac{\partial c}{\partial t} + \frac{\partial c}{\partial x}\frac{dx}{dt} + \frac{\partial c}{\partial y}\frac{dy}{dt} + \frac{\partial c}{\partial z}\frac{dz}{dt} \quad (2.1\text{-}6)$$

The velocity of the observer in the boat has the velocity components of dx/dt, dy/dt, and dz/dt.

3. *The Substantial Time Derivative.* Another useful type of time derivative is obtained if the observer floats along at the velocity **v** of the stream. Now noting the change in concentration with respect to time or the derivative that follows the motion, Dc/Dt is

$$\frac{Dc}{Dt} = \frac{\partial c}{\partial t} + v_x\frac{\partial c}{\partial x} + v_y\frac{\partial c}{\partial y} + v_z\frac{\partial c}{\partial z} = \frac{\partial c}{\partial t} + (\mathbf{v}\cdot\nabla c) \quad (2.1\text{-}7)$$

where v_x, v_y, and v_z are the velocity components of the stream velocity **v**.

2.1D Introduction to Scalars, Vectors, and Differential Operators

1. *Scalars.* In transport phenomena certain physical properties have certain mathematical representations. Such properties as temperature, length, volume, time, energy, and concentration are called scalars. These scalars have magnitude but no direction and are often thought of as zero-order tensors.

Certain common mathematical algebraic laws hold for the algebra of scalars but do not necessarily always hold for vectors. These are

$$bc = cb$$

$$b(cd) = (bc)d$$

$$b(c+d) = bc + bd$$

2. *Vectors.* When a physical property has both magnitude and direction we call it a vector. For example, velocity, force, momentum, and acceleration are considered vectors. They also are regarded as first-order tensors.

The reader should realize that the vector operations to be discussed can also be done by using the rectangular components of the vectors. However, using vectors is more efficient and less time-consuming. All vectors are represented by boldface symbols in this text.

In Fig. 2.1-1 we show a vector; the arrow represents the direction of the vector and the length represents the magnitude. The vector **B** is represented by its three projections on the x, y, and z axes. We will let unit vectors be represented by **i**, **j**, and **k** along the axes as shown. Then the magnitude of **B** is

$$B = |\mathbf{B}| = \sqrt{B_x^2 + B_y^2 + B_z^2} \quad (2.1\text{-}8)$$

Also the vector is represented by

$$\mathbf{B} = \mathbf{i}B_x + \mathbf{j}B_y + \mathbf{k}B_z \quad (2.1\text{-}9)$$

Certain useful vector operations will be summarized below without any proof.

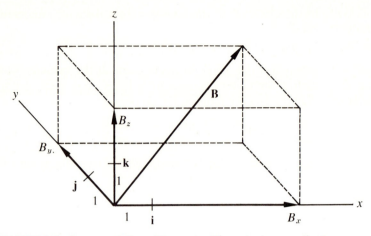

FIGURE 2.1-1 Representation of the vector **B** in cartesian coordinates

(a) *Addition.* To illustrate the addition of two vectors **B** + **C** as

$$\mathbf{B} + \mathbf{C} = \mathbf{D} \tag{2.1-10}$$

we give a geometrical representation in Fig. 2.1-2a. This parallelogram construction is often used for addition. Subtraction of two vectors, **B** − **C**, is shown in Fig. 2.1-2b. The following commutative and associative laws hold:

$$(\mathbf{B} + \mathbf{C}) = (\mathbf{C} + \mathbf{B}) \tag{2.1-11}$$
$$(\mathbf{B} + \mathbf{C}) + \mathbf{D} = \mathbf{B} + (\mathbf{C} + \mathbf{D}) \tag{2.1-12}$$

(b) *Multiplication by a scalar.* In multiplying a scalar quantity s by a vector, the magnitude of the vector is changed but the direction is, of course, not altered. The following laws hold, where r and s are scalars:

$$r\mathbf{B} = \mathbf{B}r \tag{2.1-13}$$
$$(rs)\mathbf{B} = r(s\mathbf{B}) \tag{2.1-14}$$
$$r\mathbf{B} + s\mathbf{B} = (r+s)\mathbf{B} \tag{2.1-15}$$

FIGURE 2.1-2 (a) Addition of vectors, **B** + **C**; (b) subtraction of vectors, **B** − **C**

(c) *Dot product or scalar product.* The scalar or dot product of two vectors can be represented as

$$(\mathbf{B}\cdot\mathbf{C}) = BC\cos\varphi_{BC} \tag{2.1-16}$$

where φ_{BC} is the angle between the two vectors and is less than 180°. This dot product can be considered as a scalar quantity that represents the magnitude of one of the vectors **B** multiplied by the projection of the other vector **C** upon **B**, or vice versa. These laws hold:

$$(\mathbf{B}\cdot\mathbf{C}) = (\mathbf{C}\cdot\mathbf{B}) \tag{2.1-17}$$
$$\mathbf{B}\cdot(\mathbf{C}+\mathbf{D}) = (\mathbf{B}\cdot\mathbf{C}) + (\mathbf{C}\cdot\mathbf{D}) \tag{2.1-18}$$
$$(\mathbf{B}\cdot\mathbf{C})\mathbf{D} \neq \mathbf{B}(\mathbf{C}\cdot\mathbf{D}) \tag{2.1-19}$$

(d) *Cross or vector product.* The vector or cross product is

$$(\mathbf{B}\times\mathbf{C}) = \mathbf{D} \tag{2.1-20}$$

The resultant vector **D** is normal to the plane of **B** and **C**, and its direction is such that the three vectors form a right-handed system. The magnitude of the vector product is

$$|(\mathbf{B}\times\mathbf{C})| = |\mathbf{D}| = BC\sin\varphi_{BC} \tag{2.1-21}$$

Second-order tensors τ arise primarily in momentum transfer processes and have nine components. The mathematical operations will not be given here since they are discussed in detail elsewhere (B1).

3. *Differential Operations with Scalars and Vectors.* Certain important differential operations will be summarized below for the reader's convenience.

The gradient or "grad" of a scalar field is

$$\nabla\rho = \mathbf{i}\frac{\partial\rho}{\partial x} + \mathbf{j}\frac{\partial\rho}{\partial y} + \mathbf{k}\frac{\partial\rho}{\partial z} \tag{2.1-22}$$

where ρ is a scalar such as density, and **i**, **j**, and **k** are unit vectors along the x, y, and z axes, respectively.

The divergence or "div" of a vector **v** is

$$(\nabla\cdot\mathbf{v}) = \frac{\partial v_x}{\partial x} + \frac{\partial v_y}{\partial y} + \frac{\partial v_z}{\partial z} \tag{2.1-23}$$

where the vector **v** is a function of three variables v_x, v_y, v_z.

The Laplacian of a scalar field is

$$(\nabla^2\rho) = \frac{\partial^2\rho}{\partial x^2} + \frac{\partial^2\rho}{\partial y^2} + \frac{\partial^2\rho}{\partial z^2} \tag{2.1-24}$$

The Laplacian of a vector **v** such as velocity is

$$(\nabla^2\mathbf{v}) = \nabla\cdot\nabla\mathbf{v} = \frac{\partial^2\mathbf{v}}{\partial x^2} + \frac{\partial^2\mathbf{v}}{\partial y^2} + \frac{\partial^2\mathbf{v}}{\partial z^2} \tag{2.1-25}$$

Some other operations sometimes used are

$$\nabla rs = r\nabla s + s\nabla r \tag{2.1-26}$$

$$(\nabla \cdot s\mathbf{v}) = (\nabla s \cdot \mathbf{v}) + s(\nabla \cdot \mathbf{v}) \tag{2.1-27}$$

$$(\mathbf{v} \cdot \nabla s) = v_x \frac{\partial s}{\partial x} + v_y \frac{\partial s}{\partial y} + v_z \frac{\partial s}{\partial z} \tag{2.1-28}$$

2.2 EQUATION OF CONTINUITY FOR A PURE FLUID

In Chapter 1 an elementary general property balance for unsteady-state was made in the x direction only to give Eq. (1.1-9).

$$\frac{\partial \psi}{\partial t} = \delta \frac{\partial^2 \psi}{\partial x^2} \tag{1.1-9}$$

In this equation we did not consider the y and z directions, nor did we consider convective motion; we considered only motion by diffusion in a mixture. In the present derivation we will consider only a pure fluid, and later in this chapter we will consider a mixture.

We consider in Fig. 2.2-1 a small volume element $\Delta x\, \Delta y\, \Delta z$, which is fixed in space. The fluid is flowing through this volume element, and the

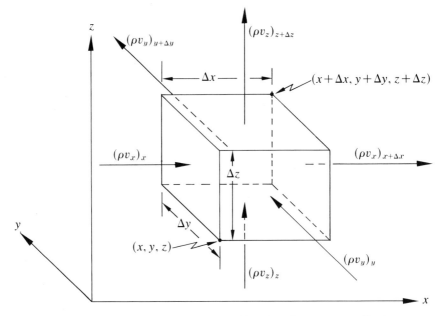

FIGURE 2.2-1 Mass balance for a pure fluid and a volume $\Delta x\, \Delta y\, \Delta z$ fixed in space

2.2 Equation of Continuity for a Pure Fluid

mass contained in this volume varies with time. The mass balance for a pure fluid with a concentration of ρ g mass/cm³ is

$$\text{(rate of mass in)} - \text{(rate of mass out)} = \text{(rate of accumulation)}. \qquad (2.2\text{-}1)$$

First considering only the x direction as given in Fig. 2.2-1, the rate of mass entering the face at x, which has an area of $\Delta y \, \Delta z$, is equal to $(\rho v_x)_x \, \Delta y \, \Delta z$. The rate of mass leaving the face at $x + \Delta x$ is $(\rho v_x)_{x+\Delta x} \, \Delta y \, \Delta z$. That entering and leaving in the y direction and in the z direction is also shown in Fig. 2.2-1.

To obtain the rate of accumulation of mass in the volume $\Delta x \, \Delta y \, \Delta z$ we multiply the time rate of change of ρ times the volume:

$$\text{rate of accumulation} = \frac{\partial \rho}{\partial t} \Delta x \, \Delta y \, \Delta z \qquad (2.2\text{-}2)$$

Substituting all of these expressions into Eq. (2.2-1), we obtain

$$\Delta y \, \Delta z [(\rho v_x)_x - (\rho v_x)_{x+\Delta x}] + \Delta x \, \Delta z [(\rho v_y)_y - (\rho v_y)_{y+\Delta y}]$$
$$+ \Delta x \, \Delta y [(\rho v_z)_z - (\rho v_z)_{z+\Delta z}] = \frac{\partial \rho}{\partial t} \Delta x \, \Delta y \, \Delta z \qquad (2.2\text{-}3)$$

If we divide both sides of Eq. (2.2-3) by $\Delta x \, \Delta y \, \Delta z$ and take the limit as each of these segments approaches zero, we obtain our final equation of continuity or conservation of mass for a pure fluid:

$$\frac{\partial \rho}{\partial t} = -\left(\frac{\partial (\rho v_x)}{\partial x} + \frac{\partial (\rho v_y)}{\partial y} + \frac{\partial (\rho v_z)}{\partial z}\right) = -(\nabla \cdot \rho \mathbf{v}) \qquad (2.2\text{-}4)$$

The vector notation on the right side comes from the fact $\mathbf{v}$ is a vector and is given in Eq. (2.1-23), where it is a function of v_x, v_y, and v_z. Hence Eq. (2.2-4) tells us how density ρ changes at a fixed point as a result of changes in the mass velocity vector $\rho \mathbf{v}$. The units of this vector $\rho \mathbf{v}$ are (g mass/cm³)(cm/sec) or g mass/sec-cm², which is the mass flux.

We can rearrange Eq. (2.2-4) into another form by carrying out the actual partial differentiation.

$$\frac{\partial \rho}{\partial t} = -\rho\left(\frac{\partial v_x}{\partial x} + \frac{\partial v_y}{\partial y} + \frac{\partial v_z}{\partial z}\right) - \left(v_x \frac{\partial \rho}{\partial x} + v_y \frac{\partial \rho}{\partial y} + v_z \frac{\partial \rho}{\partial z}\right) \qquad (2.2\text{-}5)$$

Rearranging this,

$$\frac{\partial \rho}{\partial t} + v_x \frac{\partial \rho}{\partial x} + v_y \frac{\partial \rho}{\partial y} + v_z \frac{\partial \rho}{\partial z} = -\rho\left(\frac{\partial v_x}{\partial x} + \frac{\partial v_y}{\partial y} + \frac{\partial v_z}{\partial z}\right) \qquad (2.2\text{-}6)$$

The left-hand side of Eq. (2.2-6) is the same as Eq. (2.1-7), since ρ and c are scalars; hence, Eq. (2.2-6) becomes

$$\frac{D\rho}{Dt} = -\rho\left(\frac{\partial v_x}{\partial x} + \frac{\partial v_y}{\partial y} + \frac{\partial v_z}{\partial z}\right) = -\rho(\nabla \cdot \mathbf{v}) \qquad (2.2\text{-}7)$$

As discussed before, this is the substantial time derivative and it gives the rate of change of ρ one observes when one is going at the same velocity as the fluid.

Often in engineering, and especially in dealing with liquids that are relatively incompressible, the density ρ is essentially constant. Then

$$(\nabla \cdot \mathbf{v}) = \left(\frac{\partial v_x}{\partial x} + \frac{\partial v_y}{\partial y} + \frac{\partial v_z}{\partial z}\right) = 0 \tag{2.2-8}$$

This means $D\rho/Dt = 0$.

In mass transport phenomena we are generally interested in the equation of continuity for species A and B and not in the continuity equation for a pure fluid. Equation (2.2-4) was derived to introduce the reader to the principles and methods to be used for the multicomponent balances that follow later in this chapter.

2.3 EQUATION OF MOTION FOR A PURE FLUID

2.3A Equation for All Fluids

In this section, since we are interested in the equation of motion only for its use in solving some convective mass transport problems, we will not derive this equation in detail. Only a brief outline will be given.

The equation of motion is really the equation for the conservation of momentum. We can write this in general as

$$\text{(momentum rate in)} - \text{(momentum rate out)} + (\Sigma \text{ forces acting on system})$$
$$= \text{(rate of momentum accumulation)} \tag{2.3-1}$$

We will make a balance on an element as in Fig. 2.2-1. Considering only the x component in our balance—that is, the x direction only—the convective momentum in at x is $(\rho v_x v_x)_x \Delta y \Delta z$. The quantity (ρv_x) is the concentration as momentum/cm³ in, and it is multiplied by v_x to give momentum convective flux as momentum/sec-cm². This momentum flux leaves at $x + \Delta x$ as $(\rho v_x v_x)_{x+\Delta x} \Delta y \Delta z$.

We also have the x component of momentum entering the y face by convection as $(\rho v_y v_x)_y \Delta x \Delta z$ and leaving at $y + \Delta y$ as $(\rho v_y v_x)_{y+\Delta y} \Delta x \Delta z$. A similar term can be written for the z face. Hence, the net x component for convective flow of momentum is

$$[(\rho v_x v_x)_x - (\rho v_x v_x)_{x+\Delta x}] \Delta y \Delta z + [(\rho v_y v_x)_y - (\rho v_y v_x)_{y+\Delta y}] \Delta x \Delta z$$
$$+ [(\rho v_z v_x)_z - (\rho v_z v_x)_{z+\Delta z}] \Delta x \Delta y \tag{2.3-2}$$

The rate of x component of momentum that enters the surface at x by molecular transport is $(\tau_{xx})_x \Delta y \Delta z$, and a similar term leaves at $x + \Delta x$.

2.3 Equation of Motion for a Pure Fluid

At the y and z faces similar terms can be written. These rates of molecular transport of momentum are made up of shear stresses and normal stresses. Details are given elsewhere (B1, B2, B3).

The fluid pressure force acting on the element in the x direction is $(p_x - p_{x+\Delta x}) \Delta y \Delta z$ and the gravitational force is $\rho g_x \Delta x \Delta y \Delta z$. The rate of accumulation of the momentum is

$$\Delta x \Delta y \Delta z \frac{\partial(\rho v_x)}{\partial t} \tag{2.3-3}$$

Combining the terms for convective flux, molecular transport flux, pressure and gravitational forces, and accumulation in Eq. (2.3-1) and converting to partial differentials, we obtain for the x component of momentum

$$\frac{\partial(\rho v_x)}{\partial t} = -\left(\frac{\partial(\rho v_x v_x)}{\partial x} + \frac{\partial(\rho v_y v_x)}{\partial y} + \frac{\partial(\rho v_z v_x)}{\partial z}\right)$$

$$-\left(\frac{\partial \tau_{xx}}{\partial x} + \frac{\partial \tau_{yx}}{\partial y} + \frac{\partial \tau_{zx}}{\partial z}\right) - \frac{\partial p}{\partial x} + \rho g_x \tag{2.3-4}$$

Similar equations can be written for the y and z components, and the final result is

$$\frac{\partial(\rho \mathbf{v})}{\partial t} = -[\nabla \cdot \rho \mathbf{v}\mathbf{v}] - \nabla - [\nabla \cdot \boldsymbol{\tau}] + \rho \mathbf{g} \tag{2.3-5}$$

where $\rho \mathbf{v}\mathbf{v}$ is the convective momentum flux tensor composed of nine components and $\boldsymbol{\tau}$ the stress tensor composed of nine components.

We can use Eq. (2.2-4), which is the continuity equation, and Eq. (2.1-7) for v_x, which is the definition of the substantial derivative Dv_x/Dt, and obtain a new form of Eq. (2.3-4):

$$\rho \frac{Dv_x}{Dt} = -\left(\frac{\partial \tau_{xx}}{\partial x} + \frac{\partial \tau_{yx}}{\partial y} + \frac{\partial \tau_{zx}}{\partial z}\right) + \rho g_x - \frac{\partial p}{\partial x} \tag{2.3-6}$$

Writing equations for the y and z components similar to Eq. (2.3-6) and adding, we obtain the final general equation of motion for a pure fluid:

$$\rho \frac{D\mathbf{v}}{Dt} = -[\nabla \cdot \boldsymbol{\tau}] - \nabla p + \rho \mathbf{g} \tag{2.3-7}$$

2.3B Equation for Newtonian Fluids

We are usually interested in fluids in mass transport that are Newtonian in character—that is, those that follow Newton's first law. This means the viscosity is constant and not dependent on the rate of shear. The stresses $\tau_{xx}, \tau_{yx}, \tau_{zx}$, and so on given in Eq. (2.3-7) can be related to the velocity gradients and the viscosity, μ (B1, B2). After these substitutions are made, the equation for the x component of momentum becomes

$$\rho \frac{Dv_x}{Dt} = \frac{\partial}{\partial x}\left[2\mu \frac{\partial v_x}{\partial x} - \frac{2}{3}\mu(\nabla \cdot \mathbf{v})\right] + \frac{\partial}{\partial y}\left[\mu\left(\frac{\partial v_x}{\partial y} + \frac{\partial v_y}{\partial x}\right)\right]$$
$$+ \frac{\partial}{\partial z}\left[\mu\left(\frac{\partial v_z}{\partial x} + \frac{\partial v_x}{\partial z}\right)\right] - \frac{\partial p}{\partial x} + \rho g_x \quad (2.3\text{-}8)$$

In many cases, especially for liquids, the density and the viscosity are reasonably constant, and we obtain the Navier-Stokes equation for the x component. Since $(\nabla \cdot \mathbf{v}) = 0$, we obtain

$$\rho\left(\frac{\partial v_x}{\partial t} + v_x\frac{\partial v_x}{\partial x} + v_y\frac{\partial v_x}{\partial y} + v_z\frac{\partial v_x}{\partial z}\right) = \mu\left(\frac{\partial^2 v_x}{\partial x^2} + \frac{\partial^2 v_x}{\partial y^2} + \frac{\partial^2 v_x}{\partial z^2}\right) - \frac{\partial p}{\partial x} + \rho g_x$$
(2.3-9)

If we write similar equations for the y and z components, add the three equations, and convert to the substantial derivative,

$$\rho \frac{D\mathbf{v}}{Dt} = -\nabla p + \rho \mathbf{g} + \mu \nabla^2 \mathbf{v} \quad (2.3\text{-}10)$$

2.3C Use of the Equations of Continuity and Motion

As stated earlier, the purpose and use of the general equations of change are that they can be applied to any specific problem. For a given problem, the unneeded terms are simply discarded and the remaining equations used in the solution. It is necessary, of course, to know the initial conditions and the boundary conditions before the equation can be solved. In some cases terms in the general equations of change may be very small and are assumed as zero in order to simplify the solution. An example will be given to illustrate the general method. The reader is referred to other excellent texts for many other examples (B1, B2).

Example 2.3-1 Use of the Equations of Change for Flow in a Pipe

A liquid in laminar flow is flowing in the x direction in a horizontal and circular pipe at steady-state. The fluid is assumed to be incompressible and the viscosity is constant. It is desired to obtain an equation relating the velocity with radial position.

2.3 Equation of Motion for a Pure Fluid

Solution

Figure 2.3-1 shows the flow of the fluid in the x direction. We will call the z direction horizontal and the y direction vertical. We start by writing the equation of continuity:

$$\frac{\partial \rho}{\partial t} + v_x \frac{\partial \rho}{\partial x} + v_y \frac{\partial \rho}{\partial y} + v_z \frac{\partial \rho}{\partial z} = -\rho \left(\frac{\partial v_x}{\partial x} + \frac{\partial v_y}{\partial y} + \frac{\partial v_z}{\partial z} \right) \quad (2.2\text{-}6)$$

The last term is $-\rho(\nabla \cdot \mathbf{v})$, which is equal to zero as in Eq. (2.2-8) for constant density.

Since we are at steady-state, $\partial \rho / \partial t = 0$. Also, since flow is in the x direction only, v_y and v_z are zero in Eq. (2.2-6). The final result, after these simplifications, is

$$\frac{\partial v_x}{\partial x} = 0 \quad (2.3\text{-}11)$$

We could also have reached the conclusions above by making an elementary balance on a small volume element for steady-state. This would also have stated that input in the x direction is the same as the output.

Next we write down the Navier-Stokes equation for constant density and flow in the x direction:

$$\rho \left(\frac{\partial v_x}{\partial t} + v_x \frac{\partial v_x}{\partial x} + v_y \frac{\partial v_x}{\partial y} + v_z \frac{\partial v_x}{\partial z} \right) = \mu \left(\frac{\partial^2 v_x}{\partial x^2} + \frac{\partial^2 v_x}{\partial y^2} + \frac{\partial^2 v_x}{\partial z^2} \right) - \frac{\partial p}{\partial x} + \rho g_x \quad (2.3\text{-}9)$$

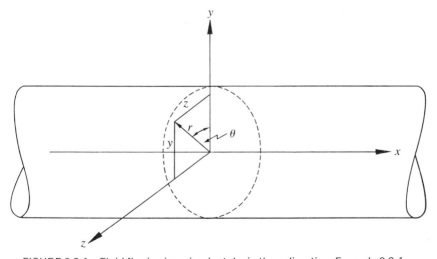

FIGURE 2.3-1 Fluid flowing in a circular tube in the x-direction, Example 2.3-1

As before, we set v_y and $v_z = 0$. Also $\partial v_x/\partial x = 0$ and $\partial v_x/\partial t = 0$. Hence, Eq. (2.3-9) finally becomes

$$\frac{dp}{dx} = \text{const.} = \mu\left(\frac{\partial^2 v_x}{\partial y^2} + \frac{\partial^2 v_x}{\partial z^2}\right) \tag{2.3-12}$$

In obtaining Eq. (2.3-12) we combine the p and ρg_x and call them p, which is the combined effect of gravitational force and static pressure. The units of p are mass/length-sec². Since v_x is not a function of x and p is independent of y and z, $dp/dx = $ constant.

In order to integrate Eq. (2.3-12) it is convenient to convert Eq. (2.3-12) into cylindrical coordinates:

$$x = x, \qquad z = r \sin \theta, \qquad y = r \cos \theta, \tag{2.3-13}$$

where r is the radial distance from the center and θ is the angle between the y axis and r as in Fig. 2.3-1. Performing the differentiations on Eq. (2.3-13) and combining the results with Eq. (2.3-12), we obtain

$$\frac{dp}{dx} = \text{const.} = \mu\left(\frac{1}{r}\frac{\partial v_x}{\partial r} + \frac{\partial^2 v_x}{\partial r^2} + \frac{1}{r^2}\frac{\partial^2 v_x}{\partial \theta^2}\right) \tag{2.3-14}$$

Since the flow is symmetrical about the x axis, $\partial v_x/\partial \theta$ and $\partial^2 v_x/\partial \theta^2 = 0$ and we have an ordinary differential equation:

$$\frac{dp}{dx} = \mu\left(\frac{1}{r}\frac{dv_x}{dr} + \frac{d^2 v_x}{dr^2}\right) = \frac{\mu}{r}\frac{d}{dr}\left(r\frac{dv_x}{dr}\right) \tag{2.3-15}$$

Integrating Eq. (2.3-15) once and letting $dv_x/dr = 0$ at $r = 0$ (symmetry at the center),

$$\frac{1}{2\mu}\frac{dp}{dx}r = \frac{dv_x}{dr}$$

Integrating again and letting $v_x = 0$ at $r = r_0$ (surface of the tube),

$$v_x = \frac{1}{4\mu}\frac{dp}{dx}(r^2 - r_0^2) \tag{2.3-16}$$

Usually we are interested in the maximum velocity $v_{x_{max}}$ at the center where $r = 0$. Then,

$$v_{x_{max}} = -\frac{1}{4\mu}\frac{dp}{dx}(r_0^2) \tag{2.3-17}$$

Substituting Eq. (2.3-17) into (2.3-16),

$$v_x = v_{x_{max}}\left(1 - \frac{r^2}{r_0^2}\right) \tag{2.3-18}$$

This final equation shows a parabolic velocity profile.

2.4 EQUATION OF ENERGY CHANGE FOR A PURE FLUID

2.4A The General Equation

In Section 2.2 we derived a general equation of continuity for a pure fluid and in Section 2.3 a general equation of motion. In this section we will outline the derivation of a general equation of energy change: the conservation-of-energy equation.

The real basis for any balance of energy is the first law of thermodynamics. If we consider an element of volume $\Delta x \Delta y \Delta z$ and write an energy balance, we get

(energy rate in) − (energy rate out) − (external rate of work done by system)

$$= \text{(rate of energy accumulation)} \quad (2.4\text{-}1)$$

Just as in momentum transfer, energy can be transported in and out by convection and by molecular transport, which is called conduction. The energy in the fluid consists of two types. The kinetic energy $\rho v^2/2$ is the energy that is associated with the observable fluid motion, where v is the mass average velocity of the fluid. The second type is the internal energy $\rho \hat{U}$ associated with motions and interactions of the individual molecules themselves. This energy is dependent on the local temperature and physical properties, such as density of the fluid.

The total energy, then, for a $\Delta x \Delta y \Delta z$ volume, is

$$(\rho \hat{U} + \tfrac{1}{2}\rho v^2) \, \Delta x \, \Delta y \, \Delta z \quad (2.4\text{-}2)$$

where $\hat{U}$ is the internal energy per unit mass.

The rate of accumulation then is simply the partial of Eq. (2.4-2) with respect to time t. The total energy coming in by convection in the x direction at point x is

$$\Delta y \, \Delta z \, (v_x(\rho \hat{U} + \tfrac{1}{2}\rho v^2))_x \quad (2.4\text{-}3)$$

The total energy coming in by conduction in the x direction is

$$\Delta y \, \Delta z \, (q_x)_x \quad (2.4\text{-}3a)$$

where q_x is the x component of the heat conduction flux vector **q**. The next work done against the gravity force g_x is

$$-\rho \, \Delta x \, \Delta y \, \Delta z \, (v_x g_x) \quad (2.4\text{-}4)$$

and the work done against the static pressure p at the inlet at x is

$$-\Delta y \, \Delta z \, (p v_x)_x \quad (2.4\text{-}5)$$

The work done at point x against the viscous forces is

$$-\Delta y \, \Delta z \, (\tau_{xx} v_x + \tau_{xy} v_y + \tau_{xz} v_z)_x \quad (2.4\text{-}6)$$

Similar terms can be written for the x direction at the position $x+\Delta x$ and also for the y and z directions. All of the terms can then be substituted into Eq. (2.4-1) and the limit taken as Δx, Δy, and $\Delta z \to 0$ to yield a final result. However, this result is not in a convenient form. So, combining it with the equation of continuity, and the equation of motion, and expressing the internal energy in terms of fluid temperature and heat capacity we obtain

$$\rho c_v \frac{DT}{Dt} = -(\nabla \cdot \mathbf{q}) - T\left(\frac{\partial p}{\partial T}\right)_V (\nabla \cdot \mathbf{v}) - (\tau:\nabla \mathbf{v}) \quad (2.4\text{-}7)$$

This is the general equation of energy change for any type of fluid, where V is $1/\rho$. Often $(\partial p/\partial T)_V$ is written as $(\partial p/\partial T)_\rho$. For details on the derivation of Eq. (2.4-7) see (B1).

2.4B Special Equations

We will write Eq. (2.4-7) for a Newtonian type fluid. Assuming a constant thermal conductivity, then using Fourier's law for molecular transport or conduction of energy,

$$(\nabla \cdot \mathbf{q}) = -k\frac{\partial^2 T}{\partial x^2} - k\frac{\partial^2 T}{\partial y^2} - k\frac{\partial^2 T}{\partial z^2} = k\nabla^2 T \quad (2.4\text{-}8)$$

The last term in Eq. (2.4-7) is $(\tau:\nabla \mathbf{v})$, which is called the viscous dissipation term. In using the equation of energy change, we usually can neglect the viscous dissipation term, except in special cases such as when high-viscosity fluids are present or large velocity gradients exist. This term will be omitted in the discussions to follow. Then, after substituting Eq. (2.4-8) into (2.4-7), Eq. (2.4-7) becomes

$$\rho c_v \left(\frac{\partial T}{\partial t} + v_x\frac{\partial T}{\partial x} + v_y\frac{\partial T}{\partial y} + v_z\frac{\partial T}{\partial z}\right)$$
$$= k\left(\frac{\partial^2 T}{\partial x^2} + \frac{\partial^2 T}{\partial y^2} + \frac{\partial^2 T}{\partial z^2}\right) - T\left(\frac{\partial p}{\partial T}\right)_V \left(\frac{\partial v_x}{\partial x} + \frac{\partial v_y}{\partial y} + \frac{\partial v_z}{\partial z}\right) \quad (2.4\text{-}9)$$

Certain special cases of Eq. (2.4-9) that occur often are as follows.
For a fluid at constant pressure, $\partial p/\partial T = 0$ and

$$\rho c_p \frac{DT}{Dt} = k\nabla^2 T \quad (2.4\text{-}10)$$

If the velocity $\mathbf{v}$ is zero, DT/Dt becomes $\partial T/\partial t$.
For a fluid at constant density, $(\nabla \cdot \mathbf{v}) = 0$ and

$$\rho c_p \frac{DT}{Dt} = k\nabla^2 T \quad (2.4\text{-}11)$$

For a solid, we have $v = 0$ and

$$\rho c_p \frac{\partial T}{\partial t} = k \nabla^2 T \tag{2.4-12}$$

If there is heat generation in the fluid by electrical or chemical means, a term q_r in calories/sec-cm^3 can be added to the right-hand side of Eqs. (2.4-8) to (2.4-12).

2.4C Use of Equation of Energy Change

In Section 2.3C we showed how to use the general equation of continuity and equation of motion to solve specific problems by simply discarding the unneeded terms in the equation. In problems involving energy changes it is often convenient to use the same approach and to simplify the equations of continuity, motion, and of energy change. Again several simple examples will illustrate the technique used. Primarily it involves selecting the terms in the equations that are comparatively very small or zero and discarding them.

Example 2.4-1 Energy and Momentum Transfer in a Circular Tube

It is desired to obtain the equations of change for a Newtonian fluid flowing in a circular tube with a constant velocity of the fluid. The inlet temperature at $z = 0$ is assumed uniform at T_0, and the direction of flow is in the z direction. We will assume that the parabolic velocity profile has been established at the point $z = 0$ and the process is steady-state. The physical properties of the fluid are assumed constant.

Solution

First we will use the continuity equation. In Example 2.3-1 for flow in a pipe Eq. (2.2-6)—the continuity equation—was shown to reduce to the following for z as the direction of flow:

$$\frac{\partial v_z}{\partial z} = 0 \tag{2.4-13}$$

The equation of motion, Eq. (2.3-9), was shown to become

$$\frac{dp}{dz} = \mu \left(\frac{1}{r} \frac{dv_z}{dr} + \frac{d^2 v_z}{dr^2} \right) \tag{2.4-14}$$

The equation above was integrated to give

$$v_z = v_{z\max}\left(1 - \frac{r^2}{r_0^2}\right) \tag{2.4-15}$$

where r_0 is the radius of the tube.

For the equation of energy change, we will use Eq. (2.4-11) for constant density:

$$\rho c_p\left(\frac{\partial T}{\partial t} + v_x\frac{\partial T}{\partial x} + v_y\frac{\partial T}{\partial y} + v_z\frac{\partial T}{\partial z}\right) = k\left(\frac{\partial^2 T}{\partial x^2} + \frac{\partial^2 T}{\partial y^2} + \frac{\partial^2 T}{\partial z^2}\right) \tag{2.4-16}$$

Since we have steady-state, $\partial T/\partial t = 0$. Also, since flow is in the z direction only, v_y and $v_x = 0$. Hence, Eq. (2.4-16) becomes

$$v_z\frac{\partial T}{\partial z} = \frac{k}{\rho c_p}\left(\frac{\partial^2 T}{\partial x^2} + \frac{\partial^2 T}{\partial y^2} + \frac{\partial^2 T}{\partial z^2}\right) \tag{2.4-17}$$

Changing the above to cylindrical coordinates for convenience, let $z = z$, $x = r\cos\theta$, and $y = r\sin\theta$. We obtain

$$v_z\frac{\partial T}{\partial z} = \frac{k}{\rho c_p}\left[\frac{1}{r}\frac{\partial}{\partial r}\left(r\frac{\partial T}{\partial r}\right) + \frac{1}{r^2}\frac{\partial^2 T}{\partial \theta^2} + \frac{\partial^2 T}{\partial z^2}\right] \tag{2.4-18}$$

Since this case is assumed symmetrical, $\partial^2 T/\partial\theta^2 = 0$. The final equation, after substitution of Eq. (2.4-15) into Eq. (2.4-18), is

$$v_{z\max}\left(1 - \frac{r^2}{r_0^2}\right)\frac{\partial T}{\partial z} = \frac{k}{\rho c_p}\left[\frac{1}{r}\frac{\partial}{\partial r}\left(r\frac{\partial T}{\partial r}\right) + \frac{\partial^2 T}{\partial z^2}\right] \tag{2.4-19}$$

The following boundary conditions can be used.

B.C. 1: $z = 0$, $r = r$, $T = T_0$

B.C. 2: $r = r_0$, $-k\dfrac{\partial T}{\partial r} = \text{constant}$ (2.4-20)

(Heat flux at the wall is assumed as a constant.)

B.C. 3: $r = 0$, T is finite

For details on the solution of the equations above see (S1).

Example 2.4-2 Heat Transfer in a Solid and Solution of the Equation of Energy Change

A solid material of thickness L is at a uniform temperature of $T_0°K$ at time $t = 0$. Suddenly the front-surface temperature of the solid at $z = 0$ is raised to a temperature of T_1 at $t = 0$ and held there and at $z = L$ at the rear to T_2 and held there. Energy transfer occurs only in the z direction. The physical properties are constant. (a) Derive the final equation to be used for unsteady-state energy transfer. (b) Do the same for steady-state and integrate the final equation.

Solution

For part (a), since the material is a solid, $v = 0$, and Eq. (2.4-12) will be used:

$$\rho c_p \frac{\partial T}{\partial t} = k\left(\frac{\partial^2 T}{\partial x^2} + \frac{\partial^2 T}{\partial y^2} + \frac{\partial^2 T}{\partial z^2}\right) \quad (2.4\text{-}12)$$

Since heat flows only in the z direction, $\partial^2 T/\partial y^2$ and $\partial^2 T/\partial x^2 = 0$. Equation (2.4-12) becomes

$$\frac{\partial T}{\partial t} = \frac{k}{\rho c_p}\frac{\partial^2 T}{\partial z^2} \quad (2.4\text{-}21)$$

The boundary and initial conditions are

$$\begin{array}{lll} \text{I.C.:} & t=0, \quad z=z, & T=T_0 \\ \text{B.C. 1:} & t=t, \quad z=0, & T=T_1 \\ \text{B.C. 2:} & t=t, \quad z=L, & T=T_2 \end{array} \quad (2.4\text{-}22)$$

The analytical solution of this problem will not be considered here, since the purpose is to show the reader how to set up the final equations from the general equations.

For part (b), we have steady-state. Then Eq. (2.4-12) becomes

$$0 = \frac{\partial^2 T}{\partial z^2} \quad (2.4\text{-}23)$$

Integrating this once,

$$\frac{dT}{dz} = K_1 \quad (2.4\text{-}24)$$

where K_1 is a constant. Integrating again,

$$T = K_1 z + K_2 \quad (2.4\text{-}25)$$

where K_2 is a constant. Substituting $T = T_1$ at $z = 0$, $K_2 = T_1$. Also, since $T = T_2$ at $z = L$, $K_1 = (T_2 - T_1)/L$. The final equation is then

$$T = \frac{(T_2 - T_1)z}{L} + T_1 \quad (2.4\text{-}26)$$

2.5 RELATION BETWEEN DIFFERENT TYPES OF FLUXES AND FICK'S LAW IN BINARY SYSTEMS

Table 1.7-1 lists the different types of velocities and concentrations that often occur in binary systems. The velocity v is the mass average velocity of the stream. This velocity, measured by a pitot tube, is the usual one used in momentum transfer. The velocity v^* is the molar

96 Transport Phenomena and Equations of Change

average velocity of the whole stream as given in Table 1.7-1. For a pure fluid the mass average velocity and the molar average velocity are the same, but not for a multicomponent mixture, as the table shows. We are dealing here not with the instantaneous velocity of each individual species but with the macroscopic velocity at which a species travels.

Many different types of fluxes can be defined, depending on whether the flux is relative to a fixed point or to a given type of velocity.

The two types of fluxes relative to fixed coordinates are $\mathbf{n}_A = \rho_A \mathbf{v}_A$, which is the mass flux in g mass A/sec-cm^2, and $\mathbf{N}_A = c_A \mathbf{v}_A$, which is the molar flux in g mole A/sec-cm^2. These are given in Table 2.5-1.

Table 2.5-1 Fluxes in Binary Systems[a]

	Mass Flux	Molar Flux
Relative to fixed coordinates	$\mathbf{n}_A = \rho_A \mathbf{v}_A$	$\mathbf{N}_A = c_A \mathbf{v}_A$
Relative to molar average velocity $\mathbf{v}^*$	$\mathbf{j}_A^* = \rho_A (\mathbf{v}_A - \mathbf{v}^*)$	$\mathbf{J}_A^* = c_A (\mathbf{v}_A - \mathbf{v}^*)$
Relative to mass average velocity $\mathbf{v}$	$\mathbf{j}_A = \rho_A (\mathbf{v}_A - \mathbf{v})$	$\mathbf{J}_A = c_A (\mathbf{v}_A - \mathbf{v})$

[a]For definitions of velocities and concentrations see Table 1.7-1.
SOURCE: R. B. Bird, W. E. Stewart, and E. N. Lightfoot, *Transport Phenomena*. New York: John Wiley & Sons, Inc., 1960. With permission.

Table 2.5-2 Fluxes and Relation between Fluxes

Sum of Fluxes
$\mathbf{N}_A + \mathbf{N}_B = c\mathbf{v}^*$
$\mathbf{J}_A + \mathbf{J}_B = c(\mathbf{v}^* - \mathbf{v})$
$\mathbf{J}_A^* + \mathbf{J}_B^* = 0$
$\mathbf{n}_A + \mathbf{n}_B = \rho \mathbf{v}$
$\mathbf{j}_A + \mathbf{j}_B = 0$
$\mathbf{j}_A^* + \mathbf{j}_B^* = \rho(\mathbf{v} - \mathbf{v}^*)$
Derived Relations
$\mathbf{N}_A = \mathbf{J}_A^* + c_A \mathbf{v}^*$
$\mathbf{N}_A = \mathbf{J}_A + c_A \mathbf{v}$
$\mathbf{N}_A = \mathbf{n}_A / M_A$
$\mathbf{J}_A = \mathbf{j}_A / M_A$
$\mathbf{J}_A = (M_B/M) \mathbf{J}_A^*$
$\mathbf{j}_A = \mathbf{n}_A - w_A (\mathbf{n}_A + \mathbf{n}_B)$
$\mathbf{j}_A^* = (M/M_B) \mathbf{j}_A$
$\mathbf{j}_A^* = \mathbf{J}_A^* M_A$
$\mathbf{J}_A^* = \mathbf{N}_A - x_A (\mathbf{N}_A + \mathbf{N}_B)$

SOURCE: R. B. Bird, W. E. Stewart, and E. N. Lightfoot, *Transport Phenomena*. New York: John Wiley & Sons, Inc., 1960. With permission.

2.5 Relation between Fluxes and Fick's Law

The common fluxes relative to the molar average velocity $\mathbf{v}^*$ are $\mathbf{j}_A^* = \rho_A(\mathbf{v}_A - \mathbf{v}^*)$, which is the mass flux in g mass A/sec-cm², and $\mathbf{J}_A^* = c_A(\mathbf{v}_A - \mathbf{v}^*)$, which is the molar flux in g mole A/sec-cm².

The fluxes relative to the mass average velocity $\mathbf{v}$ are $\mathbf{j}_A = \rho_A(\mathbf{v}_A - \mathbf{v})$, which is the mass flux in g mass A/sec-cm², and $\mathbf{J}_A = c_A(\mathbf{v}_A - \mathbf{v})$, which is the molar flux in g mole A/sec-cm². Table 2.5-2 summarizes the sum of the fluxes and the relationships between the fluxes.

Example 2.5-1 Proof of Molar Flux Equations

Prove the following relation from Table 2.5-2:

$$\mathbf{j}_A + \mathbf{j}_B = 0 \tag{2.5-1}$$

Solution

Substitute $\rho_A(\mathbf{v}_A - \mathbf{v})$ for $\mathbf{j}_A$ from Table 2.5-1 and $\rho_B(\mathbf{v}_B - \mathbf{v})$ for $\mathbf{j}_B$.

$$\rho_A \mathbf{v}_A - \rho_A \mathbf{v} + \rho_B \mathbf{v}_B - \rho_B \mathbf{v} = 0 \tag{2.5-2}$$

Rearranging,

$$\rho_A \mathbf{v}_A + \rho_B \mathbf{v}_B - \mathbf{v}(\rho_A + \rho_B) = 0 \tag{2.5-3}$$

From Table 1.7-1, $\mathbf{v} = (1/\rho)(\rho_A \mathbf{v}_A + \rho_B \mathbf{v}_B)$. Substituting this into Eq. (2.5-3), the identity is proved, since $(\rho_A + \rho_B) = \rho$.

Example 2.5-2 Sum of Molar Fluxes

Prove the following equation:

$$\mathbf{N}_A + \mathbf{N}_B = c\mathbf{v}^* \tag{2.5-4}$$

Solution

We substitute $\mathbf{N}_A = c_A \mathbf{v}_A$ and $\mathbf{N}_B = c_B \mathbf{v}_B$ from Table 2.5-1 into Eq. (2.5-4).

$$c_A \mathbf{v}_A + c_B \mathbf{v}_B = c\mathbf{v}^* \tag{2.5-5}$$

From Table 1.7-1 the definition of $\mathbf{v}^*$ is $(c_A \mathbf{v}_A + c_B \mathbf{v}_B)/c$, and the identity is proved.

Originally Fick's law was defined as

$$\mathbf{J}_A^* = -cD_{AB} \nabla x_A \tag{2.5-6}$$

where the molar flux by diffusion relative to the molar average velocity $\mathbf{v}^*$ is proportional to the concentration gradient. An important Eq. (2.5-7) is also the molar flux $\mathbf{N}_A$ relative to stationary coordinates, which has been given previously as Eq. (1.7-7).

Transport Phenomena and Equations of Change

Table 2.5-3 Different Forms of Fick's Law

$\mathbf{J}_A^* = -cD_{AB}\nabla x_A$	(a)
$\mathbf{J}_A^* = -\dfrac{\rho^2}{cM_AM_B}D_{AB}\nabla w_A$	(b)
$\mathbf{j}_A = -\rho D_{AB}\nabla w_A$	(c)
$\mathbf{j}_A = -\dfrac{c^2 M_A M_B}{\rho}D_{AB}\nabla x_A$	(d)

$$\mathbf{N}_A = -cD_{AB}\nabla x_A + x_A(\mathbf{N}_A + \mathbf{N}_B) \qquad (2.5\text{-}7)$$

where the flux $\mathbf{N}_A$ is the total molar flux of A relative to stationary coordinates and equals the bulk or convective molar flux of A, $x_A(\mathbf{N}_A + \mathbf{N}_B)$, plus the molar diffusion flux $\mathbf{J}_A^*$ of A relative to the molar average velocity.

Various forms of Fick's law, all equivalent, are given in Table 2.5-3. The D_{AB} is the same in all of the forms for a binary mixture. The most important form of Fick's law used most by engineers is Eq. (2.5-6), given above, which is the molar diffusion flux of species A. Also, the flux relationship in Eq. (2.5-7), where the molar flux $\mathbf{N}_A$ is given relative to fixed coordinates, is often used in calculations in processes where measurements are most conveniently made relative to this stationary position.

Example 2.5-3 Different Forms of Fick's Law

Using Eq. (c) in Table 2.5-3,

$$\mathbf{j}_A = -\rho D_{AB}\nabla w_A \qquad (c)$$

prove Eq. (d):

$$\mathbf{j}_A = -\dfrac{c^2}{\rho}M_A M_B D_{AB}\nabla x_A \qquad (d)$$

Solution

The definition of w_A from Table 1.7-1 is

$$w_A = \dfrac{x_A M_A}{x_A M_A + (1-x_A)M_B} \qquad (2.5\text{-}8)$$

Differentiating both sides,

$$\nabla w_A = \dfrac{[x_A M_A + (1-x_A)M_B]M_A\nabla x_A - x_A M_A(M_A\nabla x_A - M_B\nabla x_A)}{[x_A M_A + (1-x_A)M_B]^2} \qquad (2.5\text{-}9)$$

Canceling out the like terms,

$$\nabla w_A = \frac{M_A M_B \nabla x_A}{M^2} \quad (2.5\text{-}10)$$

where $M = x_A M_A + x_B M_B$ = average molecular weight. Substituting Eq. (2.5-10) into (c),

$$\mathbf{j}_A = -\rho D_{AB} \frac{M_A M_B}{M^2} \nabla x_A \quad (2.5\text{-}11)$$

From Table 1.7-1, $M = \rho/c$. Substituting this into Eq. (2.5-11), the resulting equation is the same as Eq. (d).

$$\mathbf{j}_A = -\frac{\rho D_{AB} M_A M_B}{\rho^2/c^2} \nabla x_A = -\frac{c^2}{\rho} M_A M_B D_{AB} \nabla x_A \quad (2.5\text{-}12)$$

2.6 EQUATIONS OF CONTINUITY FOR A BINARY MIXTURE

2.6A The General Equation

In Chapter 1 different problems on ordinary molecular mass transport were solved by setting up individual mass balances for each individual problem. In this section we will generalize the mass balances for more than one component. We will derive the equations of continuity for each of the two chemical species in a binary fluid mixture (B1). Then we will show how, by discarding the terms that are zero or negligible, we can solve each special situation that arises.

As in the balance made for a pure component we make a mass balance on an element $\Delta x\, \Delta y\, \Delta z$ fixed in space as shown in Fig. 2.6-1. The general mass balance equation will now include a generation term from chemical reaction:

(rate of mass A in) − (rate of mass A out) + (rate of generation of mass A)
$$= \text{(rate of accumulation of mass } A) \quad (2.6\text{-}1)$$

The rate of mass A coming in in the x direction relative to stationary coordinates is $(n_{Ax})_x\, \Delta y\, \Delta z$ and that leaving is $(n_{Ax})_{x+\Delta x}\, \Delta y\, \Delta z$. This is the total flux of A due to diffusion and convective flow. A similar term can be written for the y direction and the z direction.

The rate of chemical production of A is defined as r_A g mass A generated/sec-cm^3 volume and the total rate is $r_A\, \Delta x\, \Delta y\, \Delta z$. The rate of accumulation of mass A is $(\partial \rho_A / \partial t)\, \Delta x\, \Delta y\, \Delta z$. After substituting the terms into Eq. (2.6-1) and taking the limit as Δx, Δy, and Δz approach zero,

$$\frac{\partial \rho_A}{\partial t} + \left(\frac{\partial n_{Ax}}{\partial x} + \frac{\partial n_{Ay}}{\partial y} + \frac{\partial n_{Az}}{\partial z} \right) = r_A \quad (2.6\text{-}2)$$

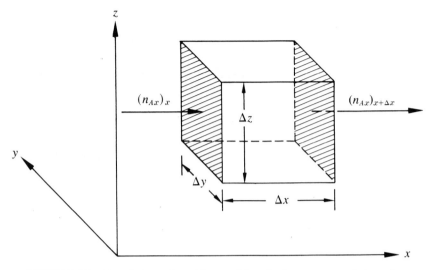

FIGURE 2.6-1 Mass balance for a binary mixture (only x-component shown)

Or, in vector notation,

$$\frac{\partial \rho_A}{\partial t} + (\nabla \cdot \mathbf{n}_A) = r_A \tag{2.6-3}$$

We can write a similar equation for component B.

$$\frac{\partial \rho_B}{\partial t} + (\nabla \cdot \mathbf{n}_B) = r_B \tag{2.6-4}$$

If we add Eqs. (2.6-3) and (2.6-4) and note from Table 2.5-2 that $\mathbf{n}_A + \mathbf{n}_B = \rho \mathbf{v}$ and that $r_A + r_B = 0$, we obtain the equation of continuity, which is identical to Eq. (2.2-4).

$$\frac{\partial \rho}{\partial t} + (\nabla \cdot \rho \mathbf{v}) = 0 \tag{2.6-5}$$

Usually we are interested in molar fluxes, so, dividing both sides of Eq. (2.6-3) by M_A,

$$\frac{\partial c_A}{\partial t} + (\nabla \cdot \mathbf{N}_A) = R_A \tag{2.6-6}$$

where R_A is the rate of generation in g mole A/sec-cm³.

The flux terms $\mathbf{n}_A$ and $\mathbf{N}_A$ represent the total flux of A with respect to stationary coordinates and include the sum of the diffusion flux plus the convective flux. In the following we will consider only the Fickian flux due to concentration differences, as given in Table 2.5-3. We will neglect other types such as pressure diffusion, thermal diffusion, and so on.

Using Eq. (2.6-6) first, we will convert this to a form that includes Fick's law. From Table 2.5-2,

$$\mathbf{N}_A = \mathbf{J}_A^* + c_A \mathbf{v}^* \tag{2.6-7}$$

2.6 Equations of Continuity for a Binary Mixture

Substituting Eq. (a) for $\mathbf{J}_A^*$ from Table 2.5-3 into Eq. (2.6-7),

$$\mathbf{N}_A = -cD_{AB}\nabla x_A + c_A\mathbf{v}^* \tag{2.6-8}$$

Finally, substituting Eq. (2.6-8) into (2.6-6) and rearranging,

$$\frac{\partial c_A}{\partial t} + (\nabla \cdot c_A\mathbf{v}^*) - (\nabla \cdot cD_{AB}\nabla x_A) = R_A \tag{2.6-9}$$

Equation (2.6-3) can also be converted to give

$$\frac{\partial \rho_A}{\partial t} + (\nabla \cdot \rho_A\mathbf{v}) - (\nabla \cdot \rho D_{AB}\nabla w_A) = r_A \tag{2.6-10}$$

Equations (2.6-9) and (2.6-10) are the final general equations that hold for variable ρ, c, and D_{AB}.

2.6B Special Equations

In many cases in actual engineering practice Eqs. (2.6-9) and (2.6-10) can be simplified for certain physical conditions.

1. *Equation for Constant c and D_{AB}.* Often with gases the total pressure P is constant and, hence, since $P = c/RT$, c total is constant. Then, starting with Eq. (2.6-9) for constant temperature and substituting $\nabla x_A = \nabla c_A/c$ for constant c, we obtain

$$\frac{\partial c_A}{\partial t} + c_A(\nabla \cdot \mathbf{v}^*) + (\mathbf{v}^* \cdot \nabla c_A) - D_{AB}\nabla^2 c_A = R_A \tag{2.6-11}$$

2. *Equation for Constant ρ and D_{AB}.* Often in dilute liquid solutions the mass density ρ and D_{AB} can be considered constant. Using the fact that for constant ρ, $\nabla w_A = \nabla \rho_A/\rho$ and also that $(\nabla \cdot \mathbf{v}) = 0$, substituting these into Eq. (2.6-10), and dividing both sides by M_A, we obtain

$$\frac{\partial c_A}{\partial t} + (\mathbf{v} \cdot \nabla c_A) - D_{AB}\nabla^2 c_A = R_A \tag{2.6-12}$$

Since the left-hand side of Eq. (2.6-12) can be written as the substantial derivative, Eq. (2.6-12) becomes for no reaction

$$\frac{Dc_A}{Dt} = D_{AB}\nabla^2 c_A \tag{2.6-13}$$

The energy transfer equation given previously, Eq. (2.4-11), for constant ρ and k is

$$\frac{DT}{Dt} = \frac{k}{\rho c_p}\nabla^2 T \tag{2.6-14}$$

These two equations are very similar and are the basis for many analogies drawn between mass and energy transport in flowing fluids. Analytical solutions derived for heat transport can be used for mass transport and vice versa by a simple change in nomenclature.

3. *Equation for Zero Velocity.* Assuming no chemical changes, when diffusion is occurring in solids or in liquids where v can be assumed as essentially zero, then Eq. (2.6-12) becomes

$$\frac{\partial c_A}{\partial t} = D_{AB} \nabla^2 c_A \qquad (2.6\text{-}15)$$

This same equation can be obtained from Eq. (2.6-11) for equimolar counterdiffusion where $v^* = 0$. Equation (2.6-15) is called Fick's second law of diffusion. This equation, as the reader can see above, was derived from the general equation of continuity for a binary mixture, Eq. (2.6-9) or (2.6-10), by setting certain physical restrictions and removing those terms that are zero. Equation (2.6-15) is the same as Eq. (1.1-9), which was derived by making a mass balance on a differential element.

2.6C Use of Equations of Continuity for a Binary Mixture

As discussed earlier, specific physical problems involving mass transport can be solved by setting up differential mass balances or by simplifying the equation of continuity for a binary mixture. Several simple examples will be given to introduce the reader to the methods used. Other examples will be given in later chapters.

Example 2.6-1 Equimolar Counterdiffusion of *A* and *B*

Gases *A* and *B* are contained in an apparatus similar to that in Fig. 2.6-2 and are counterdiffusing equimolar and steady-state in a constant-pressure system. The diffusion path given in Fig. 2.6-2 is δ cm long and

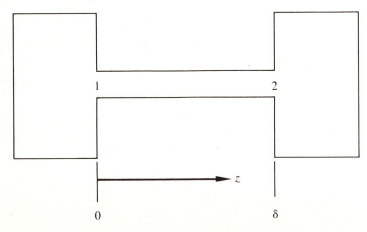

FIGURE 2.6-2 Equimolar counterdiffusion of *A* and *B* for Example 2.6-1

2.6 Equations of Continuity for a Binary Mixture

the partial pressures p_{A1} and p_{B1} at point 1 and p_{A2} and p_{B2} at point 2 are constant. Derive the integrated equation for the concentration profile of c_A.

Solution

Starting with the general Eq. (2.6-11) for constant c and D_{AB},

$$\frac{\partial c_A}{\partial t} + c_A(\nabla \cdot \mathbf{v}^*) + (\mathbf{v}^* \cdot \nabla c_A) - D_{AB}\nabla^2 c_A = R_A \qquad (2.6\text{-}11)$$

we can set $\partial c_A/\partial t = 0$ for steady-state and $R_A = 0$ for no chemical reaction. Also, for equimolar counterdiffusion, $\mathbf{v}^* = 0$. Hence, Eq. (2.6-11) becomes

$$\nabla^2 c_A = 0 \qquad (2.6\text{-}16)$$

Since diffusion is occurring only in the z direction,

$$\nabla^2 c_A = \frac{\partial^2 c_A}{\partial x^2} + \frac{\partial^2 c_A}{\partial y^2} + \frac{\partial^2 c_A}{\partial z^2} = \frac{d^2 c_A}{dz^2} = 0 \qquad (2.6\text{-}17)$$

Integrating Eq. (2.6-17) once,

$$\frac{dc_A}{dz} = K_1 \qquad (2.6\text{-}18)$$

where K_1 is a constant. Integrating again,

$$c_A = K_1 z + K_2 \qquad (2.6\text{-}19)$$

At $z = 0$, $c_A = c_{A1}$. Hence, $K_2 = c_{A1}$. At $z = \delta$, $c_A = c_{A2}$, and $K_1 = (c_{A2} - c_{A1})/\delta$. The final equation is

$$c_A = \frac{(c_{A2} - c_{A1})z}{\delta} + c_{A1} \qquad (2.6\text{-}20)$$

This same result could have been obtained by simply using Fick's law for J_A^* and integrating.

Example 2.6-2 Diffusion of A through Stagnant B

Repeat Example 2.6-1 but for the case of A diffusing through stagnant B.

Solution

Writing the general Eq. (2.6-6),

$$\frac{\partial c_A}{\partial t} + (\nabla \cdot \mathbf{N}_A) = R_A \qquad (2.6\text{-}6)$$

For steady-state $\partial c_A/\partial t = 0$ and $R_A = 0$ for no chemical reaction. Hence, for diffusion in the z direction only,

$$\frac{dN_{Az}}{dz} = 0 \qquad (2.6\text{-}21)$$

which means that N_{Az} is a constant for steady-state diffusion. From Tables 2.5-2 and 2.5-3 we obtain for diffusion in the z direction

$$N_{Az} = -cD_{AB}\frac{dx_A}{dz} + x_A(N_{Az} + N_{Bz}) \qquad (2.6\text{-}22)$$

Rearranging Eq. (2.6-22) for $N_{Bz} = 0$,

$$N_{Az} = -\frac{cD_{AB}}{1-x_A}\frac{dx_A}{dz} \qquad (2.6\text{-}23)$$

Substituting Eq. (2.6-23) into (2.6-21) and letting c and D_{AB} be constant,

$$d\left(\frac{1}{1-x_A}\frac{dx_A}{dz}\right) = 0 \qquad (2.6\text{-}24)$$

Integrating once,

$$\frac{1}{1-x_A}\frac{dx_A}{dz} = K_1 \qquad (2.6\text{-}25)$$

where K_1 is a constant. Integrating again,

$$-\ln(1-x_A) = K_1 z + K_2 \qquad (2.6\text{-}26)$$

For $z = 0$, $x_A = x_{A1}$ and for $z = \delta$, $x_A = x_{A2}$. Solving for K_1 and K_2 and rearranging,

$$\ln\left(\frac{1-x_A}{1-x_{A1}}\right) = \frac{z}{\delta}\ln\left(\frac{1-x_{A2}}{1-x_{A1}}\right) \qquad (2.6\text{-}27)$$

Hence, a plot of x_A versus z is not a straight line.

To calculate the flux, N_{Az}, Eq. (2.6-21) says that N_{Az} is constant. This means we can integrate Eq. (2.6-23) to obtain

$$N_{Az} = \frac{cD_{AB}}{\delta}\ln\left(\frac{1-x_{A2}}{1-x_{A1}}\right) \qquad (2.6\text{-}28)$$

This can be converted into Eq. (1.7-23) which was derived for the same physical situation.

2.7 EQUATIONS OF ENERGY CHANGE AND CONTINUITY FOR A BINARY MIXTURE

Often energy and mass transport occur simultaneously. This means that the diffusion of mass can affect the total energy flux. To derive this equation we start by defining an energy flux vector **e** with respect to a

2.7 Equations of Energy Change and Continuity

fixed coordinate system (B1).

$$\mathbf{e} = \rho(\hat{U} + \tfrac{1}{2}v^2)\mathbf{v} + \mathbf{q} + p\mathbf{v} + [\boldsymbol{\tau} \cdot \mathbf{v}] \tag{2.7-1}$$

where **q** is the multicomponent energy flux relative to the mass average velocity **v**. If only one component is present, **q** is simply Fourier's law for conduction. Radiation will be neglected in this discussion.

If several species are present, **q** includes the heat by conduction, by the Dufour effect which is often small and will not be considered here, and by interdiffusion of individual species with different enthalpies. Hence, **q** becomes

$$\mathbf{q} = \sum_{i=1}^{n} \bar{H}_i \mathbf{J}_i - k\nabla T \tag{2.7-2}$$

where $\bar{H}_i$ is the partial molal enthalpy of species i.

Equation (2.7-1) can be further simplified by dropping $[\boldsymbol{\tau} \cdot \mathbf{v}]$ and $(\tfrac{1}{2}\rho v^2)\mathbf{v}$ when they are small. Also, from Table 2.5-2,

$$\mathbf{N}_i = \mathbf{J}_i + c_i \mathbf{v} \tag{2.7-3}$$

Making these substitutions and upon further simplification, the final equation for simultaneous mass and energy transfer is

$$\mathbf{e} = -k\nabla T + \sum_{i=1}^{n} \mathbf{N}_i \bar{H}_i \tag{2.7-4}$$

Often $\bar{H}_i$ for an ideal gas can be represented by

$$\bar{H}_i = \tilde{c}_{pi}(T - T_0) = M_i c_{pi}(T - T_0) \tag{2.7-5}$$

where $\tilde{c}_{pi}$ is the heat capacity/mole and T_0 a reference temperature.

Equation (2.7-4) is often used to solve problems such as condensation of vapors in an air stream or absorption in a tower with large heat effects.

Example 2.7-1 Simultaneous Heat and Mass Transfer

A hot condensable water vapor (A) is in an inert air stream (B) adjacent to a cold condenser tube at $z = 0$ where the water vapor condenses. It is assumed that diffusion of A is through a stagnant gas film of thickness δ cm as shown in Fig. 2.7-1. We will assume steady-state diffusion and neglect the variation in physical properties with temperature. We will consider only conductive heat transfer and mass transfer. (a) Solve for the energy flux and mass flux. (b) Solve for the energy flux if no mass transfer is occurring and compare parts (a) and (b). This is similar to an example in Bird et al. (B1).

Solution

Since we are at steady-state, N_{Az} and e_z are constant. First, for part (a) for the mass transfer only, component A is diffusing. In Example 2.6-2 for A diffusing through stagnant B we showed that the general Eq. (2.6-6)

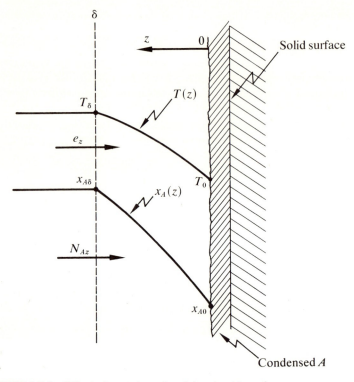

FIGURE 2.7-1 Effect of mass transfer of A on heat transfer

became

$$\frac{dN_{Az}}{dz} = 0 \tag{2.7-6}$$

Thus, for $N_{Bz} = 0$, from Tables 2.5-2 and 2.5-3,

$$N_{Az} = -cD_{AB}\frac{dx_A}{dz} + x_A(N_{Az} + 0) \tag{2.7-7}$$

On rearranging,

$$N_{Az} = -\frac{cD_{AB}}{1 - x_A}\frac{dx_A}{dz} \tag{2.7-8}$$

Integrating between $z = 0$ and $z = \delta$, we obtain an equation similar to Eq. (2.6-28) for no energy transfer.

$$N_{Az} = \frac{cD_{AB}}{\delta} \ln\left(\frac{1 - x_{A\delta}}{1 - x_{A0}}\right) \tag{2.7-9}$$

This means the flux of A is unaffected by the simultaneous heat transfer.

For the energy flux, we write Eq. (2.7-4) for components A and B,

$$e_z = -k\frac{dT}{dz} + N_{Az}\bar{H}_A + N_{Bz}\bar{H}_B \tag{2.7-10}$$

However, $N_{Bz} = 0$. Substituting Eq. (2.7-5) for A into (2.7-10), we obtain

$$e_z = -k\frac{dT}{dz} + N_{Az}\tilde{c}_{pA}(T - T_0) \tag{2.7-11}$$

Since we are at steady-state, e_z is constant. Rearranging Eq. (2.7-11) and integrating,

$$\int_{T_0}^{T_\delta} \frac{dT}{[e_z - N_{Az}\tilde{c}_{pA}(T - T_0)]} = -\frac{1}{k}\int_0^\delta dz \tag{2.7-12}$$

$$\ln\left(\frac{e_z - N_{Az}\tilde{c}_{pA}(T_\delta - T_0)}{e_z}\right) = \frac{N_{Az}\tilde{c}_{pA}\delta}{k} \tag{2.7-13}$$

Solve for e_z,

$$e_z = \frac{N_{Az}\tilde{c}_{pA}(T_\delta - T_0)}{1 - \exp(N_{Az}\tilde{c}_{pA}\delta/k)} \tag{2.7-14}$$

When no mass transfer is occurring in part (b), Eq. (2.7-10) becomes

$$e_z = -k\frac{dT}{dz} \tag{2.7-15}$$

Integrating this for steady-state,

$$e_{z(N_A=0)} = \frac{-k(T_\delta - T_0)}{\delta} \tag{2.7-16}$$

To compare the energy fluxes for mass transfer occurring and not occurring, we will take the ratio

$$\frac{e_z}{e_{z(N_A=0)}} = \frac{N_{Az}\tilde{c}_{pA}(T_\delta - T_0)}{1 - \exp(N_{Az}\tilde{c}_{pA}\delta/k)}\left(\frac{1}{-k(T_\delta - T_0)/\delta}\right)$$

$$= \frac{-N_{Az}\tilde{c}_{pA}\delta/k}{1 - \exp(N_{Az}\tilde{c}_{pA}\delta/k)} \tag{2.7-17}$$

2.8 MASS FLUXES FOR MULTICOMPONENT SYSTEMS IN TERMS OF TRANSPORT PROPERTIES

In discussing the mass flux of a species i in a multicomponent mixture of two or more components, we have thus far concerned ourselves with mass fluxes due to concentration gradients. This, of course, is a simplification used to introduce the subject of mass fluxes. The most important contribution to the mass flux is that due to a concentration gradient, and this is often termed ordinary diffusion. However, according to the rigorous kinetic theory of gases (H2) and also according to the thermodynamics of irreversible processes there will be a contribution to the total mass flux because of each driving force in the system.

The driving forces for mass flux are ordinary concentration, pressure, external forces, and one due to the temperature gradient or temperature driving force. These fluxes can be expressed as (B1)

$$\mathbf{j}_i = \mathbf{j}_{i(x)} + \mathbf{j}_{i(p)} + \mathbf{j}_{i(g)} + \mathbf{j}_{i(T)} \qquad (2.8\text{-}1)$$

where $\mathbf{j}_{i(x)}$ is the mass flux due to ordinary concentration diffusion, $\mathbf{j}_{i(p)}$ pressure diffusion, $\mathbf{j}_{i(g)}$ forced diffusion, and $\mathbf{j}_{i(T)}$ thermal diffusion.

Pressure diffusion occurs if a pressure gradient is imposed on the species, such as in a mixture in a centrifuge. Very high-pressure gradients will cause a net flux of the ith species. This specialized topic will not be considered further here.

In forced diffusion we are concerned with a system containing different ionic species where an external force such as an electrical field is applied. The force on each ionic species is proportional to the product of the ionic charge and the local electrical field strength, and a separation occurs. We will not be concerned further with this type of diffusion, since this is a specialized field in electrolytic processes; for further discussion the reader is referred to the many excellent texts in physical chemistry on this subject. If gravity force acts only on the species, all the forces are the same on each species. Bird *et al.* (B1, p. 577) discuss this case.

The last type of flux is due to thermal diffusion, whereby a mass flux is induced owing to a temperature gradient along the diffusion path. In 1911 Enskog (G1) predicted theoretically that the flow of heat or a temperature gradient in a fluid mixture of uniform composition will set up a concentration difference between the hot and cold parts. However, Soret in 1879 had found such an effect experimentally in liquids. Chapman then extended the theory using the original theories of Enskog. A good presentation of the theory using this classical kinetic theory of gases is given elsewhere (H2), and Denbigh (D1) discusses thermal diffusion and the thermodynamics of irreversible processes. Extensive compilations of experimental data are also available (A1, B4, B5, H1, G1, S2, T1).

2.9 DIMENSIONAL ANALYSIS OF EQUATIONS OF CHANGE

Since we have presented the basic equations of change for mass, energy, and momentum in the previous sections of this chapter, we will consider them now by using dimensional analysis. Using the basic differential equations of change, we will show how we can obtain various important dimensionless numbers—the Reynolds number and others—that are useful in correlating and predicting transport phenomena not only in laminar but also in turbulent flow.

Assuming constant physical properties such as ρ, μ, D_{AB}, c_p, this allows us to simplify considerably our analysis. First we select a characteristic velocity V for the flow system. For a packed bed this could be the average velocity; for an empty tube, an average velocity also. A characteristic length $\bar{D}$ is selected. For a packed bed of spheres this would be the sphere diameter. These choices are somewhat arbitrary. For pressure we pick an arbitrary pressure p_0 as a reference, so we will consider $(p-p_0)$ instead of p. Likewise for temperature we use $T-T_0$ for T. For concentration we use $x_A - x_{A0}$, where x_{A0} is a convenient reference. We will assume the range of composition change is small so that ρ and c are essentially constant.

We will limit our discussion to forced convection, excluding natural convection. Using the symbols as defined by others (B1), we define a dimensionless velocity $\mathbf{v}^*$ as

$$\mathbf{v}^* = \frac{\mathbf{v}}{V} \tag{2.9-1}$$

For pressure we define a dimensionless term as

$$p^* = \frac{p - p_0}{\rho V^2} \tag{2.9-2}$$

For time,

$$t^* = \frac{tV}{\bar{D}} \tag{2.9-3}$$

For temperature,

$$T^* = \frac{T - T_0}{T_1 - T_0} \tag{2.9-4}$$

Also, dimensionless coordinates

$$x^* = \frac{x}{\bar{D}}, \quad y^* = \frac{y}{\bar{D}}, \quad z^* = \frac{z}{\bar{D}} \tag{2.9-5}$$

Some of the mathematical operations used are

$$\nabla = \frac{1}{\bar{D}} \nabla^* \tag{2.9-6}$$

$$\nabla^2 = \frac{1}{\bar{D}^2} \nabla^{*2} \tag{2.9-7}$$

$$Dt = \frac{\bar{D}}{V} Dt^* \tag{2.9-8}$$

The equations of change we will consider are Eqs. (2.2-8), (2.3-10), and (2.4-10) for a pure fluid, and Eq. (2.6-13) for a binary fluid mixture.

$$(\nabla \cdot \mathbf{v}) = 0 \quad \text{(continuity)} \quad (2.9\text{-}9)$$

$$\rho \frac{D\mathbf{v}}{Dt} = -\nabla p + \rho \mathbf{g} + \mu \nabla^2 \mathbf{v} \quad \text{(motion)} \quad (2.9\text{-}10)$$

$$\rho c_p \frac{DT}{Dt} = k \nabla^2 T \quad \text{(energy)} \quad (2.9\text{-}11)$$

$$\frac{Dx_A}{Dt} = D_{AB} \nabla^2 x_A \quad \text{(continuity of } A\text{)} \quad (2.9\text{-}12)$$

where the dissipation term in Eq. (2.9-11) is neglected.

Next we will convert Eqs. (2.9-9) to (2.9-12) by using the dimensionless variables defined in Eqs. (2.9-1) to (2.9-8). For the equation of motion we obtain on substitution

$$\frac{\rho V^2}{\bar{D}} \frac{D\mathbf{v}^*}{Dt^*} = -\frac{\rho V^2}{\bar{D}} \nabla^* p^* + \rho \mathbf{g} + \frac{\mu V}{\bar{D}^2} \nabla^{*2} \mathbf{v}^* \quad (2.9\text{-}13)$$

On rearranging Eq. (2.9-13) and performing similar substitutions in the other equations, we obtain

$$(\nabla^* \cdot \mathbf{v}^*) = 0 \quad \text{(continuity)} \quad (2.9\text{-}14)$$

$$\frac{D\mathbf{v}^*}{Dt^*} = -\nabla^* p^* + \frac{1}{N_{Fr}} \frac{\mathbf{g}}{g} + \frac{1}{N_{Re}} \nabla^{*2} \mathbf{v}^* \quad \text{(motion)} \quad (2.9\text{-}15)$$

$$\frac{DT^*}{Dt^*} = \frac{1}{N_{Re} N_{Pr}} \nabla^{*2} T^* \quad \text{(energy)} \quad (2.9\text{-}16)$$

$$\frac{Dx_A^*}{Dt^*} = \frac{1}{N_{Re} N_{Sc}} \nabla^{*2} x_A^* \quad \text{(continuity of } A\text{)} \quad (2.9\text{-}17)$$

where the dimensionless numbers are defined as

$$N_{Re} = \frac{\bar{D} V \rho}{\mu} = \frac{\text{(inertia force)}}{\text{(viscous force)}} \quad \text{(Reynolds)} \quad (2.9\text{-}18)$$

$$N_{Pr} = \frac{c_p \mu}{k} = \frac{\text{(momentum diffusivity)}}{\text{(heat diffusivity)}} \quad \text{(Prandtl)} \quad (2.9\text{-}19)$$

$$N_{Sc} = \frac{\mu}{\rho D_{AB}} = \frac{\text{(momentum diffusivity)}}{\text{(mass diffusivity)}} \quad \text{(Schmidt)} \quad (2.9\text{-}20)$$

$$N_{Fr} = \frac{V^2}{g\bar{D}} = \frac{\text{(inertia force)}}{\text{(gravity force)}} \quad \text{(Froude)} \quad (2.9\text{-}21)$$

Hence, in correlating data for momentum transfer the Reynolds and Froude numbers are important. In energy transfer both the Reynolds number and Prandtl number play a role. In mass transport the Reynolds and Schmidt numbers are important. Bird *et al.* (B1) and Bennett and Meyers (B2) give additional details.

This method shown above of obtaining the important dimensionless groups from the basic differential equations is generally the preferred one. However, in some cases one is not able to formulate the basic differential equations because the system may be too complex or obscure. Then the Buckingham pi theorem can be used, whereby one writes down a set of variables necessary to specify the physical problem. By this method the dimensionless numbers are obtained. The disadvantage of this method is that it does not actually select the variables nor give their relative importance. The Buckingham theorem will be discussed in Chapter 6, and a good discussion is available in McAdams (M1).

PROBLEMS

2-1 Equation of Motion and Substantial Derivative

Using Eq. (2.3-4) below, obtain the substantial derivative as in Eq. (2.3-6).

$$\frac{\partial(\rho v_x)}{\partial t} = -\left(\frac{\partial(\rho v_x v_x)}{\partial x} + \frac{\partial(\rho v_y v_x)}{\partial y} + \frac{\partial(\rho v_z v_x)}{\partial z}\right)$$
$$- \left(\frac{\partial \tau_{xx}}{\partial x} + \frac{\partial \tau_{yx}}{\partial y} + \frac{\partial \tau_{zx}}{\partial z}\right) - \frac{\partial p}{\partial x} + \rho g_x \quad (2.3\text{-}4)$$

$$\rho \frac{Dv_x}{Dt} = -\left(\frac{\partial \tau_{xx}}{\partial x} + \frac{\partial \tau_{yx}}{\partial y} + \frac{\partial \tau_{zx}}{\partial z}\right) + \rho g_x - \frac{\partial p}{\partial x} \quad (2.3\text{-}6)$$

[*Hints*: First, using the definition of the substantial derivative, obtain Dv_x/Dt. Then perform the differentiation of Eq. (2.3-4). Then combine the two resulting equations. Finally, use this result and the continuity equation to eliminate terms to obtain Eq. (2.3-6).]

2-2 Derivation of Navier-Stokes Equation

Derive the Navier-Stokes equation below for the *x* component.

$$\rho\left(\frac{\partial v_x}{\partial t} + v_x\frac{\partial v_x}{\partial x} + v_y\frac{\partial v_x}{\partial y} + v_z\frac{\partial v_x}{\partial z}\right) = \mu\left(\frac{\partial^2 v_x}{\partial x^2} + \frac{\partial^2 v_x}{\partial y^2} + \frac{\partial^2 v_x}{\partial z^2}\right) - \frac{\partial p}{\partial x} + \rho g_x \quad (2.3\text{-}9)$$

[*Hints*: Use Eq. (2.3-8) and carry out the detailed steps of differentiation. Collect terms and note that $(\nabla \cdot \mathbf{v}) = 0$ for a fluid of constant density.]

2-3 Equation of Motion for an Ideal Fluid

In some situations the viscous terms in the Navier-Stokes equation that contain the viscosity may be small compared to the other inertial terms. If we set $\mu = 0$, then we obtain the equation of change for an ideal fluid called Euler's equation. For the *x* component of momentum this becomes

$$\rho\left(\frac{\partial v_x}{\partial t} + v_x\frac{\partial v_x}{\partial x} + v_y\frac{\partial v_x}{\partial y} + v_z\frac{\partial v_x}{\partial z}\right) = \rho g_x - \frac{\partial p}{\partial x} \quad (1)$$

(a) Derive Eq. (1) from Eq. (2.3-8).
(b) Also derive Eq. (2) by writing Eq. (1) for the y and z components and combining.

$$\rho \frac{D\mathbf{v}}{Dt} = -\nabla p + \rho \mathbf{g} \qquad (2)$$

2-4 Laminar Flow between Two Flat Plates and Analog Computer Solution

It is desired to derive the equation for steady-state laminar flow of a Newtonian liquid with constant density between flat, parallel plates of infinite width and at a large distance from the inlet. Call x the direction of flow and the distance between the plates in the y direction $y = 2y_0$. Choose the coordinates as $y = 0$ at a point midway between the two plates.

(a) Using the equation of continuity, what is the final form of this equation?
(b) Derive the equation for v_x as a function of y using the equation of motion.
(c) Do the same for v_x as a function of $v_{x_{max}}$ and y.
(d) Draw the analog computer diagram to solve the second-order differential equation. Show initial conditions.

Ans.: (a) $\partial v_x/\partial x = 0$

(b) $v_x = \frac{1}{2\mu}\frac{dp}{dx}(y^2 - y_0^2)$

(c) $v_x = v_{x_{max}}[1 - (y/y_0)^2]$

2-5 Average Velocity in a Circular Conduit

Using Eq. (2.3-16) for the velocity in a circular conduit as a function of radius r,

$$v_x = \frac{1}{4\mu}\frac{dp}{dx}(r^2 - r_0^2) \qquad (2.3\text{-}16)$$

and the maximum-velocity equation

$$v_x = v_{x_{max}}\left(1 - \frac{r^2}{r_0^2}\right) \qquad (2.3\text{-}18)$$

derive the equation for the average velocity, v_{ave} at steady-state. [*Hint*: This average is calculated by summing up all the velocities over the tube cross section and dividing by the cross-sectional area.]

Ans.: $v_{x_{ave}} = v_{x_{max}}/2$

2-6 Solution of Equation of Energy Change for a Cylinder

A material that is cylindrical and solid and has constant physical properties is at an original uniform temperature of T_1. At the surface the temperature is suddenly raised to temperature T_2 and held there. Since the solid is at a higher temperature, heat generation occurs inside the solid at a uniform rate of q_r calories per sec per cm^3 volume. After the solid has reached a steady-state temperature, derive the equation that relates the temperature with the position r

in the cylinder. It is assumed that there is no heat flow in the axial direction. The radius is r_0. [*Hints*: Start with Eq. (2.4-18) and add the heat-generation term to the right-hand side. Cancel out terms that are zero and integrate the ordinary differential equation.]

$$Ans.: \quad T - T_2 = \frac{q_r}{4k}(r_0^2 - r^2)$$

2-7 Sum of Fluxes and Relation of Fluxes

Prove the following equations:

(a) Table 2.5-2 gives

$$\mathbf{n}_A + \mathbf{n}_B = \rho \mathbf{v}$$

Convert it to the equation below.

$$\mathbf{J}_A^* + \mathbf{J}_B^* = 0$$

(b) Prove the following:

$$\mathbf{j}_A^* + \mathbf{j}_B^* = \rho(\mathbf{v} - \mathbf{v}^*)$$

2-8 Forms of Fick's Law

Show that the following form of Fick's law is valid:

$$c(\mathbf{v}_A - \mathbf{v}_B) = -\frac{cD_{AB}}{x_A x_B} \nabla x_A$$

2-9 Equation of Continuity for a Binary Mixture

Derive Eq. (2.6-10) from the general Eq. (2.6-3):

$$\frac{\partial \rho_A}{\partial t} + (\nabla \cdot \rho_A \mathbf{v}) - (\nabla \cdot \rho D_{AB} \nabla w_A) = r_A \quad (2.6\text{-}10)$$

$$\frac{\partial \rho_A}{\partial t} + (\nabla \cdot \mathbf{n}_A) = r_A \quad (2.6\text{-}3)$$

2-10 Different Form of Equation of Continuity

Starting with Eq. (2.6-3),

$$\frac{\partial \rho_A}{\partial t} + (\nabla \cdot \mathbf{n}_A) = r_A \quad (2.6\text{-}3)$$

convert this equation to that below for constant ρ.

$$\frac{\partial \rho_A}{\partial t} + (\mathbf{v} \cdot \nabla \rho_A) - (\nabla \cdot D_{AB} \nabla \rho_A) = r_A$$

2-11 Solid Dissolution to a Fluid in Laminar Flow

In Fig. P2-11 a fluid is flowing as a film at steady-state past a solid A, which is slightly soluble and dissolves. The velocity profile is fully developed and is parabolic and at steady-state. At a given point the fluid contacts the soluble

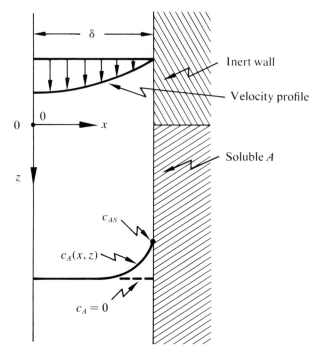

FIGURE P2-11 Diffusion to a fluid in laminar flow for Problem 2-11

portion, whose solubility is c_{AS}. Derive the partial differential equation relating c_A with x and z using the general equations of change.

[*Hints*: The parabolic velocity profile is

$$v_z(x) = v_{z\text{max}}\left(1 - \frac{x^2}{\delta^2}\right)$$

Assume the density is constant and use Eq. (2.6-12) discarding the terms that are zero. Assume diffusion in the z direction is negligible compared to the bulk or convective flow. Also assume no chemical reaction.]

$$\text{Ans.:}\quad v_{z\text{max}}\left(1 - \frac{x^2}{\delta^2}\right)\frac{\partial c_A}{\partial z} = D_{AB}\frac{\partial^2 c_A}{\partial x^2}$$

2-12 Simultaneous Energy and Mass Transfer

In Example 2.7-1 only component A was diffusing. For the case when both A and B are diffusing do as follows.

(a) Derive the integrated flux equations for N_{Az} and N_{Bz}.
(b) Derive the equation for the energy flux.

$$\text{Ans.: (b)}\quad e_z = \frac{(N_{Az}\tilde{c}_{pA} + N_{Bz}\tilde{c}_{pB})(T_\delta - T_0)}{1 - \exp\left[(\delta/k)(N_{Az}\tilde{c}_{pA} + N_{Bz}\tilde{c}_{pB})\right]}$$

2-13 Dimensionless Numbers from Basic Differential Equations

Starting with basic equations of energy and mass change obtain the following.

$$\frac{DT^*}{Dt^*} = \frac{1}{N_{Re}N_{Pr}} \nabla^{*2} T^* \qquad (2.9\text{-}16)$$

$$\frac{Dx_A^*}{Dt^*} = \frac{1}{N_{Re}N_{Sc}} \nabla^{*2} x_A^* \qquad (2.9\text{-}17)$$

NOTATION

(Boldface symbols are vectors or tensors.)

- a constant
- a scalar
- $\mathbf{B}$ vector
- b scalar
- $\mathbf{C}$ vector
- c scalar
- c total concentration, g mole/cm^3
- c_A concentration of A, g mole A/cm^3
- c_v heat capacity at constant volume per unit mass, cm^2/sec^2-°K
- c_{pi} heat capacity of species i at constant pressure per unit mass, cm^2/sec^2-°K or cal/g mass-°K
- $\tilde{c}_{pi}$ heat capacity of species i at constant pressure per mole, cm^2-g mass/sec^2-°K-g mole or cal/g mole-°K
- d scalar
- $\mathbf{D}$ vector
- D substantial derivative symbol
- $\bar{D}$ characteristic dimension, cm
- D_{AB} binary diffusivity for system A-B, cm^2/sec
- e volts on computer
- $\mathbf{e}$ total energy flux vector relative to stationary coordinates, g mass/sec^3 or cal/sec-cm^2
- g_c gravitational conversion factor, 980 g mass-cm/(g force-sec^2)
- $\mathbf{g}$ gravitational acceleration, cm/sec^2
- $\bar{H}_i$ partial molal enthalpy of species i, g mass-cm^2/sec^2-g mole or cal/g mole
- $\mathbf{i}$ unit vector along x axis
- $\mathbf{j}_i$ mass flux vector of species i relative to mass average velocity, g mass i/sec-cm^2
- $\mathbf{j}$ unit vector along y axis

$\mathbf{j}_i^*$	mass flux vector of species i relative to molar average velocity, g mass i/sec-cm²
$\mathbf{J}_i$	molar flux vector of species i relative to mass average velocity, g mole i/sec-cm²
$\mathbf{J}_i^*$	molar flux vector of species i relative to molar average velocity, g mole i/sec-cm²
K_1, K_2	constants
k	thermal conductivity, g mass-cm/sec³-°K or cal-cm/sec-cm²-°K
$\mathbf{k}$	unit vector along z axis
k_1'	heterogeneous first-order reaction velocity constant, cm/sec
k'	homogeneous first-order reaction velocity constant, sec⁻¹
k_c	mass transfer coefficient, cm/sec
L	distance, cm
M	average molecular weight, g mass/g mole
M_A	molecular weight of species A, g mass A/g mole
$\mathbf{N}_A$	molar flux vector of A relative to stationary coordinates, g mole A/sec-cm²
N_{Az}	z component of flux vector $\mathbf{N}_A$, g mole A/sec-cm²
$\mathbf{n}_A$	mass flux vector of A relative to stationary coordinates, g mass A/sec-cm²
N_{Fr}	Froude number, $V^2/g\bar{D}$, dimensionless
N_{Pr}	Prandtl number, $c_p\mu/k$, dimensionless
N_{Re}	Reynolds number, $\bar{D}V\rho/\mu$, dimensionless
N_{Sc}	Schmidt number, $\mu/\rho D_{AB}$, dimensionless
P	component P
p	constant
p	pressure, atm or g mass/sec²-cm
p_A	partial pressure of A, atm
p_B	partial pressure of B, atm
$\mathbf{q}$	energy flux vector relative to mass average velocity, g mass/sec³ or cal/sec-cm²
q_r	energy generated per unit time and volume, cal/sec-cm³ or g mass/cm-sec³
r	radial distance, cm
r	scalar
R_A	rate of generation, g mole A/sec-cm³
r_A	rate of generation, g mass A/sec-cm³
R	universal gas law constant, 82.06 cm³-atm/g mole-°K or units of g mass-cm²/sec²-°K-g mole
s	scalar
t	time, sec
T	temperature, °K or °R
$\hat{U}$	internal energy per unit mass, cm²/sec²
V	characteristic velocity, cm/sec
V	volume per unit mass or $1/\rho$, cm³/g mass
$\mathbf{v}$	mass average velocity vector relative to stationary coordinates, cm/sec
$\mathbf{v}^*$	molar average velocity vector relative to stationary coordinates, cm/sec
$\mathbf{v}_A$	velocity vector of A relative to stationary coordinates, cm/sec

v_x x component of velocity vector $\mathbf{v}$, cm/sec
w_A mass fraction of A
w_B mass fraction of B
x distance in x direction, cm
x_A mole fraction of A
x_B mole fraction of B
y distance in y direction, cm
z distance in z direction, cm

Greek Letters

α thermal diffusivity $= k/\rho c_p$, cm²/sec
α time scale factor for $\tau = \alpha t$
β amplitude scale factor for $e = \beta x$
δ distance of diffusion path, cm
δ diffusivity, cm²/sec
∇ "del" operator
θ angle in cylindrical or spherical coordinates, radians
μ viscosity, g mass/cm-sec
ρ density, g mass/cm³
ρ_A concentration of A, g mass A/cm³
τ viscous stress tensor, g mass/cm-sec²
τ_{yx} component of stress tensor, flux of x momentum in y direction, g mass/cm-sec²
τ analog computer time, sec
φ angle, radians
ψ concentration of a property per unit volume

Subscripts

A component A
B component B
ave   average
(g) forced diffusion
L liquid
max   maximum
P component P
i component i
(p) pressure diffusion
(T) thermal diffusion
x, y, z x, y, and z direction
(x) concentration diffusion
δ point at distance δ
0 reference point
1 beginning of diffusion path
2 end of diffusion path

Superscripts

* dimensionless form in Eqs. (2.9-1) to (2.9-17)

REFERENCES

(A1) B. E. Atkins, R. E. Bastick, and T. L. Ibbs, *Proc. Roy. Soc. (London)* **A172**, 142 (1939).

(B1) R. B. Bird, W. E. Stewart, and E. N. Lightfoot, *Transport Phenomena.* New York: John Wiley & Sons, Inc., 1960.

(B2) C. O. Bennett and J. E. Myers, *Momentum, Heat, and Mass Transfer.* New York: McGraw-Hill, Inc., 1962.

(B3) R. B. Bird, *Advances in Chemical Engineering*, vol. I. New York: Academic Press, Inc., 1956.

(B4) R. E. Bastick, H. R. Heath, and T. L. Ibbs, *Proc. Roy. Soc. (London)* **A173**, 543 (1939).

(B5) H. Brown, *Phys. Rev.* **58**, 661 (1940).

(D1) K. G. Denbigh, *The Thermodynamics of the Steady-State.* New York: John Wiley & Sons, Inc., 1951.

(G1) K. E. Grew and T. L. Ibbs, *Thermal Diffusion in Gases.* London: Cambridge University Press, 1952.

(H1) H. R. Heath, T. L. Ibbs, and N. E. Wild, *Proc. Roy. Soc. (London)* **A178**, 380 (1941).

(H2) J. O. Hirschfelder, C. F. Curtiss, and R. B. Bird, *Molecular Theory of Gases and Liquids.* New York: John Wiley & Sons, Inc., 1954.

(M1) W. H. McAdams, *Heat Transmission*, 3d ed. New York: McGraw-Hill, Inc., 1954.

(S1) R. Siegel, E. M. Sparrow, and T. M. Hallman, *Appl. Sci. Research* **A7**, 386 (1958).

(S2) R. L. Saxton, E. L. Dougherty, and H. G. Drickamer, *Jr. Chem. Phys.* **22**, 1166 (1954).

(T1) L. J. Tichacek, W. S. Kmak, and H. G. Drickamer, *J. Phys. Chem.* **60**, 660 (1956).

3 Molecular Mass Transport Phenomena in Liquids

3.1 MOLECULAR TRANSPORT EQUATIONS FOR LIQUIDS

As discussed in Chapter 1, the molecules in a liquid are packed much more closely together than in gases. Because of the liquid's increased density, the resistance to diffusion is much greater. Also, because of the closer spacing of the molecules, the attraction between molecules plays an important part in the diffusion process.

Unlike the kinetic theory of gases, which is well developed, the kinetic theory of liquids is only partially developed. As a result we write the equations for liquids by analogy to those for gases.

3.1A General Equation

By analogy to Eq. (1.7-8) for gases the equation can also be written for a binary liquid A and B, where

$$-c\,dx_A = \beta_{AB} c_A c_B (v_A - v_B)\,dz \qquad (3.1\text{-}1)$$

where β_{AB} is a proportionality constant for molecules A and B in the liquid, c_A is the concentration of A in the solution, g mole A/cm^3, c is the total concentration, $c_A + c_B$, g mole $A+B/\text{cm}^3$. In the gaseous diffusion equation (1.7-8), c is constant for constant total pressure P. Hence, integration of Eq. (1.7-8) or (1.7-3) presents no problem. However, for liquids c often varies, since densities of solutions and/or the volume/mole of a solute in a liquid vary some with concentration.

Substituting $N_A = c_A v_A$, $N_B = c_B v_B$ and $c_B = c - c_A$

$$-c\,dx_A = \beta_{AB}[(c-c_A)N_A - c_A N_B]\,dz \tag{3.1-2}$$

Rearranging,

$$N_A = -\frac{1}{\beta_{AB}}\frac{dx_A}{dz} + \frac{c_A}{c}(N_A + N_B) \tag{3.1-3}$$

Defining the diffusivity for liquids as

$$D_{AB} = \frac{1}{\beta_{AB}c} \tag{3.1-4}$$

Eq. (3.1-3) becomes

$$N_A = -cD_{AB}\frac{dx_A}{dz} + \frac{c_A}{c}(N_A + N_B) \tag{3.1-5}$$

This equation is the same as Eq. (1.7-3) which is the general equation for all fluids.

To integrate Eq. (3.1-5) requires the assumption that D_{AB} and c are constant. Both of these may vary with concentration. However, we usually use average experimental values of D_{AB} and use the average value of c at both ends of the diffusion path between points 1 and 2. Integrating between points 1 and 2 at steady-state,

$$N_A = \frac{N_A}{N_A + N_B}\frac{D_{AB}}{z}c_{\text{ave}}\ln\left(\frac{x_{A2} - N_A/(N_A + N_B)}{x_{A1} - N_A/(N_A + N_B)}\right) \tag{3.1-6}$$

This equation is similar to Eq. (1.7-17) for fluids except for the c_{ave}, which is defined (T2) as

$$c_{\text{ave}} = \left(\frac{\rho}{M}\right)_{\text{ave}} \tag{3.1-7}$$

where ρ is the density of the solution and M the molecular weight of the solution. Generally c_{ave} is calculated by the linear average

$$c_{\text{ave}} = \frac{\frac{\rho_1}{M_1} + \frac{\rho_2}{M_2}}{2} \tag{3.1-8}$$

In order to use Eq. (3.1-6), the relationship between the fluxes N_A and N_B must be known or set by the physical characteristics of the system.

3.1B Diffusion of A through Stagnant B

An example of the case where A is diffusing and B is stagnant and does not diffuse arises in the situation where a dilute propionic acid (A)-water (B) solution is contacted with toluene. Only the propionic acid (A) diffuses

3.1 Molecular Transport Equations for Liquids 121

into the toluene phase; the water (B) does not diffuse, since it is insoluble in toluene. The toluene-water interface is a barrier to diffusion of B. Hence, $N_B = 0$. This case is often approximated in industry (T1).

To derive the equation at steady-state for this case, set $N_B = 0$ in Eq. (3.1-6).

$$N_A = \frac{D_{AB}c_{ave}}{z} \ln\left(\frac{1-x_{A2}}{1-x_{A1}}\right) \qquad (3.1\text{-}9)$$

Since the term in the ln may often be close to 1.0, we can set $x_{A1} + x_{B1} = x_{A2} + x_{B2} = 1.0$ and Eq. (3.1-9) becomes

$$N_A = \frac{D_{AB}c_{ave}}{zx_{BM}}(x_{A1} - x_{A2}) \qquad (3.1\text{-}10)$$

where

$$x_{BM} = \frac{x_{B2} - x_{B1}}{\ln(x_{B2}/x_{B1})} \qquad (3.1\text{-}11)$$

Equation (3.1-10) is preferred, since the concentrations of the diffusing solute x_{A1} and x_{A2} are usually accurately known.

Example 3.1-1 Diffusion of Ethanol in Water

A thin film 0.4 cm thick of an ethanol (A)-water (B) solution is in contact at 20°C at one surface with an organic liquid in which water is insoluble so B will not diffuse. The concentration of ethanol at the interface or point 2 is 6.8 wt percent ethanol and 16.8 wt percent at the other side of the film, or point 1, and the densities from Perry (P2) are 0.9881 g mass/cm^3 and 0.9728 g mass/cm^3, respectively. The diffusivity of the ethanol (T1) is 0.74×10^{-5} cm^2/sec. Calculate the steady-state flux N_A.

Solution

The weight percent of ethanol is 16.8 and the mole fraction is

$$x_{A1} = \frac{16.8/46.05}{16.8/46.05 + 83.2/18.02} = \frac{0.365}{0.365 + 4.62} = 0.0732$$

Hence, $x_{B1} = 1 - 0.0732 = 0.9268$. In a similar calculation for 6.8 wt percent ethanol, $x_{A2} = 0.0277$ and $x_{B2} = 0.9723$. To calculate the molecular weight at point 1,

$$M_1 = \frac{100}{0.365 + 4.62} = 20.07 \text{ g mass/g mole}$$

Also $M_2 = 18.75$ g mass/g mole. Then to calculate c_{ave} from Eq. (3.1-8),

$$c_{ave} = \frac{\frac{\rho_1}{M_1} + \frac{\rho_2}{M_2}}{2} = \frac{\frac{0.9728}{20.07} + \frac{0.9881}{18.75}}{2} = \frac{0.0485 + 0.0527}{2} = 0.0506$$

Substituting into Eq. (3.1-11),

$$x_{BM} = \frac{x_{B2} - x_{B1}}{\ln(x_{B2}/x_{B1})} = \frac{0.9723 - 0.9268}{\ln(0.9723/0.9268)} = 0.949$$

To calculate N_A, use Eq. (3.1-10), since $N_B = 0$.

$$N_A = \frac{D_{AB} c_{ave}}{z x_{BM}} (x_{A1} - x_{A2}) = \frac{(0.74 \times 10^{-5})(0.0506)}{0.4(0.949)} (0.0732 - 0.0277)$$

$$= 4.48 \times 10^{-8} \text{ g mole ethanol/sec-cm}^2$$

Example 3.1-2 Diffusion and Effect of Flux Ratios

The following is a modification of an example given by Treybal (T1) for a drop of a methyl ethyl ketone-water solution dissolving in water.

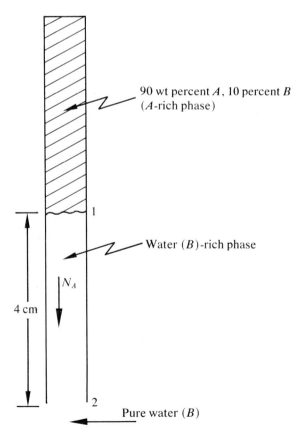

FIGURE 3.1-1 Diffusion of methyl ethyl ketone (A) in water (B) for Example 3.1-2

3.1 Molecular Transport Equations for Liquids

The rate of dissolution of a solution of methyl ethyl ketone or MEK (A) saturated with water (B), where the concentrations are 90.0 wt percent MEK and 10.0 wt percent water at 25°C, is being studied. The saturated solution is dissolving into a water phase. The saturated MEK-water solution is inside and, because of its lower density, at the top of a long, narrow test tube inverted in a stream of flowing pure water. The space between the bottom of this MEK-water solution interface and the open end of the tube is filled with the water phase, and this distance is 4.0 cm. Diffusion of A and B is occurring in this 4.0 cm distance shown in Fig. 3.1-1.

Pure water is flowing continuously outside by the open end of the tube. Calculate the fluxes of A and B through the 4 cm of water at steady-state to the pure water. The saturation solubility of MEK in water is 24.4 wt percent MEK (density = 0.962 g mass/cm³). Treybal (T1) gives an estimated value for D_{AB} as 0.975×10^{-5} cm²/sec.

Solution

Calling point 1 the interface between the two layers, the equilibrium saturation concentration of MEK in the water is 24.4 wt percent or

$$x_{A1} = \frac{24.4/72.0}{24.4/72.0 + 75.6/18.02} = \frac{0.339}{0.339 + 4.19} = 0.0747$$

Hence, $M_1 = 100/4.529 = 22.05$ and $\rho_1 = 0.962$. At point 2 at the end of the diffusion path 4.0 cm away at the open end of the tube, for pure flowing water $x_{A2} = 0$, $M_2 = 18.02$, and $\rho_2 = 1.000$. Then $\rho_1/M_1 = 0.962/22.05 = 0.0436$ and $\rho_2/M_2 = 1.00/18.02 = 0.0555$. $c_{ave} = (0.0436 + 0.0555)/2 = 0.0496$. In order to use Eq. (3.1-6) it is necessary to know the relation between N_A and N_B.

The saturated solution of MEK contains 90 percent MEK (A) or

$$x_A = \frac{90.0/72.0}{90.0/72.0 + 10/18.02} = 0.692$$

and $x_B = 1 - 0.692 = 0.308$. If some A diffuses out of this MEK-water phase, then some of the B must also leave to keep the solution saturated. Thus, the ratio of loss of A to B must be $N_A/N_B = 0.692/0.308 = 2.25$. Or, $N_B = (1/2.25)N_A = 0.445 N_A$. Substituting these values into Eq. (3.1-6),

$$N_A = \frac{N_A}{N_A + N_B} \frac{D_{AB}}{z} c_{ave} \ln\left(\frac{x_{A2} - N_A/(N_A + N_B)}{x_{A1} - N_A/(N_A + N_B)}\right)$$

$$= \frac{N_A}{N_A + 0.445 N_A} \frac{0.975 \times 10^{-5}}{4.0} (0.0496) \ln\left(\frac{0 - N_A/(N_A + 0.445 N_A)}{0.0747 - N_A/(N_A + 0.445 N_A)}\right)$$

$$= 9.54 \times 10^{-9} \text{ g mole } A/\text{sec-cm}^2$$

$$N_B = 0.445(9.54 \times 10^{-9}) = 4.25 \times 10^{-9} \text{ g mole } B/\text{sec-cm}^2$$

3.1C Equimolar Counterdiffusion of A and B

In the case of steady-state equimolar counterdiffusion, $N_A = -N_B$ and Eq. (3.1-5) becomes, upon integration,

$$N_A = \frac{D_{AB}}{z}(c_{A1} - c_{A2}) = \frac{D_{AB}}{z}c_{\text{ave}}(x_{A1} - x_{A2}) \qquad (3.1\text{-}12)$$

3.2 MOLECULAR DIFFUSION COEFFICIENTS FOR LIQUIDS

3.2A Experimental Determination of Diffusion Coefficients

A number of different experimental methods are used to obtain diffusion coefficients for liquids. In one method unsteady-state diffusion in a long capillary tube is carried out and the diffusivity determined from the concentration profile. In a very common method relatively dilute and slightly more concentrated solutions are placed on opposite sides of a porous membrane made of sintered glass as shown in Fig. 3.2-1. Molecular diffusion takes place in the narrow passageways of the pores, where convection is dampened. The two compartments on either side of the membrane are stirred. The effective diffusion length of the pores is first obtained by calibrating with a solute of known diffusivity. Gordon (G1) and Bidstrup and Geankoplis (B1) discuss this method in detail.

To derive the equation for this method shown in Fig. 3.2-1, the assumption of quasi-steady-state in the diffusion path is made. This means

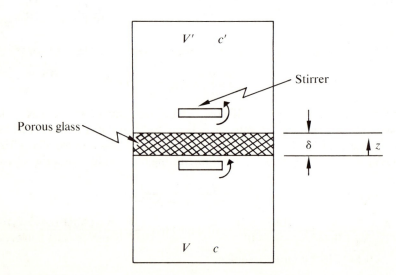

FIGURE 3.2-1 Diffusion cell for liquids

3.2 Molecular Diffusion Coefficients for Liquids

that the concentration gradient in the porous glass can be given by

$$\frac{\partial c}{\partial z} \cong \frac{c'-c}{K_1 \delta} \tag{3.2-1}$$

where K_1 is a constant > 1.0 to correct for the fact that the actual length of path is not δ but is greater. Next we assume dilute solutions so we can neglect the effect of the bulk flow term $x_A(N_A + N_B)$ in Eq. (3.1-5). Then,

$$N_A = J_A^* = -\epsilon D_{AB} \frac{dc}{K_1 dz} = \epsilon D_{AB} \frac{c-c'}{K_1 \delta} \tag{3.2-2}$$

where D_{AB} is assumed constant for dilute solutions and ϵ is the fraction of the area that is open for diffusion.

Making a mole balance of component A on the upper chamber, rate in = rate out + rate of accumulation, and

$$\frac{\epsilon A}{K_1 \delta} D_{AB}(c-c') = 0 + V' \frac{dc'}{dt} \tag{3.2-3}$$

where A is area, cm².

Assuming that $V = V'$ and making a balance on the lower chamber,

$$\frac{\epsilon A}{K_1 \delta} D_{AB}(c-c') = -V \frac{dc}{dt} \tag{3.2-4}$$

Adding Eqs. (3.2-3) and (3.2-4),

$$\frac{2\epsilon A}{K_1 \delta} D_{AB}(c-c') = V \frac{dc'-dc}{dt} = -V \frac{d(c-c')}{dt} \tag{3.2-5}$$

Rearranging and integrating,

$$\int_{c_0-c_0'}^{c-c'} \frac{d(c-c')}{c-c'} = \frac{-2\epsilon A}{K_1 \delta V} D_{AB} \int_0^t dt$$

where $c = c_0$ and $c' = c_0'$ at $t = 0$. Integrating,

$$\ln\left(\frac{c_0-c_0'}{c-c'}\right) = \frac{2\epsilon A}{K_1 \delta V} D_{AB} t = \beta D_{AB} t \tag{3.2-6}$$

where $\beta = 2\epsilon A/(K_1 \delta V)$, a cell constant.

Example 3.2-1 Cell Diffusion Constant

In run 1 by Bidstrup and Geankoplis (B1) to determine the cell constant at 25.0°C using dilute KCl in water the following data were obtained: $t = 2.229 \times 10^5$ sec, $c' = 0.0195$ g mole A/liter, $c_0' = 0.000$, $c = 0.0805$, $c_0 = 0.1000$. The diffusivity of KCl in dilute solution is 1.87×10^{-5} cm²/sec. Calculate the cell constant β.

Solution

Substituting the data into Eq. (3.2-6),

$$\ln\left(\frac{0.100-0}{0.0805-0.0195}\right) = \beta(1.87 \times 10^{-5})(2.229 \times 10^5)$$

Solving, $\beta = 0.1186$ cm^{-2}.

3.2B Experimental Data

The experimental diffusivity data for binary mixtures in the liquid phase are given in Table 3.2-1. Most of the data are for dilute solutions of the solute in the solvent. The diffusivities of solutes in liquids vary

Table 3.2-1 Diffusion Coefficients of Liquids

Solute	Solvent	Temp., °C	Solute Conc., g mole/liter	Diffusivity, (cm²/sec)(10⁵)	Ref.
CO_2	Water	25.0	0	2.00	(V1)
NH_3	Water	5	3.5	1.24	(I1)
		12	1.0	1.64	(I1)
		15	1.0	1.77	(I1)
Ethanol	Water	10	3.75	0.50	(I1)
		10	0.05	0.83	(I1)
		10	0	0.84	(I1)
		15	0	1.00	(I1)
		25	0	1.24	(J1)
Methanol	Water	15	0	1.28	(I1)
n-Propanol	Water	15	0	0.87	(I1)
n-Butanol	Water	15	0	0.77	(I1)
Formic acid	Water	25	0.05	1.52	(B1)
Acetic acid	Water	9.7	0.05	0.769	(B1)
		25	0.05	1.26	(B1)
Propionic acid	Water	25	0.05	1.01	(B1)
Butyric acid	Water	25	0.05	0.92	(B1)
HCl	Water	10	9	3.3	(I1)
		10	2.5	2.5	(I1)
Benzoic acid	Water	25	0	1.21	(C2)
Acetone	Water	25	0	1.28	(A1)
Acetic acid	Acetone	25	0	3.31	(C3)
Acetic acid	Benzene	25	0	2.09	(C3)
Ethanol	Benzene	15	0	2.25	(J1)
Formic acid	Benzene	25	0	2.28	(C3)
Benzene	Chloroform	15	0	2.51	(J1)
Ethanol	Chloroform	15	0	2.20	(J1)
Water	Ethanol	25	0	1.13	(H1)
Toluene	n-Hexane	25	0	4.21	(C3)
KCl	Water	25	0.05	1.870	(P1)
KCl	Ethylene glycol	25	0.05	0.119	(P1)
O_2	Water	18	0	1.98	(I1)
		25	0	2.41	(V1)
H_2	Water	25	0	6.3	(V1)

markedly with concentration, contrary to the case in gases where the effect of concentration was very small. Hence, most data reported for liquids are for relatively dilute solutions.

As noted in the table, the diffusivities are quite small and in the range of about 0.5×10^{-5} to 5×10^{-5} cm^2/sec for relatively nonviscous liquids. Since most gases at 1 atm pressure have values of about 0.5 cm^2/sec, liquid diffusivities are smaller by about a factor of 10^{-5} to 10^{-4}.

3.2C Prediction of Diffusivities for Binary Solutions

Since the theory for diffusion in liquids is not very well established as yet, the equations for predicting diffusivities in liquids are by necessity semiempirical. However, we shall review briefly the basic theoretical equations from which the final semiempirical equations are derived.

1. *Basic Theoretical Equations for Dilute Solutions.* The Stokes-Einstein (E1) equation, one of the first theories, was derived for a very large spherical solute particle (A) diffusing through a liquid solvent (B) of small particles. The Stokes' law was used to describe the drag on the solute particle as it moved. The equation is

$$D_{AB} = \frac{kt}{6\pi\mu_B r_A} \qquad (3.2\text{-}7)$$

where r_A (W1), the radius of the molecule, is $(3V_A/4\pi N)^{1/3}$. This means that D_{AB} is inversely proportional to the viscosity μ_B and the molar volume of the solute to the one-third power or $V_A^{1/3}$. This equation applies very well to very large unhydrated molecules of about 1000 molecular weight and greater in low-molecular-weight solvents (R1) or where the molar volume V_A of the solute is above about 500 cm^3/g mole (W1).

The theory of absolute reaction rates by Eyring (T1) has also been used to predict viscosity, thermal diffusivity, and molecular diffusion coefficients. The liquid is treated as a latticelike array of molecules, which is imperfect because of holes or vacancies. The solute molecules carrying momentum, heat, or mass migrate or jump from one hole or equilibrium position to another. As the molecule migrates, it actually carries momentum, heat or thermal energy, and mass — a different species of material. However, heat and momentum can also be transported by an additional mechanism by actual collision with adjacent molecules without migrating to a "hole." Hence, the mass diffusivity will be the smallest of the three.

For example, in water at 0°C the self-diffusivity of water in water, D_{AA}, is 1.35×10^{-5} cm^2/sec, thermal diffusivity α is 142×10^{-5} cm^2/sec, and momentum diffusivity μ/ρ is 1790×10^{-5} cm^2/sec. Hence, the actual migration mechanism is small compared to the collision mechanism in liquids.

128 Molecular Mass Transport Phenomena in Liquids

The final equations of the absolute reaction rate theory parallel those by the Stokes-Einstein method and do not predict the diffusivities for liquids well. Arnold (D1) applied the kinetic theory of gases to the liquid state and predicted the diffusivity to be proportional to $1/\mu_B^{0.5}$.

2. *Estimation of Diffusion Coefficients in Dilute Solutions.* The Wilke-Chang (W1) correlation can be used for most general purposes where the solute (A) is dilute in the solvent (B):

$$D_{AB} = 7.4(10^{-8})(\varphi M_B)^{1/2} \frac{T}{\mu_B V_A^{0.6}} \qquad (3.2\text{-}8)$$

where M_B is the molecular weight of B, μ_B is the viscosity of B in cp, V_A is the solute molar volume at the boiling point (Table 1.6-4), and φ is an "association parameter" of the solvent, where φ is 2.6 for water, 1.9 methanol, 1.5 ethanol, 1.0 benzene, 1.0 ether, 1.0 heptane, 1.0 other unassociated solvents. When molar volumes are above about 400 cm³/g mole in low-molecular-weight solvents, the Stokes-Einstein equation should be used. When water is the solute, the values predicted from Eq. (3.2-8) should be multiplied by a factor of 1/2.3 (R1). Equation (3.2-8) predicts the diffusivities with a mean deviation of about 10 to 15 percent for aqueous solutions but only to about 25 percent in nonaqueous solvents. Data indicate that the temperature correction of D_{AB} proportional to T/μ_B may overcorrect. The equation should be used with caution outside the temperature range of 5 to 40°C (W1). Perkins and Geankoplis (P1) state that for highly viscous liquids of hydrocarbons and glycerols the viscosity exponent should be less than 1.0 and between 0.5 and 1.0.

Othmer and Thakar (O1) give a simple equation for solutes (A) in water (B), which is as accurate as Eq. (3.2-8):

$$D_{AB} = \frac{14.0(10^{-5})}{\mu_B^{1.1} V_A^{0.6}} \qquad (3.2\text{-}9)$$

where μ_B is again in cp. This equation is often preferred for water as the solvent.

Example 3.2-2 Estimation of Liquid Diffusivity

It is desired to estimate the diffusivity of dilute acetic acid in water at 9.7°C and 25.0°C and compare them with the experimental values using the two methods: (a) Wilke-Chang method, (b) Othmer-Thakar method. (c) Also correct the experimental value at 9.7°C to 25.0°C using both methods.

Solution

From Table A.2-5 the viscosities of water at 9.7°C and 25.0°C are 1.319 and 0.8937 cp, respectively. From Table 1.6-4 for acetic acid, $V_A = 2(14.8) + 4(3.7) + 12.0 + 7.4 = 63.8$. $\varphi = 2.6$ for water, $M_B = 18$, $T = 282.7°K$. Substituting into Eq. (3.2-8) for 9.7°C,

$$D_{AB} = \frac{7.4(10^{-8})[2.6(18)]^{1/2}(282.7)}{1.319(63.8)^{0.6}} = 0.898 \times 10^{-5} \text{ cm}^2/\text{sec}$$

Substituting into Eq. (3.2-9) for 9.7°C,

$$D_{AB} = \frac{14.0(10^{-5})}{(1.319)^{1.1}(63.8)^{0.6}} = 0.853 \times 10^{-5} \text{ cm}^2/\text{sec}$$

Substituting into Eq. (3.2-8) for 25.0°C or 298°K,

$$D_{AB} = \frac{7.4(10^{-8})[2.6(18)]^{1/2}(298)}{0.8937(63.8)^{0.6}} = 1.40 \times 10^{-5} \text{ cm}^2/\text{sec}$$

Substituting into Eq. (3.2-9) for 298°K,

$$D_{AB} = \frac{14.0(10^{-5})}{(0.8937)^{1.1}(63.8)^{0.6}} = 1.31 \times 10^{-5} \text{ cm}^2/\text{sec}$$

The experimental values are 0.769×10^{-5} and 1.26×10^{-5} cm^2/sec at 9.7 and 25.0°C, respectively. Hence, both methods predict the values to within 17 percent [Ans. to (a) and (b)].

To correct the experimental value at 9.7 to 25.0°C, Eq. (3.2-8) predicts

$$D_{AB} = \left(\frac{298}{282.7}\right)^{1.0}\left(\frac{1.319}{0.8937}\right)^{1.0}(0.769 \times 10^{-5}) = 1.20 \times 10^{-5} \text{ cm}^2/\text{sec}$$

Equation (3.2-9) predicts

$$D_{AB} = \left(\frac{1.319}{0.8937}\right)^{1.1}(0.769 \times 10^{-5}) = 1.18 \times 10^{-5} \text{ cm}^2/\text{sec}$$

In this case it is more accurate to correct the experimental value at 9.7°C than to estimate a completely new diffusivity at 25.0°C [Ans. to (c)].

Reddy and Doraiswamy (R1) modified Eq. (3.2-8) and eliminated the association parameter replacing it by the square root of the solvent molar volume. They proposed two equations. For $V_B/V_A \leq 1.5$:

$$D_{AB} = 10(10^{-8})M_B^{1/2}\frac{T}{\mu_B V_A^{1/3} V_B^{1/3}} \qquad (3.2\text{-}10)$$

for $V_B/V_A > 1.5$:

$$D_{AB} = 8.5(10^{-8})M_B^{1/2}\frac{T}{\mu_B V_A^{1/3} V_B^{1/3}} \qquad (3.2\text{-}11)$$

where V_B is the solvent molar volume from Table 1.6-4.

Equation (3.2-10) represents the data to within 13.5 percent; and Eq. (3.2-11) represents the data to within 18 percent, which includes six systems in which water diffuses into an organic solvent. These two equations are useful when an association parameter is not available for use in Eq. (3.2-8).

3. *Estimation of Diffusion Coefficients for Concentrated Binary Solutions.* Cullinen (C1) derived the following by modifying the Eyring theory of absolute reaction rates for a concentrated mixture:

$$D_{AB} = (D_{AB}^0)^{x_B} (D_{BA}^0)^{x_A} \left(1 + \frac{\partial \ln \gamma_A}{\partial \ln x_A}\right) \qquad (3.2\text{-}12)$$

where D_{AB}^0 is the diffusion coefficient for dilute A in B and D_{BA}^0 for dilute B in A and γ_A is the activity coefficient of A in the concentrated mixture. This equation checks experimental data very well for nonideal and nonassociating mixtures. It should not be used for viscous mixtures.

3.2D Prediction of Diffusivities for Multicomponent Solutions

Only a few diffusivity data on ternary liquid systems have been published. Burchard and Toor (B2) studied a concentrated ternary hydrocarbon system and found that the diffusion in a ternary system could be represented by Fick's law of

$$J_A^* = -D_{Am} \frac{dc_A}{dz} \qquad (3.2\text{-}13)$$

where D_{Am} is a molal average diffusion coefficient of the three binary coefficients in dilute solutions. However, this is for a nonviscous relatively ideal system of toluene, chlorobenzene, and bromobenzene; it cannot be used for viscous or nonideal systems.

Perkins and Geankoplis (P1) derived an equation for a ternary system for diffusion of a dilute species A in a concentrated mixture of B and C. The equation is

$$D_{Am}^0 \eta_m^{0.8} = x_B D_{AB}^0 \eta_B^{0.8} + x_C D_{AC}^0 \eta_C^{0.8} \qquad (3.2\text{-}14)$$

where D_{Am}^0 is the diffusivity of dilute A in a mixture of B and C of mole fraction compositions x_B and x_C, η_m is the viscosity of the mixture, η_B is the viscosity of pure B, and η_C is the viscosity of pure C. This equation represents quite well the diffusivities of the system acetic acid in ethanol-water, which is highly nonideal with an average deviation of 8.5 percent, the system KCl in ethylene glycol-water, which has a high viscosity up to 18 cp within 14.6 percent, CO_2 in ethanol-water within 22.2 percent, and several hydrocarbon systems within 0.5 to 8.2 percent. If the diffusivities of the dilute solute A in the pure solvent D_{AB}^0 and D_{AC}^0 are not available, they can be estimated by Eq. (3.2-8) and Eq. (3.2-9). Equation (3.2-13) can be used to predict the flux, since c_A is small and the cross-flux terms contributed by N_B and N_C will be small and can be neglected. The case where dilute A is diffusing in a mixture of solvents is often approximated in industrial diffusion processes.

Example 3.2-3 Prediction of Multicomponent Diffusivity

Perkins and Geankoplis (P1) give the following experimental data for the diffusion of dilute acetic acid (A) in ethanol (B)-water (C) mixtures at 25°C. The D^0_{AB} for dilute A in pure $B = 1.032 \times 10^{-5}$ cm²/sec, η_B of pure ethanol = 1.096 cp; D^0_{AC} for dilute A in pure $C = 1.295 \times 10^{-5}$ cm²/sec, η_C of pure water = 0.8937 cp; D_{AB} for dilute A in a solution of 0.207 mole fraction ethanol − 0.793 mole fraction water = 0.5706×10^{-5} cm²/sec, η_m of solution = 2.350 cp.

Using Eq. (3.2-14) predict D^0_{Am} in the mixture and compare with the experimental value given.

Solution

For Eq. (3.2-14),

$$D^0_{Am}\eta_m^{0.8} = x_B D^0_{AB}\eta_B^{0.8} + x_C D^0_{AC}\eta_C^{0.8}$$

$$D^0_{Am}(2.350)^{0.8} = 0.207(1.032 \times 10^{-5})(1.096)^{0.8} + 0.793$$

$(1.295 \times 10^{-5})(0.8937)^{0.8}$. $D^0_{Am} = 0.590 \times 10^{-5}$ cm²/sec

Experimental value is 0.5706×10^{-5} cm²/sec.

3.3 MOLECULAR TRANSPORT AND CHEMICAL REACTION IN LIQUIDS

3.3A Steady-State Reaction and Diffusion

If the solute diffusing and reacting in the liquid is dilute, then Eq. (3.3-1) can be used to predict the flux in the solution:

$$N_{Az} = -D_{AB}\frac{dc_A}{dz} \tag{3.3-1}$$

where the D_{AB} is estimated using Eqs. (3.2-8) to (3.2-11). This is the same as the flux equation used for homogeneous reaction in the gas phase when the gaseous component A is dilute in the gas stream. These equations were derived in Section 1.8B, Chapter 1, for the cases of first-order reaction, empirical reaction rate, and variable diffusivity. These methods and solutions will also hold for molecular diffusion and chemical reaction in liquids when the solute is dilute.

We will consider a case similar to that for Eqs. (1.8-8) to (1.8-11) and discussed by Bird et al. (B3), where homogeneous reaction is occurring in a phase at steady-state. In the present case pure gas A is being contacted by pure liquid C as shown in Fig. 3.3-1. The concentration of species A at the surface of the liquid is constant at c_{A1}. The irreversible first-order reaction occurring is

$$A + C \rightarrow D \tag{3.3-2}$$

The solute A and the product D are dilute in the liquid C phase, so that we can use Eq. (3.3-1) for component A diffusing in the liquid phase. Writing the material balance on A,

rate in + rate of generation = rate out + rate of accumulation (3.3-3)

The first-order reaction rate per cm³ volume is

$$\text{rate of generation} = -k'c_A \tag{3.3-4}$$

where k' is the reaction velocity constant, sec⁻¹. Substituting into Eq. (3.3-3) for steady-state,

$$N_{Az}|_z(1) - k'c_A(1)\,\Delta z = N_{Az}|_{z+\Delta z}(1) + 0 \tag{3.3-5}$$

where the cross-sectional area is taken as 1 cm². Next we divide through

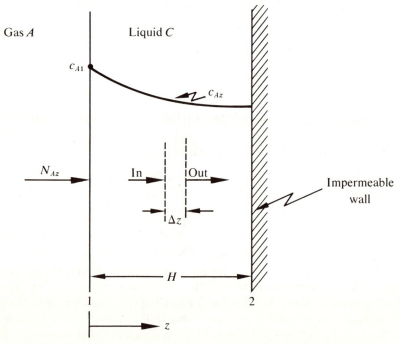

FIGURE 3.3-1 Homogeneous chemical reaction in a liquid at steady-state

3.3 Molecular Transport and Reaction in Liquids

by Δz and let Δz approach zero. As stated, N_{Az} is written for a dilute solution as Eq. (3.3-1). Substituting Eq. (3.3-1) for A diffusing in C into Eq. (3.3-5) and rearranging,

$$\frac{d^2c_A}{dz^2} - \frac{k'}{D_{AC}} c_A = 0 \qquad (3.3\text{-}6)$$

The boundary conditions are different from Eq. (1.8-10) and are

B.C. 1: $\quad z = 0, \quad c_A = c_{A1}$
B.C. 2: $\quad z = H, \quad N_{Az} = 0 \quad$ or $\quad -D_{AC} dc_A/dz = 0 \qquad (3.3\text{-}7)$

The last B.C. 2 occurs since the point at $z = H$ is impermeable to A.

Equation (3.3-6) is a linear second-order differential equation and its solution is

$$c_A = c_{A1}\left(\frac{\cosh a(1 - z/H)}{\cosh a}\right) \qquad (3.3\text{-}8)$$

where $a = \sqrt{k'H^2/D_{AC}}$. To obtain the concentration of A in the liquid at the point $z = H$,

$$c_{Az} = \frac{c_{A1} \cosh 0}{\cosh a} = \frac{c_{A1}}{\cosh a} \qquad (3.3\text{-}9)$$

The total flux of A entering the liquid is equal to N_{Az} at the point $z = 0$ or

$$N_{A(z=0)} = -D_{AC}\left(\frac{dc_A}{dz}\right)_{z=0} \qquad (3.3\text{-}10)$$

First dc_A/dz is obtained from Eq. (3.3-8) as

$$\frac{dc_A}{dz} = \frac{c_{A1}}{\cosh a}\left(\frac{-a}{H}\right) \sinh a\left(1 - \frac{z}{H}\right) \qquad (3.3\text{-}11)$$

Substituting $z = 0$ into Eq. (3.3-11),

$$\left(\frac{dc_A}{dz}\right)_{z=0} = \frac{c_{A1}}{\cosh a}\left(\frac{-a}{H}\right) \sinh a = -c_{A1}\left(\frac{a}{H} \tanh a\right) \qquad (3.3\text{-}12)$$

Substituting Eq. (3.3-12) into (3.3-10),

$$N_{A(z=0)} = \frac{D_{AC} c_{A1} a}{H} \tanh a \qquad (3.3\text{-}13)$$

As an alternate and often preferable method we will derive Eq. (3.3-6) by using the general equation of continuity for a binary mixture. Writing Eq. (2.6-12) for constant ρ and D_{AC},

$$\frac{\partial c_A}{\partial t} + (\mathbf{v} \cdot \nabla c_A) - D_{AC} \nabla^2 c_A = R_A \qquad (3.3\text{-}14)$$

The first term $\partial c_A/\partial t = 0$ for steady-state. Also, since we are assuming dilute solutions and neglecting the bulk flow term, then $v = 0$. The second term of Eq. (3.3-14) is thus zero. For R_A we write

$$R_A = -k'c_A \tag{3.3-15}$$

where R_A is the rate of generation, g mole A/sec-cm³. Substituting Eq. (3.3-15) into (3.3-14) and writing the equation for the z direction, we obtain

$$D_{AC}\frac{d^2c_A}{dz^2} = k'c_A \tag{3.3-16}$$

This, of course, is identical to Eq. (3.3-6).

3.3B Unsteady-State Reaction and Diffusion in a Semi-Infinite Medium

Next we consider a case presented by Danckwerts (D2) whereby a dilute solute A is absorbed in a pure liquid B at the surface $x = 0$, where the surface concentration is kept constant at c_{A0}. The dilute solute A reacts by a first-order mechanism, so

$$A + B \rightarrow C \tag{3.3-17}$$

and the rate of generation is $-k'c_A = R_A$.

Starting with the general equation of continuity for a binary mixture, Eq. (2.6-12) becomes as follows for a dilute mixture where $v \cong 0$ and diffusion is in the x direction:

$$\frac{\partial c_A}{\partial t} = D_{AB}\frac{\partial^2 c_A}{\partial x^2} - k'c_A \tag{3.3-18}$$

The solution will be considered a semiinfinite medium, so at $x \to \infty$ the concentration of solute A will be considered zero at all times. The initial and boundary conditions are

$$\begin{array}{llll} \text{I.C.:} & \text{at } t = 0, & c_A = 0 & \text{for } x > 0 \\ \text{B.C. 1:} & \text{at } x = 0, & c_A = c_{A0} & \text{for } t > 0 \\ \text{B.C. 2:} & \text{at } x = \infty, & c_A = 0 & \text{for } t > 0 \end{array} \tag{3.3-19}$$

The solution to Eq. (3.3-18) has been given by Danckwerts (D2) as follows:

$$\frac{c_A}{c_{A0}} = \frac{1}{2}\exp\left(-x\sqrt{\frac{k'}{D_{AB}}}\right) \cdot \text{erfc}\left(\frac{x}{2\sqrt{tD_{AB}}} - \sqrt{k't}\right)$$

$$+ \frac{1}{2}\exp\left(x\sqrt{\frac{k'}{D_{AB}}}\right) \cdot \text{erfc}\left(\frac{x}{2\sqrt{tD_{AB}}} + \sqrt{k't}\right) \tag{3.3-20}$$

where erfc is 1-error function discussed in Chapter 5, Section 5.3, and is

$$\text{erfc } z = 1 - \text{erf } z = 1 - \frac{2}{\sqrt{\pi}} \int_0^z e^{-y^2} \, dy \tag{3.3-21}$$

In order to obtain the flux N_A at $x = 0$, we must set

$$N_{A(x=0)} = -D_{AB} \left(\frac{\partial c_A}{\partial x} \right)_{x=0} \tag{3.3-22}$$

Differentiating Eq. (3.3-20) with respect to x and setting $x = 0$, we obtain the flux at point $x = 0$ at a given time t.

$$N_{A(x=0)} = c_{A0} \sqrt{k' D_{AB}} \left(\text{erf } \sqrt{k't} + \frac{e^{-k't}}{\sqrt{\pi k't}} \right) \tag{3.3-23}$$

Next, to obtain the total amount absorbed up to time t, we can obtain the integral of $N_{A(x=0)} \, dt$ between $t = 0$ and $t = t$:

$$Q = c_{A0} \sqrt{\frac{D_{AB}}{k'}} \left[\left(k't + \frac{1}{2} \right) \text{erf } \sqrt{k't} + \sqrt{\frac{k't}{\pi}} e^{-k't} \right] \tag{3.3-24}$$

where Q is the total g mole A absorbed/cm² up to time t.

When $k't$ becomes large, erf $\sqrt{k't} \cong 1.0$ and Eqs. (3.3-20), (3.3-23), and (3.3-24) become

$$c_A \cong c_{A0} \exp\left(-x \sqrt{\frac{k'}{D_{AB}}} \right) \tag{3.3-25}$$

$$N_{A(x=0)} \cong c_{A0} \sqrt{k' D_{AB}} \tag{3.3-26}$$

$$Q \cong c_{A0} \sqrt{k' D_{AB}} \left(t + \frac{1}{2k'} \right) \tag{3.3-27}$$

When $k't > 4$, the error in Eqs. (3.3-26) and (3.3-27) is less than 2 percent. When $k't$ is very small, then we obtain

$$N_{A(x=0)} \cong c_{A0} \sqrt{\frac{D_{AB}}{\pi t}} (1 + k't) \tag{3.3-28}$$

$$Q \cong 2 c_{A0} \sqrt{\frac{D_{AB} t}{\pi}} \left(1 + \frac{k't}{3} \right) \tag{3.3-29}$$

The equations above are quite useful in cases where absorption at the surface of a stagnant fluid or a solid occurs and then unsteady-state diffusion and reaction occurs in the fluid or solid. Many actual cases are approximated by this case. Often the reaction rate constant can be approximated by a pseudo first-order reaction rate constant over the concentration range used.

3.4 INTRODUCTION TO THERMODYNAMICS OF IRREVERSIBLE PROCESSES AND ONSAGER'S RECIPROCAL RELATIONS

As we noted in Chapter 2 when discussing thermal diffusion, the thermodynamics of irreversible processes state that there will be a contribution to a total flux because of each driving force in the system. Hence, in a multicomponent system and in the absence of pressure or forced diffusion, the mass flux of a species will depend primarily on the concentration gradient (ordinary diffusion) and on the temperature gradient (thermal diffusion or Soret effect). Similarly the energy flux depends primarily on the temperature gradient and also on the concentration gradient, which is the Dufour effect.

The Onsager reciprocal relations give us an equation relating the two coupled effects of Soret and Dufour and also give us information on the forms of the flux vectors.

Thermodynamics and the first and second laws are essentially concerned with equilibrium processes. Onsager tried to explain irreversible processes in terms of thermodynamics by his assumption of microscopic reversibility where the system is not far from equilibrium. He used this assumption in proof of his reciprocal relations.

Experimentally it has been shown to a reasonable degree that the rates of transport processes are proportional to the gradients of suitable potentials. For example, in Fourier's law the energy flux is proportional to the temperature gradient. In Fick's law the mass flux is proportional to the concentration gradient. These are called linear laws and phenomenological laws, where

$$\mathbf{J} = L\mathbf{X} \tag{3.4-1}$$

where L is a scalar phenomenological coefficient or rate constant and $\mathbf{X}$ is a generalized driving force or potential difference or affinity. Suppose we have a system where two processes take place simultaneously and their rates are $\mathbf{J}_1$ and $\mathbf{J}_2$ and the forces are $\mathbf{X}_1$ and $\mathbf{X}_2$. It is assumed that $\mathbf{J}_1$ is a linear function of both forces by Onsager's linear relations:

$$\mathbf{J}_1 = L_{11}\mathbf{X}_1 + L_{12}\mathbf{X}_2 \tag{3.4-2}$$
$$\mathbf{J}_2 = L_{21}\mathbf{X}_1 + L_{22}\mathbf{X}_2 \tag{3.4-3}$$

For three simultaneous processes,

$$\begin{aligned}\mathbf{J}_1 &= L_{11}\mathbf{X}_1 + L_{12}\mathbf{X}_2 + L_{13}\mathbf{X}_3 \\ \mathbf{J}_2 &= L_{21}\mathbf{X}_1 + L_{22}\mathbf{X}_2 + L_{23}\mathbf{X}_3 \\ \mathbf{J}_3 &= L_{31}\mathbf{X}_1 + L_{32}\mathbf{X}_2 + L_{33}\mathbf{X}_3\end{aligned} \tag{3.4-4}$$

Hence, we can write a general equation as

$$\mathbf{J}_i = \sum_k L_{ik}\mathbf{X}_k \tag{3.4-5}$$

3.4 Introduction to Irreversible Thermodynamics

If Eqs. (3.4-2) and (3.4-3) represent the simultaneous irreversible transfer of energy and mass flux, then L_{11} is related to thermal conductivity and describes the direct effects due to the temperature gradient, and L_{22} is related to ordinary mass diffusivity. Also L_{12} and L_{21} are the interference or cross coefficients relating thermal diffusion and Dufour diffusion and give coupling.

Onsager further showed by this theorem of reciprocity relation that, provided the proper choices of the form of fluxes and affinities are made, the following holds:

$$L_{12} = L_{21} \tag{3.4-6}$$

Or, for more than two forces,

$$L_{ik} = L_{ki} \quad (i \neq k), \quad i, k = 1, \ldots, n \tag{3.4-7}$$

This means that for thermal diffusion there is symmetry between the effect of the diffusion force on the flux of energy and the effect of thermal force on the flux of mass.

We can choose for the forces a number of types. For example, for Fourier's law we could select dT or $d(1/T)$. These changes merely alter the form of the scalar coefficient L and its relation to the conventional definition of thermal conductivity. Thus, for a choice of $d(1/T) = -dT/T^2$, L is changed by a factor of $-1/T^2$.

Onsager uses an entropy balance on the system to derive the form of the forces $\mathbf{X}_i$. Diffusion and conduction of heat are irreversible processes, where entropy S is created, and the amounts depend on the magnitudes of $\mathbf{J}_i$ and forces $\mathbf{X}_i$. As for any other property, one can make a property balance of entropy on a system: input plus generation equals output plus accumulation. One chooses the forces in such a way that when the flux $\mathbf{J}_i$ is multiplied by the force, the sums are θ, the rate of creation of entropy per unit volume, multiplied by the temperature.

$$T\theta = \mathbf{J}_1 \cdot \mathbf{X}_1 + \mathbf{J}_2 \cdot \mathbf{X}_2 + \cdots = \sum_i \mathbf{J}_i \cdot \mathbf{X}_i \tag{3.4-8}$$

Often this is written as

$$\theta = \sum_i \mathbf{J}_i \cdot \mathbf{x}_i \tag{3.4-9}$$

where $\mathbf{x}_i = \mathbf{X}_i/T$. Many authors use different symbols with $\mathbf{X}$ in place of $\mathbf{x}$, and so on; the reader is cautioned to determine the meaning of the symbols.

When the forces have been chosen so as to satisfy Eq. (3.4-8), then Onsager proved that the coupling coefficients in Eq. (3.4-7) are equal by using the principle of microscopic reversibility.

For isothermal diffusion it can be shown that the rate of creation of entropy $(dS/dt)_{\text{irr}}$ is

$$\left(\frac{dS}{dt}\right)_{\text{irr}} = -\frac{1}{T}\frac{dG}{dt} \tag{3.4-10}$$

where $-dG/dt$ is the rate of decrease of Gibbs free energy. Since the change in chemical potential μ_i of a species can be related to Eq. (3.4-10), the equation below can be derived.

$$T\theta = -\sum_i \mathbf{J}_i \cdot \nabla \mu_i \qquad (3.4\text{-}11)$$

Hence, for a nonideal system the chemical potential gradient is the thermodynamic driving force and not the concentration gradient.

Details of some of the derivations above are lengthy, and the excellent texts by Hirschfelder, Curtiss, and Bird (H2), Denbigh (D3), and De Groot (D4) should be consulted for further details.

The theories above are useful in systems where interactions occur between processes. They also give information as to the form of the flux vectors. This is important in liquids and other dense fluids or solids where the forms of the flux vectors would otherwise be unknown. However, for dilute gases the rigorous kinetic theory of gases gives us results consistent with those from the thermodynamics of irreversible processes.

Onsager's theory does not show us whether the equations of motion are correct. This must be determined by experiment. His theory only gives the reciprocal relations between the coefficients if the equations are correct. His approach does not give the equations for the transport properties in terms of the molecular properties.

The values of L's must be determined experimentally. The reciprocal relations are used to eliminate the coefficient $L_{ik} = L_{ki}$ and thus to give us certain physical relations that can be tested by experiment. For uses of these theories in multicomponent diffusion in dense-phase systems see Burchard and Toor (B2), Fujita and Gosting (F1), and Kirkaldy et al. (K1).

PROBLEMS

3-1 Diffusion of Components in a Binary Liquid

A thin film 0.2 cm thick of an ammonia (A)-water (B) solution is in contact at 5°C at one surface with an organic liquid at this interface. The concentration of ammonia in the organic is constant and such that the equilibrium concentration of ammonia in water at this surface is 2.0 wt percent ammonia (density 0.9917 g mass/cm^3) and the concentration is 10 wt percent ammonia (density 0.9617 g mass/cm^3) at the other end of the film 0.2 cm away. The diffusivity of NH$_3$ in water is 1.24×10^{-5} cm^2/sec. Water and the organic are insoluble in each other
  (a) Calculate the flux N_A at steady-state.
  (b) Calculate the flux N_B. Explain.
  (c) Calculate the concentration x_A at various points in the path and plot x_A versus distance z up to 0.2 cm.

3-2 Diffusion of HCl in a Solution at Steady-State

In a thin film 0.3 cm thick of an HCl (A)-water (B) solution at 10°C, the concentration of HCl at one end of the film at point 1 is 12.0 wt percent HCl (density 1.0607 g mass/cm^3) and, at point 2, 6.0 wt percent HCl (density 1.0303 g

mass/cm^3) at the other end of the film. The diffuvisity of the HCl in water is 2.5×10^{-5} cm^2/sec. Assume steady-state.
   (a) Calculate the flux of HCl assuming that water does not diffuse.
   (b) Calculate the flux of HCl and the flux of water assuming that they are related by $N_A = 2N_B$.

3-3 Prediction of Binary Diffusion Coefficient

It is desired to predict the diffusivity of acetone in water at 25°C and 50°C.
   (a) Predict it using the Wilke-Chang equation.
   (b) Predict it using the Othmer-Thakar equation.
   (c) The experimental value at 25°C is 1.28×10^{-5} cm^2/sec. Using this experimental value, predict the value at 50°C using the temperature correction factors derived from the Wilke-Chang and Othmer-Thakar equations.

3-4 Prediction of Diffusivity of KCl in Ethylene Glycol

The diffusivity of KCl in water at 25°C is 1.87×10^{-5} cm^2/sec. It is desired to predict the diffusivity of KCl in ethylene glycol at 25°C. Predict it using the Wilke-Chang equation and the data in water. The viscosity of ethylene glycol is 18.09 cp. [*Hint*: Assume the association factor of ethylene glycol is close to that for ethanol. Is it necessary to predict the value of V_A, the solute molar volume of KCl? The experimental value is given in Table 3.2-1 for comparison.]

Ans.: 0.130×10^{-5} cm^2/sec

3-5 Diffusion in a Ternary Liquid Solution

Using the data in Example 3.2-3, predict the diffusivity D^0_{Am} of dilute acetic acid (*A*) in a mixture of ethanol (*B*)-water (*C*) containing a mole fraction 0.370 mole fraction ethanol and 0.630 mole fraction water at 25.0°C. The viscosity of this mixture is 2.24 cp and the experimental value of D^0_{Am} is 0.597×10^{-5} cm^2/sec.
   (a) Predict the D^0_{Am} using the data above and Example 3.2-3.
   (b) Predict the value of D^0_{Am} as above but assume that the values for dilute *A* in pure solvent *B* or D^0_{AB} and also D^0_{AC} are not available. [*Hint*: Use the Wilke-Chang equation.]
   (c) Predict the flux of *A* at 25°C when it is diffusing steady-state in this mixture through a distance of 0.20 cm and the concentration of *A* at one boundary is 0.010 mole fraction and 0.0010 mole fraction at the other boundary. [*Hint*: To calculate the flux, you can use Eq. (3.2-13). The density of the mixture is 0.897 g mass/cm^3.]

3-6 Derivation of Equation for Diaphragm Cell Apparatus

Derive the equation for the diffusion cell given in Fig. 3.2-1 but for volumes V and V' being unequal.

$$\text{Ans.: } \ln\left(\frac{c_0 - c'_0}{c - c'}\right) = \frac{2\epsilon A}{K_1 \delta}\left(\frac{1}{V} + \frac{1}{V'}\right) D_{AB} t$$

3-7 Experimental Diffusivity of Acetic Acid in Water

In a run at 25.0°C to determine the diffusivity of dilute acetic acid in water, Bidstrup and Geankoplis (B1) obtained the following data. $c'_0 = 0.0002$, $c = 0.0851$, $c' = 0.0134$, $c_0 = 0.0984$ for $t = 2.133 \times 10^5$ sec. The cell constant is the same as in Example 3.2-1. What is the diffusivity of acetic acid?

Ans.: $D_{AB} = 1.25 \times 10^{-5}$ cm^2/sec

3-8 Homogeneous Chemical Reaction and Analog Computer Solution

For the same case as pictured in Fig. 3.3-1 for a homogeneous chemical reaction in a liquid, but where the reaction velocity constant k' is not a constant but is $k' = (K_1 + K_2 \log c_A)$, solve the final equation using the analog computer as follows. (K_1 and K_2 are constants.)

(a) First solve for the highest derivative, which will be

$$\frac{d^2 c_A}{dz^2} = \frac{K_1}{D_{AC}} c_A + \frac{K_2 c_A}{D_{AC}} \log c_A \tag{3.3-6a}$$

(b) Draw the analog computer diagram.

(c) Make a sketch of the expected computer output results of c_A versus z. Give the initial conditions and state how to use them. What should the slope of the c_A versus z line be at $z = H$?

3-9 Unsteady-State Diffusion and Reaction

A solute A is diffusing unsteady-state into a semiinfinite medium of pure B and undergoes a first-order reaction with the pure B phase. Solute A is dilute in the medium. Calculate the concentration c_A versus distance x from the surface for $t = 1.0 \times 10^5$ sec at points $x = 0, 0.1, 0.2, 0.4,$ and 1.0 cm and plot the data. The physical data are $D_{AB} = 1 \times 10^{-5}$ cm²/sec, $k' = 1 \times 10^{-4}$ sec⁻¹, $c_{A0} = 1.0 \times 10^{-3}$ g mole A/cm³.

NOTATION

(Boldface symbols are vectors.)

a	dimensionless constant $= \sqrt{k' H^2/D_{AB}}$
A	cross-sectional area, cm²
c_A	concentration of A, g mole A/cm³ or lb mole A/ft³
c	total concentration, g mole/cm³ or lb mole/ft³
c	concentration of component A, g mole/cm³
D_{AB}	diffusivity of A in B, cm²/sec or ft²/hr
D^0_{AB}	diffusivity of dilute A in B, cm²/sec or ft²/hr
D^0_{BA}	diffusivity of dilute B in A, cm²/sec or ft²/hr
D^0_{AC}	diffusivity of dilute A in C, cm²/sec or ft²/hr
D_{Am}	molal average diffusivity of A in a mixture, cm²/sec or ft²/hr
D^0_{Am}	diffusivity of dilute A in a mixture, cm²/sec or ft²/hr
e	dependent variable on analog computer, volts
G	Gibbs free energy
H	distance, cm
J^*_A	flux of A relative to the molar average velocity, g mole A/sec-cm² or lb mole A/hr-ft²
J	generalized flux
K_1	constant

k Boltzmann constant
k' homogeneous first order reaction velocity constant, sec^{-1}
L phenomenological coefficient
M molecular weight, g mass/g mole or lb mass/lb mole
N Avogadro's number, molecules/mole
$\mathbf{N}_A$ flux of A relative to a fixed point, g mole A/sec-cm² or lb mole A/hr-ft²
P total pressure, atm abs.
Q total absorbed, g mole A/cm²
R_A rate of generation, g mole A/sec-cm³
r_A radius of molecule A, cm
S entropy of a system
T absolute temperature, °K
t time, sec
$\mathbf{v}_A$ velocity of A relative to stationary coordinates, cm/sec or ft/hr
V volume, cm³
V_A molar volume of liquid at normal boiling point, cm³/g mole (see Table 1.6-4)
$\mathbf{v}^*$ molar average velocity, cm/sec
x_A mole fraction of A
x_{BM} ln mean of x_{B1} and x_{B2}
$\mathbf{x}$ generalized force $= \mathbf{X}/T$
$\mathbf{X}$ generalized force
z distance in z direction, cm or ft

Greek Letters

α analog computer time scale factor in $\tau = \alpha t$
β cell constant, cm^{-2}
β analog computer amplitude scale factor
β_{AB} proportionality constant for diffusion of A in B
γ_A activity coefficient of A
δ distance, cm
ϵ void fraction, dimensionless
η viscosity, cp
θ rate of entropy created per unit volume
μ chemical potential
μ viscosity of solution, g mass/cm-sec or in some cases cp
ρ density, g mass/cm³ or lb mass/ft³
τ analog computer time, sec
φ association parameter in Eq. (3.2-8)

Subscripts

A compound A
ave   average
B compound B
C compound C

142 Molecular Mass Transport Phenomena in Liquids

m mixture
x x direction
z z direction
1 beginning of diffusion path
2 end of diffusion path

REFERENCES

(A1) D. K. Anderson, J. R. Hall, and A. L. Babb, *J. Phys. Chem.* **62**, 404 (1958).
(B1) D. E. Bidstrup and C. J. Geankoplis, *J. Chem. Eng. Data* **8**, 170 (1963).
(B2) J. K. Burchard and H. L. Toor, *J. Phys. Chem.* **66**, 2015 (1962).
(B3) R. B. Bird, W. E. Stewart, and E. N. Lightfoot, *Transport Phenomena*. New York: John Wiley & Sons, Inc., 1960.
(C1) H. T. Cullinen, *Ind. Eng. Chem. Fund.* **5**, 281 (1966).
(C2) S. Y. Chang, M.S. thesis, Massachusetts Institute of Technology (1959).
(C3) Pin Chang and C. R. Wilke, *J. Phys. Chem.* **59**, 592 (1955).
(D1) G. A. Davies, A. B. Ponter, and K. Craine, *Can. J. Chem. Eng.* **45**, 372 (1967).
(D2) P. V. Danckwerts, *Trans. Faraday Soc.* **46**, 300 (1950).
(D3) K. G. Denbigh, *The Thermodynamics of the Steady State*. New York: John Wiley & Sons, Inc., 1951.
(D4) S. R. De Groot, *Thermodynamics of Irreversible Processes*. New York: Interscience Publishers, 1951.
(E1) A. Einstein, *Ann. Physik* **19**, 371 (1906).
(F1) H. Fujita and L. Gosting, *J. Am. Chem. Soc.* **78**, 1099 (1956).
(G1) A. R. Gordon, *Ann. N.Y. Acad. Sci.* **46**, 235 (1945).
(H1) B. R. Hammond and R. H. Stokes, *Trans. Faraday Soc.* **49**, 890 (1953).
(H2) J. O. Hirschfelder, C. F. Curtiss, and R. B. Bird, *Molecular Theory of Gases and Liquids*. New York: John Wiley & Sons, Inc., 1954.
(I1) National Research Council, *International Critical Tables*, vol. V. New York: McGraw-Hill, Inc., 1929.
(J1) P. A. Johnson and A. L. Babb, *Chem. Revs.* **56**, 387 (1956).
(K1) J. S. Kirkaldy, J. E. Lane, and G. R. Mason, *Can. J. Phys.* **41**, 2174 (1963).
(O1) D. F. Othmer and M. S. Thakar, *Ind. Eng. Chem.* **45**, 589 (1953).
(P1) L. R. Perkins and C. J. Geankoplis, *Chem. Eng. Sci.* **24**, 1035 (1969).
(P2) J. H. Perry, *Chemical Engineers' Handbook*, 4th ed. New York: McGraw-Hill, Inc., 1963.
(R1) K. A. Reddy and L. K. Doraiswamy, *Ind. Eng. Chem. Fund.* **6**, 77 (1967).
(T1) R. E. Treybal, *Liquid Extraction*, 2d ed. New York: McGraw-Hill, Inc., 1963.
(T2) R. E. Treybal, *Mass-Transfer Operations*, 2d ed. New York: McGraw-Hill, Inc., 1968.
(V1) J. E. Vivian and C. J. King, *AIChEJ* **10**, 220 (1964).
(W1) C. R. Wilke and Pin Chang, *AIChEJ* **1**, 264 (1955).

4 Mass Transport Phenomena in Solids

4.1 STEADY-STATE MASS TRANSPORT EQUATIONS FOR SOLIDS

4.1A Introduction and General Types of Diffusion in Solids

The diffusion of gases, liquids, and solids in solids is quite important in mass transfer operations. For example, in processes such as leaching of solids, drying of solids, absorption and catalytic reaction in solid catalysts, separation of fluids by membranes, and treating of metals at high temperatures by gases, a fluid is diffusing through a solid matrix or porous solid.

In general we can broadly classify these processes into two types of diffusion: diffusion that is not a function of the actual solid structure, and diffusion in porous solids where the actual structure and interstices are important. These two types will be discussed below.

4-1B Diffusion That Does Not Depend on Structure

1. *Derivation of Equations.* This type of diffusion occurs when the fluid or solute diffusing actually dissolves in the solid to form a homogeneous solution—for example, in leaching, where the solid contains a large amount of water and a solute is diffusing through this solution, or in the diffusion of zinc through copper, where solid solutions are formed. The diffusion of gaseous hydrogen and oxygen through rubber

can be classified here, since equations of similar type can be used to predict the diffusion rate. Even though the actual mechanism of the movement of the solute molecules is quite complex, simplified equations are generally used to describe the overall diffusion.

The equation used is Eq. (1.7-5), where the bulk flow term, even if present, is neglected.

$$N_A = -D_{AB}\frac{dc_A}{dz} \tag{4.1-1}$$

where D_{AB} is the diffusivity of A through solid B and usually is a constant independent of pressure for diffusion through solids. Integration of this between points 1 and 2 for a solid slab at steady-state is

$$N_A = \frac{D_{AB}(c_{A1} - c_{A2})}{z} \tag{4.1-2}$$

If the cross-sectional area A is not constant, then Eq. (4.1-3) must be integrated, where A will be a function of z.

$$\frac{\bar{N}_A}{A} = -D_{AB}\frac{dc_A}{dz} \tag{4.1-3}$$

If diffusion is proceeding through a cylinder wall of inner radius r_1 and outer r_2 and length L, then

$$\frac{\bar{N}_A}{2\pi rL} = -D_{AB}\frac{dc_A}{dr} \tag{4.1-4}$$

On integration,

$$\bar{N}_A = D_{AB}(c_{A1} - c_{A2})\frac{2\pi L}{\ln(r_2/r_1)} \tag{4.1-5}$$

Example 4.1-1 Leakage of CO_2 through Rubber

It is desired to calculate the diffusion at 25°C of CO_2 at 2.0 atm pressure of pure CO_2 in a container which has a 2.0 cm² flat plug of vulcanized rubber closing an opening in the container. The rubber is 3.0 cm thick. From Barrer (B1) the solubility of the CO_2 gas is 0.90 cm³ gas (at N.T.P. of 0°C and 1 atm) per cm³ rubber per atm partial pressure of CO_2. The diffusivity is 0.11×10^{-5} cm²/sec which is assumed independent of pressure.

Solution

The solubility is proportional to the pressure as in Henry's law. So, solubility = $0.90(2) = 1.80$ cm³ CO_2/cm³ rubber measured at standard conditions of 0°C and 1 atm. The equilibrium concentration c_{A1} at the inside surface of the rubber is $1.80/22,414 = 8.05 \times 10^{-5}$ g mole CO_2/cm^3 rubber. The concentration at the outer surface $c_{A2} = 0$ if it is assumed that the resistance to diffusion in the outside is negligible and the pressure of CO_2 outside is zero. Substituting into Eq. (4.1-2),

$$N_A(A) = \frac{2.0(0.11 \times 10^{-5})(8.05 \times 10^{-5} - 0)}{3.0}$$

$$= 5.90 \times 10^{-11} \text{ g mole } CO_2/\text{sec}$$

2. *Experimental Diffusivities.* At present the knowledge of the theory of the solid state is not sufficient to predict the diffusivities in solids. Several theories have predicted that the log of the diffusivity should be a linear function of $1/T$ or that the energy of activation is a constant for certain solids and metals. However, the actual values predicted are off several or more orders of magnitude. Hence, generally experimental values are used. The diffusivity is often dependent markedly on concentration. Often the data given do not give the concentration range over which the experimental value was obtained and is valid. Hence, care must be exercised in using the data.

Experimental data are tabulated for a few systems in Table 4.1-1. Systems included are gases diffusing in solids and solids diffusing in solids. The diffusivities in the solids can be assumed to be independent of pressure.

The effect of temperature has been shown in many cases experimentally to follow the equation

$$D_{AB} = D_0 e^{-E/RT} \quad (4.1\text{-}6)$$

where D_0 is a constant and E is the activation energy in cal/g mole. For example, for Bi diffusing in Pb the equation is

$$D_{AB} = 7.7(10^{-3})e^{-18,600/RT}$$

A plot of $\ln D_{AB}$ versus $1/T$ gives a straight line, and the slope is related to E. Often, if only two experimental points at two temperatures are available, values at other temperatures can be estimated this way.

Example 4.1-2 Temperature Effect on Diffusivity

Calculate the diffusivity for Hg in Pb at 100°C when $D_{AB} = 3.6 (10^{-1})e^{-19,000/RT}$.

Solution

Substituting $T = 373°K$ and $R = 1.987$ cal/g mole-°K into the equation above, $D_{AB} = 6.9 \times 10^{-12}$ cm^2/sec.

Table 4.1-1 Diffusivities in Solids

Solute (A)	Solid (B)	T, °C	D_{AB}, Diffus. Coeff., cm²/sec	Solubility, S, cc solute at N.T.P. cc solid-atm	Permeability, P_M, cc solute at N.T.P. sec-cm² C.S.-cm Hg/mm
H_2	Vulc. rubber	25	0.85 (10^{-5})	0.040	0.045 (10^{-6})
O_2		25	0.21 (10^{-5})	0.070	0.020 (10^{-6})
N_2		25	0.15 (10^{-5})	0.035	0.0071 (10^{-6})
CO_2		25	0.11 (10^{-5})	0.90	0.132 (10^{-6})
H_2	Vulc. neoprene	0	0.037 (10^{-5})	0.065	
		17	0.103 (10^{-5})	0.051	
		27	0.180 (10^{-5})	0.053	
		46.5	0.481 (10^{-5})	0.050	
Air	Newspaper	25			0.047
Air	English leather	25			0.02–0.09
H_2O	Wax	23			0.021 (10^{-6})
H_2O	Cellophane	38			0.12–0.24 (10^{-6})
He	Pyrex glass	0			0.37 (10^{-11})
		20			0.64 (10^{-11})
		100			2.64 (10^{-11})
Air	Porcelain	25			0.106 (10^{-11})
He	SiO_2	20	2.4–5.5 (10^{-10})	0.01	
H_2	Ni	85	1.16 (10^{-8})	0.202	
		125	3.4 (10^{-8})	0.194	
		165	10.5 (10^{-8})	0.192	
H_2	Fe	20	2.59 (10^{-9})		
CO	Ni	950	4.0 (10^{-8})		
Bi	Pb	20	1.1 (10^{-16})		
Cd	Cu	20	2.7 (10^{-15})		
Al	Cu	20	1.3 (10^{-30})		

SOURCE: Data from Barrer (B1), pp. 136, 137, 141, 222, 223, 229, 275, 409, 410, 417, 418, 420, 441, 443.

3. *Diffusivities and Permeabilities in Solids.* Often the data for diffusion of gases in solids are not given as diffusivities and solubilities but as permeabilities, P_M, in cm³ of gas at N.T.P. (0°C and 1 atm press) diffusing per second per cm² cross-sectional area through a solid 1.0 mm thick under a pressure difference of 1 cm Hg. Some data are given in Table 4.1-1. This permeability can be related to the diffusivity as follows. The experiment is done with 1 cm Hg partial pressure of gas A on one side and zero partial pressure on the other side. Hence, in Eq. (4.1-2), $c_{A2} = 0$ at the outside. The value of c_{A1} is the concentration at the inside face, which is the concentration in the solid at the interface in equilibrium with the gas phase at 1 cm Hg pressure. The solubility S in the solid is directly proportional to the pressure by Henry's law. So, for 1 cm Hg pressure or $\frac{1}{76}$ atm the concentration c_{A1} in g mole A/cm³ is

$$c_{A1} = \frac{S \text{ cm}^3 \text{ gas}}{\text{cm}^3 \text{ solid-atm}} \left(\frac{1}{22,414 \frac{\text{cm}^3}{\text{g mole}}} \right) (\tfrac{1}{76} \text{ atm}) \qquad (4.1\text{-}7)$$

In the permeability measurement, the $z = 0.1$ cm. Also, N_A is related to the permeability constant P_M as follows, since P_M is cm³ at N.T.P. per sec per cm².

$$N_{AM} = \frac{P_M}{22,414} \qquad (4.1\text{-}8)$$

Note that N_{AM} is the flux only for 1 cm Hg pressure difference, and 0.10 cm thickness. Combining Eqs. (4.1-2), (4.1-7), and (4.1-8),

$$N_{AM} = \frac{P_M}{22,414} = \frac{D_{AB}}{0.10}\left(\frac{S}{22,414(76)} - 0\right) \qquad (4.1\text{-}9)$$

Hence,

$$D_{AB} = \frac{P_M L}{\Delta p_A S} \qquad (4.1\text{-}10)$$

where D_{AB} is constant and not dependent on p_A, L is thickness in cm used to determine P_M, Δp_A is pressure difference in atm used to determine P_M, and S solubility at 1 atm. Note that if the permeability P'_M were determined for a pressure difference of 1.0 atm and 1.0 cm thickness, then $P'_M = D_{AB} S$. Also, $P'_M = 7.60 P_M$. To calculate the flux N_A for any thickness and any Δp_A from the permeability at steady-state,

$$N_A = \frac{P_M}{22,414}\left(\frac{0.1}{L}\right)(76 \Delta p_A) \qquad (4.1\text{-}11)$$

Example 4.1-3 *Relation between Permeability and Diffusivity*

Hydrogen is diffusing through vulcanized rubber 2.0 cm thick. The partial pressure of hydrogen inside is 1.5 atm and 0 outside. Using the data from Table 4.1-1, calculate (a) diffusivity, D_{AB}, from the permeability

P_M and solubility S, and compare with the value given in Table 4.1-1 at 25°C; (b) the flux N_A of H_2 at 1.5 atm.

Solution

Using Eq. (4.1-10) the Δp_A is $\frac{1}{76}$ atm and $L = 0.1$ cm for the given P_M.

$$D_{AB} = \frac{P_M L}{\Delta p_A S} = \frac{0.045(10^{-6})(0.10)}{(\frac{1}{76})(0.040)}$$

$$= 0.85 \times 10^{-5} \text{ cm}^2/\text{sec} \quad \text{[Ans. to (a)]}$$

This is the same value for D_{AB} as in Table 4.1-1.

To calculate the flux at 1.50 atm for a solubility S of 0.040 from Eq. (4.1-7),

$$c_{A1} = \frac{0.040(1.50)}{22,414} = 2.68 \times 10^{-6} \text{ g mole } A/\text{cm}^3 \text{ rubber}$$

$c_{A2} = 0$ outside.

Using Eq. (4.1-2),

$$N_A = \frac{D_{AB}(c_{A1} - c_{A2})}{z} = \frac{0.85 \times 10^{-5}(2.68 \times 10^{-6} - 0)}{2.0}$$

$$= 1.14 \times 10^{-11} \text{ g mole } A/\text{sec-cm}^2$$

Instead of using Eq. (4.1-2), we could have calculated the flux N_A directly from Eq. (4.1-11) using P_M.

$$N_A = \frac{0.045(10^{-6})}{22,414}\left(\frac{0.1}{2.0}\right)(76)(1.5-0)$$

$$= 1.14 \times 10^{-11} \text{ g mole } A/\text{sec-cm}^2 \quad \text{[Ans. to (b)]}$$

4. *Classification of Solids by Type of Permeability.* The solids in Table 4.1-1 can be grouped into four general classes or groups according to their physical structure and the type of permeability occurring of gases through the solids.

(a) *Glasses and crystals.* In structures or solids like SiO_2 or Pyrex glass the holes are very small—in the angstrom range. The diffusion occurs by complicated processes of Knudsen diffusion, grain boundary flow, and surface flow. The experimental measurements for most gases usually have partial pressure differences of 1 cm Hg up to 1 atm. But for H_2 it was done up to 800 atm. Equations (4.1-2) to (4.1-11) hold. One should not use the experimental D_{AB} or P_M data to predict diffusion above 1 or 2 atm unless they have actually been measured at high pressures, such as H_2 gas. The D_{AB} is constant with pressure in these ranges.

(b) *Metals.* Metals such as Ni, Cd, and so on, where gases such as H_2 and He are diffusing in them, are in this group. Measurements have been made up to 112 atm or more. Equation (4.1-6) holds. However,

Eq. (4.1-10) says P_M is proportional to Δp_A. In this group it is found that P_M is proportional to $\sqrt{\Delta p_A}$. Hence, if the permeability is given, the value of Δp_A actually used in the original experiment must be known in order to predict the P_M at another pressure. However, one can use Eq. (4.1-2) with D_{AB} and solubility if the solubility is known at the given pressure. The solubility does not necessarily follow Henry's law. In Table 4.1-1 for Ni and SiO_2 the solubilities and D_{AB} values are given only for 1 atm pressure.

(c) *Organic solids.* In this group are waxes, rubbers, cellulose, and some papers. The diffusion up to about 5 atm absolute partial pressure with pressure differences below 5 atm follows Henry's law and Eqs. (4.1-2) to (4.1-11). The diffusion of one gas, say H_2, is independent of the other gases present, such as O_2 and N_2. Hence, the partial pressures of H_2 are used in the Δp_A.

(d) *Coarse membranes.* This includes fiberboard, some papers, and leathers. In many cases they obey Eqs. (4.1-2) to (4.1-11), and in some the permeability is not quite directly proportional to Δp_A. In these cases one must use the permeability values with caution when trying to predict flows. The exact conditions under which the experimental values were determined must then be known.

Additional experimental data for permeation in polymers are available in a monograph by Crank and Park (C4) who show that there is a substantial structure or morphology contribution to rates of diffusion.

4.2 DIFFUSION IN POROUS SOLIDS THAT DEPENDS ON STRUCTURE

4.2A Introduction and Diffusion of Liquids

In Section 4.1B, where the diffusion was through solids that did not depend on structure, we used Fick's law and treated the solid as a uniform material. We were not concerned with the actual structure but used an effective diffusivity D_{AB}.

In this section we are concerned with porous solids that have interconnected voids or pores in the solid. The diffusion is greatly affected by the size and type of voids. Figure 4.2-1 shows a sketch of a cross section of such a porous solid.

Suppose the voids are filled with liquid water and a concentration of salt in water at boundary 1 is c_{A1} and at point 2 is 0. Then salt diffuses through the water to point 2. The salt must take a very tortuous path, which is unknown and is greater than $z_2 - z_1$ by a factor k_t called tortuosity. This k_t must be determined experimentally. The equation can be written using Eq. (3.3-1) for diffusion of salt in water as follows at steady-state:

$$N_A = \frac{\epsilon}{k_t^2} D_{AB} \frac{c_{A1} - c_{A2}}{z_2 - z_1} \quad (4.2\text{-}1)$$

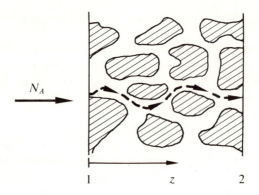

FIGURE 4.2-1 Sketch of a typical porous solid

where ϵ is the open void fraction and D_{AB} is the diffusivity of salt (A) in water (B). The one k_t corrects for the path longer than $z_2 - z_1$, and the second k_t corrects for the fact that many openings or pores at the surface are at an angle with the surface. The experimental values of k_t vary from 1.2 to 2.5. Often the terms are lumped together, so

$$D_{A_{\text{effec}}} = \frac{\epsilon}{k_t^2} D_{AB} \tag{4.2-2}$$

Example 4.2-1 Diffusion of a Liquid in a Porous Solid

A porous sintered silica solid filled with water at 25°C is 0.3 cm thick. At one face the concentration of KCl is held constant at 0.10 g mole/liter. Fresh running water flows rapidly by the other face. What is the diffusion rate? The ϵ is 0.30 and k_t is 2.1.

Solution

The diffusivity of KCl is 1.87×10^{-5} cm²/sec. The value of $c_{A1} = 0.10/1000 = 1.0 \times 10^{-4}$ g mole/cm³. $c_{A2} = 0$. By Eq. (4.2-1),

$$N_A = \frac{\epsilon}{k_t^2} D_{AB} \frac{c_{A1} - c_{A2}}{z_2 - z_1} = \frac{0.30}{(2.1)^2} \frac{(1.87 \times 10^{-5})(1.0 \times 10^{-4} - 0)}{(0.30)}$$
$$= 4.23 \times 10^{-10} \text{ g mole KCl/sec-cm}^2$$

Equations (4.2-1) and (4.2-2) also hold for molecular diffusion of gases through the porous structure if the pores are quite large. This will be discussed in more detail in the following sections.

4.2B Diffusion of Gases in Pores

Since the pores of porous solids are often very small, the diffusion of gases may depend on the diameter of the pores or capillaries. Hence, different mechanisms of diffusion can occur. We first define a mean free path, λ

$$\lambda = \frac{3.2\mu}{P}\sqrt{\frac{RT}{2\pi g_c M}} \tag{4.2-3}$$

where λ is mean free path in cm, μ is viscosity in poises, P is pressure in g mass/cm^2 = atm (1033.2), T is temperature in °K, $g_c = 980$ g mass-cm/g force-sec^2, M = molecular weight g mass/g mole, and $R = 84{,}780$ g force-cm/°K-g mole. The mean free path is the average distance a molecule travels before it collides with another molecule. Next we will consider the different types of diffusion that can occur in the pores and are dependent on pore size and λ. It should be emphasized that the total pressure on either side in these cases is the same, but partial pressures of A and B may be different.

1. *Knudsen Diffusion.* In Fig. 4.2-2 a gas molecule A is shown diffusing through a capillary of d diameter. The total pressure P on both sides is constant, but $p_{A1} > p_{A2}$. In Fig. 4.2-2a the mean free path is very

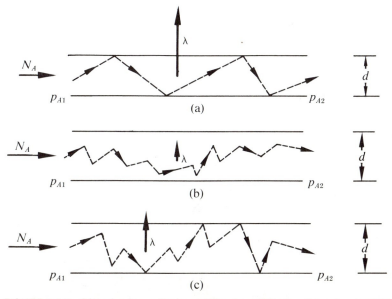

FIGURE 4.2-2 Diffusion in capillaries: (a) Knudsen diffusion, (b) molecular (Fick's) diffusion, (c) transition diffusion

large compared to the diameter d. Hence, the molecule collides with the walls, and the diameter d is important. This is called Knudsen diffusion; the Knudsen diffusivity is independent of pressure and is calculated by

$$D_{KA} = 9.7(10^{-3})\bar{r}\sqrt{\frac{T\,°K}{M_A}} \qquad (4.2\text{-}4)$$

where $\bar{r}$ is the radius in cm of the pore and D_{KA} the diffusivity of A in cm²/sec. The flux equation for a length L cm at steady-state is

$$N_A = -D_{KA}\frac{dc_A}{dz} = \frac{-D_{KA}}{RT}\frac{dp_A}{dz} = \frac{D_{KA}P}{RTL}(x_{A1} - x_{A2}) \qquad (4.2\text{-}5)$$

The diffusion of A is completely independent of B, since A collides with the walls and not with B.

When the Knudsen number N_{Kn} defined as $\lambda/2r \geqslant 10/1$, then the diffusion is primarily Knudsen, and Eq. (4.2-5) will predict the diffusion to within about 10 percent error. As the Knudsen number gets larger, this error decreases, since the diffusion approaches more closely the Knudsen type.

If two components A and B are diffusing, inspection of Eqs. (4.2-4) and (4.2-5) shows that

$$N_A \propto \frac{1}{\sqrt{M_A}} \qquad (4.2\text{-}6)$$

2. *Molecular Diffusion.* In Fig. 4.2-2b, where the pore diameter is very large compared to the mean free path λ or where $N_{Kn} \leqslant \frac{1}{100}$, then molecular-type diffusion predominates. The molecule-to-molecule collisions are the most important, the molecule-to-wall collisions have practically no effect. The ordinary law of molecular diffusion (Fickian type) holds, and for $N_{Kn} \cong \frac{1}{100}$ will predict the diffusion to within about 10 percent. As the Knudsen number gets smaller, the error becomes less and less, because any effect of molecule-to-wall collisions becomes more insignificant.

The equation for molecular diffusion that has been derived in Chapters 1 and 2 is for one-dimensional-diffusion:

$$N_A = \frac{-D_{AB}P\,dx_A}{RT\,dz} + x_A(N_A + N_B) \qquad (4.2\text{-}7)$$

Defining a flux ratio factor α as

$$\alpha = 1 + \frac{N_B}{N_A} \qquad (4.2\text{-}8)$$

we can combine Eqs. (4.2-7) and (4.2-8) as follows:

$$N_A = \frac{-D_{AB}P\,dx_A}{(1-\alpha x_A)RT\,dz} = \frac{-D_{AB}\,dp_A}{(1-\alpha x_A)RT\,dz} \qquad (4.2\text{-}9)$$

4.2 Diffusion in Porous Solids That Depends on Structure

We should remember that N_B can be the negative of N_A in using α in Eq. (4.2-9). If the diffusion is equimolar, $N_B = -N_A$ and $\alpha = 0$. Integration of Eq. (4.2-9) gives for steady-state,

$$N_A = \frac{D_{AB}P}{\alpha RTL} \ln\left(\frac{1-\alpha x_{A2}}{1-\alpha x_{A1}}\right) \tag{4.2-10}$$

3. *Transition Region Diffusion.* When the mean free path λ and the diameter $2\bar{r}$ are intermediate in size between the two limits above for Knudsen or molecular diffusion, then transition or mixed type diffusion occurs as shown in Fig. 4.2-2c. In this diffusion regime, both molecule-to-molecule and molecule-to-wall collisions are important in the diffusion of the molecule being considered.

The transition-region diffusion equation can be derived in several ways. It will be derived below using a momentum balance in a manner similar to that of Rothfeld (R1) and Scott and Dullien (S1). For molecule-to-wall collisions of the component A in diffusing the length of the capillary, Eq. (4.2-5) can be solved for dp_A and multiplied by $g_c A$ to give the loss in momentum in length dz due to molecule-to-wall collisions:

$$-(dp_A)_{\text{Knud}}(g_c A) = N_A \frac{RT}{D_{KA}} dz(g_c A) \tag{4.2-11}$$

where A is the cross-sectional area of the capillary in cm^2.

Equation (4.2-9) can also be multiplied by $g_c A$ to give the momentum loss due to molecule-to-molecule collisions:

$$-(dp_A)_{\text{Molec}}(g_c A) = N_A \frac{RT}{D_{AB}}(1-\alpha x_A) dz(g_c A) \tag{4.2-12}$$

Now the total momentum loss due to both types of collisions is the sum of Eqs. (4.2-11) and (4.2-12), or

$$-dp_A(g_c A) = N_A \frac{RT}{D_{KA}} dz(g_c A) + N_A \frac{RT}{D_{AB}}(1-\alpha x_A) dz(g_c A) \tag{4.2-13}$$

Rearranging,

$$N_A = -\left[\frac{1}{(1-\alpha x_A)/D_{AB} + 1/D_{KA}}\right] \frac{P \, dx_A}{RT \, dz} \tag{4.2-14}$$

One can consider the bracketed terms to be a transition region diffusivity, D_{NA}, which depends slightly on concentration, x_A. Integrating Eq. (4.2-14) between the limits $z = 0$, $x_A = x_{A1}$ and $z = L$, $x_A = x_{A2}$, at steady-state,

$$N_A = \frac{D_{AB}P}{\alpha RTL} \ln\left(\frac{1-\alpha x_{A2} + D_{AB}/D_{KA}}{1-\alpha x_{A1} + D_{AB}/D_{KA}}\right) \tag{4.2-15}$$

An equation similar to the above can also be written for B. This derivation assumes that the viscous shear between the gas and the wall is negligible.

154 Mass Transport Phenomena in Solids

The reader should note that since the molecules concerned are relatively far apart, the gas or fluid is not considered a continuum; hence the continuum fluid-dynamics equations do not apply here.

Equation (4.2-15) is valid over the entire transition region, so it should reduce to the pure Knudsen diffusion equation at low pressures and to pure molecular diffusion at higher pressures. Taking Eq. (4.2-14) and multiplying the numerator and denominator by D_{KA},

$$N_A = -\frac{1}{(1-\alpha x_A)D_{KA}/D_{AB}+1} D_{KA} \frac{P\,dx_A}{RT\,dz} \qquad (4.2\text{-}16)$$

Now D_{KA} is independent of pressure by Eq. (4.2-4), and D_{AB} is proportional to $1/P$. Hence, $D_{KA}/D_{AB} \propto P$. At low pressures, then, $D_{KA}/D_{AB} \ll 1$, and Eq. (4.2-16) becomes Eq. (4.2-5) as expected. At high pressures, $D_{KA}/D_{AB} \gg 1$, and Eq. (4.2-16) becomes Eq. (4.2-9) for molecular diffusion.

Example 4.2-2 Knudsen and Transition Region Diffusion

A nitrogen (A) and helium gas (B) mixture at 25°C is diffusing through a capillary in an open system 10.0 cm long with a diameter of 5 microns. In this open system the gases flow by both ends of the capillary. The mole fractions are constant at point 2 of $x_{A2} = 0.2$, and at point 1, $x_{A1} = 0.8$. For an open system the flux ratio $N_A/N_B = -\sqrt{M_B/M_A}$. Viscosity of $N_2 = 0.018$ cp. Also D_{AB} molecular is 0.698 cm²/sec at 1.0 atm, which is an average value. (a) Calculate λ for total pressures of 0.001, 0.01, and 10.0 atm for pure N_2 at these pressures. Also calculate the ratio $\lambda/2\bar{r}$ or Knudsen number. (b) Calculate D_{KA}. (c) Calculate N_A for the gas mixture, using the transition region equation at steady-state.

Solution

For Eq. (4.2-3), $\mu = 0.00018$ poise, and $P = 0.001$ atm.

$$\lambda = \frac{3.2\mu}{P}\sqrt{\frac{RT}{2\pi g_c M}} = \frac{3.2(0.00018)}{0.001(1033.3)}\sqrt{\frac{84,780(298)}{2\pi(980)(28.02)}}$$

$$= 6.75 \times 10^{-3} \text{ cm or } 67.5 \text{ microns}$$

The Knudsen number is $\lambda/2\bar{r} = 67.5/5.0 = 13.5/1$. Hence the diffusion is predominantly Knudsen. For 0.01 atm, $\lambda = 6.75$ microns, $\lambda/2\bar{r} = 1.35/1$. Hence, the diffusion is transition. For 10.0 atm, $\lambda/2\bar{r} = 0.00135/1$. Hence, the diffusion is mainly molecular [Ans. to (a)].

To calculate D_{KA}, Eq. (4.2-4) gives

$$D_{KA} = 9.7(10^3)\bar{r}\sqrt{\frac{T}{M}} = 9.7(10^3)\frac{5(10^{-4})}{2}\sqrt{\frac{298}{28.02}}$$

$$= 7.90 \text{ cm}^2/\text{sec at all pressures} \qquad [\text{Ans. to (b)}]$$

To calculate the flux, the D_{AB} for N_2-He at 1 atm and 25°C is 0.698 cm²/sec. At 0.001 atm, $D_{AB} = 0.698/0.001$. The α is $1 + N_B/N_A$.

$$\frac{N_B}{N_A} = -\sqrt{\frac{M_A}{M_B}} = -\sqrt{\frac{28.02}{4.003}} = -2.645$$

Now N_B and N_A are fluxes in opposite directions, so $\alpha = 1 - 2.645 = -1.645$. Substituting into Eq. (4.2-15),

$$N_A = \frac{(0.698/0.001)(0.001)}{(-1.645)(82.06)(298)(10)} \ln\left(\frac{1 + 1.645(0.2) + (0.698/0.001)/7.90}{1 + 1.645(0.8) + (0.698/0.001)/7.90}\right)$$

$$= 1.89 \times 10^{-8} \text{ g mole } A/\text{sec-cm}^2 \text{ at } P = 0.001 \text{ atm}$$

Using Eq. (4.2-5) would have given an answer close to the above. For 0.01 atm, using Eq. (4.2-15),

$$N_A = 1.62 \times 10^{-7} \text{ at } P = 0.01 \text{ atm}$$

For 10.0 atm, using Eq. (4.2-15),

$$N_A = 0.955 \times 10^{-6} \text{ at } P = 10.0 \text{ atm} \qquad [\text{Ans. to (c)}]$$

If a plot of Eq. (4.2-15) is made using constant values of x_{A1}, x_{A2}, and D_{KA} with total pressure P varying, then a characteristic curve is obtained where the flux increases with pressure. Using the data from Example 4.2-2, Eq. (4.2-15) is plotted on Fig. 4.2-3. The flux increases with P and

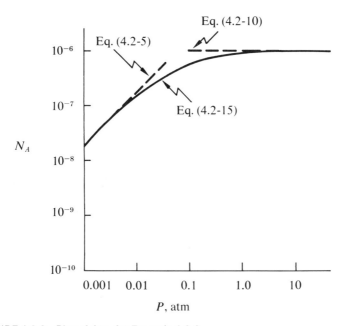

FIGURE 4.2-3 Plot of data for Example 4.2-2

levels off at high values of P. At low pressures the diffusion is essentially Knudsen and the slope approaches that of Eq. (4.2-5), which is a 45-degree line. At high pressures the transition region curve approaches Eq. (4.2-10) for molecular diffusion, which is a horizontal line. The transition region covers a fairly wide pressure range.

Using a smaller-diameter capillary, D_{KA} would be smaller and a line plotted on Fig. 4.2-3 would start out at lower values of N_A but be approximately parallel to the existing line at low pressures. This line at high pressures would asymptotically approach the existing horizontal line, since molecular diffusion is not affected by capillary radius.

4.2C   Flux Ratios for Diffusion of Gases

Dullien and Scott (D3) discuss in detail the kinds of systems that occur with diffusion of gases through channels or porous solids. They show that in molecular diffusion in a closed system with constant total pressure P on either side of the diffusion path, equimolar counterdiffusion must occur. This system is shown in Fig. 4.2-4a. This means

$$\sum_i N_i = 0 \qquad (4.2\text{-}17)$$

For two components, $N_A = -N_B$. The reason, of course, is that if one side had a greater total pressure, gas would flow to the other side.

Another system quite common in experiments to measure diffusivities in porous solids or capillaries is the open system shown in Fig. 4.2-4b. Here each of the two different gas streams flows by the end of the tube. Dullien and Scott (D3) and Rothfeld (R1) show that the fluxes must be related by the following for the Knudsen, molecular, and transition regions:

$$\sum_i N_i \sqrt{M_i} = 0 \qquad (4.2\text{-}18)$$

For a binary system this reduces to Eq. (4.2-19):

$$N_A \sqrt{M_A} = -N_B \sqrt{M_B} \qquad (4.2\text{-}19)$$

In some cases, however, the flux ratios are determined by other means. For example, if a chemical reaction occurs at one end of the diffusion path, then the flux ratios at steady-state are determined by the stoichiometry of the reaction.

4.2D   Multicomponent Diffusion in the Transition Region

Just as for multicomponent molecular diffusion in a closed system discussed in Chapters 1 and 2, multicomponent diffusion in the transition region in open systems is quite complex. Rothfeld (R1) and Scott and

4.2 Diffusion in Porous Solids That Depends on Structure

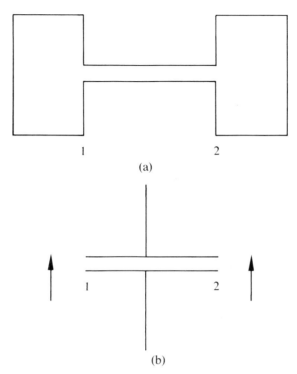

FIGURE 4.2-4 Types of diffusion systems: (a) closed system at constant total pressure, (b) open system at constant total pressure

Dullien (S1) used a momentum balance and derived the following for a multicomponent system:

$$\frac{-P\,dx_i}{RT\,dz} = \frac{N_i}{D_{Ki}} + \sum_j \frac{x_j N_i - x_i N_j}{D_{ij}} \qquad (4.2\text{-}20)$$

Cunningham and Geankoplis (C2) integrated the equation above for three components. They showed theoretically that a diffusion barrier could exist whereby the flux of a component can be zero even though a concentration gradient exists for this component and even though there are no actual physical barriers to diffusion. This seems to occur because of the interactions between the various diffusing species.

For the three component equations above, which have been integrated (C2), actual numerical values are difficult to obtain. The solutions are trial-and-error. Remick and Geankoplis (R2) obtained some numerical solutions for these equations and also found osmotic diffusion occurring where a component diffused with a zero concentration gradient. They

158   Mass Transport Phenomena in Solids

also proposed (R2) a simple method using two binary equations to bracket and quickly approximate the flux values for the three component equations.

4.2E   Forced Flow of Gases in Pores due to Total Pressure Differences

Unlike the cases in Section 4.2B, where Knudsen, molecular, or transition diffusion occurred and where the total pressure was the same on both ends of the pores, in this case the total pressures differ; hence, we will have forced or hydrodynamic flow. If the pore sizes in a porous solid or a single capillary are very large, then we can have ordinary Poiseuille flow. If the total pressure difference is quite large, then we could even have turbulent flow, which is not as common. In Poiseuille flow the flow is proportional to the total ΔP or pressure difference and not the partial pressure difference.

The ordinary Poiseuille or laminar flow equation is valid for a Reynolds number less than 2100. This equation can be converted to a form that includes N_A as follows.

The laminar flow equation in terms of ΔH in centimeters of fluid pressure drop is

$$\Delta H = \frac{4fv^2 L}{2g_c D} \quad (4.2\text{-}21)$$

where f is the Fanning friction factor and D the diameter. For laminar flow $f = 16\mu/(Dv\rho)$. Substituting this value for f into Eq. (4.2-21) and also noting that $\Delta H = (p_1 - p_2)/\rho$,

$$(p_1 - p_2) = \frac{32vL\mu}{g_c D^2} \quad (4.2\text{-}22)$$

Since $v = N_A RT/p_{\text{ave}}$, Eq. (4.2-22) becomes, after substitution for v and rearrangement,

$$N_A = \frac{\bar{r}^2 g_c p_{\text{ave}} (p_1 - p_2)}{8\mu LRT} \quad (4.2\text{-}23)$$

where $p_{\text{ave}} = (p_1 + p_2)/2$. If the flow is through a porous solid, Eq. (4.2-23) is modified by multiplying it by the porosity and a tortuosity correction factor as in Eq. (4.2-1). Equation (4.2-23) holds for a pure gas A.

Inspection of Eq. (4.2-23) shows that as the temperature is increased, the viscosity of the gas also increases and the flow decreases if in Poiseuille flow. However, if diffusion also occurs in solid solution, the solubility is affected by temperature, and the overall affect may be complex.

If the pores are very small, then the gas molecules hit the wall and behave as though in the Knudsen region. Then each gas acts independently, and the "flows" of A and of B are calculated by the Knudsen diffusion equations (4.2-4) and (4.2-5).

If partial pressure differences of A exist at both ends of the capillary along with total pressure differences, then forced flow and diffusion occur simultaneously. Such complex cases are beyond the scope of this text.

4.2F Diffusion and Reaction in Porous Solids

1. *Introduction.* For diffusion in porous catalytic type solids, such as silica-alumina, various physical models have been worked out to calculate the pore radii and the effective path lengths. See Wakao and Smith (W1), Henry, Cunningham, and Geankoplis (H1), and Cunningham and Geankoplis (C1) for a summary.

Catalytic reactions in porous solids are often used to speed up chemical reactions. We can visualize such a solid as in Fig. 4.2-1, where there are many pores inside the solid of irregular shape. For a reactant A in a bulk gas stream to react, it must first diffuse from the main gas stream to the mouth of a pore. Then it diffuses in the pore, where it finally reacts on the catalyst surface. The process is then reversed as the product diffuses out of the pore and into the main gas stream. Other phenomena also occur, such as adsorption on the solid surface and surface layer flow, which will not be considered here, since these topics belong in advanced treatments. First we will consider the diffusion and reaction inside a single pore.

2. *Diffusion and Reaction in a Pore.* We consider an idealized single cylindrical pore as given in Fig. 4.2-5. The single pore extends a distance of L cm into the catalyst pellet. The solute A diffuses from the bulk gas stream at a concentration c_{Ag} through the gas film, where its concentration is c_{AS} at the pore mouth. Some reacts on the exterior catalyst surface, which we will assume is negligible here, and the rest diffuses into the pore to react.

Assuming for simplicity a first-order irreversible reaction of

$$A \rightarrow B$$

and also the rate equation of

$$-\frac{1}{S}\frac{dN'_A}{dt} = k_S c_A \tag{4.2-24}$$

where k_S is the reaction velocity constant in cm/sec on the catalyst surface S cm^2 and N'_A is g mole A diffusing. Next we define an "effective diffusivity" D_A for species A in the catalyst pore. This diffusivity is a complicated function of the pore geometry and structure and often must be measured experimentally or predicted. It contains the Knudsen and molecular diffusivity for the transition region diffusion mentioned in Section 4.2B of this chapter.

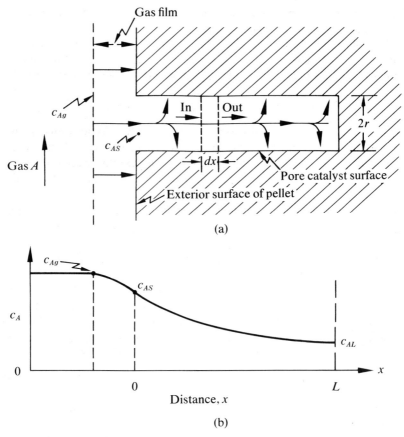

FIGURE 4.2-5 Idealized model for diffusion and reaction in a single pore: (a) pore diffusion path, (b) concentration versus distance plot

Making a mole balance on A for an element of the pore dx long,

(rate of input) − (rate of output) + (rate of generation)

$$= 0 \text{ (no accumulation)}. \quad (4.2\text{-}25)$$

The input occurs by diffusion into the element. Thus Eq. (4.2-25) becomes for steady-state

$$-\left[\frac{d}{dx}\left(\frac{-D_A dc_A}{dx}\right)\right](\pi r^2) - k_S c_A (2\pi r) = 0 \quad (4.2\text{-}26)$$

Rearranging,

$$\frac{d^2 c_A}{dx^2} - \frac{2k_S}{D_A r} c_A = 0 \quad (4.2\text{-}27)$$

4.2 Diffusion in Porous Solids That Depends on Structure

Often the reaction velocity constant is expressed in terms of a catalyst volume or as k, rather than as k_S for the surface. These are related by the following for a cylindrical pore:

$$k = \frac{2}{r} k_S \qquad (4.2\text{-}28)$$

The boundary conditions for Eq. (4.2-27) are

B.C. 1: at $x = 0$, $c_A = c_{AS}$

B.C. 2: at $x = L$, $\dfrac{dc_A}{dx} = 0$ (no diffusion) $\qquad (4.2\text{-}29)$

Upon integration Eq. (4.2-27) becomes

$$c_A = \frac{c_{AS} \cosh m(L-x)}{\cosh mL} \qquad (4.2\text{-}30)$$

where $m = \sqrt{2k_S/D_A r} = \sqrt{k/D_A}$. This equation relates the concentration c_A to distance x, as shown in Fig. 4.2-5b.

If the pore has no resistance to diffusion, this would be equivalent to slicing open the pore and having all its surface available for reaction at the pore mouth. This means the reaction rate for no pore diffusion resistance would be $k_S c_{AS}$ instead of $k_S c_A$ as in Eq. (4.2-24). If we obtain a mean value of $\bar{c}_A$ from Eq. (4.2-30), then $k_S \bar{c}_A$ is a measure of the reaction rate with pore diffusion resistance. Finally, the ratio $k_S \bar{c}_A / k_S c_{AS}$ is the ratio of the average reaction rate in a pore/the maximum reaction rate with pore diffusion absent. This ratio is termed the effectiveness factor, ϵ, and becomes

$$\epsilon = \frac{\tanh mL}{mL} \qquad (4.2\text{-}31)$$

A plot of this equation is shown in Fig. 4.2-6. It shows that for low values of mL, ϵ is close to 1 and pore diffusion resistance is small. A small value of $mL = \sqrt{k/D_A} L$ means either slow reaction (small k), fast diffusion (large D_A), or short pore (small L). At large values of mL pore diffusion resistance is quite important.

In Fig. 4.2-6 a line is also given for a spherical catalyst particle, where the mL on the abcissa is $mR/3$, where R is the radius of the particle, and for a cylindrical catalyst pellet, where mL is $mR/2$ with R the cylinder radius. Little difference is found between these geometries as discussed by Aris (A1) and shown in Fig. 4.2-6.

The total or average reaction rate inside the complete pore of Fig. 4.2-5 is then

$$-\frac{1}{S}\frac{dN'_A}{dt} = k_S \bar{c}_A = k_S c_{AS} \epsilon \qquad (4.2\text{-}32)$$

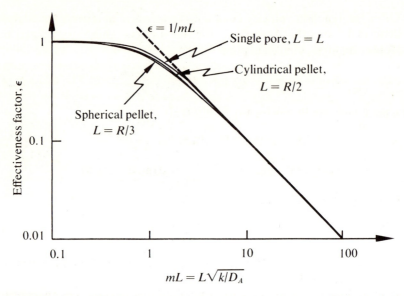

FIGURE 4.2-6 Effectiveness factor for catalyst particles of different shapes. (From O. Levenspiel, *Chemical Reaction Engineering* New York: John Wiley & Sons, Inc., 1962. With permission.)

or

$$-\frac{1}{\text{pore volume}} \frac{dN'_A}{dt} = kc_{AS}\epsilon \qquad (4.2\text{-}33)$$

Example 4.2-3 Diffusion in a Single Pore and Reaction

Gas A is diffusing from the main gas stream into a single catalyst pore as in Fig. 4.2-5. The pore length $L = 0.50$ cm, the reaction velocity constant k is 80 sec^{-1}, and the mean diffusivity $D_A = 0.2$ cm²/sec. Calculate the effectiveness factor.

Solution

Substituting into the equation for mL,

$$mL = \sqrt{\frac{k}{D_A}} L = \sqrt{\frac{80}{0.20}} (0.50) = 10.0$$

From Fig. 4.2-6, the effectiveness factor $\epsilon = 0.10$. Or, substituting into Eq. (4.2-31),

$$\epsilon = \frac{\tanh (10)}{10} = \frac{1.00}{10} = 0.10$$

3. *Overall Reaction Rates in a Catalyst Pellet.* To obtain the overall reaction rate in a catalyst pellet we must combine the equation for diffusion through the exterior of the pellet with the one for diffusion inside the pellet. First, for diffusion from the main gas stream at concentration c_{Ag} to the mouth of the pore where $c_A = c_{AS}$ (see Fig. 4.2-5),

$$-\frac{1}{S_{ex}}\frac{dN'_A}{dt} = k_g(c_{Ag} - c_{AS}) \tag{4.2-34}$$

where S_{ex} is the gross exterior surface area of the pellet in cm^2 and k_g is the gas phase mass transfer coefficient in cm/sec (see Section 6.2). We neglect the reaction rate on S_{ex}. For the diffusion and reaction in the catalyst pores,

$$-\frac{1}{V_p}\frac{dN'_A}{dt} = k c_{AS} \epsilon \tag{4.2-35}$$

where V_p is the catalyst particle volume and k is based on that volume. Combining Eqs. (4.2-34) and (4.2-35) to eliminate c_{AS},

$$-\frac{1}{V_p}\frac{dN'_A}{dt} = \frac{1}{1/k\epsilon + V_p/k_g S_{ex}} c_{Ag} \tag{4.2-36}$$

For a spherical particle of radius R, $V_p/S_{ex} = R/3$.

Hence, in Eq. (4.2-36) we have the sum of two resistances in the demoninator. The first is the pore diffusion and reaction resistance, the second the exterior mass transfer resistance. In catalytic processes either can be limiting. For further details see Levenspiel (L2) and Smith (S3) who discuss the various physical models used for pore diffusion combined with reaction inside the pores.

4.3 DIFFUSION AND REACTION IN SOLIDS

Often the diffusing solute in a solid reacts with the solid as it diffuses. This occurs in many industrial situations. For example, nitrogen gas diffuses in iron and reacts, forming metal nitrides. The gas H_2S diffuses into silver to form Ag_2S on the outside. Oxygen diffuses to iron and forms iron oxides on the outside, so another phase forms. A few cases will be considered here. For a more complete discussion, see Jost (J1).

4.3A Diffusion and Reaction in a Single Phase

In this case the solute diffuses into the solid and reacts with the solid. The product stays in solution in the solid, and a second phase does not form. This case is identical to the case for gases and liquids where A was dilute and diffused and reacted in the phase. Fick's law was combined with the reaction rate. Those equations hold here.

4.3B Diffusion and Reaction Where a Second Phase Forms

As an example of this situation, H_2S or other sulfur gases diffuse to the surface of the silver and Ag_2S forms at the surface; hence a second phase forms at the surface. Figure 4.3-1 shows the phase being formed as described by Jost (J1). The compound being formed as phase II has a larger volume than Ag, and a protective dense layer is formed. The sulfur must diffuse through this layer to attack the silver, and this is a slow process. We assume that the chemical reaction rate is very fast so that it does not limit the overall process. The limiting step is the diffusion. Hence, N_A is proportional to the rate of increase in thickness z of the layer, or $N_A \propto \partial z/\partial t$. But also from Fick's law, $N_A \propto 1/z$ for quasi-steady-state. Equating both N_A values,

$$N_A = k_1 \frac{\partial z}{\partial t} = k_2 \frac{1}{z} \qquad (4.3\text{-}1)$$

where k_1 and k_2 are constants. Hence,

$$\frac{\partial z}{\partial t} = \frac{k_3}{z} \qquad (4.3\text{-}2)$$

Integrating between 0 and z and 0 and t,

$$z^2 = 2k_3 t \qquad (4.3\text{-}3)$$

This is called the quadratic law. Hence, the layer gets thicker with time and slows down the process, or forms a "protective layer."

Another example where a layer forms outside but offers little protection involves an oxide formed with a smaller density than the original

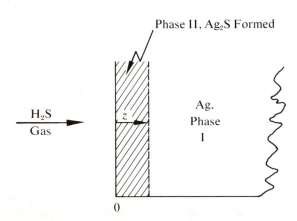

FIGURE 4.3-1 Tarnishing of silver

4.3 Diffusion and Reaction in Solids

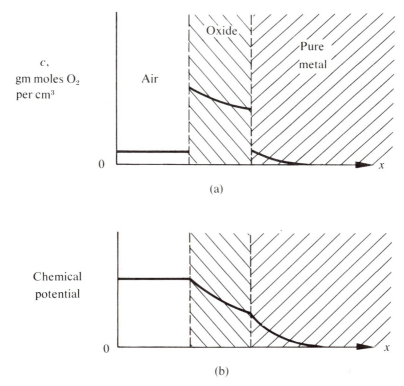

FIGURE 4.3-2 Oxidation of a metal: (a) concentration profile, (b) chemical potential profile

metal. Hence, large cracks or pores are formed and the diffusion is not slowed appreciably.

Often a protective oxide is formed outside a metal. Oxygen gas diffuses from the air to the metal surface, where the oxide is formed rapidly. As diffusion proceeds, the oxide layer gets thicker, as in the tarnishing of silver. In many cases the rate-determining step (the slowest step in the overall process) is the diffusion. Figure 4.3-2 shows the phases present during this oxidation and diffusion process (S2).

The concentration profile in Fig. 4.3-2a shows abrupt changes at the interfaces. At each interface the concentrations in the two phases are in equilibrium. They are related by the distribution coefficient. If a plot of chemical potential versus distance is made as in Fig. 4.3-2b, the curve is continuous and does not show abrupt changes at the interface. This occurs because at the interface both chemical potentials are equal to each other.

4.4 STEADY-STATE DIFFUSION IN TWO DIMENSIONS

4.4A   Derivation of Basic Equation and Analytical Solution

For no generation or convection Eq. (2.6-15) was given in Chapter 2 for unsteady-state transport of mass in three directions and can be written as follows:

$$\frac{\partial c_A}{\partial \theta} = D_{AB}\left(\frac{\partial^2 c_A}{\partial x^2} + \frac{\partial^2 c_A}{\partial y^2} + \frac{\partial^2 c_A}{\partial z^2}\right) \quad (4.4\text{-}1)$$

When steady-state diffusion is occurring in only one direction,

$$\frac{\partial^2 c_A}{\partial x^2} = 0 \quad (4.4\text{-}2)$$

This, of course, can be integrated, and it is really Fick's first law.

When diffusion is occurring at steady-state in two directions, then we obtain the celebrated Laplace equation:

$$\frac{\partial^2 c_A}{\partial x^2} + \frac{\partial^2 c_A}{\partial y^2} = 0 \quad (4.4\text{-}3)$$

Equation (4.4-3) can also be obtained from the general Eq. (2.6-12). This is a very useful equation for diffusion in two directions.

There are a number of methods to solve Laplace's equation for steady-state diffusion. It can be solved analytically using the method of separation of variables, which gives the result as an infinite Fourier series. This method will be discussed very briefly here to indicate the nature of the results. In the next chapter the separation-of-variables method will be discussed in detail.

Figure 4.4-1 shows a three-dimensional rectangular slab that is impermeable to diffusion in the z direction — that is, the two parallel z faces are insulated. The slab has a length $y \to \infty$ and a width $x = L$. Diffusion occurs in the x and y directions only in this semi-infinite strip. The following equations hold, where c is used in place of c_A:

$$\frac{\partial^2 c}{\partial x^2} + \frac{\partial^2 c}{\partial y^2} = 0 \quad (4.4\text{-}4)$$

B.C. 1: $c = 0$ at $x = 0$
B.C. 2: $c = 0$ at $x = L = \pi$
B.C. 3: $c = 0$ at $y = \infty$ (4.4-5)
I.C.: $c = f(x) = 1.0$ at $y = 0$, $0 < x < L$

Hence, the concentration c is 1.0 at the face where $y = 0$ and is 0 at the other two faces and a large distance away at $y = \infty$. To solve this we assume that c is a function of the product of two separate functions

4.4 Steady-State Diffusion in Two Dimensions

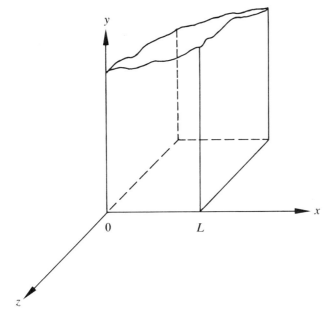

FIGURE 4.4-1 Rectangular slab with diffusion in the x and y directions

$X(x)$ and $Y(y)$, where $X(x)$ is a function only of x and $Y(y)$ a function of y:

$$c(x, y) = X(x)Y(y) \qquad (4.4\text{-}6)$$

Differentiating Eq. (4.4-6) twice with respect to x and also twice with respect to y and substituting the results into Eq. (4.4-4),

$$\frac{1}{X}\frac{\partial^2 X}{\partial x^2} = -\frac{1}{Y}\frac{\partial^2 Y}{\partial y^2} \qquad (4.4\text{-}7)$$

Since both sides are independent of each other, they can be set equal to a constant. This then gives two ordinary differential equations, which can be integrated and substituted back into Eq. (4.4-6). The final result is an infinite Fourier series:

$$c(x, y) = \frac{4}{\pi}\left(\frac{e^{-1y}}{1}\sin 1x + \frac{e^{-3y}}{3}\sin 3x + \frac{e^{-5y}}{5}\sin 5x + \cdots\right) \qquad (4.4\text{-}8)$$

With the equation above we can calculate c at any position $0 < x < \pi$ and $0 < y < +\infty$.

Mickley et al. (M1, p. 265) show how to solve the equation below for diffusion in three directions:

$$\frac{\partial^2 c_A}{\partial x^2} + \frac{\partial^2 c_A}{\partial y^2} + \frac{\partial^2 c_A}{\partial z^2} = 0 \qquad (4.4\text{-}9)$$

The result is a double series of orthogonal functions in a Fourier series form.

In general, many steady-state problems for diffusion in two and three dimensions cannot be solved analytically. The boundary conditions may not be constant but vary with position, and/or the geometry may not be a flat slab or a cylinder but may be irregular. As a result, because of the mathematical difficulty, various numerical methods have been developed. In the next section one such important method will be taken up in detail.

4.4B Numerical Methods for Solution of Steady-State Diffusion

1. *General Theory.* The solution of the partial differential Eq. (4.4-3), which is called an elliptical form, can be obtained numerically by several different methods. The solution of this equation is termed a boundary-value problem. It could be considered to represent the value approached by the solution of a corresponding initial-value problem after an infinite amount of time has passed. Hence, the solution of (4.4-3),

$$\frac{\partial^2 c_A}{\partial x^2} + \frac{\partial^2 c_A}{\partial y^2} = 0 \qquad (4.4\text{-}3)$$

can be obtained by solving the unsteady-state initial-value problem for a very long time (M1):

$$\frac{\partial c_A}{\partial t} = D_{AB}\left(\frac{\partial^2 c_A}{\partial x^2} + \frac{\partial^2 c_A}{\partial y^2}\right) \qquad (4.4\text{-}10)$$

This suggests setting up Eq. (4.4-10) as an initial-value problem using finite-difference methods and obtaining the result for a very large time. However, this process is very time-consuming and not very suitable for general use. A much more useful approach is to use finite differences and apply a relaxation method.

In the relaxation method the partial differentials are set up as finite differences. Using Eq. (4.4-3), the finite difference of $\partial^2 c/\partial x^2$ can be expressed as follows:

$$\frac{\partial^2 c}{\partial x^2} = \frac{\partial(\partial c/\partial x)}{\partial x} = \frac{\dfrac{c_{n+1,m} - c_{n,m}}{\Delta x} - \dfrac{c_{n,m} - c_{n-1,m}}{\Delta x}}{\Delta x}$$

$$= \frac{c_{n+1,m} - 2c_{n,m} + c_{n-1,m}}{(\Delta x)^2} \qquad (4.4\text{-}11)$$

where the index m stands for a given value of y, $m+1$ stands for $y + 1\,\Delta y$, and n is the index indicating the position of c on the x scale. For example, $c_{n+1,m}$ is the concentration at a position $1\,\Delta x$ to the right of x and $c_{n-1,m}$ at a position $1\,\Delta x$ backwards from x. This is shown in Fig. 4.4-2.

4.4 Steady-State Diffusion in Two Dimensions

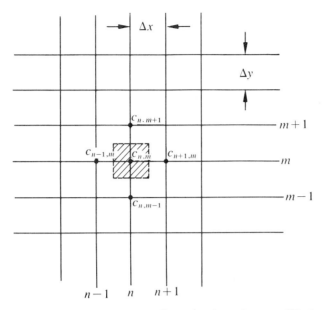

FIGURE 4.4-2 Concentrations for two-dimensional steady-state diffusion

The finite difference of $\partial^2 c / \partial y^2$ is

$$\frac{\partial^2 c}{\partial y^2} = \frac{c_{n,m+1} - 2c_{n,m} + c_{n,m-1}}{(\Delta y)^2} \qquad (4.4\text{-}12)$$

Substituting Eqs. (4.4-11) and (4.4-12) into (4.4-3), and setting $(\Delta x)^2 = (\Delta y)^2$,

$$\frac{c_{n+1,m} - 2c_{n,m} + c_{n-1,m}}{(\Delta x)^2} + \frac{c_{n,m+1} - 2c_{n,m} + c_{n,m-1}}{(\Delta x)^2} = 0 \qquad (4.4\text{-}13)$$

Rearranging,

$$c_{n,m+1} + c_{n,m-1} + c_{n+1,m} + c_{n-1,m} - 4c_{n,m} = 0 \qquad (4.4\text{-}14)$$

This final equation states that in a square grid pattern and at steady-state the concentration at a given point, $c_{n,m}$, is equal to the arithmetic average of the concentrations at the four surrounding points.

Eq. (4.4-14) can be applied in a relatively simple manner to an example of a square grid given in Fig. 4.4-3. At the four boundaries of this grid the concentrations are all known or fixed. The internal concentrations are unknown. For example, for point (3, 3) Eq. (4.4-14) becomes

$$c_{2,3} + c_{4,3} + c_{3,4} + c_{3,2} - 4c_{3,3} = 0 \qquad (4.4\text{-}15)$$

with all five concentrations being unknown. For point 4, 2 we obtain

$$c_{3,2} + c_{4,3} - 4c_{4,2} = -c_{4,1} - c_{5,2} \qquad (4.4\text{-}16)$$

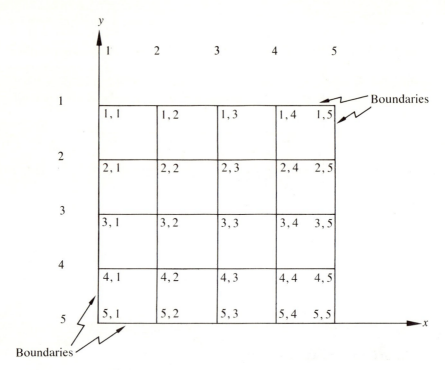

FIGURE 4.4-3 Square grid with boundary concentrations fixed

This equation has two known values on the right-hand side. Hence, we can write N equations and N unknowns with one equation for each internal unknown point. In Fig. 4.4-3 there are nine equations and nine unknowns.

In order to keep the truncation errors small we must take very small values of Δx. This, of course, leads to very large values of N. This means a large number of simultaneous linear algebraic equations to be solved directly on the digital computer using the methods discussed in Chapter 1 of this text. Often the capabilities of the computer may be exceeded, so various iteration or relaxation procedures are used.

A relaxation method which employs an iteration process due to Liebmann (L1) and is suitable for hand or machine computation will be described in the next section.

For those interested, Lapidus (L1) gives more sophisticated and efficient methods such as the successive-overrelaxation method which requires less iterations and the Peaceman–Rackford method. All of these methods are stable because of the iterative nature of the process. A recent procedure sometimes used is the Monte Carlo method which involves a set of rules based on random numbers.

2. *Relaxation Method.* This method uses Eq. (4.4-14) for its solution. However, to offer the reader a better understanding of this method, Eq.

4.4 Steady-State Diffusion in Two Dimensions

(4.4-14) will be rederived as follows. Referring to Fig. 4.4-2 again, moles of A are being transferred in the x and y directions but not in the z direction into the paper. Let us assume a thickness z of the square grid shown. We start by making a mole balance of solute A on the shaded square grid with $\Delta x = \Delta y$. The moles of A diffuse into the solid from four directions. The moles/sec going to the shaded square, which we will call $N_{n,m}$, will be equal to

$$N_{n,m} = D_{AB}(z\,\Delta y)\frac{c_{n-1,m} - c_{n,m}}{\Delta x} - D_{AB}(z\,\Delta y)\frac{c_{n,m} - c_{n+1,m}}{\Delta x}$$
$$+ D_{AB}(z\,\Delta x)\frac{c_{n,m-1} - c_{n,m}}{\Delta y} - D_{AB}(z\,\Delta x)\frac{c_{n,m} - c_{n,m+1}}{\Delta y} \quad (4.4\text{-}17)$$

Canceling out the Δy and Δx terms and rearranging,

$$\theta_{n,m} = c_{n-1,m} + c_{n+1,m} + c_{n,m+1} + c_{n,m-1} - 4c_{n,m} \quad (4.4\text{-}18)$$

where $\theta_{n,m} = N_{n,m}/(D_{AB}z)$, which is called a "residual."

Since $\theta_{n,m} = 0$ under steady-state conditions, we obtain the following, which is identical to Eq. (4.4-14) but rearranged:

$$c_{n,m} = \frac{c_{n-1,m} + c_{n+1,m} + c_{n,m+1} + c_{n,m-1}}{4} \quad (4.4\text{-}19)$$

Equations (4.4-18) and (4.4-19) are the final equations to be used, and their use will be illustrated in the example that follows.

Example 4.4-1 Steady-State Diffusion in a Hollow Rectangular Chamber

Figure 4.4-4 shows a cross section of a hollow rectangular chamber through which steady-state diffusion is occurring only in the x and y directions. The chamber is $z = 20$ cm long. Since it is symmetrical, only one-fourth of the chamber (shaded portion) will be used in the calculations. The size of each grid will be selected as 1.0 cm $\times$ 1.0 cm. The concentration at all of the inside surfaces of the hollow chamber is held constant at 6.00×10^{-3} g mole A/cm^3 and that at the outside surface at 3.00×10^{-3}. The diffusivity $D_{AB} = 1.0 \times 10^{-5}$ cm^2/sec. It is desired to calculate the concentrations at each point on the grid and also the total flux diffusing through the chamber walls for a chamber 1 cm long. This problem is similar to one for heat transfer given by Jakob and Hawkins (J2).

Solution

To simplify calculations, all of the concentrations will be multiplied by 10^5 before use. Thus, the inside concentration is 600 and the outside 300. These are shown in Fig. 4.4-4, where $c_{1,1}$, $c_{2,1}$, $c_{3,1}$, $c_{4,1}$, $c_{5,1}$, $c_{5,2}$, $c_{5,3}$, $c_{5,4}$, and $c_{5,5}$ are all 300. The inside points, $c_{1,3}$ and so on, are all 600.

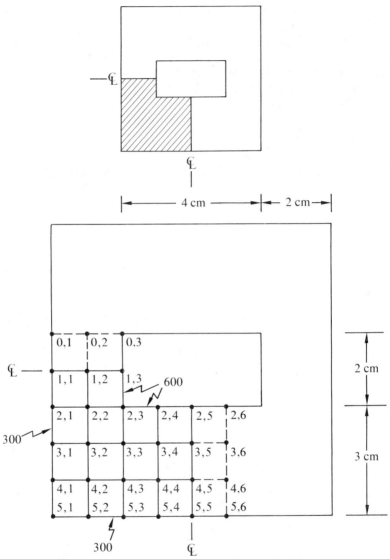

FIGURE 4.4-4 Square grid pattern for hollow chamber for Example 4.4-1

To speed up the calculations, preliminary estimates will be made for the internal points for the first approximation. (For a digital computer solution they could all be assumed the same at, say, 450). These estimates of concentration are for point $(1, 2) = 450$, $(2, 2) = 400$, $(3, 2) = 400$, $(3, 3) = 400$, $(3, 4) = 450$, $(3, 5) = 500$, $(4, 2) = 325$, $(4, 3) = 350$, $(4, 4) = 375$, $(4, 5) = 400$.

4.4 Steady-State Diffusion in Two Dimensions

First Approximation. One can start the calculation of the interior points at any point. However, it is usually better to select a point that is next to the boundaries. Starting with point (1, 2), we will calculate the residual $\theta_{1,2}$ by Eq. (4.4-18), with the point (0, 2) being the same as (2, 2).

$$\theta_{1,2} = (1, 1) + (1, 3) + (0, 2) + (2, 2) - 4(1, 2) \qquad (4.4\text{-}20)$$

Substituting the assumed and known values,

$$\theta_{1,2} = 300 + 600 + 400 + 400 - 4(450) = -100$$

This indicates that point (1, 2) is not at steady-state.

Next, we "relax" $\theta_{1,2}$ to zero and calculate point (1, 2) from Eq. (4.4-19), which becomes

$$(1, 2) = \frac{(1, 1) + (1, 3) + (0, 2) + (2, 2)}{4}$$

$$= \frac{300 + 600 + 400 + 400}{4} = 425 \qquad (4.4\text{-}21)$$

This new value of (1, 2) of 425 replaces the old value of 450 and will be used to calculate the other points.

Proceeding as before, we calculate $\theta_{2,2}$:

$$\theta_{2,2} = (2, 1) + (2, 3) + (1, 2) + (3, 2) - 4(2, 2)$$
$$= 300 + 600 + 425 + 400 - 4(400) = 125$$

Relaxing $\theta_{2,2}$ to zero, the new value of (2, 2) is

$$(2, 2) = \frac{(2, 1) + (2, 3) + (1, 2) + (3, 2)}{4} = \frac{300 + 600 + 425 + 400}{4} = 431$$

All calculated fractional values are rounded off to the nearest whole number. This is continued for (3, 2), ...

$$\theta_{3,2} = 300 + 400 + 431 + 325 - 4(400) = -144$$

Relaxing,

$$(3, 2) = 364$$
$$\theta_{3,3} = 364 + 450 + 600 + 350 - 4(400) = 164$$
$$(3, 3) = 441$$
$$\theta_{3,4} = 441 + 500 + 600 + 375 - 4(450) = 116$$
$$(3, 4) = 479$$
$$\theta_{3,5} = 479 + 479 + 600 + 400 - 4(500) = -42$$
$$(3, 5) = 489$$
$$\theta_{4,2} = 300 + 350 + 364 + 300 - 4(325) = 14$$
$$(4, 2) = 329$$
$$\theta_{4,3} = 329 + 375 + 441 + 300 - 4(350) = 45$$

(4, 3) = 361
$\theta_{4,4} = 361 + 400 + 479 + 300 - 4(375) = 40$
(4, 4) = 385
$\theta_{4,5} = 385 + 385 + 489 + 300 - 4(400) = -41$
(4, 5) = 390

We have now made one approximation or sweep across the grid map, and we start our second approximation using the new values just calculated.

Second Approximation

$\theta_{1,2} = 300 + 600 + 431 + 431 - 4(425) = 62$
(1, 2) = 440
$\theta_{2,2} = 300 + 600 + 440 + 364 - 4(431) = -20$
(2, 2) = 426
$\theta_{3,2} = 300 + 441 + 426 + 329 - 4(364) = 40$
(3, 2) = 374
$\theta_{3,3} = 374 + 479 + 600 + 361 - 4(441) = 50$
(3, 3) = 453
$\theta_{3,4} = 453 + 489 + 600 + 385 - 4(479) = 11$
(3, 4) = 482
$\theta_{3,5} = 482 + 482 + 600 + 390 - 4(489) = -2$
(3, 5) = 489
$\theta_{4,2} = 300 + 361 + 453 + 300 - 4(329) = 98$
(4, 2) = 353
$\theta_{4,3} = 353 + 385 + 453 + 300 - 4(361) = 47$
(4, 3) = 373
$\theta_{4,4} = 373 + 390 + 482 + 300 - 4(385) = 5$
(4, 4) = 386
$\theta_{4,5} = 386 + 386 + 489 + 300 - 4(390) = 1$
(4, 5) = 390

A third, fourth, and fifth approximation were each carried out. The fourth and fifth values were all within 1 of each other, so the problem was assumed to have converged. Below the third and final values are tabulated.

To calculate the total flux through the chamber we use the drawing in Fig. 4.4-5 for a small section of the slab. Calculating the flux from the inside wall 1 Δx or 1 Δy into the solid, we can write for element (2, 4) to (3, 4)

$$N = \frac{AD_{AB}}{\Delta x}(c_{2,4} - c_{3,4}) = \frac{\Delta x\,(1)D_{AB}}{\Delta x}(c_{2,4} - c_{3,4})$$

$$= D_{AB}(c_{2,4} - c_{3,4}) \qquad (4.4\text{-}22)$$

where the area A is Δx times 1 cm deep. We consider the concentration at (2, 4) to be the same or constant for $\Delta x/2$ distance to either side. Note

Point	Third Approximation θ	c	Final Values c
(1, 2)	−8	438	441
(2, 2)	8	428	432
(3, 2)	38	383	384
(3, 3)	26	460	461
(3, 4)	8	484	485
(3, 5)	2	490	490
(4, 2)	56	339	340
(4, 3)	−7	371	372
(4, 4)	1	386	387
(4, 5)	2	391	391

that the flux for $(2, 5)$ to $(3, 5)$ should be multiplied by $\frac{1}{2}$ because of symmetry. Hence, for four duplicate parts of the one-fourth of the solid given in Fig. 4.4-5, the total flux is the sum of the five diffusion paths:

$$N_I = 4D_{AB}[\tfrac{1}{2}(c_{1,3} - c_{1,2}) + (c_{2,3} - c_{2,2}) + (c_{2,3} - c_{3,3})$$

$$+ (c_{2,4} - c_{3,4}) + \tfrac{1}{2}(c_{2,5} - c_{3,5})] \quad (4.4\text{-}23)$$

$$N_I = 4(1 \times 10^{-5})[\tfrac{1}{2}(600 - 441) + (600 - 432) + (600 - 461)$$
$$+ (600 - 485) + \tfrac{1}{2}(600 - 490)](10^{-5})$$
$$= 4(1 \times 10^{-5})(556.5)(10^{-5}) = 2.23 \times 10^{-7} \text{ g mole } A/\text{sec}$$

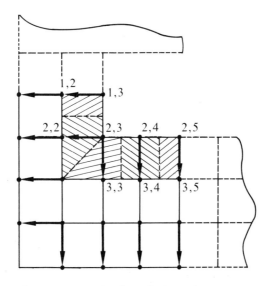

FIGURE 4.4-5 Drawing to calculate flux in each element

176 Mass Transport Phenomena in Solids

The total flux can also be calculated using the driving forces at the outside layer, as shown in Fig. 4.4-5. Thus

$$N_{II} = 4(1 \times 10^{-5})[\tfrac{1}{2}(441-300) + (432-300) + (384-300) + (340-300)$$
$$+ (340-300) + (372-300) + (387-300) + \tfrac{1}{2}(391-300)]10^{-5}$$
$$= 2.28 \times 10^{-7} \text{ g mole } A/\text{sec}$$

Using an average value,

$$N_{\text{ave}} = \frac{(2.23 + 2.28)10^{-7}}{2} = 2.25 \times 10^{-7} \text{ g mole } A/\text{sec}$$

If a finer grid had been used, the two values would have agreed more closely.

For other cases, where convection occurs at the boundaries, the boundaries are irregular in shape, the boundaries are insulated, and/or the physical properties change, the reader is referred to more detailed discussions in other references (D1, J1, L1, M1).

3. *Digital Computer Solution.* This type of a problem is ideally suited to machine computation; an example is given below.

Example 4.4-2 Steady-State Diffusion Using the Digital Computer

Repeat Example 4.4-1 using the digital computer. Use a grid spacing and size of $\tfrac{1}{2} \times \tfrac{1}{2}$ cm. For the first guess assume all interior concentrations are the average of the interior and exterior, or $(600 + 300)/2 = 450$.

Solution

The grid pattern for $\tfrac{1}{2}$-cm squares is shown in Fig. 4.4-6 for one-fourth of the hollow chamber, as in Fig. 4.4-4. As in the previous example, the first approximation will be started at point (2, 2). Then (2, 3), (2, 4), (3, 2), (3, 3), and so on will be computed. Note that the numbering corresponds to a matrix numbering system. The computer is programmed to calculate the residuals for each point and then the new "relaxed" concentration.

When all of the residuals in a complete sweep or approximation are less than 0.4, then the program is completed — that is, when

$$|\theta_{n,m}| < 0.4 \tag{4.4-24}$$

The digital Fortran program for this example is given in Table 4.4-1. The comment statements should enable the reader to follow the program. In the statement "C: set matrix geometry control values" and the statements that follow, a full 10×10 matrix is set up. However, it is cut off at column 5 and row 4 in the upper part. In statements 12 to 15 all of the

4.4 Steady-State Diffusion in Two Dimensions 177

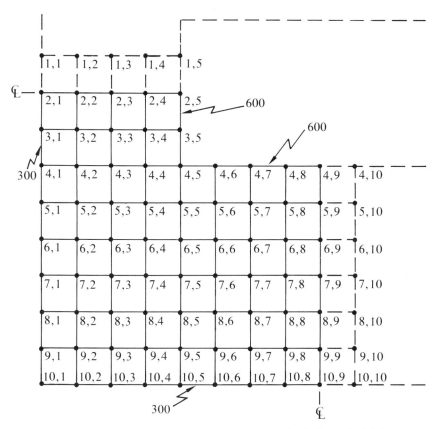

FIGURE 4.4-6 Square grid pattern for computer solution in Example 4.4-2

Table 4.4-1 Fortran Program, Example 4.4-2

```
C     EXAMPLE PROBLEM
C
C     CALCULATION OF CONCENTRATION AND MASS FLUX IN A SOLID
C
C     CALCULATION USES THE RELAXATION METHOD
C
C     ALL CONCENTRATIONS SCALED BY MULTIPLICATION BY 1.0E+05
C
1     FORMAT('1EXAMPLE PROBLEM - CALCULATION OF CONCENTRATION AND'/'OMAS
     1S FLUX IN A SOLID BY USE OF THE RELAXATION METHOD'/'OTHE DIFFUSIVI
     2TY IS',E16.8/'OTHE INSIDE  SURFACE CONCENTRATION IS',F8.2/'OTHE OU
     3TSIDE SURFACE CONCENTRATION IS',F8.2//)
2     FORMAT('OTHE TOTAL NUMBER OF ITERATIONS OF THE RELAXATION METHOD'/
     1'ONEEDED TO FIND THE FOLLOWING VALUES ARE',I6//)
3     FORMAT(9(F7.2)/)
4     FORMAT('OTHE FLUX CROSSING THE INSIDE  SURFACE IS',E16.8/'OTHE FLU
     1X CROSSING THE OUTSIDE SURFACE IS',E16.8/'OTHE AVERAGE FLUX IS',E1
     26.8)
5     FORMAT('1END OF JOB')
6     FORMAT(5(F7.2)/)
      DIMENSION T(10,10)
      REAL INCONC
      IO=6
```

Table 4.4-1 (continued)

```
        EPS=0.4
        DAB=1.0E-05
C       SET MATRIX GEOMETRY CONTROL VALUES
        NCOL=10
        NROW=10
        MCOL=5
        MROW=4
C       SET DO LOOP CONTROL VALUES
        KA=MROW+1
        KB=MROW-1
        KC=MCOL+1
        KD=MCOL-1
        KF=NROW-1
        KH=NCOL-1
        KI=NCOL-2
C       SET INITIAL BOUNDARY VALUE CONCENTRATIONS
        INCONC=600.
        OUCONC=300.
        AVCONC=(INCONC+OUCONC)/2.
C       SET INITIAL MATRIX ELEMENT VALUES
C       SET OUTSIDE SURFACE VALUES
        DO 10 I=1,NROW
10      T(I,1)=OUCONC
        DO 11 I=2,NCOL
11      T(NROW,I)=OUCONC
C       SET INSIDE SURFACE VALUES
        DO 12 I=1,MROW
12      T(I,MCOL)=INCONC
        DO 13 I=KC,NCOL
13      T(MROW,I)=INCONC
C       SET REMAINING INITIAL VALUES
        DO 14 I=1,MROW
        DO 14 J=2,KD
14      T(I,J)=AVCONC
        DO 15 I=KA,KF
        DO 15 J=2,NCOL
15      T(I,J)=AVCONC
        IT=0
        WRITE (IO,1) DAB,INCONC,OUCONC
C       RELAXATION LOOP
16      IT=IT+1
        IER=0
        DO 18 I=2,MROW
        DO 18 J=2,KD
        RES=ABS(T(I,J-1)+T(I,J+1)+T(I-1,J)+T(I+1,J)-4.*T(I,J))
        IF(EPS-RES)17,17,18
17      IER=IER+1
18      T(I,J)=(T(I,J-1)+T(I,J+1)+T(I-1,J)+T(I+1,J))/4.0
        DO 40 I=KA,KF
        DO 40 J=2,KH
        RES=ABS(T(I,J-1)+T(I,J+1)+T(I-1,J)+T(I+1,J)-4.*T(I,J))
        IF(EPS-RES)19,19,20
19      IER=IER+1
20      T(I,J)=(T(I,J-1)+T(I,J+1)+T(I-1,J)+T(I+1,J))/4.0
        IF(J-KI) 40,41,40
41      T(I,NCOL)=T(I,J)
40      CONTINUE
        IF(IER)21,21,24
24      DO 25 J=2,KD
25      T(1,J)=T(3,J)
        GO TO 16
21      CONTINUE
C       OUTPUT CONCENTRATION VALUES
        WRITE (IO,2) IT
        DO 22 I=2,KB
22      WRITE (IO,6) (T(I,J),J=1,MCOL)
        DO 23 I=MROW,NROW
23      WRITE (IO,3) (T(I,J),J=1,KH)
C       CALCULATE FLUX
        FLUXI=(T(2,MCOL)-T(2,MCOL-1))/2.
        DO 30 I=3,MROW
```

Table 4.4-1 (continued)

```
30      FLUXI=FLUXI+(T(I,MCOL)-T(I,MCOL-1))
        DO 31 J=MCOL,KI
31      FLUXI=FLUXI+(T(MROW,J)-T(MROW+1,J))
        FLUXI=FLUXI+(T(MROW,NCOL-1)-T(MROW+1,NCOL-1))/2.
        FLUXI=FLUXI*DAB*4.0
        FLUXO=(T(2,2)-T(2,1))/2.
        DO 32 I=3,KF
32      FLUXO=FLUXO+(T(I,2)-T(I,1))
        DO 33 J=2,KI
33      FLUXO=FLUXO+(T(NROW-1,J)-T(NROW,J))
        FLUXO=FLUXO+(T(NROW-1,NCOL-1)-T(NROW,NCOL-1))/2.0
        FLUXO=FLUXO*DAB*4.0
C       RESCALE FLUX BY 1.0E-05
        SF=1.0E-05
        FLUXI=FLUXI*SF
        FLUXO=FLUXO*SF
        AVFLUX=(FLUXI+FLUXO)/2.0
C       OUTPUT FLUX VALUES
        WRITE (IO,4) FLUXI,FLUXO,AVFLUX
        WRITE (IO,5)
        STOP
        END
```

Table 4.4-2 Results for Example 4.4-2

```
EXAMPLE PROBLEM - CALCULATION OF CONCENTRATION AND
MASS FLUX IN A SOLID BY USE OF THE RELAXATION METHOD
THE DIFFUSIVITY IS   0.99999997E-05
THE INSIDE  SURFACE CONCENTRATION IS   600.00
THE OUTSIDE SURFACE CONCENTRATION IS   300.00


THE TOTAL NUMBER OF ITERATIONS OF THE RELAXATION METHOD
NEEDED TO FIND THE FOLLOWING VALUES ARE     30

300.00 367.57 438.55 515.89 600.00
300.00 365.82 435.33 512.47 600.00
300.00 360.30 424.39 498.63 600.00 600.00 600.00 600.00 600.00
300.00 350.91 403.22 457.58 509.92 530.20 538.95 542.74 543.80
300.00 340.02 379.87 418.40 451.74 471.79 482.73 488.06 489.63
300.00 329.18 357.67 384.23 406.66 422.29 431.91 436.97 438.51
300.00 318.95 337.25 354.05 368.19 378.64 385.47 389.23 390.39
300.00 309.31 318.25 326.40 333.28 338.49 341.99 343.96 344.58
300.00 300.00 300.00 300.00 300.00 300.00 300.00 300.00 300.00

THE FLUX CROSSING THE INSIDE  SURFACE IS   0.21489808E-06
THE FLUX CROSSING THE OUTSIDE SURFACE IS   0.21689488E-06
THE AVERAGE FLUX IS   0.21589648E-06
```

180   Mass Transport Phenomena in Solids

initial values are stored in the matrix. The statement after number 15 sets the number of iterations counter to 0. Statement 16 starts the relaxation loop calculations, which are made up of two loops or types, one for the top part of the matrix and one for the bottom part. During the relaxation calculation the absolute value of the residual is compared to epsilon of 0.4 in the statements before numbers 17 and 19. If it is greater than 0.4, a number 1 is added to an error counter. At the end of the pass at statement 40, if the error counter is 1 or higher, then the complete calculation is repeated for another relaxation starting at statement 16 again.

The results are tabulated in Table 4.4-2. The average flux is 2.1589×10^{-7} g mole A/sec, which is slightly different than 2.25×10^{-7} calculated for Example 4.4-1 with a coarser grid pattern.

4.5   DETERMINATION OF DIFFUSIVITIES IN SOLIDS

4.5A   Steady-State Methods

1. *Constant Diffusivity.* There are numerous experimental methods to measure experimentally the diffusivity D_{AB} in solids using steady-state methods. If the solute A is a gas, then a permeability experiment is usually done, as discussed in Section 4.1. In this method the solid as a thin diaphragm separates two chambers containing different pressures of the solute gas A. The flow of gas or flux N_A through the thin solid is measured and the D_{AB} obtained from Eq. (4.1-2) or (4.5-3) below. This works well if the diffusivity is reasonably high, but when the diffusion is very slow it is very difficult to measure a flux, and unsteady-state methods are preferred. For liquids such as urea in water diffusing through agar gel this steady-state method is also convenient.

If the diffusivity is constant and if the solution is dilute, Eq. (4.1-1) can be used, and

$$N_A = -D_{AB}\frac{dc_A}{dz} \qquad (4.5\text{-}1)$$

Integrating between 0 and z and c_{A1} and c_{Az},

$$N_A \int_0^z dz = -D_{AB} \int_{c_{A1}}^{c_{Az}} dc_A \qquad (4.5\text{-}2)$$

$$N_A = D_{AB}\frac{c_{A1} - c_{Az}}{z - 0} \qquad (4.5\text{-}3)$$

Integrating between 0 and δ and c_{A1} and c_{A2},

$$N_A = D_{AB}\frac{c_{A1} - c_{A2}}{\delta - 0} \qquad (4.5\text{-}4)$$

Equating Eq. (4.5-3) to (4.5-4) and rearranging with c_A used for c_{Az},

$$c_A = c_{A1} + \frac{(c_{A2} - c_{A1})z}{\delta} \qquad (4.5\text{-}5)$$

4.5 Determination of Diffusivities in Solids

which is a straight line when c_A is plotted versus z for a constant D_{AB}. When the diffusion experiment is done with a solid, the concentration versus z data are obtained by slicing the solid into thin slices and experimentally measuring the concentrations. If the data give a straight line, D_{AB} is constant. Then measurement of N_A, δ, c_{A1}, and c_{A2} will give D_{AB}.

2. *Variable Diffusivity.* In many cases for diffusion in solids the diffusivity varies with concentration. This is often particularly true in diffusion of solutes in metal systems. In a steady-state experiment the following equation can be written for constant flux by rearranging Eq. (4.5-1):

$$D_{AB} = \frac{-N_A}{dc_A/dz} \tag{4.5-6}$$

Hence, if the flux and concentration profile are both obtained in an experiment, a plot of c_A versus z can be made. If the line is straight, as shown in Eq. (4.5-5), the diffusivity is constant. If it is curved, D_{AB} varies and its value at a given concentration may be obtained from the slope dc_A/dz at this point and Eq. (4.5-6).

This method was described by Darkin (D2) in obtaining the diffusivity of carbon in iron-carbon alloys. A carburizing gas was passed inside a hollow cylinder of an iron-carbon alloy and a decarburizing gas outside. After steady-state had been reached, thin slices were removed successively and analyzed for carbon. To analyze the data, Eq. (4.1-4) for a cylinder can be rearranged as

$$\overline{N}_A = -2\pi L D_{AB} \frac{dc_A}{dr/r} = -2\pi L D_{AB} \frac{dc_A}{d\ln r} \tag{4.1-4a}$$

$$D_{AB} = -\frac{\overline{N}_A}{2\pi L} \frac{1}{dc_A/(d\ln r)} \tag{4.5-7}$$

Hence, D_{AB} is inversely proportional to the slope of a plot of c_A versus $\ln r$. If D_{AB} is constant, the line should be linear. Data for 1000°C are shown in Fig. 4.5-1. The line is quite curved, and the diffusivity varies considerably with concentration. There is at least a threefold change in slope or diffusivity. In some other extreme cases in metals the variation is 100-fold or more (D2).

If it is assumed D_{AB} is a linear function of concentration, where $D_{AB} = (k_1 + k_2 c_A)$, then, inserting this into Eq. (4.5-1) and integrating between $z = 0$ and $c_A = c_{A1}$ and $z = z$ and $c_A = c_A$,

$$N_A = -(k_1 + k_2 c_A)\frac{dc_A}{dz} \tag{4.5-8}$$

$$z = \frac{k_1}{N_A}(c_{A1} - c_A) + \frac{k_2}{2N_A}(c_{A1}^2 - c_A^2) \tag{4.5-9}$$

where k_1 and k_2 are constants. Hence, a plot of c_A versus z is curved.

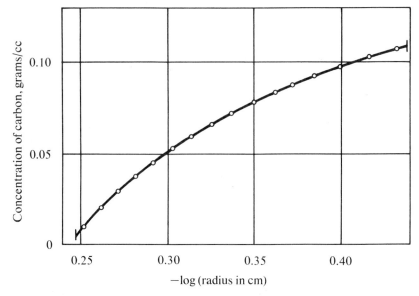

FIGURE 4.5-1 Steady-state diffusion of carbon at 1000°C through a hollow iron cylinder (From L. S. Darkin and K. W. Gurry, *Physical Chemistry of Metals*. New York: McGraw-Hill, Inc., 1953. With permission.)

4.5B Unsteady-State Methods

1. *Constant Diffusivity.* In many cases the determination of diffusivity in solids is more readily done using unsteady-state methods. Since diffusion in solids is quite slow, in steady-state experiments it is difficult to measure the flux and/or to wait long enough to insure steady-state. These transient methods also apply to diffusion in liquids.

A common method used is to perform an unsteady-state experiment in a semiinfinite medium. The solid initially has a uniform composition c_0 and the front face is suddenly subjected to a constant concentration at the surface of c_1, which is held there. Diffusion occurs but does not penetrate all the way in the solid. Starting with Fick's second law for constant D,

$$\frac{\partial c}{\partial t} = D\frac{\partial^2 c}{\partial x^2} \qquad (4.5\text{-}10)$$

I.C.: at $x = x$, $t = 0$, $c = c_0$
B.C. 1: at $x = 0$, $t = t$, $c = c_1$
B.C. 2: at $x = \infty$, $t = t$, $c = c_0$

One method to solve the above is to make the substitution $y = x/t^{1/2}$.

4.5 Determination of Diffusivities in Solids

Then,

$$\left(\frac{\partial y}{\partial t}\right)_x = -\frac{1}{2}\frac{x}{t^{3/2}} = -\frac{1}{2}(yt^{1/2})\frac{1}{t^{3/2}} = -\frac{y}{2t} \tag{4.5-11}$$

$$\left(\frac{\partial c}{\partial t}\right)_x = \frac{\partial c}{\partial y}\frac{\partial y}{\partial t} = \frac{\partial c}{\partial y}\left(\frac{-y}{2t}\right) \tag{4.5-12}$$

In a similar manner,

$$\left(\frac{\partial^2 c}{\partial x^2}\right)_t = \frac{1}{t}\left(\frac{\partial^2 c}{\partial y^2}\right)_t \tag{4.5-13}$$

Substituting Eqs. (4.5-13) and (4.5-12) into (4.5-10) and calling $p = \partial c/\partial y$,

$$-\frac{p}{2}y = D\frac{dp}{dy} \tag{4.5-14}$$

This is an ordinary differential equation and, upon integrating once,

$$D\ln p = \frac{-y^2}{4} + D\ln A \tag{4.5-15}$$

where A is a constant. Solving for p,

$$p = \frac{dc}{dy} = Ae^{-y^2/(4D)} \tag{4.5-16}$$

Letting $\lambda = y/(2\sqrt{D})$, then $d\lambda = dy/(2\sqrt{D})$. Substituting this value for dy into Eq. (4.5-16) and integrating using B.C. 1,

$$\int_{c_1}^{c} dc = c - c_1 = A2\sqrt{D}\int_{0}^{x/2\sqrt{Dt}} e^{-\lambda^2}\,d\lambda \tag{4.5-17}$$

Setting $x = \infty$ and $c = c_0$ from B.C. 2, the value of the integral, which is really $(\sqrt{\pi}/2)$ times the error function, is $(\sqrt{\pi}/2)(1)$. Hence, $A = (c_0 - c_1)/\sqrt{\pi D}$ and Eq. (4.5-17) becomes

$$\frac{c_1 - c}{c_1 - c_0} = \frac{2}{\sqrt{\pi}}\int_{0}^{x/2\sqrt{Dt}} e^{-\lambda^2}\,d\lambda = \mathrm{erf}\left(\frac{x}{2\sqrt{Dt}}\right) \tag{4.5-18}$$

where erf is the error function tabulated in Table 5.3-2. Equation (4.5-18) can be rearranged to give

$$\frac{c - c_0}{c_1 - c_0} = 1 - \mathrm{erf}\left(\frac{x}{2\sqrt{Dt}}\right) \tag{4.5-19}$$

In the next chapter this same equation will be derived in a different manner using the Laplace transform.

Experimental data of c versus x can be used along with Eq. (4.5-19) to determine D. This is done by plotting the experimental data as

$(c-c_0)/(c_1-c_0)$ versus x at constant t on linear graph paper. Then by trial and error select a value of D and plot the predicted curve using Eq. (4.5-19). Use the value of D giving the closest matching curve.

Closer investigation of Eq. (4.5-19) reveals some interesting observations. Suppose a family of curves is plotted as $(c-c_0)/(c_1-c_0)$ versus x at different constant values of t. For the same concentration c, for time t_1, we will read x_1 off the curve and x_2 for t_2. Then from Eq. (4.5-19)

$$\frac{x_1}{2\sqrt{Dt_1}} = \frac{x_2}{2\sqrt{Dt_2}} \qquad (4.5\text{-}20)$$

Hence, a given composition moves away from the plane $x=0$ at a rate proportional to $\sqrt{Dt}$ as shown in Fig. 4.5-2. This is similar to the quadratic law mentioned earlier in this chapter.

To determine the total amount of material that has diffused per unit area from $t=0$ to $t=t$, we evaluate the area under the c versus x curve as

$$\int_0^\infty (c-c_0)\, dx \quad \text{or} \quad 2\sqrt{Dt} \int_0^\infty (c-c_0)\, d\left(\frac{x}{2\sqrt{Dt}}\right) \qquad (4.5\text{-}21)$$

Substituting Eq. (4.5-19) into (4.5-21),

$$\int_0^\infty (c-c_0)\, dx = 2\sqrt{Dt}\,(c_1-c_0) \int_0^\infty \left[1-\operatorname{erf}\left(\frac{x}{2\sqrt{Dt}}\right)\right] d\left(\frac{x}{2\sqrt{Dt}}\right) \qquad (4.5\text{-}22)$$

From tables by Crank (C3), the definite integral has a value of $1/\sqrt{\pi}$. Hence,

$$\text{total amount diffusing} = \frac{2}{\sqrt{\pi}} \sqrt{Dt}\,(c_1-c_0) \qquad (4.5\text{-}23)$$

This gives another method to evaluate D by obtaining the area from the experimental data and solving for D in Eq. (4.5-23).

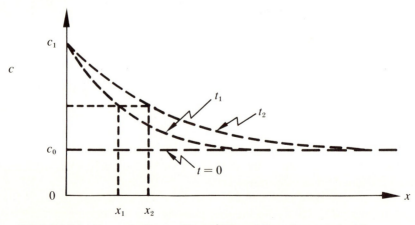

FIGURE 4.5-2 Concentration profiles at different times for a constant D

4.5 Determination of Diffusivities in Solids

2. Variable Diffusivity. The determination of the diffusivity experimentally when it is a function of concentration is much more complex than when it is constant. It is usually done using unsteady-state methods, which is relatively straightforward, but the analysis of the data is complicated and sometimes uncertain.

An example will be considered. To do an ordinary diffusion experiment with metals pure Cu is electroplated onto an alloy of Cu-Al. Diffusion is allowed to proceed in a furnace. Then thin sections are machined off and analyzed and the data plotted as shown in Fig. 4.5-3.

Shewman (S2) and Crank (C3) show that the diffusivity is related to c as follows:

$$D(c') = -\frac{1}{2t}\left(\frac{dx}{dc}\right)_{c'}\int_0^{c'} x\, dc \qquad (4.5\text{-}24)$$

The interface, of course, is not known, since both Cu and Al atoms have counterdiffused and there has not necessarily been equimolar counterdiffusion. Hence, bulk flow also occurs. The interface is selected by making the two crosshatched areas equal. Then to obtain the diffusivity $D(c')$ at the concentration c', a tangent is drawn to the curve at c' to obtain the slope $(dx/dc)_{c'}$. The area of $\int x\, dc$ is obtained from $c = 0$ to c'. The values are finally substituted into Eq. (4.5-24). This is done at various values of c' to obtain various values of $D(c')$.

This method to obtain the diffusion coefficient (S2) is called the Boltzmann-Matano method and describes the overall diffusion and bulk motion of both elements in the mixture. It does not describe the mobility of each type of atom. This D is called the chemical $\tilde{D}$ and is the same for each element in the binary.

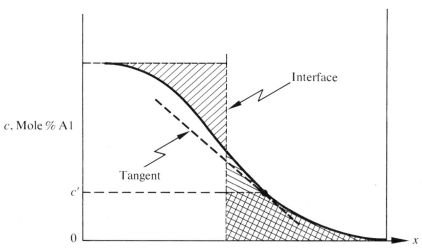

FIGURE 4.5-3 Typical diffusion data in a binary metal alloy

186 Mass Transport Phenomena in Solids

In the newer techniques radiotracers are used. Two alloys of the same composition are welded together. One alloy has a low concentration of one element component 1 as a radiotracer. Then the Matano method is used to obtain D_1^*, which is called the self-diffusion coefficient of element 1. This is repeated for D_2^*. Then, the following has been shown to hold theoretically:

$$\tilde{D} = (D_1^* N_2 + D_2^* N_1)\left(1 + \frac{\partial \ln \gamma_1}{\partial \ln N_1}\right) \qquad (4.5\text{-}25)$$

where N_1 is mole fraction of element 1 and γ_1 its activity coefficient. Experimental data check this equation. The reader is referred elsewhere (D2, S2) for more details on ordinary diffusion and on other types, such as grain boundary diffusion and so on.

PROBLEMS

4-1 Diffusion Through a Membrane

Hydrogen gas at 17°C and 0.010 atm pressure is diffusing through a thin membrane of vulcanized neoprene 0.05 cm thick. The pressure of H_2 on the other side is zero. Calculate the flux at steady-state assuming that the only resistance to diffusion is in the membrane.

4-2 Diffusivity and Permeability

Helium gas at 2.0 atm pressure is contained in a metal vessel at 20°C. A window of SiO_2 0.20 cm thick and 1.0 cm² area is used to view the contents in the vessel. The pressure of He gas outside can be assumed as zero. To be conservative use D_{AB} as $5.5(10^{-10})$ cm²/sec.
(a) Calculate the loss of He gas per hour at steady-state.
(b) Calculate the permeability constant P_M in cc gas at N.T.P./(sec-cm²) for 1 mm thickness and 1.0 cm Hg pressure difference.

$$\textit{Ans.:} \quad \text{(b)} \ P_M = 7.23(10^{-13}) \ \frac{\text{cc at N.T.P.}}{\text{sec-cm}^2\text{-cm Hg/mm}}$$

4-3 Diffusivity and Temperature

The equation for the diffusivity of Ag in Pb is $D_{AB} = 7.5(10^{-2})e^{-15,200/RT}$. Calculate the values of D_{AB} at 25, 100, 300, and 500°C and make a plot on graph paper that gives a straight line.

4-4 Loss from a Neoprene Tube

Hydrogen gas at 2.0 atm is flowing in a neoprene tube 0.3 cm I.D. and 1.0 cm O.D. at 27°C. Calculate the leakage of H_2 through the walls of a tube 3.0 ft long.

4-5 Diffusion Through Membranes in Series

Helium gas at 2.0 atm pressure and 20°C is diffusing through a thickness of Pyrex glass 0.2 cm thick and then through a thickness of SiO_2 0.3 cm thick to the air. Assume no resistance at the interface between the two solids. Calculate the loss per cm^2 area per sec. [*Hint*: Let p_{Ai} = the pressure of He gas in the space between the two layers. Write the flux equations for each layer. Then solve for the unknowns N_A and p_{Ai}.]

4-6 Breathing of Cellophane

Cellophane is being used to keep food moist at 38°C. Calculate the loss of water vapor in g mass/day for a wrapping 0.010 cm thick and 2000 cm^2 when the pressure of water vapor inside is 10 mm Hg and the air outside contains water vapor at a pressure of 5 mm Hg. Use the larger permeability in Table 4.1-1.

4-7 Diffusion in a Porous Solid

A porous fritted glass disk of 2.0 cm^2 area is filled with water at 25°C and is 0.15 cm thick. At one face a solution containing acetic acid at a concentration of 0.15 g mole/liter is agitated. Fresh water flows by the other side. The void fraction ϵ is 0.27 and the tortuosity k_t is 2.4. Calculate the loss in g mass acetic acid/day and the effective diffusivity D_{A_effec}.

4-8 Diffusion of Gases in Fine Pores

A mixture of nitrogen gas (A) and helium (B) at 25°C is diffusing through a capillary 100 cm long in an open system with a diameter of 10 microns. The mole fractions are constant at $x_{A1} = 1.0$ and $x_{A2} = 0$. See Example 4.2-2 for physical properties.

(a) Calculate the Knudsen diffusivity D_{KA} and D_{KB} at the total pressures of 0.001, 0.1, and 10.0 atm.
(b) Calculate the flux N_A at steady-state at these pressures.
(c) Plot N_A versus P on log-log paper. What are the limiting lines at lower pressures and very high pressures? Calculate and plot these lines.

4-9 Catalyst Effectiveness Factor and Size of Pellet

A gas A is diffusing from the main gas stream into a spherical catalyst pellet with radius $R = 0.5$ cm. The reaction velocity constant $k = 100$ sec^{-1} and the mean diffusivity $D_A = 0.75$ cm^2/sec in the pores.

(a) Calculate the catalyst effectiveness factor.
(b) To reduce the effect of pore diffusion a catalyst pellet of $R = 0.20$ cm is being considered. What will be the new effectiveness factor? Will reducing the size still further help substantially?

4-10 Analog Computer Solution for Pore Reaction

Draw the analog computer diagram to solve Eq. (4.2-27) and the *IC* in Eq. (4.2-29). Do not scale but call $x = \tau$ computer time and $c_A = e$ volts. Show the initial conditions on all integrators. Make an expected plot of e versus τ.

Explain how to start the run and how we know when the solution is reached. [*Hint*: The *IC* on the first integrator with an output of dc_A/dx is trial and error. What should its output be at $x = L$?]

4–11 Numerical Methods for Steady-State Diffusion

Repeat Example 4.4-1 using the relaxation method, except that the inside surface concentration is held constant at 8.00×10^{-3} g mole/cm^3 and the outside at 1.00×10^{-3} g mole/cm^3.

4–12 Numerical Methods and Digital Computer

Repeat Example 4.4-1 using the digital computer. Use square grids 0.25×0.25 cm and compare the results with Example 4.4-1.

4–13 Steady-State Diffusion by Numerical Methods

A hollow square-shaped chamber made of agar gel is described in Fig. P4-13. Steady-state diffusion is occurring only in the x and y directions. The chamber is 100 cm long in the z direction. The concentration at all of the inside faces is held constant at 4.0×10^{-4} g mole urea/cm^3. The outside of the chamber is immersed in pure fast-running water, so no surface resistance can be assumed. The diffusivity of urea is 4.70×10^{-6} cm^2/sec. Use the relaxation method to calculate the steady-state concentrations and the loss of urea in g mole/sec. Select a grid size of 1.0×1.0 cm and do the calculation by hand. To simplify the calculations multiply all of the figures by 10^6 before starting. Do not do over four approximations.

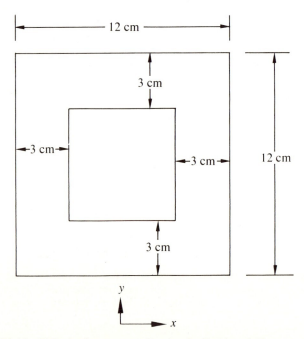

FIGURE P4-13 Diffusion in a square chamber for Problem 4-13

4-14 Steady-State Diffusion Using Digital Computer

Repeat Problem 4-13 but use a grid size of $\frac{1}{2} \times \frac{1}{2}$ cm. Program this on the digital computer.

4-15 Integration of Fick's Second Law for Steady-State

Using Fick's second law,

$$\frac{\partial c_A}{\partial t} = D_{AB} \frac{\partial^2 c_A}{\partial z^2},$$

integrate this for steady-state and obtain the following equation:

$$c_A = c_{A1} + \frac{(c_{A2} - c_{A1})z}{\delta}.$$

The B.C. are when $z = 0$, $c_A = c_{A1}$ and when $z = \delta$, $c_A = c_{A2}$.

4-16 Determination of Experimental Diffusivity

An unsteady-state experiment was performed to determine the diffusivity of solute A in a solid. The solid initially had a uniform composition of $c_0 = 1.00 \times 10^{-4}$ g mole A/cm^3. The front face was suddenly subjected to a constant concentration $c_1 = 11.0 \times 10^{-4}$ g mole A/cm^3. After 1.0×10^6 sec the sample was analyzed and the following data obtained of concentration c versus position x.

c, g mole A/cm^3	x, cm
11.0×10^{-4}	0
8.4×10^{-4}	0.05
6.00×10^{-4}	0.10
3.90×10^{-4}	0.15
1.75×10^{-4}	0.25

Assuming the diffusivity is constant, use the experimental data and obtain D by two methods. [*Hint*: See Section 4.5B for the two methods.]

NOTATION

A constant
A cross-sectional area perpendicular to flux, cm^2 or ft^2
$\bar{c}_A$ mean value of c_A
c_A concentration of A, g mole A/cm^3 or lb mole A/ft^3
c concentration, g mole/cm^3
D_1^* self-diffusion coefficient of component 1, cm^2/sec or ft^2/hr
D_A diffusivity of A in a solid, cm^2/sec or ft^2/hr
$\tilde{D}$ chemical diffusivity, cm^2/sec or ft^2/hr
D diameter, cm or ft
D_{AB} diffusivity, cm^2/sec or ft^2/hr
$D_{A\text{effec}}$ effective diffusivity $= \epsilon D_{AB}/k_t^2$, cm^2/sec or ft^2/hr

D_{KA} Knudsen diffusivity, cm^2/sec or ft^2/hr
D_0 constant in solid diffusivity equation
E activation energy, cal/g mole
e dependent variable on analog computer, volts
f Fanning friction factor, dimensionless
g_c 980 g mass-cm/g force-sec^2
ΔH pressure drop, cm of fluid
k first order reaction velocity constant defined in Eq. (4.2-28), sec^{-1}
k_g mass transfer coefficient, cm/sec or ft/hr
k_s first order reaction velocity constant at surface, cm/sec
k_t tortuosity, dimensionless
k_1, k_2, k_3 constants
L distance, cm or ft
m $\sqrt{k/D_A}$, cm^{-1}
M_A molecular weight of A, g mass A/g mole or lb mass A/lb mole
$N_{n,m}$ flux at point n, m in g mole/sec
N'_A g mole of A diffusing
$\bar{N}_A$ flux of A relative to a fixed point, g mole A/sec or lb mole A/hr
N_A flux of A relative to a fixed point, g mole A/sec-cm^2 or lb mole A/hr-ft^2
N_{AM} flux for a 0.10 cm thickness and 1.0 cm Hg pressure difference, g mole A/sec-cm^2
N_{Kn} Knudsen number $= \lambda/2r$, dimensionless
N_1 mole fraction of component 1
p_A partial pressure of A, atm
p total pressure, atm
P total pressure, g mass/cm^2 or atm
P_M permeability, cc gas at N.T.P. (0°C, 1 atm) diffusing per sec per cm^2 cross-sectional area through a solid 1.0 mm thick with a pressure difference of 1.0 cm Hg
P'_M permeability, cc gas at N.T.P. (0°C, 1 atm) diffusing per sec per cm^2 cross-sectional area through a solid 1.0 cm thick with a pressure difference of 1.0 atm
R universal gas law constant, 84,780 g force-cm/°K-g mole
R universal gas law constant, 82.06 cm^3-atm/g mole-°K
R universal gas law constant, 1.987 cal/g mole-°K
R radius, cm or ft
r radius, cm or ft
$\bar{r}$ mean radius, cm
S catalyst surface, cm^2
S solubility, cc solute at N.T.P. (0°C, 1 atm)/cc solid-atm
t time, sec
T absolute temperature, °K or °R
v velocity, cm/sec or ft/sec
V_p volume of catalyst particle, cm^3
$X(x)$ variable dependent only on x

x_A mole fraction of A
x distance in x direction, cm or ft
$Y(y)$ variable dependent only on y
y distance in y direction, cm or ft
z distance in z direction, cm or ft

Greek Letters

α flux ratio, $1 + N_B/N_A$
α analog computer time scale factor in $\tau = \alpha t$
β analog computer amplitude scale factor
γ_1 activity coefficient of component 1, dimensionless
ϵ void fraction, dimensionless
ϵ effectiveness factor defined by Eq. (4.2-31)
θ time, sec or hr
$\theta_{n,m}$ residual at point n, m in Eq. (4.4-18)
λ mean free path, cm
μ viscosity in poises, g mass/cm-sec or lb mass/ft-hr
τ analog computer time, sec

Subscripts

A component A
ave average
B component B
ex exterior surface
g bulk gas phase
i component $1, 2, 3, \ldots, i$
j component $1, 2, 3, \ldots, j$
n position $0, 1, 2, 3, \ldots, n$
m position $0, 1, 2, 3, \ldots, m$
0 value at $\theta = 0$
S surface
x, y, z in the x, y, or z direction
1 beginning of diffusion path
2 end of diffusion path
θ time

REFERENCES

(A1) R. Aris, *Chem. Eng. Sci.* **6**, 265 (1957).
(B1) R. M. Barrer, *Diffusion in and Through Solids*. London: Cambridge University Press, 1941.
(C1) R. S. Cunningham and C. J. Geankoplis, *Ind. Eng. Chem. Fund.* **7**, 535 (1968).
(C2) *Ibid.*, **7**, 429 (1968).

(C3) J. Crank, *The Mathematics of Diffusion*. Oxford: Clarendon Press, 1956.
(C4) J. Crank and G. S. Park, *Diffusion in Polymers*. New York: Academic Press, Inc., 1968.
(D1) G. M. Dusinberre, *Heat Transfer Calculations by Finite Differences*. Scranton, Pa: International Textbook Company, 1961.
(D2) L. S. Darkin and R. W. Gurry, *Physical Chemistry of Metals*. New York: McGraw-Hill, Inc., 1953.
(D3) F. A. L. Dullien and D. S. Scott, *Chem. Eng. Sci.* **17**, 771 (1962).
(H1) J. P. Henry, R. S. Cunningham, and C. J. Geankoplis, *Chem. Eng. Sci.* **22**, 11 (1967).
(J1) W. Jost, *Diffusion in Solids, Liquids, Gases*. New York: Academic Press, Inc., 1960.
(J2) M. Jacob and G. A. Hawkins, *Elements of Heat Transfer and Insulation*, 2d ed. New York: John Wiley & Sons, Inc., 1950.
(L1) L. Lapidus, *Digital Computation for Chemical Engineers*. New York: McGraw-Hill, Inc., 1962.
(L2) O. Levenspiel, *Chemical Reaction Engineering*. New York: John Wiley & Sons, Inc., 1962.
(M1) H. S. Mickley, T. K. Sherwood, and C. E. Reed, *Applied Mathematics in Chemical Engineering*, 2d ed. New York: McGraw-Hill, Inc., 1957.
(R1) L. B. Rothfeld, *AIChEJ*. **9**, 19 (1963).
(R2) R. S. Remick and C. J. Geankoplis, *Ind. Eng. Chem. Fund.* **9**, 206 (1970).
(S1) D. S. Scott and F. A. L. Dullien, *AIChEJ*. **8**, 113 (1962).
(S2) P. G. Shewman, *Diffusion in Solids*. New York: McGraw-Hill, Inc., 1963.
(S3) J. M. Smith, *Chemical Engineering Kinetics*, 2d ed. New York: McGraw-Hill, Inc., 1970.
(W1) N. Wakao and J. M. Smith, *Chem. Eng. Sci.* **17**, 825 (1962).

5 Unsteady-State Diffusion

5.1 DERIVATION OF BASIC EQUATIONS FOR UNSTEADY-STATE DIFFUSION

In Chapter 4, Section 4.4, the following equation was given for unsteady-state transport of mass:

$$\frac{\partial c_A}{\partial \theta} = D_{AB}\left(\frac{\partial^2 c_A}{\partial x^2} + \frac{\partial^2 c_A}{\partial y^2} + \frac{\partial^2 c_A}{\partial z^2}\right) = D_{AB}\nabla^2 c_A \qquad (4.4\text{-}1)$$

where c_A is the concentration of A in g mole A/cm^3 and D_{AB} is molecular diffusivity, cm^2/sec. This equation is only for molecular diffusion in a phase and not for any convective flow, which usually is not present or is neglected in solids, and also for no generation. This same equation was also derived for mass transfer from the continuity Equation (2.6-15).

The equation above will be rederived for mass transfer for variable diffusivities in all three directions. In wood, for example, moisture diffuses differently with the grain than across the grain. First draw an element of a cube as in Fig. 5.1-1. Then, writing

rate of input $+ 0 =$ rate of output $+$ rate of accumulation (5.1-1)

we can write the value for the rate of accumulation as

rate of accumulation $=$ (time rate of change of A in volume) (5.1-2)

rate of accumulation $= \Delta x\, \Delta y\, \Delta z\, \dfrac{\partial c_A}{\partial \theta}$ (5.1-3)

Unsteady-State Diffusion

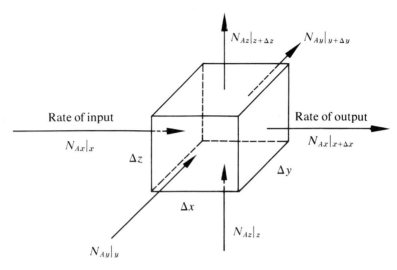

FIGURE 5.1-1 Unsteady-state diffusion in a solid

Substituting Eq. (5.1-3) and the values for the inputs and outputs from Fig. 5.1-1 multiplied by the cross-sectional areas into Eq. (5.1-1),

$$N_{Ax}|_x \, \Delta y \, \Delta z + N_{Ay}|_y \, \Delta x \, \Delta z + N_{Az}|_z \, \Delta x \, \Delta y$$
$$= N_{Ax}|_{x+\Delta x} \, \Delta y \, \Delta z + N_{Ay}|_{y+\Delta y} \, \Delta x \, \Delta z + N_{Az}|_{z+\Delta z} \, \Delta x \, \Delta y + \Delta x \, \Delta y \, \Delta z \frac{\partial c_A}{\partial \theta}$$

(5.1-4)

Dividing by $\Delta x \, \Delta y \, \Delta z$ and letting Δx, Δy, and Δz approach zero,

$$-\left(\frac{\partial N_{Ax}}{\partial x} + \frac{\partial N_{Ay}}{\partial y} + \frac{\partial N_{Az}}{\partial z}\right) = \frac{\partial c_A}{\partial \theta} \qquad (5.1\text{-}5)$$

Substituting the equations for the fluxes,

$$\frac{\partial c_A}{\partial \theta} = \frac{\partial (D_{Ax} \partial c_A / \partial x)}{\partial x} + \frac{\partial (D_{Ay} \partial c_A / \partial y)}{\partial y} + \frac{\partial (D_{Az} \partial c_A / \partial z)}{\partial z} \qquad (5.1\text{-}6)$$

If the diffusivities are constant,

$$\frac{\partial c_A}{\partial \theta} = D_{Ax} \frac{\partial^2 c_A}{\partial x^2} + D_{Ay} \frac{\partial^2 c_A}{\partial y^2} + D_{Az} \frac{\partial^2 c_A}{\partial z^2} \qquad (5.1\text{-}7)$$

If the diffusivities are the same in all directions and constant,

$$\frac{\partial c_A}{\partial \theta} = D_A \left(\frac{\partial^2 c_A}{\partial x^2} + \frac{\partial^2 c_A}{\partial y^2} + \frac{\partial^2 c_A}{\partial z^2}\right) = D_A \nabla^2 c_A \qquad (5.1\text{-}8)$$

Equation (5.1-7) can be converted into a more useful form in the following manner. Let

$$u = x\sqrt{D/D_x}, \qquad v = y\sqrt{D/D_y}, \qquad w = z\sqrt{D/D_z} \qquad (5.1\text{-}9)$$

where D is a constant. Using the first equation from above,

$$\frac{\partial x}{\partial u} = \sqrt{\frac{D_x}{D}} \tag{5.1-10}$$

Also,

$$\frac{\partial c}{\partial u} = \frac{\partial c}{\partial x}\frac{\partial x}{\partial u} = \frac{\partial c}{\partial x}\sqrt{\frac{D_x}{D}} \tag{5.1-11}$$

Taking the second partial with respect to u of Eq. (5.1-11),

$$\frac{\partial^2 c}{\partial u^2} = \frac{\partial^2 c}{\partial x \, \partial u}\sqrt{\frac{D_x}{D}} \tag{5.1-12}$$

Also,

$$\frac{\partial c}{\partial x} = \frac{\partial c}{\partial u}\frac{\partial u}{\partial x} = \frac{\partial c}{\partial u}\sqrt{\frac{D}{D_x}} \tag{5.1-13}$$

Again taking the second partial of Eq. (5.1-13) with respect to x,

$$\frac{\partial^2 c}{\partial x^2} = \frac{\partial^2 c}{\partial x \, \partial u}\sqrt{\frac{D}{D_x}} \tag{5.1-14}$$

Substituting Eq. (5.1-14) into (5.1-12) and rearranging,

$$D_x \frac{\partial^2 c}{\partial x^2} = D\frac{\partial^2 c}{\partial u^2} \tag{5.1-15}$$

Repeating the above for v and w, the final equation is

$$\frac{\partial c}{\partial \theta} = D\left(\frac{\partial^2 c}{\partial u^2} + \frac{\partial^2 c}{\partial v^2} + \frac{\partial^2 c}{\partial w^2}\right) \tag{5.1-16}$$

Hence, this case can be solved by the same methods as used for Eq. (5.1-8).

5.2 SOLUTION OF EQUATIONS BY ANALYTICAL METHODS USING THE METHOD OF SEPARATION OF VARIABLES

5.2A Basic Derivation

The partial differential equations derived and given in Section 5.1 are linear partial differential equations. There are no general formalized analytical procedures for the solution of an arbitrary partial differential equation. The solution of ordinary differential equations involves on integration the introduction of arbitrary constants. However, the solution of partial differential equations involves the introduction of arbitrary functions.

An important property of a linear, homogeneous partial differential

196   Unsteady-State Diffusion

equation is as follows. If each of the functions $f_1, f_2, \ldots, f_n$ is a solution of a linear, homogeneous partial differential equation, then the following function f is also a solution:

$$f = \sum_{1}^{\infty} f_n \tag{5.2-1}$$

This is true provided the infinite series converges and is termwise differentiable as far as the highest derivatives appearing in the original differential equation. In engineering situations the physical situation usually enables one to establish the necessary boundary conditions to ensure a specialized solution of a given equation.

A useful and convenient method to solve unsteady-state diffusion analytically is to express the solution as an expansion in an infinite series of orthogonal functions. This method is not applicable to all partial differential equations but works well for diffusion equations that are homogeneous and linear. Often the expansion is done using a Fourier series.

Since it is not the purpose of this text to go into great mathematical detail, only a short summary of the theory and the method of solution will be given here. This should enable the reader to solve simple problems. The method of the separation of variables will be illustrated in the example that follows. Then the method of solution will be formalized.

Example 5.2-1   Unsteady-State Diffusion in a Stagnant Liquid

A dilute stagnant solution is in a tube of length $2H$ cm. The solution occupies the region between $x = 0$ and $x = 2H$, and the solute is initially present in the solution at a uniform concentration c_0 g mole solute/cm³.

At the time $t = 0$ sec the concentration at the two ends of the tube at $x = 0$ and $x = 2H$ is suddenly lowered to c_1 concentration and held there. It is desired to find the concentration c as a function of position and time, or $c(x, t)$. Diffusion occurs only in the x direction.

Solution

The concentration profile in the liquid is plotted for $c(x, 0)$ at the time $t = 0$ and for $c(x, t)$ in Fig. 5.2-1. Since the solution is assumed dilute, it will be assumed that the density ρ of the solution is constant and that D_{AB} is constant. The general equation for constant ρ and D_{AB} derived in Chapter 2 is as follows, since $(\nabla \cdot \mathbf{v}) = 0$ and $R_A = 0$ for no chemical reaction:

$$\frac{\partial c_A}{\partial t} + (\mathbf{v} \cdot \nabla c_A) = D_{AB} \nabla^2 c_A \tag{5.2-2}$$

5.2 Solution Using Method of Separation of Variables

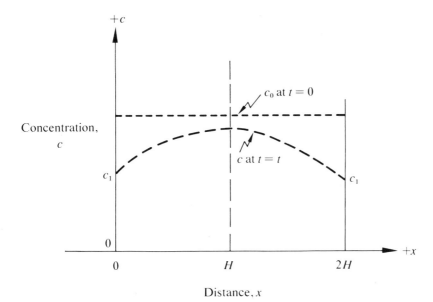

Distance, x

FIGURE 5.2-1 Unsteady-state diffusion in a stagnant liquid for Example 5.2-1

Since this is a stationary liquid with $\mathbf{v} = 0$, and also since diffusion is only in the x direction,

$$\frac{\partial c_A}{\partial t} = D_{AB} \frac{\partial^2 c_A}{\partial x^2} \qquad (5.2\text{-}3)$$

Dropping the subscripts A and B,

$$\frac{\partial c}{\partial t} = D \frac{\partial^2 c}{\partial x^2} \qquad (5.2\text{-}4)$$

Generally it is convenient in solving partial differential equations to redefine the dependent variable $c(x, t)$ so that it is a dimensionless concentration and goes between 0 and 1. Hence,

$$Y = \frac{c_1 - c}{c_1 - c_0} \qquad (5.2\text{-}5)$$

Then after substituting Eq. (5.2-5) into (5.2-4),

$$\frac{\partial Y}{\partial t} = D \frac{\partial^2 Y}{\partial x^2} \qquad (5.2\text{-}6)$$

198   Unsteady-State Diffusion

We can also redefine the boundary conditions and initial conditions as

$$\text{I.C.:} \quad \text{at } x = x, \quad t = 0, \quad Y = \frac{c_1 - c_0}{c_1 - c_0} = 1 \quad (5.2\text{-}7)$$

$$\text{B.C. 1:} \quad \text{at } x = 0, \quad t = t, \quad Y = \frac{c_1 - c_1}{c_1 - c_0} = 0 \quad (5.2\text{-}8)$$

$$\text{B.C. 2:} \quad \text{at } x = +2H, \quad t = t, \quad Y = \frac{c_1 - c_1}{c_1 - c_0} = 0 \quad (5.2\text{-}9)$$

Now we assume that the solution of Eq. (5.2-6) is of the following form, which is the product of two variables:

$$Y(x, t) = X(x) T(t) \quad (5.2\text{-}10)$$

where $X(x)$ is a function of the variable x only and $T(t)$ is a function of the variable t only. Next Eq. (5.2-10) is differentiated twice with respect to x to give

$$\frac{\partial^2 Y}{\partial x^2} = T \frac{\partial^2 X}{\partial x^2} \quad (5.2\text{-}11)$$

Then, differentiating Eq. (5.2-10) with respect to t,

$$\frac{\partial Y}{\partial t} = X \frac{\partial T}{\partial t} \quad (5.2\text{-}12)$$

After substituting Eqs. (5.2-11) and (5.2-12) into (5.2-6),

$$\frac{1}{X} \frac{\partial^2 X}{\partial x^2} = \frac{1}{DT} \frac{\partial T}{\partial t} \quad (5.2\text{-}13)$$

Now X is independent of t by definition, and so as t varies, the left-hand side of Eq. (5.2-13) remains constant. Similarly, the right-hand side of Eq. (5.2-13) remains constant as x varies. As a result, Eq. (5.2-13) must equal a constant, which can be 0, $+a^2$, or $-a^2$. It can be shown that the constant $-a^2$ does not give a trivial answer, and so

$$\frac{1}{X} \frac{\partial^2 X}{\partial x^2} = \frac{1}{DT} \frac{\partial T}{\partial t} = -a^2 \quad (5.2\text{-}14)$$

This means the above is equivalent to two separate differential equations:

$$\frac{dT}{dt} + Da^2 T = 0 \quad (5.2\text{-}15)$$

$$\frac{d^2 X}{dx^2} + a^2 X = 0 \quad (5.2\text{-}16)$$

The two differential equations above can be integrated and give

$$T = C \exp(-a^2 Dt) \quad (5.2\text{-}17)$$
$$X = A \cos ax + B \sin ax \quad (5.2\text{-}18)$$

5.2 Solution Using Method of Separation of Variables

where A, B, and C are constants. After substituting Eqs. (5.2-17) and (5.2-18) into (5.2-10),

$$Y(x, t) = C \exp(-a^2 Dt)(A \cos ax + B \sin ax) \tag{5.2-19}$$

Using B.C. 1 in Eq. (5.2-19),

$$Y(0, t) = 0 = C \exp(-a^2 Dt)(A(1) + B(0)) \tag{5.2-20}$$

However, this can be true only if $A = 0$. Then,

$$Y(x, t) = BC \exp(-a^2 Dt)(\sin ax) \tag{5.2-21}$$

Now, after substituting B.C. 2 into the equation above,

$$Y(0, t) = 0 = BC \exp(-a^2 Dt)(\sin a2H) \tag{5.2-22}$$

This equation can be equal to 0 only if $\sin a2H = 0$, or

$$a = \frac{n\pi}{2H}, \quad n = 1, 2, 3, \ldots \tag{5.2-23}$$

Substituting Eq. (5.2-23) into (5.2-21),

$$Y(x, t) = BC \exp\left(\frac{-n^2\pi^2}{4H^2} Dt\right)\left(\sin \frac{n\pi x}{2H}\right) \tag{5.2-24}$$

The equation above satisfies the boundary conditions for $n = 1, 2, 3, \ldots$. Since the original partial differential equation was linear, then by Eq. (5.2-1) an infinite sum of such terms each multiplied by a suitable constant will satisfy the original equation and the boundary conditions.

$$Y(x, t) = \sum_{n=1}^{\infty} b_n \exp\left(\frac{-n^2\pi^2 Dt}{4H^2}\right)\left(\sin \frac{n\pi x}{2H}\right) \tag{5.2-25}$$

The coefficient b_n is different for each n and must be evaluated. If the reader desired, he may omit the following proof and go directly to Section 5.2B.

Using the I.C. from Eq. (5.2-7) in the above,

$$Y(x, 0) = f(x) = 1 = \sum_{n=1}^{\infty} b_n e^{-0} \sin \frac{n\pi x}{2H} = \sum_{n=1}^{\infty} b_n \sin \frac{n\pi x}{2H} \tag{5.2-26}$$

Next we multiply both sides of Eq. (5.2-26) by $\sin(m\pi x/2H)$ and dx, where m is an integer, in order to evaluate b_n. Then we integrate between the limits of B.C. 1 and 2 or 0 and $2H$. This gives

$$\int_0^{2H} (1) \sin\left(\frac{m\pi x}{2H}\right) dx = \int_0^{2H} \sin\left(\frac{m\pi x}{2H}\right)\left[\sum_{n=1}^{\infty} b_n \sin\left(\frac{n\pi x}{2H}\right)\right] dx \tag{5.2-27}$$

The integral of the sum of many terms may be expressed as the sum of

200 Unsteady-State Diffusion

the integrals.

$$\int_0^{2H} \sin\left(\frac{m\pi x}{2H}\right) dx = \sum_{n=1}^{\infty} b_n \int_0^{2H} \sin\left(\frac{m\pi x}{2H}\right) \sin\left(\frac{n\pi x}{2H}\right) dx \quad (5.2\text{-}28)$$

Integrating the right-hand term,

$$\int_0^{2H} \sin\left(\frac{m\pi x}{2H}\right) \sin\left(\frac{n\pi x}{2H}\right) dx = \left[\frac{\sin\left((m-n)\frac{\pi x}{2H}\right)}{2(m-n)\frac{\pi}{2H}} - \frac{\sin\left((m+n)\frac{\pi x}{2H}\right)}{2(m+n)\frac{\pi}{2H}}\right]_0^{2H}$$
$$(5.2\text{-}29)$$

The terms on the right-hand side are zero except when $m = n$, since m and n are integers. So, setting $m = n$ in Eq. (5.2-29) and integrating,

$$\int_0^{2H} \left[\sin\left(\frac{n\pi x}{2H}\right)\right]^2 dx = H \quad (5.2\text{-}30)$$

Hence, substituting Eq. (5.2-30) into (5.2-28) and setting $m = n$,

$$\int_0^{2H} \sin\left(\frac{n\pi x}{2H}\right) dx = \sum_{n=1}^{\infty} b_n H \quad (5.2\text{-}31)$$

Solving for b_n,

$$b_n = \frac{1}{H} \int_0^{2H} \sin\left(\frac{n\pi x}{2H}\right) dx = -\frac{1}{H}\left[\frac{2H}{n\pi} \cos\left(\frac{n\pi x}{2H}\right)\right]_0^{2H}$$

$$= \frac{2}{n\pi}(1 - (-1)^n) \quad (5.2\text{-}32)$$

Using the value of b_n from Eq. (5.2-32) in (5.2-25), we obtain the final equation:

$$Y(x,t) = \frac{2}{\pi} \sum_{n=1}^{\infty} \left(\frac{1-(-1)^n}{n}\right) \exp\left(\frac{-n^2\pi^2 Dt}{4H^2}\right) \sin\left(\frac{n\pi x}{2H}\right)$$

$$= \frac{4}{\pi}\left[\frac{1}{1}\exp\left(\frac{-1^2\pi^2 Dt}{4H^2}\right) \sin\left(\frac{1\pi x}{2H}\right) + \frac{1}{3}\exp\left(\frac{-3^2\pi^2 Dt}{4H^2}\right) \sin\left(\frac{3\pi x}{2H}\right)\right.$$

$$\left. + \frac{1}{5}\exp\left(\frac{-5^2\pi^2 Dt}{4H^2}\right) \sin\left(\frac{5\pi x}{2H}\right) + \cdots\right] \quad (5.2\text{-}33)$$

The next section gives a method to evaluate the coefficients rapidly.

5.2B Formalized Method Using Euler Formulas

A widely used expansion of orthogonal functions is an infinite Fourier series, where

$$f(x) = a_0 + \sum_{n=1}^{\infty} (a_n \cos nx + b_n \cos nx) \qquad (5.2\text{-}34)$$

The Euler formulas can be used to determine the coefficients.

$$a_0 = \frac{1}{2}\frac{1}{L}\int_{-L}^{L} f(x)\,dx \qquad (5.2\text{-}35)$$

$$a_n = \frac{1}{L}\int_{-L}^{L} f(x) \cos\left(\frac{n\pi x}{L}\right) dx \qquad (5.2\text{-}36)$$

$$b_n = \frac{1}{L}\int_{-L}^{L} f(x) \sin\left(\frac{n\pi x}{L}\right) dx \qquad (5.2\text{-}37)$$

The factor $1/L$ in front of the integrals is 2/length of the interval in the boundary conditions. For the above, this is $2/[L-(-L)] = 2/(2L) = 1/L$. If the interval of integration were 0 to L, the factor in front would be $2/L$.

If a certain special case arises where $a_0 = 0$ and $a_n = 0$, a sine series is obtained. For $b_n = 0$ a cosine series is obtained.

These formulas will be used to obtain b_n for Example 5.2-1. Using Eqs. (5.2-7) to (5.2-9) in (5.2-37),

$$b_n = \frac{2}{2H}\int_0^{2H} f(x) \sin\left(\frac{n\pi x}{2H}\right) dx = \frac{1}{H}\int_0^{2H} (1) \sin\left(\frac{n\pi x}{2H}\right) dx$$

$$= \left[\frac{-2H}{n\pi H} \cos\left(\frac{n\pi x}{2H}\right)\right]_0^{2H} = \frac{2}{n\pi}[1 - (-1)^n] \qquad (5.2\text{-}38)$$

The results given above for b_n are the same as those in Eq. (5.2-32).

5.2C Unsteady-State Charts

Many cases for unsteady-state mass transfer and/or heat transfer have been solved in the literature. For the case of mass transfer only in the x direction to both sides of a slab of thickness $2x_1$, where the slab is at an initial uniform concentration of c_{A0} at $\theta = 0$ and all x values, and the two surfaces are suddenly subjected to a concentration c_{A1}, the final

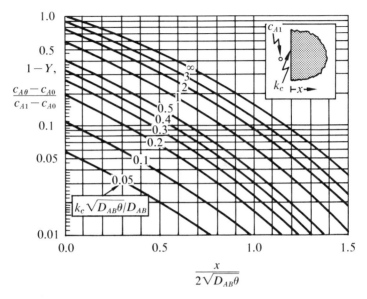

FIGURE 5.2-2 Mass transfer unsteady-state to a semi-infinite solid. (From P. J. Schneider, *Conduction Heat Transfer*. Cambridge, Mass.: Addison-Wesley Publishing Co., Inc., 1955. With permission.)

equation solved by Fourier series is similar to Eq. (5.2-33) and is

$$\frac{c_{A1} - c_{A\theta}}{c_{A1} - c_{A0}} = Y$$

$$= \frac{4}{\pi}\left(\frac{1}{1}e^{-\pi^2 X/4}\sin\frac{\pi x}{2x_1} + \frac{1}{3}e^{-9\pi^2 X/4}\sin\frac{3\pi x}{2x_1} + \frac{1}{5}e^{-25\pi^2 X/4}\sin\frac{5\pi x}{2x_1} + \cdots\right)$$

(5.2-39)

where x = distance from center of slab to a point in slab,
   $X = D\theta/x_1^2$,
   $c_{A\theta}$ = concentration at time = θ,
   Y = fraction of unaccomplished change,
   $1 - Y$ = fraction of change = $(c_{A\theta} - c_{A0})/(c_{A1} - c_{A0})$

Charts or graphs of Eq. (5.2-39) for a slab and others for spheres, cylinders, and semi-infinite solids are given in many references and are reproduced in Figs. 5.2-2 to 5.2-5.

Figure 5.2-2 is for the condition of unsteady-state diffusion to a semi-infinite solid with surface convection. If the diffusion in the solid is slow enough or if k_c is very large, the top line with $k_c\sqrt{D_{AB}\theta}/D_{AB} = \infty$ is used. The parameters to be used for heat transfer in the charts here and in the appendix are given below along with the mass transfer analogs.

Mass transfer	Heat transfer
$\dfrac{k_c}{D_{AB}}\sqrt{D_{AB}\theta}$	$\dfrac{h}{k}\sqrt{\alpha\theta}$
$\dfrac{x}{2\sqrt{D_{AB}\theta}}$	$\dfrac{x}{2\sqrt{\alpha\theta}}$
$1-Y, \dfrac{c_{A\theta}-c_{A0}}{c_{A1}-c_{A0}}$	$\dfrac{T_\theta-T_0}{T_1-T_0}$
$Y, \dfrac{c_{A1}-c_{A\theta}}{c_{A1}-c_{A0}}$	$\dfrac{T_1-T_\theta}{T_1-T_0}$
$X, \dfrac{D_{AB}\theta}{x_1^{\,2}}$	$\dfrac{\alpha\theta}{x_1^{\,2}}$
$m, \dfrac{D_{AB}}{k_c x_1}$	$\dfrac{k}{h x_1}$
$n, \dfrac{x}{x_1}$	$\dfrac{x}{x_1}$

(x is distance from center of slab, cylinder, or sphere. x_1 is one-half the thickness of slab, radius of cylinder or sphere. For a semiinfinite slab x is the distance from the surface.) Note that these charts hold only for the solid initially having a uniform concentration and the boundary condition imposed on the outside remaining constant with time. This limits these charts somewhat. In the next section methods will be discussed to handle variable initial concentrations and boundary conditions.

A set of more comprehensive unsteady-state charts are given in Appendix A.3. These charts by Heisler (H1) cover a wider range and allow easier interpolation. In these charts Y_n/Y_0 is the ratio of the value of Y at $n = n$ to the value of Y at $n = 0$.

Example 5.2-2 Unsteady-State Diffusion in a Thin Slab

A wet solid slab made of agar gel at 5°C contains a uniform concentration of 1.0×10^{-4} g mole of urea/cm^3. The slab is 0.40 in. thick. Diffusion is only through two parallel flat surfaces 0.40 in. apart. At zero time the slab is immersed in pure fast-running water so that it can be assumed there is no surface resistance to diffusion. The diffusivity of urea is 4.70×10^{-6} cm^2/sec or 18.3×10^{-6} ft^2/hr. (a) What is the concentration at the midpoint of the slab (0.20 in. from the surface) and at a distance of 0.10 in. from the surface after 10 hr? (b) If the thickness of the slab is halved, what is the midpoint concentration in 10 hr?

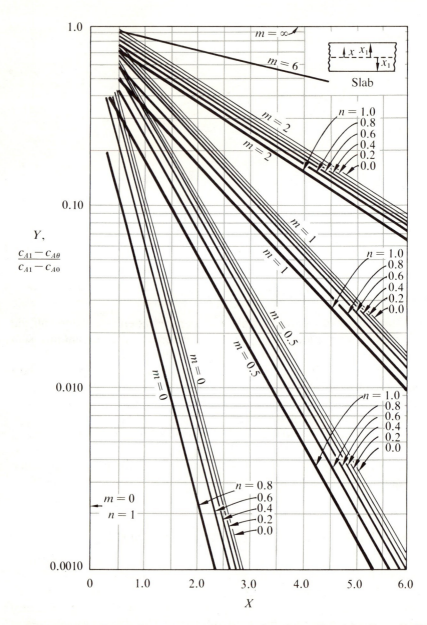

FIGURE 5.2-3 Unsteady-state mass transport in a large flat slab. [From H. P. Gurney and J. Lurie, *Ind. Eng. Chem.* **15**, 1170 (1923). With permission.]

5.2 Solution Using Method of Separation of Variables

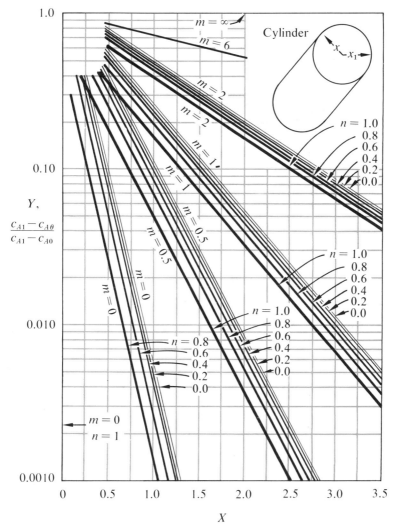

FIGURE 5.2-4 Unsteady-state mass transport in a long cylinder. [From H. P. Gurney and J. Lurie, *Ind. Eng. Chem.* **15**, 1170 (1923). With permission.]

Solution

$c_{A0} = 1.0 \times 10^{-4}$ g mole urea/cm^3. $c_{A1} = 0$ for fresh running water. $c_{A\theta}$ = concentration at time θ and distance x from slab centerline. $Y = (c_{A1} - c_{A\theta})/(c_{A1} - c_{A0}) = (0 - c_{A\theta})/(0 - 1.0 \times 10^{-4})$. $x_1 = 0.2/12 = 0.0167$ ft, $x = 0$ for center. Then $X = D\theta/x_1^2 = 18.3 \times 10^{-6}(10)/(0.0167)^2 = 0.656$. The relative position $n = x/x_1 = 0/0.0167 = 0$, and relative resistance

206 Unsteady-State Diffusion

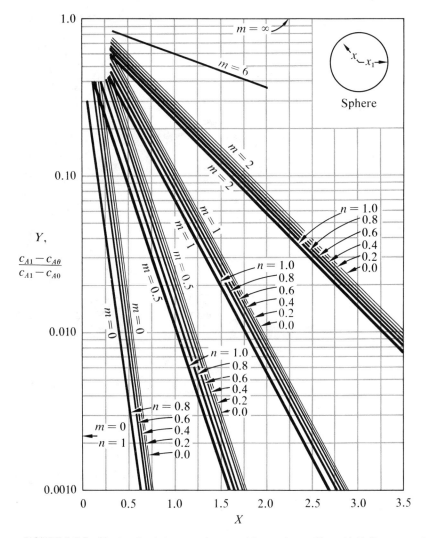

FIGURE 5.2-5 Unsteady-state mass transport in a sphere. [From H. P. Gurney and J. Lurie, *Ind. Eng. Chem.* **15**, 1170 (1923). With permission.]

$m = (D_{AB}/x_1)/k_c = 0$, since k_c is large (zero resistance). From Fig. 5.2-3, for $X = 0.656$, $Y = 0.275 = (0 - c_{A\theta})/(0 - 1.0 \times 10^{-4})$. Hence, $c_{A\theta} = 0.275 \times 10^{-4}$.

For the position $x = 0.10$ in./12 $= 0.00835$ ft, $X = 0.656$, $m = 0$, $n = 0.00835$ ft/0.0167 ft $= 0.5$. Hence, $Y = 0.172$. Then, $c_{A\theta} = 0.172 \times 10^{-4}$ [Ans. to (a)]. For half the thickness $X = 0.656/0.5^2 = 2.624$, $n = 0$, and $m = 0$. Hence, $Y = 0.0020$. Then $c_{A\theta} = 0.0020 \times 10^{-4}$ [Ans. to (b)].

5.2 Solution Using Method of Separation of Variables 207

Example 5.2-3 Unsteady-State Diffusion and Convection

Repeat Example 5.2-2, part (a), for the case where $k_c = 2.19 \times 10^{-3}$ cm/sec, using the Heisler charts in the appendix.

Solution

As before, $X = 0.656$ and $n = 0$ (center). $m = (D_{AB}/x_1)/k_c = (18.3 \times 10^{-6}/0.0167)/(2.19 \times 10^{-3}) = 0.5$. In A.3-3 for $X = 0.656$, $n = 0$, and $m = 0.5$, $Y_0 = 0.53$ (at center of plate). This fraction of unaccomplished change is greater than that for $m = 0$, as expected. For $n = 0.5$, $m = 0.5$, and $X = 0.656$, from A.3-7, the position correction factor Y_n/Y_0 is 0.855. Then $Y_n = 0.855(0.53) = 0.45$.

Note that the curves for the slab can be used only for mass transfer to two parallel faces. Newman (N1) used the principle of superposition and showed mathematically how to combine the solutions for one-dimensional momentum, heat, or mass transfer in the x, the y, and the z direction into an overall solution for simultaneous transfer in all three directions. For example, we are given a rectangular block as shown in Fig. 5.2-6 with dimensions $2x_1$, $2y_1$, and $2z_1$. Then, to calculate the Y value for transfer in the x direction,

$$Y_x = \frac{c_{A1} - c_{A\theta,x}}{c_{A1} - c_{A0}} \qquad (5.2\text{-}40)$$

where $c_{A\theta,x}$ is the concentration at time θ and position x distance from centerline of slab. Also, $n = x/x_1$, $m = (D/x_1)/k_c$ as before.

$$X_x = \frac{D\theta}{x_1^2} \qquad (5.2\text{-}41)$$

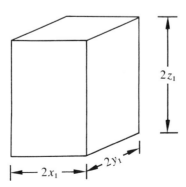

FIGURE 5.2-6 Transfer in three directions

208 Unsteady-State Diffusion

Then Y_y is defined as $(c_{A1} - c_{A\theta,y})/(c_{A1} - c_{A0})$, X_y as $D\theta/y_1^2$, and $n = y/y_1$, and so on. Similarly for Y_z and X_z. Then, for simultaneous transfer,

$$Y_{x,y,z} = (Y_x)(Y_y)(Y_z) = \frac{c_{A1} - c_{A\theta,x,y,z}}{c_{A1} - c_{A0}} \tag{5.2-42}$$

where $c_{A\theta,x,y,z}$ is the concentration at the point x, y, z from the center of the slab. Similar reasoning holds for a cylinder and its ends.

Example 5.2-4 Unsteady-State Diffusion in Three Directions

Use the conditions in Example 5.2-2 except that the slab is 0.40 in. thick in the x direction, 0.3 in. thick in the y direction, and 0.40 in. thick in the z direction and mass transfer occurs at all six faces. Calculate the concentration at the midpoint after 10 hr.

Solution

$X_x = D\theta/x_1^2 = 18.3 \times 10^{-6}(10)/(0.0167)^2 = 0.656$, $n = 0$, $m = 0$. From Fig. 5.2-3 $Y_x = 0.275$ as in the previous example. Then, $X_y = D\theta/y_1^2 = 18.3 \times 10^{-6}(10)/(0.15/12)^2 = 1.16$. Then $Y_y = 0.082$. $X_z = 0.656$, $Y_z = 0.275$. Then, by Eq. (5.2-42),

$$Y_{x,y,z} = Y_x Y_y Y_z = (0.275)(0.082)(0.275) = 6.22 \times 10^{-3}$$

$$\frac{c_{A1} - c_{A\theta,x,y,z}}{c_{A1} - c_{A0}} = \frac{0 - c_{A\theta,x,y,z}}{0 - 1.0 \times 10^{-4}} = 6.22 \times 10^{-3}$$

Hence, $c_{A\theta}$ at the center $= 6.22 \times 10^{-3}(1.0 \times 10^{-4}) = 6.22 \times 10^{-7}$.

If it is assumed that the resistance to diffusion in the surrounding fluid is negligible—that is, the mass transfer coefficient is very large—the curves given in Fig. 5.2-7 will give the total fraction unremoved, E, for slabs, cylinders, or spheres. This value of E is

$$E = \frac{c_{A\theta_{ave}} - c_{A1}}{c_{A0} - c_{A1}} \tag{5.2-43}$$

where $c_{A\theta_{ave}}$ is the average concentration in the solid at time θ.

The values of E_a, E_b, or E_c are each used for diffusion between a pair of parallel faces. For example, for diffusion in the a and b directions in a rectangular bar,

$$E = E_a E_b \tag{5.2-44}$$

For diffusion from all three sets of faces,

$$E = E_a E_b E_c \tag{5.2-45}$$

5.2 Solution Using Method of Separation of Variables

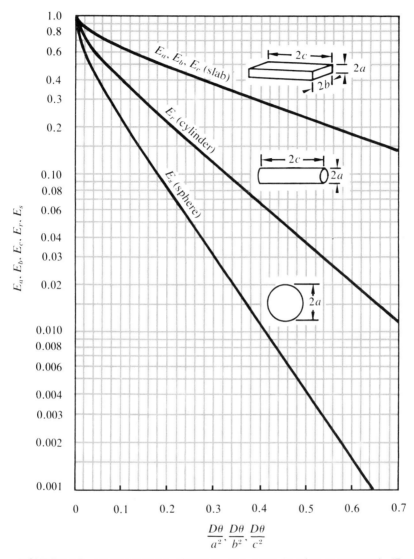

FIGURE 5.2-7 Unsteady-state diffusion and total fraction unremoved. (From R. E. Treybal, *Mass Transfer Operations*, 2d ed. New York: McGraw-Hill, Inc., 1968. With permission.)

For diffusion from a short cylinder $2c$ long and radius a,

$$E = E_c E_r \qquad (5.2\text{-}46)$$

5.3 SOLUTION OF EQUATIONS BY ANALYTICAL METHODS USING THE LAPLACE TRANSFORM

5.3A Basic Theory of Laplace Transform

In general, partial differential equations in transport processes are usually solved by the following methods: separation of variables (discussed in the previous section), numerical methods (to be discussed later in this chapter), complex variables, and operational calculus or Laplace transform.

The Laplace transform is very useful in solving linear partial differential diffusion equations. This method introduces the constants of integration early in the analysis and evaluates them automatically. The Laplace transform removes the time variable in the diffusion equation, leaving an ordinary differential equation, which can be easily solved.

The Laplace transform can be defined in the following manner. Suppose $f(t)$ is a known function of t defined for all positive values of t. Then the Laplace transform, $f(p)$, of $f(t)$ is defined as

$$f(p) = \mathscr{L}[f(t)] = \int_0^\infty e^{-pt} f(t)\, dt \qquad (5.3\text{-}1)$$

In this formula the function is multiplied by $e^{-pt}\, dt$ and integrated. The term p is a number sufficiently large to make the integral converge. It can be a complex number with its real part large, but for the present discussion it can be thought of as simply a positive number.

The Laplace transforms of simple functions are easily obtained. Several examples will be considered.

Given the function $f(t) = 1.0$, then

$$\mathscr{L}[f(t)] = f(p) = \int_0^\infty e^{-pt}(1)\, dt = \left.\frac{e^{-pt}}{-p}\right]_0^\infty = 0 - \frac{e^{-p(0)}}{-p} = \frac{1}{p} \qquad (5.3\text{-}2)$$

Given the function $f(t) = e^{at}$, then

$$f(p) = \int_0^\infty e^{-pt}(e^{at})\, dt = \int_0^\infty e^{-(p-a)t}\, dt = \left.\frac{e^{-(p-a)t}}{-(p-a)}\right]_0^\infty = \frac{1}{p-a} \qquad (5.3\text{-}3)$$

Given the function $f(t) = \sin bt$, then

$$f(p) = \int_0^\infty e^{-pt}(\sin bt)\, dt = \left.\frac{e^{-pt}}{p^2+b^2}(-p\sin bt - b\cos bt)\right]_0^\infty = \frac{b}{p^2+b^2} \qquad (5.3\text{-}4)$$

Given two functions $f(t)$ and $g(t)$, the Laplace of the sum of these two functions is

$$\mathscr{L}[f(t) + g(t)] = \mathscr{L}[f(t)] + \mathscr{L}[g(t)] \qquad (5.3\text{-}5)$$

5.3 Solution of Equations Using Laplace Transform

The obtaining of the Laplace of a derivative of a function where $f'(t) = d[f(t)]/dt$ is quite important.

$$\mathscr{L}[f'(t)] = \int_0^\infty e^{-pt} f'(t)\, dt \qquad (5.3\text{-}6)$$

This can be integrated by parts.

$$\int_0^\infty U\, dV = UV\Big|_0^\infty - \int_0^\infty V\, dU \qquad (5.3\text{-}7)$$

Setting $U = e^{-pt}$ and $dV = f'(t)\, dt$, then $dU = -pe^{-pt}\, dt$ and $V = f(t)$. Substituting into Eq. (5.3-7),

$$\mathscr{L}[f'(t)] = \int_0^\infty e^{-pt} f'(t)\, dt = e^{-pt} f(t)\Big|_0^\infty + p\int_0^\infty e^{-pt} f(t)\, dt \qquad (5.3\text{-}8)$$

$$\mathscr{L}[f'(t)] = 0 - f(0) + p\mathscr{L}[f(t)]$$
$$= pf(p) - f(0)$$

where $f(0)$ is the value of $f(t)$ when $t = 0$ (a boundary condition).

The Laplace of a second derivative $f''(t)$ is obtained in a similar manner:

$$\mathscr{L}[f''(t)] = p^2 f(p) - pf(0) - f'(0) \qquad (5.3\text{-}9)$$

where $f'(0)$ is the value of $f'(t)$ at $t = 0$.

Very extensive tables of Laplace transforms have been worked out and are available in many common mathematics handbooks—Perry (P1) and others. An abbreviated table of transforms occurring frequently in diffusion problems is given in Table 5.3-1.

To obtain the inverse transform—that is, to reverse the steps and obtain $f(t)$ from $f(p)$—one goes to the tables and obtains the $f(t)$

Table 5.3-1 Short Table of Laplace Transforms

	$f(t)$	$f(p)$
(1)	1	$1/p$
(2)	t	$1/p^2$
(3)	t^n (n a positive integer)	$n!/p^{n+1}$
(4)	$1/\sqrt{\pi t}$	$1/\sqrt{p}$
(5)	e^{at}	$1/(p-a)$
(6)	$\sin \omega t$	$\omega/(p^2 + \omega^2)$
(7)	$\cos \omega t$	$p/(p^2 + \omega^2)$
(8)	$\sinh at$	$a/(p^2 - a^2)$
(9)	$\cosh at$	$p/(p^2 - a^2)$
(10)	erfc $(k/2\sqrt{t})$	$(1/p)e^{-k\sqrt{p}}, k \geq 0$
(11)	$t \cos at$	$(p^2 - a^2)/(p^2 + a^2)^2$

corresponding to the $f(p)$. To illustrate this a simple differential equation will be solved by Laplace transform methods.

Given the following equation and boundary conditions, it is desired to solve the equation:

$$f''(t) - 3f'(t) + 2f(t) = 0 \qquad (5.3\text{-}10)$$

when $t = 0, f(t) = 0;\ t = 0, f'(t) = 1.0$

First the Laplace is taken of both sides.

$$[p^2 f(p) - pf(0) - f'(0)] - 3[pf(p) - f(0)] + 2[f(p)] = 0 \qquad (5.3\text{-}11)$$

The first term in the brackets is the Laplace of the second derivative. Substituting 1.0 for $f'(0)$ and 0 for $f(0)$,

$$p^2 f(p) - 1 - 3pf(p) + 2f(p) = 0 \qquad (5.3\text{-}12)$$

Solving for $f(p)$,

$$f(p) = \frac{1}{(p-1)(p-2)} \qquad (5.3\text{-}13)$$

Using the method of partial fractions, the above is converted to

$$f(p) = \frac{-1}{(p-1)} + \frac{1}{(p-2)} \qquad (5.3\text{-}14)$$

Next, taking the inverse of both sides, the left-hand term becomes $f(t)$; the next term, using Table 5.3-1, becomes $-e^{-t}$; and similarly the last term e^{2t}. Hence,

$$f(t) = -e^t + e^{2t} \qquad (5.3\text{-}15)$$

Of course this differential equation could have been solved easily by other methods, but it was solved in the manner above to illustrate the Laplace transform method.

5.3B Solution of Unsteady-State Diffusion by Laplace Transform

The Laplace transform method has some of its greatest utility in solution of unsteady-state diffusion equations. The method used will be illustrated by a simple example.

Example 5.3-1 Unsteady-State Diffusion in a Semiinfinite Medium

Let us consider the case of unsteady-state diffusion in a semiinfinite medium extending from $x = 0$ to $x \to \infty$. The solid medium for $x > 0$ is shown in Fig. 5.3-1 and has an initial uniform concentration profile of $c_1 = $ constant at $t = 0$. At a given time $t = 0$, the boundary is suddenly

5.3 Solution of Equations Using Laplace Transform

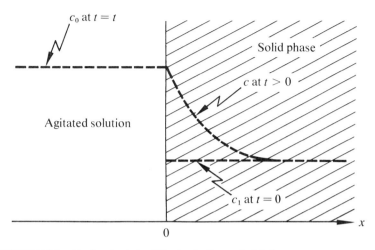

FIGURE 5.3-1 Unsteady-state diffusion in Example 5.3-1

raised to $c = c_0 =$ constant and held there. This is done by having a large volume of highly agitated solution of concentration c_0. The convection coefficient k_c is so high that there is no resistance in the solution to diffusion. Solve the partial diffential equation using the Laplace transform.

Solution

The original partial differential equation for diffusion in the solid is

$$\frac{\partial c}{\partial t} = D\frac{\partial^2 c}{\partial x^2} \tag{5.3-16}$$

I.C.: at $x = x$, $t = 0$, $c = c_1$
B.C. 1: at $x = 0$, $t = t$, $c = c_0$
B.C. 2: at $x \to \infty$, $t = t$, $c = c_1$

First we take the Laplace of $\partial c/\partial t$. This is simply the Laplace of a derivative, and we will use the symbol $c(x, p)$ instead of $f(p)$, since it still depends on the value of x selected.

$$\mathscr{L}\left(\frac{\partial c}{\partial t}\right) = pc(x, p) - f(0) = pc(x, p) - c_1 \tag{5.3-17}$$

Note that the I.C. is used here for $f(0)$, which equals c_1.

Next we write out the formula for the Laplace of $D\ \partial^2 c/\partial x^2$ as

$$\mathscr{L}\left(D\frac{\partial^2 c}{\partial x^2}\right) = D\int_0^\infty e^{-pt}\frac{\partial^2 c(x, t)}{\partial x^2}\, dt \tag{5.3-18}$$

The order of integration and differention in Eq. (5.3-18) can be interchanged.

214 Unsteady-State Diffusion

$$D \int_0^\infty e^{-pt} \frac{\partial^2 c(x,t)}{\partial x^2} dt = D \frac{\partial^2}{\partial x^2} \left[\int_0^\infty e^{-pt} c(x,t)\, dt \right] = D \frac{\partial^2 c(x,p)}{\partial x^2}$$
(5.3-19)

The term in the square brackets is simply the $\mathscr{L}[c(x,t)]$. Substituting Eqs. (5.3-17) and (5.3-19) into (5.3-16), we obtain the following ordinary differential equation with the time variable removed. On rearranging,

$$\frac{d^2 c(x,p)}{dx^2} - \frac{p}{D} c(x,p) = \frac{-c_1}{D}$$
(5.3-20)

The general solution of this nonhomogeneous differential equation is

$$c(x,p) = A_1 e^{-\sqrt{p/D}\, x} + A_2 e^{\sqrt{p/D}\, x} + \frac{c_1}{p}$$
(5.3-21)

Before the boundary conditions can be used to solve for the integration constants A_1 and A_2, they must also be transformed. Taking the Laplace of both sides of B.C. 1 and B.C. 2,

B.C. 1: $\quad \mathscr{L}[c(0,t)] = c(0,p) = \mathscr{L}(c_0) = \dfrac{c_0}{p}$

B.C. 2: $\quad \mathscr{L}[\underset{x\to\infty}{c(x,t)}] = \underset{x\to\infty}{c(x,p)} = \mathscr{L}(c_1) = \dfrac{c_1}{p}$
(5.3-22)

Using B.C. 2 for $x \to \infty$, Eq. (5.3-21) becomes

$$\underset{x\to\infty}{c(x,p)} = \frac{c_1}{p} = A_1(0) + A_2 e^\infty + \frac{c_1}{p}$$
(5.3-23)

Hence, A_2 must be 0. Using B.C. 1 in Eq. (5.3-21),

$$c(0,p) = \frac{c_0}{p} = A_1(1) + \frac{c_1}{p}$$
(5.3-24)

Thus, $A_1 = (c_0 - c_1)/p$ and Eq. (5.3-21) becomes

$$c(x,p) = (c_0 - c_1) \frac{e^{-\sqrt{p/D}\, x}}{p} + \frac{c_1}{p}$$
(5.3-25)

Finally the inverse transform is taken of both sides. The left-hand side becomes $c(x,t)$. The middle term is the same as formula (10) in Table 5.3-1, where k is $x/\sqrt{D}$, and this gives erfc $(x/2\sqrt{Dt})$. The inverse of the right-hand term is just c_1. The final equation is

$$c(x,t) = c_1 + (c_0 - c_1)\, \text{erfc}\left(\frac{x}{2\sqrt{Dt}}\right)$$
(5.3-26)

where erfc $(x/2\sqrt{Dt}) = 1 - \text{erf}\,(x/2\sqrt{Dt})$

5.3 Solution of Equations Using Laplace Transform

Table 5.3-2 Table of Error Function

$$\text{erf } z = \frac{2}{\sqrt{\pi}} \int_0^z e^{-\eta^2} d\eta$$

z	erf z	erfc z	z	erf z	erfc z
0	0	1.0	1.1	0.880205	0.119795
0.05	0.056372	0.943628	1.2	0.910314	0.089686
0.1	0.112463	0.887537	1.3	0.934008	0.065992
0.15	0.167996	0.832004	1.4	0.952285	0.047715
0.2	0.222703	0.777297	1.5	0.966105	0.033895
0.25	0.276326	0.723674	1.6	0.976348	0.023652
0.3	0.328627	0.671373	1.7	0.983790	0.016210
0.35	0.379382	0.620618	1.8	0.989091	0.010909
0.4	0.428392	0.571608	1.9	0.992790	0.007210
0.45	0.475482	0.524518	2.0	0.995322	0.004678
0.5	0.520500	0.479500	2.1	0.997021	0.002979
0.55	0.563323	0.436677	2.2	0.998137	0.001863
0.6	0.603856	0.396144	2.3	0.998857	0.001143
0.65	0.642029	0.357971	2.4	0.999311	0.000689
0.7	0.677801	0.322199	2.5	0.999593	0.000407
0.75	0.711156	0.288844	2.6	0.999764	0.000236
0.8	0.742101	0.257899	2.7	0.999866	0.000134
0.85	0.770668	0.229332	2.8	0.999925	0.000075
0.9	0.796908	0.203092	2.9	0.999959	0.000041
0.95	0.820891	0.179109	3.0	0.999978	0.000022
1.0	0.842701	0.157299			

The term $\text{erf }(x/2\sqrt{Dt})$ is the infinite series error function, and numerical values are tabulated in standard tables. An abbreviated compilation is given in Table 5.3-2. This error function has the following properties.

$$\text{erfc } z = 1 - \text{erf } z, \text{ erf } (-z) = -\text{erf } z, \text{ erf } (0) = 0, \text{ erf } (\infty) = 1 \quad (5.3\text{-}27)$$

Equation (5.3-26) can be rearranged to give

$$Y = \frac{c_0 - c}{c_0 - c_1} = \text{erf}\left(\frac{x}{2\sqrt{Dt}}\right) \quad (5.3\text{-}28)$$

This equation above corresponds to the line marked ∞ in Fig. 5.2-2, where $(k_c/D_{AB})\sqrt{D_{AB}\theta}$ is very large because of k_c being very large. For example, for $(x/2\sqrt{Dt}) = 0.5$, $\text{erf}(x/2\sqrt{Dt}) = 0.5205$ from Table 5.3-2. This means $Y = 0.5205$ or $1 - Y = 0.4795$, which checks the value of $1 - Y$ from Fig. 5.2-2 of 0.48.

5.4 INTRODUCTION TO NUMERICAL METHODS USING THE EXPLICIT METHOD WITH THE DIGITAL COMPUTER

As discussed in the beginning of this chapter, the partial differential equations can usually be solved analytically if they are linear and the boundary conditions are constant with time. Then the unsteady-state charts are often used (only, however, if the initial concentration is uniform with position). However, when the boundary conditions are not constant but vary with time and/or the initial conditions are not constant with position, then numerical methods must be used.

In Chapter 2 and the beginning of this chapter the partial differential equation relating C_A with position x and time θ, often called a parabolic type equation, was given as follows:

$$\frac{\partial C_A}{\partial \theta} = D_{AB} \frac{\partial^2 C_A}{\partial x^2} \qquad (5.4\text{-}1)$$

where D_{AB} is the diffusivity, which is constant. The partial derivative on the left can be expressed as a finite difference:

$$\frac{\partial C_A}{\partial \theta} = \frac{_{\theta+\Delta\theta}C_{Ax} - {_\theta}C_{Ax}}{\Delta \theta} \qquad (5.4\text{-}2)$$

where ${_\theta}C_{Ax}$ is the concentration of A at position x at time θ and ${_{\theta+\Delta\theta}}C_{Ax}$ is the concentration at the same position x at time $\theta + \Delta\theta$.

The second partial on the right can be written as

$$\frac{\partial^2 C_A}{\partial x^2} = \frac{\partial(\partial C_A/\partial x)}{\partial x} = \frac{\left(\frac{_\theta C_{Ax+\Delta x} - {_\theta}C_{Ax}}{\Delta x}\right) - \left(\frac{_\theta C_{Ax} - {_\theta}C_{Ax-\Delta x}}{\Delta x}\right)}{\Delta x}$$

$$= ({_\theta}C_{Ax+\Delta x} - 2{_\theta}C_{Ax} + {_\theta}C_{Ax-\Delta x})/(\Delta x)^2 \qquad (5.4\text{-}3)$$

where ${_\theta}C_{Ax+\Delta x}$ is the concentration at the position Δx to the right or forward from x and ${_\theta}C_{Ax-\Delta x}$ the concentration backwards a distance of Δx from x.

Substituting Eqs. (5.4-2) and (5.4-3) into (5.4-1) and rearranging,

$$_{\theta+\Delta\theta}C_{Ax} = \frac{1}{M}[{_\theta}C_{Ax+\Delta x} + (M-2){_\theta}C_{Ax} + {_\theta}C_{Ax-\Delta x}] \qquad (5.4\text{-}4)$$

where $M = (\Delta x)^2/(D_{AB} \Delta\theta)$ and is a constant. In many texts β is used for $1/M$ as follows:

$$\frac{1}{M} = \beta = \frac{D_{AB}\Delta\theta}{(\Delta x)^2} \qquad (5.4\text{-}5)$$

Note that in Eq. (5.4-4) the concentration ${_{\theta+\Delta\theta}}C_{Ax}$ at position x and at a new time $\theta + \Delta\theta$ is calculated from three points given at time θ or the starting time. This is called the explicit method, because the concentration

5.4 Numerical Explicit Method and Digital Computer

at a new time can be calculated explicitly from the concentrations at the previous time.

For stability and convergence M must be ≥ 2 or $\beta \leq \frac{1}{2}$. Stability means that the errors in the numerical solution do not grow exponentially as the solution proceeds but damp out. Convergence means that the solution of the difference equation approaches the exact solution of the partial differential equation as $\Delta\theta$ and Δx go to zero with β or M fixed. Using smaller sizes of the intervals $\Delta\theta$ and Δx increases the accuracy in general but greatly increases the number of calculations required. Hence, a digital computer is ideally suited for such computations. The program is relatively simple, and the storage requirements are modest.

If $M = 2$, then the well-known Schmidt method is obtained from Eq. (5.4-4):

$$_{\theta+\Delta\theta}C_{Ax} = \frac{_{\theta}C_{Ax+\Delta x} + {_{\theta}C_{Ax-\Delta x}}}{2} \tag{5.4-6}$$

An alternate method will be given for deriving the equations for a slab where the boundary conditions at the front and rear faces are different. Figure 5.4-1 shows a slab with the position x in Eq. (5.4-4) corresponding to n in the figure. The figure shows the concentration profile of A, called C, at a given time instant θ. Making a mole balance on A on the half of the slab to the left and right of n, then the rate of moles in − rate of moles out = the rate of accumulation in $\Delta\theta$ hours:

$$\frac{D_{AB}A({_{\theta}C_{n-1}} - {_{\theta}C_n})}{\Delta x} - \frac{D_{AB}A({_{\theta}C_n} - {_{\theta}C_{n+1}})}{\Delta x} = \frac{A\Delta x({_{\theta+\Delta\theta}C_n} - {_{\theta}C_n})}{\Delta\theta} \tag{5.4-7}$$

where A is cross-sectional area and $_{\theta+\Delta\theta}C_n$ is the concentration at point n one $\Delta\theta$ later. Rearranging,

$$_{\theta+\Delta\theta}C_n = \frac{1}{M}[{_{\theta}C_{n+1}} + (M-2){_{\theta}C_n} + {_{\theta}C_{n-1}}] \tag{5.4-8}$$

which is the same as Eq. (5.4-4).

For the boundary condition at the front for the case where there is a finite resistance at the surface and the concentration of the fluid outside is C_a, then the mass entering at the surface is $k_c A({_{\theta}C_a} - {_{\theta}C_0})$, so a balance on the outside half element is

$$k_c A({_{\theta}C_a} - {_{\theta}C_0}) - \frac{D_{AB}A({_{\theta}C_0} - {_{\theta}C_1})}{\Delta x} = \frac{(A\Delta x/2)({_{\theta+\Delta\theta}C_{0.25}} - {_{\theta}C_{0.25}})}{\Delta\theta} \tag{5.4-9}$$

where $_{\theta}C_{0.25}$ is the concentration at the midpoint of the $0.5\Delta x$ outside slab. As an approximation, the concentration at the surface can be used to replace that of $0.25\Delta x$. Then, rearranging,

$$_{\theta+\Delta\theta}C_0 = \frac{1}{M}[2N{_{\theta}C_a} + [M - (2N+2)]{_{\theta}C_0} + 2{_{\theta}C_1}] \tag{5.4-10}$$

218 Unsteady-State Diffusion

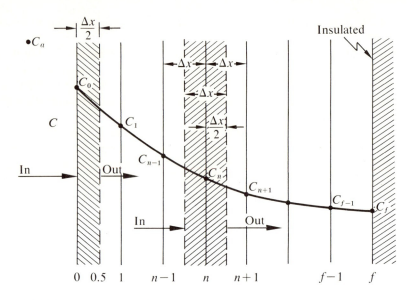

FIGURE 5.4-1 Unsteady-state diffusion in a slab

where $N = k_c \Delta x / D_{AB}$, k_c = mass transfer coefficient, cm/sec. Note that M must have a value such that $M \geq (2N+2)$. For the first time increment, to calculate C_0 one should use a C_a of $(C_a + C_0)/2$. For succeeding $\Delta \theta$ values the value of C_a should be used.

If the C_a is a variable with θ, a new value of C_a can be used for each $\Delta \theta$ used and it appears as $_\theta C_a$ in Eq. (5.4-10).

To derive the equation for the boundary condition where the rear face is insulated, a balance is made on the rear $\frac{1}{2} \Delta x$ slab just as on the front $\frac{1}{2} \Delta x$ slab. The resulting equation is the same as Eqs. (5.4-9) and (5.4-10), but $k_c = 0$ or $N = 0$; hence,

$$_{\theta+\Delta\theta}C_f = \frac{(M-2)_\theta C_f + 2_\theta C_{f-1}}{M} \qquad (5.4\text{-}11)$$

The value of M must be ≥ 2. This equation is the same as Eq. (5.4-8), with $_\theta C_{n+1}$ being set equal to $_\theta C_{n-1}$ because of symmetry.

To use Eqs. (5.4-8), (5.4-10), and (5.4-11) for a given problem, the same values of M, $\Delta \theta$, and Δx must be used throughout. Another form of Eq. (5.4-10) can be derived to be used if N gets too large and, hence, M may be inconveniently too large. This equation is derived by neglecting the accumulation in the half slab in Eq. (5.4-9). Then the value of M is not restricted by the N value. The equation is

$$_{\theta+\Delta\theta}C_0 = \frac{N}{N+1} {_{\theta+\Delta\theta}C_a} + \frac{1}{1+N} {_{\theta+\Delta\theta}C_1} \qquad (5.4\text{-}12)$$

The derivations for Eqs. (5.4-9), (5.4-10), and (5.4-12) were made assuming that the distribution coefficient between the liquid concentration and the solid concentration is 1.0 — that is, the concentration just within the surface of the solid at C_0 is the same as that in the liquid just adjacent to the solid. This may not be so (Crank, C1, p. 53), and then the concentration C_{L0} just in the adjacent liquid is K times C_0 in the adjacent solid:

$$C_{L0} = KC_0 \tag{5.4-13}$$

Hence, the driving force across the liquid fluid for mass transfer is not $({}_\theta C_a - {}_\theta C_0)$ as before but is $({}_\theta C_a - K{}_\theta C_0)$. This means the "in" term in Eq. (5.4-9) is

$$\text{in} = k_c A ({}_\theta C_a - K{}_\theta C_0) \tag{5.4-14}$$

Dividing and multiplying by K,

$$\text{in} = (Kk_c) A \left(\frac{{}_\theta C_a}{K} - {}_\theta C_0 \right) \tag{5.4-15}$$

This means that wherever the term k_c appears alone in Eq. (5.4-10), which is in N, a value of (Kk_c) should be used. Hence,

$$N = \frac{Kk_c \, \Delta x}{D_{AB}} \tag{5.4-16}$$

If K is 1.0, then the original definition holds for $N = (1)k_c \, \Delta x/D_{AB}$. Also, wherever C_a appears, the term C_a/K should be used in Eq. (5.4-10) to give

$$_{\theta+\Delta\theta}C_0 = \frac{1}{M}\left[\frac{2N_\theta C_a}{K} + (M - (2N+2)){}_\theta C_0 + 2{}_\theta C_1 \right] \tag{5.4-17}$$

where $N = Kk_c \, \Delta x/D_{AB}$. The same conditions hold for Eq. (5.4-12), which becomes

$$_{\theta+\Delta\theta}C_0 = \frac{N}{N+1} \frac{{}_{\theta+\Delta\theta}C_a}{K} + \frac{1}{1+N} {}_{\theta+\Delta\theta}C_1 \tag{5.4-18}$$

where $N = Kk_c \, \Delta x/D_{AB}$.

The discussion above concerning the distribution coefficient also holds for the use of the unsteady-state charts given in Figs. 5.2-2 through 5.2-5 and in Appendix A.3. The value of C_{A1}, which is the bulk liquid concentration outside of the slab or solid, should be divided by K and C_{A1}/K used in Y.

$$Y = \frac{(C_{A1}/K) - C_{A\theta}}{(C_{A1}/K) - C_{A0}} \tag{5.4-19}$$

Also,

$$m = \frac{D_{AB}/x_1}{Kk_c} \tag{5.4-20}$$

Example 5.4-1 Numerical Solution for Unsteady-State Mass Transfer

Unsteady-state mass transfer is occurring from a slab 4.0 cm thick with the rear face insulated and the front face subjected to a flowing fluid whose concentration C_a varies with time θ in sec as follows:

$$C_a = 10(10^{-3}) + 2(10^{-7})\theta \text{ g mole } A/\text{cm}^3$$

The diffusivity is $D_{AB} = 1 \times 10^{-4}$ cm²/sec, and the mass transfer coefficient of the flowing liquid to the surface is $k_c = 1.0(10^{-3})$ cm/sec and is constant. The original concentration in the slab is nonuniform but is 5.0×10^{-3} g mole A/cm^3 at the rear face and 1.0×10^{-3} at the front face with a linear variation in between. Solve for the concentration profile in 6.94 hr. The distribution coefficient K defined by Eq. (5.4-13) is 1.0.

Solution

In Fig. 5.4-2 the concentration profile is shown for four slabs of $\Delta x = 1.0$ cm each at time $\theta = 0$ sec, where $C_0 = 1(10^{-3})$, $C_1 = 2(10^{-3})$, $C_2 = 3(10^{-3})$, $C_3 = 4(10^{-3})$, $C_4 = 5(10^{-5})$. For $\theta = 0$, $C_a = 10(10^{-3}) + 2(10^{-7})$

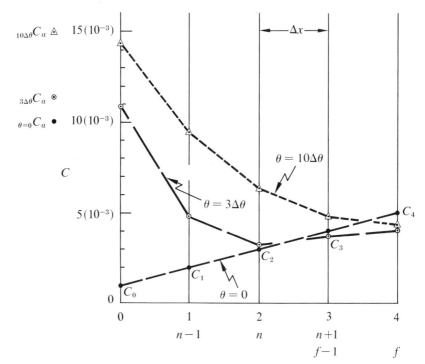

FIGURE 5.4-2 Concentration profile for finite difference, Example 5.4-1

5.4 Numerical Explicit Method and Digital Computer

$(0) = 10(10^{-3})$ g mole/cm^3 and is shown in Fig. 5.4-2. To calculate $\Delta\theta$, set $M = 4$.

$$M = 4 = \frac{(\Delta x)^2}{D_{AB}\Delta\theta} = \frac{(1.0)^2}{1 \times 10^{-4}\Delta\theta}$$

Hence, $\Delta\theta = 2.50(10^3)$ sec or 0.694 hr. This means that 6.94/0.694 or 10 increments of $\Delta\theta$ in seconds will be needed. To calculate N,

$$N = \frac{k_c \Delta x}{D_{AB}} = \frac{1(10^{-3})(1.0)}{1 \times 10^{-4}} = 10.0$$

Since M must be $\geq (2N+2)$, Eq. (5.4-10) will not be used, but Eq. (5.4-12) will be used to calculate the values of C_0 at the outside of the face. Substituting into Eq. (5.4-12) for C_0,

$$_{\theta+\Delta\theta}C_0 = \frac{N}{N+1}{_{\theta+\Delta\theta}C_a} + \frac{1}{1+N}{_{\theta+\Delta\theta}C_1} = \frac{10}{10+1}{_{\theta+\Delta\theta}C_a} + \frac{1}{1+10}{_{\theta+\Delta\theta}C_1}$$

$$= 0.909{_{\theta+\Delta\theta}C_a} + 0.091{_{\theta+\Delta\theta}C_1}$$

Substituting into Eq. (5.4-8) for slabs C_1, C_2 and C_3,

$$_{\theta+\Delta\theta}C_n = \tfrac{1}{4}[_{\theta}C_{n+1} + (4-2)_{\theta}C_n + {_\theta}C_{n-1}]$$

$$= 0.250{_\theta}C_{n+1} + 0.500{_\theta}C_n + 0.250{_\theta}C_{n-1}$$

Substituting into Eq. (5.4-11) for slab C_4, which is insulated,

$$_{\theta+\Delta\theta}C_4 = {_{\theta+\Delta\theta}C_f} = \frac{(4-2)}{4}{_\theta}C_f + \frac{2{_\theta}C_{f-1}}{4} = 0.50{_\theta}C_4 + 0.50{_\theta}C_3$$

The final equations are tabulated below.

$$_{\theta+\Delta\theta}C_0 = 0.909{_{\theta+\Delta\theta}C_a} + 0.091{_{\theta+\Delta\theta}C_1}$$
$$_{\theta+\Delta\theta}C_1 = 0.250{_\theta}C_2 + 0.500{_\theta}C_1 + 0.250{_\theta}C_0$$
$$_{\theta+\Delta\theta}C_2 = 0.250{_\theta}C_3 + 0.500{_\theta}C_2 + 0.250{_\theta}C_1$$
$$_{\theta+\Delta\theta}C_3 = 0.250{_\theta}C_4 + 0.500{_\theta}C_3 + 0.250{_\theta}C_2$$
$$_{\theta+\Delta\theta}C_4 = 0.50{_\theta}C_4 + 0.50{_\theta}C_3$$

To start the calculations we must start to calculate C_1 first at $\theta + \Delta\theta$, since C_0 at $\theta + \Delta\theta$ requires a knowledge of C_1 at $\theta + \Delta\theta$. Substituting for C_1,

$$_{\theta+\Delta\theta}C_1 = 0.250(3)(10^{-3}) + 0.500(2)(10^{-3}) + 0.250(1)(10^{-3}) = 2.0(10^{-3})$$

Next calculate C_0 at $\theta + \Delta\theta$. For the first increment $\Delta\theta$, we first calculate $_{\theta+\Delta\theta}C_a = 10(10^{-3}) + 2(10^{-7})(2.50)(10^3) = 10.5(10^{-3})$. However, for the first time increment we will use instead of C_a,

$$\frac{C_a + C_0}{2} = \frac{10.5(10^{-3}) + 1(10^{-3})}{2} = 5.75(10^{-3})$$

222 Unsteady-State Diffusion

Substituting for C_0,

$$_{\theta+\Delta\theta}C_0 = 0.909(5.75)(10^{-3}) + 0.091(2.0)(10^{-3}) = 5.41(10^{-3})$$

Continuing and substituting for C_2, and so on,

$$_{\theta+\Delta\theta}C_2 = 0.250(4)(10^{-3}) + 0.500(3)(10^{-3}) + 0.250(2)(10^{-3}) = 3.0(10^{-3})$$
$$_{\theta+\Delta\theta}C_3 = 0.250(5)(10^{-3}) + 0.500(4)(10^{-3}) + 0.250(3)(10^{-3}) = 4.0(10^{-3})$$
$$_{\theta+\Delta\theta}C_4 = 0.500(5)(10^{-3}) + 0.500(4)(10^{-3}) = 4.5(10^{-3})$$

For $2\,\Delta\theta$,

$$_{\theta+2\Delta\theta}C_a = 10(10^{-3}) + 2(10^{-7})(2)(2.50)(10^3) = 11.0(10^{-3})$$
$$_{\theta+2\Delta\theta}C_1 = 0.250(3.0)(10^{-3}) + 0.500(2.0)(10^{-3}) + 0.250(5.41)(10^{-3})$$
$$= 3.11(10^{-3})$$
$$_{\theta+2\Delta\theta}C_0 = 0.909(11.0)(10^{-3}) + 0.091(3.11)(10^{-3}) = 10.28(10^{-3})$$
$$_{\theta+2\Delta\theta}C_2 = 0.250(4.0)(10^{-3}) + 0.500(3.0)(10^{-3}) + 0.250(2.0)(10^{-3})$$
$$= 3.0(10^{-3})$$
$$_{\theta+2\Delta\theta}C_3 = 0.250(4.5)(10^{-3}) + 0.500(4.0)(10^{-3}) + 0.250(3.0)(10^{-3})$$
$$= 3.88(10^{-3})$$
$$_{\theta+2\Delta\theta}C_4 = 0.500(4.5)(10^{-3}) + 0.500(4.0)(10^{-3}) = 4.25(10^{-3})$$

For $3\,\Delta\theta$,

$$_{\theta+3\Delta\theta}C_a = 10(10^{-3}) + 2(10^{-7})(3)(2.50)(10^3) = 11.5(10^{-3})$$
$$_{\theta+3\Delta\theta}C_1 = 0.250(3.00)(10^{-3}) + 0.500(3.11)(10^{-3}) + 0.250(10.28)(10^{-3})$$
$$= 4.87(10^{-3})$$
$$_{\theta+3\Delta\theta}C_0 = 0.909(11.5)(10^{-3}) + 0.091(4.87)(10^{-3}) = 10.89(10^{-3})$$
$$_{\theta+3\Delta\theta}C_2 = 0.250(3.88)(10^{-3}) + 0.500(3.00)(10^{-3}) + 0.250(3.11)(10^{-3})$$
$$= 3.25(10^{-3})$$
$$_{\theta+3\Delta\theta}C_3 = 0.250(4.25)(10^{-3}) + 0.500(3.88)(10^{-3}) + 0.250(3.00)(10^{-3})$$
$$= 3.75(10^{-3})$$
$$_{\theta+3\Delta\theta}C_4 = 0.500(4.25)(10^{-3}) + 0.500(3.88)(10^{-3}) = 4.07(10^{-3})$$

The completion of this problem is left up to the reader, who should proceed in the same manner for $7\,\Delta\theta$ increments, for a total of $10\,\Delta\theta$ increments. The profile for $3\,\Delta\theta$ increments is plotted in Fig. 5.4-2.

The preceding example shows how a hand calculation can be done for the explicit method, which is very useful for such calculations. This is especially true for the Schmidt method with $M = 2$. To increase the accuracy more increments are required.

Example 5.4-2 Unsteady-State Mass Transfer Using the Digital Computer

Repeat Example 5.4-1, using the digital computer and the same size increments as with the explicit method. Write the Fortran IV program and obtain a numerical answer. Plot the results for $10\,\Delta\theta$ on Fig. 5.4-2.

5.4 Numerical Explicit Method and Digital Computer 223

Table 5.4-1 Fortran Program for Example 5.4-2

```
C       EXAMPLE PROBLEM - NUMERICAL SOLUTION FOR UNSTEADY STATE MASS
C          TRANSFER IN A SLAB USING THE EXPLICIT METHOD
C
C       THE FRONT SURFACE CONTACTS LIQUID WHOSE CONC IS A FUNCTION OF
C          TIME AND THE REAR SURFACE IS INSULATED
C
C
        DIMENSION YI(1000),YO(1000)
2       FORMAT(34H1EXAMPLE PROBLEM - EXPLICIT METHOD)
3       FORMAT(11H1END OF JOB)
4       FORMAT(24H0THE SPACE INTERVAL H IS,F10.3/25H0THE TIME INTERVAL K
       1 IS,F11.3/45H0THE NUMBER OF STEPS IN THE TIME DIRECTION IS,I5/
       232H0THE NUMBER OF SPACE UNKNOWNS IS,I5/19H0THE DIFFUSIVITY IS,F10.
       35/27H0THE MASS TRANSFER COEFF IS,F10.4/14H0THE LENGTH IS,F10.4)
5       FORMAT(1H0/10H0TIME(SEC),5X,2HCA,9X,4HY(0),8X,4HY(1),8X,4HY(2),8X,
       14HY(3),8X,4HY(4))
6       FORMAT(1H0,F7.0,6F12.8)
        IO=6
        KK=1
        XKC=1.0E-03
        DAB=1.0E-04
        XL=4.
C       CONSTANTS FOR CA(T)= A1 + A2*TIME
        A1=1.0E-02
        A2=2.0E-07
        H=1.0
        XK=2500.
        NT=10
        NY=(XL/H+1.0)*1.00001
C       INSULATED EDGE ADDS ONE TO NY
        YO(1)=1.0E-03
        YO(NY)=5.0E-03
C       INITIAL PROFILE IS LINEAR
        NYM1=NY-1
        RNYM1=1./FLOAT(NYM1)
        DIFF=YO(NY)-YO(1)
        DO 36 I=2,NYM1
36      YO(I)=YO(1)+DIFF*FLOAT(I-1)*RNYM1
        WRITE(IO,2)
        WRITE(IO,4)H,XK,NT,NY,DAB,XKC,XL
C       AT TIME=0, CA = A1
        TIME=0.
        WRITE(IO,5)
        WRITE(IO,6)TIME,A1,(YO(I),I=1,NY)
        B=DAB*XK/(H*H)
        BB=1.-B-B
        A3=XKC*H/DAB
        A4=A3/(A3+1.)
        A5=1./(A3+1.)
        TLAST=XK*FLOAT(NT)*0.999999
45      TIME=TIME+XK
        DO 47 I=2,NYM1
47      YI(I)=B*(YO(I+1)+YO(I-1))+BB*YO(I)
        CA=A1+A2*TIME
        GO TO(49,50),KK
49      KK=2
C       FOR FIRST STEP, USE CA=(CA+YO(1))/2. TO FIND Y(1)
        YI(1)=A4*0.5*(CA+YO(1))+A5*YI(2)
        GO TO 51
50      YI(1)=A4*CA+A5*YI(2)
51      YI(NY)=B*YO(NYM1)*2.0+BB*YO(NY)
        WRITE(IO,6)TIME,CA,(YI(I),I=1,NY)
        IF(TLAST-TIME)56,56,53
53      DO 54 I=1,NY
54      YO(I)=YI(I)
        GO TO 45
56      WRITE(IO,3)
        STOP
        END
```

Solution

The Fortran IV program is given in Table 5.4-1, and the results at 10 $\Delta\theta$ or a time of 25,000 sec are given in Table 5.4-2. The comment statements should assist the reader in following the program. Some changes in nomenclature were necessary for the program but are self-explanatory. The symbols in the Fortran program are related to those of the example as follows. $\Delta\theta = $ XK, $\Delta x = $ H, $L = $ XL, $\theta = $ TIME, $C_a = $ CA $=$ A1 $+$ A2*TIME, $(M-2)/M = $ BB, $N/(N+1) = $ A4, $1/(N+1) = $ A5, $1/M = $ B, $_\theta C_n = $ Y0(I), $k_c = $ XKC.

Normally when the digital computer is used, more slab and time increments are used to obtain more accurate results. However, this example was given to allow the reader to compare more easily the digital to the hand calculation.

Table 5.4-2 Numerical Results for Examples 5.4-2 and 5.5-1 (time $=$ 25,000 sec)

	C, g mole A/cm^3	
	Explicit Method, Example 5.4-2	Implicit Method,[a] Example 5.5-1
C_a	0.01500	0.01500
C_0	0.01451	0.01452
C_1	0.00956	0.00967
C_2	0.00644	0.00667
C_3	0.00484	0.00516
C_4	0.00436	0.00472

[a]Crank-Nicolson Method

5.5 NUMERICAL METHODS USING THE IMPLICIT METHOD WITH THE DIGITAL COMPUTER

In order to minimize the computational time in lengthy problems and also to minimize the stability problems, implicit methods using different finite difference formulas have been developed. The Crank-Nicolson method, one of the most important of these, will be considered here.

In this method the second derivative is formed (M1, L1) by calculating the finite difference approximation at the known row at θ, as in Eq. (5.4-3), and also at the unknown row at $\theta + \Delta\theta$, and using the average. Using instead of θ the subscript j and instead of $\theta + \Delta\theta$ the subscript $j+1$ (one time increment later), this second derivative can be written as

$$\frac{\partial^2 C_A}{\partial x^2} = \frac{1}{2h^2}\left[(y_{i+1,j+1} - 2y_{i,j+1} + y_{i-1,j+1}) + (y_{i+1,j} - 2y_{i,j} + y_{i-1,j})\right] \quad (5.5\text{-}1)$$

where i is the index for position x, $i+1$ for $x+\Delta x$; h is the spacing in x or $h = \Delta x = x_{i+1} - x_i$; and $y = C_A$.

For the time derivative, Eq. (5.4-2) is used with different nomenclature as

$$\frac{\partial C_A}{\partial \theta} = \frac{1}{k}(y_{i,j+1} - y_{i,j}) \tag{5.5-2}$$

where k is the time increment $\Delta\theta$ and $y_{i,j+1}$ is the value of y_i at the time $j+1$ or $\theta + \Delta\theta$. Substituting Eqs. (5.5-1) and (5.5-2) into (5.4-1) and rearranging,

$$\frac{\beta}{2}y_{i+1,j+1} - (\beta+1)y_{i,j+1} + \frac{\beta}{2}y_{i-1,j+1} = -\frac{\beta}{2}y_{i+1,j} + (\beta-1)y_{i,j} - \frac{\beta}{2}y_{i-1,j} \tag{5.5-3}$$

$$\beta = \frac{D_{AB}k}{h^2} = \frac{D_{AB}\Delta\theta}{(\Delta x)^2} = \frac{1}{M} \tag{5.5-4}$$

This means that a value of $y_{i,j+1}$ cannot be calculated only from the values at row or time j but that the whole new row must be calculated simultaneously.

The stability and convergence criteria for this method are satisfied for all positive values of β or M. This means M values less than 2 can be used safely. This is an advantage in calculation of complex situations (M1).

The solution of Eq. (5.5-3) proceeds as follows. The knowns on one row j at time j are $y_{i+1,j}$, $y_{i,j}$ and $y_{i-1,j}$. The number of internal points exclusive of boundary points on a row is given as N and the number of intervals or spaces as $N+1$. Also $(N+1)h = L$ (length in x direction). To calculate $y_{i,j+1}$ at the $j+1$ row, Eq. (5.5-3) has three unknowns except if at a boundary. If this equation is used for each point on a row (N times for i goes from 1 to N), then we have N simultaneous linear algebraic equations with N unknowns—that is, N simultaneous equations to be solved for one $\Delta\theta$, which are repeated for each $\Delta\theta$. These equations can be solved by using a matrix just like the method given in Chapter 1.

The boundary condition equations used for this implicit method are the same as used for the explicit method, Eqs. (5.4-11) and (5.4-12). An alternate implicit method for convection at the surface is available elsewhere (C1, p. 199).

Example 5.5-1 Numerical Solution for Unsteady-State Using the Digital Computer (Implicit Method)

Solve Example 5.4-1 again, using the Crank-Nicolson implicit method. Use the same size increments, write the Fortran IV program, and obtain a numerical answer. Compare the results with the explicit method, Example 5.4-2. Note that generally with this implicit method much smaller increments are used to give greater accuracy.

Solution

For this case $M = 4$ and $\beta = \frac{1}{4}$. $\Delta\theta = k = $ XK (Fortran symbol) $= 2500$ sec. $\Delta x = h = $ H $= 1.0$ cm. $D_{AB} = $ DAB $= 1 \times 10^{-4}$. $k_c = $ XKC $= 1 \times 10^{-3}$. $L = $ XL $= 4$ cm. Number of intervals or spaces $= $ XL/H $= 4$. The y values are numbered starting at the front face as $y_{1,j}, y_{2,j}, \ldots, y_{NY,j}$, where NY $= 5$. The flowing liquid concentration is $C_a = $ CA $= $ A1 + A2 *TIME, where A1 $= 1 \times 10^{-2}$, A2 $= 2 \times 10^{-7}$.

The equations to be used are as follows. For the front face, Eq. (5.4-12) is valid. This becomes, after rearrangement,

$$_{\theta+\Delta\theta}C_0 = \frac{k_c \Delta x / D_{AB}}{1 + \dfrac{k_c \Delta x}{D_{AB}}} {}_{\theta+\Delta\theta}C_a + \frac{1}{1 + \dfrac{k_c \Delta x}{D_{AB}}} {}_{\theta+\Delta\theta}C_1 \qquad (5.5\text{-}5)$$

Calling position 0 above as $i = 1$ for the front face and 1 as $i = 2; j + 1$ for $\theta + \Delta\theta$; and CA for C_a, the following is obtained after rearrangement:

$$y_{1,j+1} - \frac{1}{\dfrac{k_c H}{D_{AB}} + 1} y_{2,j+1} = \text{CA} \frac{k_c H / D_{AB}}{\dfrac{k_c H}{D_{AB}} + 1} \qquad (5.5\text{-}6)$$

or

$$y_{1,j+1} - \text{A5} \cdot y_{2,j+1} = \text{CA} \frac{\text{A3}}{\text{A4}} \qquad (5.5\text{-}7)$$

where A3, A4, and A5 are constants given in Eq. (5.5-6) and in the Fortran IV program, Table 5.5-1.

Table 5.5-1 Fortran Program for Example 5.5-1

```
C       EXAMPLE PROBLEM - NUMERICAL SOLUTION FOR UNSTEADY STATE MASS
C          TRANSFER IN A SLAB USING THE IMPLICIT METHOD
C
C       THE FRONT SURFACE CONTACTS LIQUID WHOSE CONC IS A FUNCTION OF
C          TIME AND THE REAR SURFACE IS INSULATED
C
C       PROGRAM WILL HANDLE UP TO 51 SIMULTANEOUS EQUATIONS IN PRESENT
C          FORM
C
        DIMENSION A(2601),D(51),Y(51)
   2    FORMAT(34H1EXAMPLE PROBLEM - IMPLICIT METHOD)
   3    FORMAT(11H1END OF JOB)
   4    FORMAT(24H0THE SPACE INTERVAL H IS,F10.3/25H0THE TIME INTERVAL K
       1 IS,F11.3/45H0THE NUMBER OF STEPS IN THE TIME DIRECTION IS,I5/
       232H0THE NUMBER OF SPACE UNKNOWNS IS,I5/19H0THE DIFFUSIVITY IS,F10.
       35/27H0THE MASS TRANSFER COEFF IS,F10.4/14H0THE LENGTH IS,F10.4)
   5    FORMAT(1H0/10H0TIME(SEC),5X,2HCA,9X,4HY(0),8X,4HY(1),8X,4HY(2),8X,
       14HY(3),8X,4HY(4))
   6    FORMAT(1H0,F7.0,6F12.8)
        IO=6
        XKC=1.0E-03
        DAB=1.0E-04
        XL=4.
C       CONSTANTS FOR CA(T)= A1 + A2*TIME
        A1=1.0E-02
        A2=2.0E-07
        H=1.0
        XK=2500.
```

Table 5.5-1 (continued)

```
            NT=10
            NY=(XL/H+1.0)*1.00001
C           INSULATED EDGE ADDS ONE TO NY
C           ZERO MATRIX ELEMENTS
            MY=NY**2
            DO 30 I=1,MY
  30        A(I)=0
            Y(1)=1.0E-03
            Y(NY)=5.0E-03
C           INITIAL PROFILE IS LINEAR
            NYM1=NY-1
            RNYM1=1./FLOAT(NYM1)
            DIFF=Y(NY)-Y(1)
            DO 36 I=2,NYM1
  36        Y(I)=Y(1)+DIFF*FLOAT(I-1)*RNYM1
            WRITE(6,2)
            WRITE(6,4)H,XK,NT,NY,DAB,XKC,XL
            TIME=0.
            WRITE(6,5)
C           AT TIME=0, CA = A1
            WRITE(6,6)TIME,A1,(Y(I),I=1,NY)
            B=DAB*XK/(H*H)
            B1=-B-1.
            B2=B*0.5
            B3=-B2
            B4=B-1.
            TLAST=XK*FLOAT(NT)*0.999999
            A3=XKC*H/DAB
            A4=A3/(A3+1.)
            A5=1./(A3+1.)
C           CALCULATE NONZERO ELEMENTS OF COEFFICIENT MATRIX
C           FIRST TIME ONLY, USE CA = (CA+Y(1))/2. THIS AFFECTS A(1) AND D(1)
            A(1)=1.-A4*0.5
            A(NY+1)=-A5
            TIME=XK
            CA=A1+A2*TIME
            D(1)=A4*CA*0.5
  70        J=0
            K=J+NY
            L=K+NY
            DO 37 I=2,NYM1
            A(J+I)=B2
            A(K+I)=B1
            A(L+I)=B2
            J=J+NY
            K=K+NY
            L=L+NY
  37        D(I)=B3*(Y(I-1)+Y(I+1))+B4*Y(I)
            A(MY-NY)=B2+B2
            A(MY)=B1
            D(NY)=2.*B3*Y(NYM1)+B4*Y(NY)
  40        CALL SIMQ(A,D,NY,KS)
            DO 43 I=1,NY
  43        Y(I)=D(I)
            WRITE(6,6)TIME,CA,(Y(I),I=1,NY)
            IF(TLAST-TIME)65,65,53
  53        TIME=TIME+XK
            CA=A1+A2*TIME
C           RESTORE MATRIX COEFFICIENTS
C           ZERO MATRIX ELEMENTS
            DO 54 I=1,MY
  54        A(I)=0
C           CALCULATE NONZERO ELEMENTS OF COEFFICIENT MATRIX
            A(1)=1.
            A(NY+1)=-A5
            D(1)=A4*CA
            GO TO 70
  65        WRITE(6,3)
            STOP
            END
```

228 Unsteady-State Diffusion

For the internal points $i = 2, 3,$ and 4, Eq. (5.5-3) is written three times:

$$\frac{\beta}{2} y_{3,j+1} - (\beta+1) y_{2,j+1} + \frac{\beta}{2} y_{1,j+1} = -\frac{\beta}{2} y_{3,j} + (\beta-1) y_{2,j} - \frac{\beta}{2} y_{1,j} \quad (i=2)$$

$$\frac{\beta}{2} y_{4,j+1} - (\beta+1) y_{3,j+1} + \frac{\beta}{2} y_{2,j+1} = -\frac{\beta}{2} y_{4,j} + (\beta-1) y_{3,j} - \frac{\beta}{2} y_{2,j} \quad (i=3)$$

$$\frac{\beta}{2} y_{5,j+1} - (\beta+1) y_{4,j+1} + \frac{\beta}{2} y_{3,j+1} = -\frac{\beta}{2} y_{5,j} + (\beta-1) y_{4,j} - \frac{\beta}{2} y_{3,j} \quad (i=4)$$

$$(5.5-8)$$

For the insulated end, Eq. (5.4-11) holds as before. This last point is at NY $= i = 5$. This equation sets $_0C_{f-1} = {_0C_{f+1}}$ because of symmetry. Hence, in the present nomenclature

$$y_{NY-1,j} = y_{NY+1,j} \quad (5.5-9)$$

Substituting NY for i into Eq. (5.5-3), we obtain

$$\frac{\beta}{2} y_{NY+1,j+1} - (\beta+1) y_{NY,j+1} + \frac{\beta}{2} y_{NY-1,j+1}$$

$$= -\frac{\beta}{2} y_{NY+1,j} + (\beta-1) y_{NY,j} - \frac{\beta}{2} y_{NY-1,j} \quad (5.5-10)$$

Substituting Eq. (5.5-9) into (5.5-10),

$$\frac{(2)\beta}{2} y_{NY-1,j+1} - (\beta+1) y_{NY,j+1} = \frac{(2)(-\beta)}{2} y_{NY-1,j} + (\beta-1) y_{NY,j} \quad (5.5-11)$$

For NY $= 5$,

$$\frac{(2)\beta}{2} y_{4,j+1} - (\beta+1) y_{5,j+1} = \frac{2(-\beta)}{2} y_{4,j} + (\beta-1) y_{5,j}. \quad (5.5-12)$$

The final five equations to use in five unknowns are Eqs. (5.5-7), (5.5-8), and (5.5-12). The right side of each equation is constant. These equations are set up in a matrix form and solved using the standard subroutine SIMQ(A,D,NY,KS) which is available in the System 360 Scientific Subroutine Package (360A-CM-03X) Version III of IBM. This is readily available, and the subroutine can be copied and put on punch cards. In many cases it is in the computer library, and the programmer need only call for the subroutine. The numerical results for this example are tabulated in Table 5.4-2. The implicit and explicit methods differ somewhat. If smaller time and slab increments had been used, the results would have been closer together.

In this subroutine,

    A   is the coefficient matrix stored columnwise in the computer memory.

D is the vector of constants (known values on right-hand side of equations) replaced by the solution vector.
NY is the size of the vector or number of equations, 5.
KS is an indicator that tells whether the coefficient matrix is singular (not used here).

The reader is referred to Section 1.9C in Chapter 1 for a more efficient method to solve the system of five equations given using a tridiagonal type matrix.

As an alternate method for the time interval of θ to $\theta + \Delta\theta$, the value of C_a could be calculated using the average time of $\theta + \Delta\theta/2$ instead of using $\theta + \Delta\theta$.

In summary, explicit methods are generally simpler to use, but because of stability considerations, implicit methods such as the Crank-Nicolson procedure are usually preferred. For solving nonlinear partial differential equations the explicit method can be used. The implicit method can also be used where the finite difference equations are "linearized" by holding the factor or factors which cause the nonlinearity constant over a time interval.

5.6 NUMERICAL METHODS FOR OTHER PHYSICAL GEOMETRIES

5.6A Flat Slabs in Series

In some cases in unsteady-state diffusion in solids, diffusion occurs in two different solid phases in series. For the diffusion in each solid phase the equations derived in Sections 5.4 and 5.5 are valid for calculation of the points inside of each phase. However, at the interface a new equation must be used, because the physical properties on either side differ.

In the following derivation using the explicit method the assumption is made that there is no resistance at the interface. Also the distribution coefficient K between phases is 1.0. In Fig. 5.6-1 component A is diffusing through phases B and D in series. In making a mole balance on A at a given instant θ, the concentrations are shown at this time θ. In the balance on the half slab of B to the left and the half slab of D to the right of the interface i, the rate of moles in $-$ the rate of moles out $=$ the rate of accumulation in each half slab in $\Delta\theta$ hours.

$$\frac{D_{AB} A (_\theta C_{i-1} - _\theta C_i)}{\Delta x_B} - \frac{D_{AD} A (_\theta C_i - _\theta C_{i+1})}{\Delta x_D} = \frac{A \Delta x_B}{2} \frac{(_{\theta+\Delta\theta} C_i - _\theta C_i)}{\Delta\theta}$$

$$+ \frac{A \Delta x_D}{2} \frac{(_{\theta+\Delta\theta} C_i - _\theta C_i)}{\Delta\theta} \quad (5.6\text{-}1)$$

On rearranging and solving for $_{\theta+\Delta\theta}C_i$,

$$_{\theta+\Delta\theta}C_i = \frac{_\theta C_{i-1} + (M_B/2 + M_D R/2 - 1 - R)_\theta C_i + R_\theta C_{i+1}}{M_B/2 + M_D R/2} \quad (5.6\text{-}2)$$

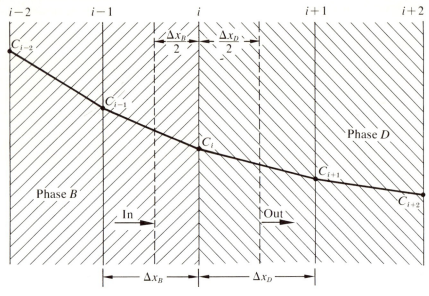

FIGURE 5.6-1 Concentration profile in two dissimilar slabs B and D in series (distribution coefficient $K = 1.0$)

where

$$M_B = \frac{(\Delta x_B)^2}{D_{AB} \Delta \theta} \quad (5.6\text{-}3)$$

$$M_D = \frac{(\Delta x_D)^2}{D_{AD} \Delta \theta} \quad (5.6\text{-}4)$$

$$R = \frac{D_{AD}}{D_{AB}} \frac{\Delta x_B}{\Delta x_D} \quad (5.6\text{-}5)$$

When $M_B = M_D$, $\Delta x_B = \Delta x_D$, and $D_{AB} = D_{AD}$, then Eq. (5.6-2) reduces to Eq. (5.4-4) for a homogeneous slab.

Equation (5.6-2) can be used in two different ways in a numerical solution. In the first method since $\Delta\theta$ is the same always in both slabs, one can select $\Delta x_B = \Delta x_D$—that is, slabs of the same thickness in both phases. Then equating $(\Delta x_B)^2$ of Eq. (5.6-3) to $(\Delta x_D)^2$ of Eq. (5.6-4),

$$M_B D_{AB} \Delta\theta = M_D D_{AD} \Delta\theta \quad (5.6\text{-}6)$$

Hence,

$$M_B = \frac{D_{AD}}{D_{AB}} M_D \quad (5.6\text{-}7)$$

This means that if Δx_B, $\Delta\theta$, and hence M_B are first set, then in this method $\Delta x_D = \Delta x_B$, and M_D is related to M_B by Eq. (5.6-7). The stability and convergence criteria still hold, however, and M_B and M_D must each be

⩾ 2.0. To calculate the concentration C at various interior points in phase B, Eq. (5.4-4) is used with M_B replacing M. Only C_i is calculated from Eq. (5.6-2). For interior points in phase D, M_D is used in Eq. (5.4-4).

In the second method $\Delta\theta$ is still the same in both slabs but M_B is selected as equal to M_D.

$$M_B = \frac{(\Delta x_B)^2}{D_{AB}\,\Delta\theta} = M_D = \frac{(\Delta x_D)^2}{D_{AD}\,\Delta\theta} \qquad (5.6\text{-}8)$$

Hence, Δx_B is related to Δx_D by

$$\Delta x_B = \sqrt{\frac{D_{AB}}{D_{AD}}}\,\Delta x_D \qquad (5.6\text{-}9)$$

The sizes of the slabs are then of unequal lengths.

Example 5.6-1 Unsteady-State Diffusion through Dissimilar Slabs in Series Using Numerical Method

Solute A is diffusing unsteady-state through slabs B and D in series, with B being 3.0 cm thick and D being 2.0 cm thick. At time $\theta = 0$, the initial concentration profile of A is linear and is 3.0×10^{-3} g mole A/cm^3 at the front face of B and 0.5×10^{-3} at the rear face of D. The concentration at the front of phase B is suddenly raised to a value of 5.0×10^{-3} g mole A/cm^3 and held there. The rear of phase D is insulated. The diffusivity $D_{AB} = 1.0 \times 10^{-4}$ cm²/sec and $D_{AD} = 1.333 \times 10^{-4}$ cm²/sec. It is desired to calculate the concentration profile in 6.94 hr (25,000 sec) using $\Delta\theta$ increments of 2500 sec.

Solution

Figure 5.6-2 shows the two slabs. Either method 1 or method 2 can be used, but it is more convenient to use $\Delta x_B = \Delta x_D$ in method 1. Solving for M_B after selecting $\Delta x_B = 1.0$ cm $= \Delta x_D$,

$$M_B = \frac{(\Delta x_B)^2}{D_{AB}\,\Delta\theta} = \frac{(1.0)^2}{1.0 \times 10^{-4} \times 2500} = 4.0 \qquad (5.6\text{-}3)$$

Substituting into Eq. (5.6-7),

$$M_D = \frac{D_{AB}}{D_{AD}} M_B = \frac{1.0 \times 10^{-4}}{1.333 \times 10^{-4}}(4.0) = 3.0$$

Using Eq. (5.6-5),

$$R = \frac{D_{AD}}{D_{AB}}\frac{\Delta x_B}{\Delta x_D} = \frac{1.333 \times 10^{-4}(1.0)}{1.0 \times 10^{-4}(1.0)} = 1.333$$

After time $\theta = 0$,

$$C_1 = C_a = 5.0 \times 10^{-3} \qquad (5.6\text{-}10)$$

232 Unsteady-State Diffusion

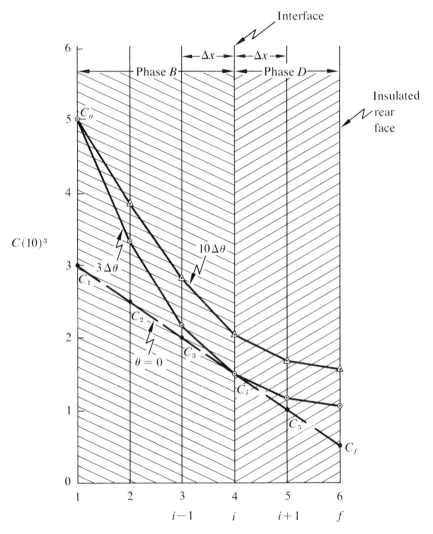

FIGURE 5.6-2   Diffusion in slabs in series, Example 5.6-1

However, for the first $\Delta\theta$ an average value of $C_1 = (5.0 \times 10^{-3} + 3.0 \times 10^{-3})/2 = 4.0 \times 10^{-3}$ will be used. In subsequent calculations $C_1 = 5.0 \times 10^{-3}$ will be used. Equation (5.4-8) will be used for calculating C_2 and C_3. Substituting $M_B = M = 4.0$,

$$_{\theta+\Delta\theta}C_n = \tfrac{1}{4}[_\theta C_{n+1} + (4-2)_\theta C_n + {_\theta C_{n-1}}]$$
$$= 0.250\,_\theta C_{n+1} + 0.500\,_\theta C_n + 0.250\,_\theta C_{n-1} \qquad (5.6\text{-}11)$$

where $n = 2, 3$.

5.6 Numerical Methods for Other Physical Geometries 233

For calculating C_i, which is C_4, Eq. (5.6-2) is used with $M_B = 4.0$, $M_D = 3.0$, and $R = 1.333$:

$$_{\theta+\Delta\theta}C_i = \frac{_\theta C_{i-1} + [4.0/2 + 3.0(1.333)/2 - 1 - 1.333]_\theta C_i + 1.333_\theta C_{i+1}}{4.0/2 + 3.0(1.333)/2}$$

$$= 0.250_\theta C_{i-1} + 0.417_\theta C_i + 0.333_\theta C_{i+1} \quad (5.6\text{-}12)$$

where $i = 4$.

To calculate C_5, Eq. (5.4-8) becomes, with $M_D = 3.0$,

$$_{\theta+\Delta\theta}C_n = \tfrac{1}{3}[_\theta C_{n+1} + (3-2)_\theta C_n + _\theta C_{n-1}]$$

$$= 0.333_\theta C_{n+1} + 0.333_\theta C_n + 0.333_\theta C_{n-1} \quad (5.6\text{-}13)$$

where $n = 5$.

For calculation of C_f, which is C_6 at the insulated boundary, Eq. (5.4-11) is used with $M_D = M = 3.0$ and

$$_{\theta+\Delta\theta}C_f = \frac{(3-2)_\theta C_f + 2_\theta C_{f-1}}{3} = 0.333_\theta C_f + 0.666_\theta C_{f-1} \quad (5.6\text{-}14)$$

where $f = 6$. The final equations (5.6-10) through (5.6-14) will be used:

$C_1 = 4.0 \times 10^{-3}(1\,\Delta\theta)$, $C_1 = C_a = 5.0 \times 10^{-3}(2\,\Delta\theta, 3\,\Delta\theta, \ldots, 10\,\Delta\theta)$

$_{\theta+\Delta\theta}C_2 = 0.25_\theta C_3 + 0.50_\theta C_2 + 0.25_\theta C_1$

$_{\theta+\Delta\theta}C_3 = 0.25_\theta C_4 + 0.50_\theta C_3 + 0.25_\theta C_2$

$_{\theta+\Delta\theta}C_4 = 0.25_\theta C_3 + 0.417_\theta C_4 + 0.333_\theta C_5$

$_{\theta+\Delta\theta}C_5 = 0.333_\theta C_6 + 0.333_\theta C_5 + 0.333_\theta C_4$

$_{\theta+\Delta\theta}C_6 = 0.333_\theta C_6 + 0.666_\theta C_5$

At $\theta = 0$, $C_1 = 3.0 \times 10^{-3}$, $C_2 = 2.5 \times 10^{-3}$, $C_3 = 2.0 \times 10^{-3}$, $C_4 = 1.5 \times 10^{-3}$, $C_5 = 1.0 \times 10^{-3}$, $C_6 = 0.5 \times 10^{-3}$.

Proceeding with the calculations for $1\,\Delta\theta$,

$C_1 = 4.0 \times 10^{-3}$ for the first $1\,\Delta\theta$.

$_{\theta+\Delta\theta}C_2 = 0.25(2.0)(10^{-3}) + 0.50(2.5)(10^{-3}) + 0.25(4.0)(10^{-3}) = 2.75 \times 10^{-3}$

$_{\theta+\Delta\theta}C_3 = 0.25(1.5)(10^{-3}) + 0.50(2.0)(10^{-3}) + 0.25(2.5)(10^{-3}) = 2.00 \times 10^{-3}$

$_{\theta+\Delta\theta}C_4 = 0.25(2.0)(10^{-3}) + 0.417(1.5)(10^{-3}) + 0.333(1.0)(10^{-3})$
$= 1.458 \times 10^{-3}$

$_{\theta+\Delta\theta}C_5 = 0.333(0.5)(10^{-3}) + 0.333(1.0)(10^{-3}) + 0.333(1.5)(10^{-3})$
$= 1.00 \times 10^{-3}$

$_{\theta+\Delta\theta}C_6 = 0.333(0.5)(10^{-3}) + 0.666(1.0)(10^{-3}) = 0.833 \times 10^{-3}$

$C_1 = C_a = 5.0 \times 10^{-3} = $ constant for all remaining $\Delta\theta$'s

$_{\theta+2\Delta\theta}C_2 = 0.25(2.00)(10^{-3}) + 0.50(2.75)(10^{-3}) + 0.25(5.0)(10^{-3})$
$= 3.125 \times 10^{-3}$

$_{\theta+2\Delta\theta}C_3 = 0.25(1.458)(10^{-3}) + 0.50(2.00)(10^{-3}) + 0.25(2.75)(10^{-3})$
$= 2.05 \times 10^{-3}$

$_{\theta+2\Delta\theta}C_4 = 0.25(2.00)(10^{-3}) + 0.417(1.458)(10^{-3}) + 0.333(1.00)(10^{-3})$
$= 1.44 \times 10^{-3}$

234 Unsteady-State Diffusion

$$_{\theta+2\Delta\theta}C_5 = 0.333(0.833)(10^{-3}) + 0.333(1.00)(10^{-3}) + 0.333(1.458)(10^{-3})$$
$$= 1.096 \times 10^{-3}$$
$$_{\theta+2\Delta\theta}C_6 = 0.333(0.833)(10^{-3}) + 0.666(1.00)(10^{-3}) = 0.944 \times 10^{-3}$$
$$C_1 = 5.0 \times 10^{-3}$$
$$_{\theta+3\Delta\theta}C_2 = 0.25(2.05)(10^{-3}) + 0.50(3.125)(10^{-3}) + 0.25(5.0)(10^{-3})$$
$$= 3.324 \times 10^{-3}$$
$$_{\theta+3\Delta\theta}C_3 = 0.25(1.44)(10^{-3}) + 0.50(2.05)(10^{-3}) + 0.25(3.125)(10^{-3})$$
$$= 2.166 \times 10^{-3}$$
$$_{\theta+3\Delta\theta}C_4 = 0.25(2.05)(10^{-3}) + 0.417(1.44)(10^{-3}) + 0.333(1.096)(10^{-3})$$
$$= 1.477 \times 10^{-3}$$
$$_{\theta+3\Delta\theta}C_5 = 0.333(0.944)(10^{-3}) + 0.333(1.096)(10^{-3}) + 0.333(1.44)(10^{-3})$$
$$= 1.160 \times 10^{-3}$$
$$_{\theta+3\Delta\theta}C_6 = 0.333(0.944)(10^{-3}) + 0.666(1.096)(10^{-3}) = 1.045 \times 10^{-3}$$

The completion of this problem is left up to the reader, who should proceed in the same manner for 7 $\Delta\theta$ increments for a total of 10 $\Delta\theta$ increments. The profile at 3 $\Delta\theta$ is plotted in Fig. 5.6-2.

Example 5.6-2 Unsteady-State Diffusion through Dissimilar Slabs in Series Using Digital Computer

Repeat Example 5.6-1, but write the Fortran IV program and obtain numerical results for 10 $\Delta\theta$ increments.

Solution

The Fortran program is given in Table 5.6-1 and the results in Table 5.6-2. The data for 10 $\Delta\theta$ are plotted in Fig. 5.6-2.

In many cases the distribution coefficient between the two solid phases B and D is not 1.0. Then

$$C_D = KC_B \qquad (5.6\text{-}15)$$

This fact must be taken into account in the derivation. Figure 5.6-3 shows such a case. Making a balance similar to Eq. (5.6-1),

$$\frac{D_{AB}A(_\theta C_{i-1} - {_\theta}C_{iB})}{\Delta x_B} - \frac{D_{AD}A(_\theta C_{iD} - {_\theta}C_{i+1})}{\Delta x_D} = \frac{A\,\Delta x_B}{2} \frac{(_{\theta+\Delta\theta}C_{iB} - {_\theta}C_{iB})}{\Delta\theta}$$
$$+ \frac{A\,\Delta x_D}{2} \frac{(_{\theta+\Delta\theta}C_{iD} - {_\theta}C_{iD})}{\Delta\theta} \qquad (5.6\text{-}16)$$

where C_{iB} is the concentration of the solute at the interface in phase B, C_{iD} the concentration of the solute at the interface in phase D, and both are related by Eq. (5.6-15).

Substituting KC_{iB} for C_{iD} into Eq. (5.6-16) and rearranging,

$$_{\theta+\Delta\theta}C_{iB} = \frac{_\theta C_{i-1} + (M_B/2 + M_D RK/2 - 1 - KR)_\theta C_{iB} + R_\theta C_{i+1}}{(M_B/2 + M_D RK/2)} \qquad (5.6\text{-}17)$$

5.6 Numerical Methods for Other Physical Geometries   235

$$C_{iD} = KC_{iB} \qquad (5.6\text{-}18)$$

To use these equations, $_{\theta+\Delta\theta}C_{iB}$ is calculated from Eq. (5.6-17), where C_{i-1} is the concentration in phase B and C_{i+1} that in D. Then C_{iD} is calculated from Eq. (5.6-18). For the new C_{i+1}, the regular equation is used for internal points. One analytical method is given in Section 5.9A.

Table 5.6-1 Fortran Program for Example 5.6-2

```
C      EXAMPLE PROBLEM - NUMERICAL METHODS FOR OTHER PHYSICAL
C         GEOMETRIES - FLAT SLABS IN SERIES - EXPLICIT METHOD
C
C      THE FRONT SURFACE CONTACTS LIQUID WHOSE CONC JUMPS AT TIME=0 TO
C         A CONSTANT VALUE - THE REAR SURFACE IS INSULATED
C
1      FORMAT('1EXAMPLE PROBLEM - FLAT SLABS IN SERIES - EXPLICIT METHOD'
      1)
2      FORMAT('1END OF JOB')
3      FORMAT('0THE SPACE INTERVAL H IS',F10.3/'0THE TIME INTERVAL XK IS'
      1,F11.3/'0THE NUMBER OF STEPS IN THE TIME DIRECTION IS',I5/'0THE NU
      2MBER OF SPACE UNKNOWNS IS',I5/'0THE DIFFUSIVITY FOR PHASE B IS',F1
      30.5/'0THE DIFFUSIVITY FOR PHASE D IS',F10.5/'0THE MASS TRANSFER MO
      4D  FOR PHASE B IS',F10.4/'0THE MASS TRANSFER MOD FOR PHASE D IS'
      5,F10.4/'0THE LENGTH IS',F10.4)
4      FORMAT('0'/'0TIME(SEC)',5X,'C(1)',8X,'C(2)',8X,'C(3)',8X,'C(4)',
      18X,'C(5)',8X,'C(6)')
5      FORMAT('0',F7.0,6F12.8)
       DIMENSION C(6),CO(6)
       IO=6
C      SET UP PROGRAM CONTROL CONSTANTS
C      IS IS THE INTERFACE BOUNDARY
       IS=4
       XL=5.
       H=1.
C      INSULATED EDGE ADDS ONE TO NCO
       NCO=(XL/H+1.)*1.00001
       XK=2500.
       NT=10
C      SET UP EQUATION CONSTANTS
C      DIFFUSIVITY OF PHASE B
       DAB=1.0E-04
C      DIFFUSIVITY OF PHASE D
       DAD=1.3333E-04
       DXB=1.0
       DXD=DXB
       XMB=4.0
       XMD=3.0
       TIME=0.0
       TLAST=XK*FLOAT(NT)*0.999999
C      INITIAL PROFILE IS LINEAR
       CC=5.0E-03
       C(1)=3.0E-03
       C(NCO)=5.0E-04
       MCO=NCO-1
       RMCO=1./FLOAT(MCO)
       DIFF=C(NCO)-C(1)
       DO 7 I=2,MCO
7      C(I)=C(1)+DIFF*FLOAT(I-1)*RMCO
       WRITE(IO,1)
       WRITE(IO,3)H,XK,NT,NCO,DAB,DAD,XMB,XMD,XL
       WRITE(IO,4)
       WRITE(IO,5)TIME,(C(I),I=1,NCO)
       R=(DAD*DXB)/(DAB*DXD)
C      CALCULATE NEW VALUES
       C(1)=(C(1)+CC)*0.5
8      CO(1)=C(1)
       DO 15 I=2,MCO
```

Table 5.6-1 (continued)

```
        IF(I-IS) 16,17,18
  16    CO(I)=(C(I+1)+(XMB-2.0)*C(I)+C(I-1))/XMB
        GO TO 15
  17    CO(I)=(C(I-1)+(XMB*.5+XMD*R*.5-1.-R)*C(I)+R*C(I+1))
        CO(I)=CO(I)/(XMB*.5+XMD*.5*R)
        GO TO 15
  18    CO(I)=(C(I+1)+(XMD-2.)*C(I)+C(I-1))/XMD
  15    CONTINUE
        CO(NCO)=((XMD-2.)*C(NCO)+2.*C(MCO))/XMD
        TIME=TIME+XK
        WRITE(IO,5)TIME,(CO(I),I=1,NCO)
        DO 20 I=2,NCO
  20    C(I)=CO(I)
        C(1)=CC
        IF(TLAST-TIME)30,30,8
  30    WRITE(IO,2)
        STOP
        END
```

Table 5.6-2 Results for Example 5.6-2

```
EXAMPLE PROBLEM - FLAT SLABS IN SERIES - EXPLICIT METHOD
THE SPACE INTERVAL H IS      1.000
THE TIME INTERVAL XK IS      2500.000
THE NUMBER OF STEPS IN THE TIME DIRECTION IS    10
THE NUMBER OF SPACE UNKNOWNS IS     6
THE DIFFUSIVITY FOR PHASE B IS     0.00010
THE DIFFUSIVITY FOR PHASE D IS     0.00013
THE MASS TRANSFER MOD FOR PHASE B IS     4.0000
THE MASS TRANSFER MOD FOR PHASE D IS     3.0000
THE LENGTH IS     5.0000
```

TIME(SEC)	C(1)	C(2)	C(3)	C(4)	C(5)	C(6)
0.	0.00300000	0.00250000	0.00200000	0.00150000	0.00100000	0.00050000
2500.	0.00400000	0.00275000	0.00200000	0.00145834	0.00100000	0.00083333
5000.	0.00500000	0.00312500	0.00205208	0.00144098	0.00109722	0.00094444
7500.	0.00500000	0.00332552	0.00216753	0.00147917	0.00116088	0.00104630
10000.	0.00500000	0.00345464	0.00228494	0.00154517	0.00122878	0.00112269
12500.	0.00500000	0.00354855	0.00239242	0.00162465	0.00129888	0.00119342
15000.	0.00500000	0.00362238	0.00248951	0.00170800	0.00137231	0.00126372
17500.	0.00500000	0.00368357	0.00257735	0.00179149	0.00144801	0.00133612
20000.	0.00500000	0.00373612	0.00265744	0.00187347	0.00152521	0.00141071
22500.	0.00500000	0.00378242	0.00273111	0.00195338	0.00160313	0.00148704
25000.	0.00500000	0.00382399	0.00279951	0.00203106	0.00168118	0.00156443

5.6 Numerical Methods for Other Physical Geometries

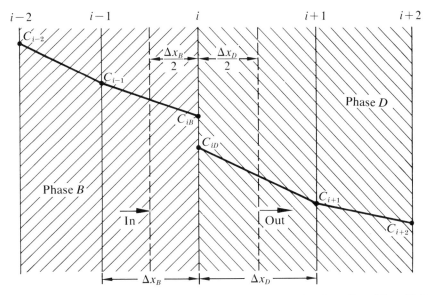

FIGURE 5.6-3 Concentration profile in two phases B and D where $C_D = KC_B$, with K being the distribution coefficient

5.6B Resistance between Flat Slabs in Series

In unsteady-state diffusion in two different solids in series, mass transfer in a fluid may be occurring in the region between the two phases. Hence, a resistance is present in series at the interface. This is shown in Fig. 5.6-4, where there is a concentration drop of $C_{nB} - C_{nD}$ at this interface.

If this gap of material of phase E is small, then it can be assumed that it has negligible capacity for accumulation of solute A in this gap or "film." Expressing the resistance in terms of a mass transfer coefficient k_c, cm/sec,

$$k_c = \frac{D_{AE}}{\delta} \tag{5.6-19}$$

where δ is the equivalent film thickness, cm, of phase E.

To derive the equation for point C_{nB} at the surface of phase B, it can be considered as the same case as Eq. (5.4-10) for mass transfer to a surface from a fluid outside of concentration C_a (which is C_{nD} in this case). Hence, rewriting Eq. (5.4-10), we obtain

$$_{\theta+\Delta\theta}C_{nB} = \frac{1}{M_B}\{2N_{B\,\theta}C_{nD} + [M_B - (2N_B + 2)]_\theta C_{nB} + 2_\theta C_{nB-1}\} \tag{5.6-20}$$

where $N_B = k_c\,\Delta x_B / D_{AB}$.

238 Unsteady-State Diffusion

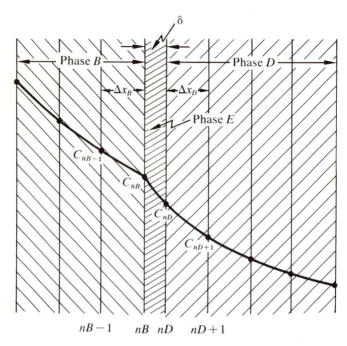

FIGURE 5.6-4 Diffusion through solids in series with a resistance at the interface

A similar situation exists for mass transfer to the surface at C_{nD} and

$$_{\theta+\Delta\theta}C_{nD} = \frac{1}{M_D}\{2N_{D\,\theta}C_{nB} + [M_D - (2N_D + 2)]_{\theta}C_{nD} + 2_{\theta}C_{nD+1}\} \quad (5.6\text{-}21)$$

where $N_D = k_c\,\Delta x_D/D_{AD}$.

If M_B or M_D becomes inconveniently too large, then Eq. (5.4-12) can be modified and used in place of (5.4-10) in the above derivations.

The derivation above was made for a value of 1.0 of the distribution coefficient K between phase B and E and between E and D.

5.7 COMBINED NUMERICAL-ANALOG COMPUTER METHOD

In the approaches used in previous sections of this chapter, both of the partial derivatives were replaced by finite differences. In the present section only one, the second partial derivative, will be replaced by a finite difference. Then the other partial derivative with respect to time remains and becomes an ordinary derivative. As a result, we have a system of first-order differential-difference equations remaining. These can be solved by the numerical techniques discussed in Chapter 1 for solving initial-value and boundary-value differential equations. Another method often used is to solve these equations on the analog computer; this is discussed elsewhere (L1).

5.8 NUMERICAL METHODS FOR HEAT OR MASS TRANSFER

The numerical methods derived in this chapter for mass transfer can easily be used for heat transfer, unsteady-state. Equations (2.4-12) and (2.6-15) are the basic equations used for unsteady-state transport of heat and mass. For heat transfer α, the thermal diffusivity in cm²/sec is substituted for D_{AB}. Temperature, $T\,°C$, is substituted for C, concentration.

For example, Eq. (5.4-8) for mass transfer becomes for heat transfer

$$_{\theta+\Delta\theta}T_n = \frac{1}{M}[_\theta T_{n+1} + (M-2)_\theta T_n + _\theta T_{n-1}] \qquad (5.8\text{-}1)$$

where $M = (\Delta x)^2/\alpha\,\Delta\theta$, $\alpha = k/(\rho c_p)$. For convection to the surface and for interphase heat transfer N is defined as $h\,\Delta x/k$ and R as $(k_D\Delta x_B)/(k_B\Delta x_D)$. The term h is the heat transfer coefficient in cal/(sec·cm²·°C) or Btu/(hr·ft²·°F).

Mention must also be made of a text by Crank (C1), who has solved many unsteady-state cases for diffusion and gives tables and graphs of these results. Carslaw and Jaeger (C2) have also done this for heat transfer.

5.9 UNSTEADY-STATE DIFFUSION IN COMPOSITE MEDIA

5.9A   Two Slabs in Series, No Interface Resistance

Diffusion in two-phase systems is quite important in mass transport phenomena. A case that occurs often is unsteady-state diffusion of a solute A from phase II to a second phase I that is immiscible with phase II. Figure 5.9-1a shows the conditions at time $t = 0$.

The solute is at a uniform concentration initially of c_I^0 in phase I and a uniform concentration of c_{II}^0 in phase II at time $t = 0$. It is assumed that Fick's second law holds in each phase. Further, it is assumed that there is no resistance at the interface and that there is equilibrium at the interface described by

$$c_{II} = Kc_I \qquad (5.9\text{-}1)$$

where K is the equilibrium distribution coefficient, which is constant. The following equations and boundary and initial conditions are given:

$$\frac{\partial c_I}{\partial t} = \frac{D_I \partial^2 c_I}{\partial x^2}, \qquad x < 0 \qquad (5.9\text{-}2)$$

240 Unsteady-State Diffusion

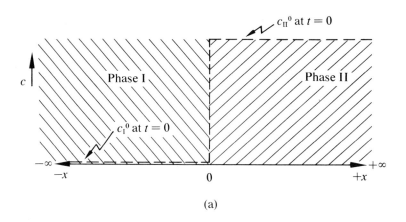

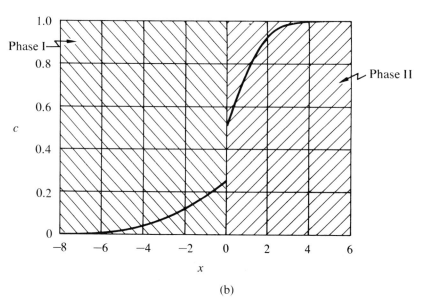

FIGURE 5.9-1 Unsteady-state diffusion in composite media: (a) initial concentration distribution at $t = 0$, (b) concentration distribution for $D_I = 4D_{II}$, $K = 2.0$, $D_{II}t = 1$. (From J. Crank, *The Mathematics of Diffusion*. Oxford: Clarendon Press, 1956. With permission.)

$$\frac{\partial c_{II}}{\partial t} = \frac{D_{II} \partial^2 c_{II}}{\partial x^2}, \quad x > 0 \quad (5.9\text{-}3)$$

$$c_{II} = Kc_I \quad \text{when } x = 0, t > 0 \quad (5.9\text{-}4)$$

$$\frac{-D_I \partial c_I}{\partial x} = \frac{-D_{II} \partial c_{II}}{\partial x} \quad \text{when } x = 0 \quad (5.9\text{-}5)$$

$$c_I = c_I^0 \quad \text{when } x = -\infty \quad (5.9\text{-}6)$$

5.9 Unsteady-State Diffusion in Composite Media

$$c_{II} = c_{II}^0 \quad \text{when } x = +\infty \quad (5.9\text{-}7)$$

$$c_I = c_I^0 \quad \text{when } t = 0, -\infty < x < 0 \quad (5.9\text{-}8)$$

$$c_{II} = c_{II}^0 \quad \text{when } t = 0, 0 < x < \infty \quad (5.9\text{-}9)$$

First taking the Laplace of Eq. (5.9-2), the result is an ordinary differential equation:

$$\frac{d^2 c_I(x,p)}{dx^2} - \left(\frac{p}{D_I}\right) c_I(x,p) = -\frac{c_I^0}{D_I} \quad (5.9\text{-}10)$$

The steps to obtain Eq. (5.9-10) are identical to those used to obtain Eq. (5.3-20). The equation above is valid only in the region x going from $-\infty$ to 0. A similar procedure can be used for Eq. (5.9-3). Then, using the boundary and initial conditions, the final two equations are

$$c_I = \frac{c_{II}^0 - K c_I^0}{K + \sqrt{D_I/D_{II}}} \left(1 + \text{erf}\,\frac{x}{2\sqrt{D_I t}}\right) + c_I^0, \quad -\infty < x < 0 \quad (5.9\text{-}11)$$

$$c_{II} = \frac{c_I^0 - c_{II}^0/K}{1/K + \sqrt{D_{II}/D_I}} \left(1 - \text{erf}\,\frac{x}{2\sqrt{D_{II} t}}\right) + c_{II}^0, \quad 0 < x < \infty \quad (5.9\text{-}12)$$

It is interesting to note that as diffusion proceeds, on setting $x = 0$ (interface), we obtain from the two equations above [since erf (0) = 0]

$$c_I = \frac{c_{II}^0 - K c_I^0}{K + \sqrt{D_I/D_{II}}} + c_I^0, \quad x = 0 \quad (5.9\text{-}13)$$

$$c_{II} = \frac{c_I^0 - c_{II}^0/K}{1/K + \sqrt{D_{II}/D_I}} + c_{II}^0, \quad x = 0 \quad (5.9\text{-}14)$$

This means the concentrations c_I and c_{II} remain constant at the interface during the diffusion.

Crank (C1) shows such a case for $D_I = 4 D_{II}$, $K = 2.0$, $c_{II}^0 = 1.0$, and $c_I^0 = 0$. Substituting these numbers into Eqs. (5.9-11) to (5.9-14), the concentration at a time where $D_{II} t = 1$ is plotted in Fig. 5.9-1b. Substituting the numerical values into Eq. (5.9-13) and (5.9-14), $c_I = \frac{1}{4}$ and $c_{II} = \frac{1}{2}$ at the interface. These values at the interface will be constant at all values of t. Section 5.6A gives a numerical method to solve this case.

An interesting case occurs when c_I^0 is greater than c_{II}^0 at $t = 0$. A brief sketch in Fig. 5.9-2 shows this case. Here $D_{II} = 2 D_I$ and $K = 1.5 = c_{II}/c_I$. The diffusion in this case is from phase I to phase II, which is the opposite direction than in Fig. 5.9-1. At the interface it would seem that the solute is diffusing "uphill." However, this is not so, since the concentrations at the interface are in equilibrium. For equilibrium at the interface, the chemical potentials are equal, and this does not imply that the concentrations are necessarily equal.

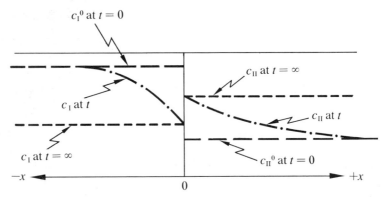

FIGURE 5.9-2 Concentration distribution for $K = 1.5$ and $D_{II} = 2D_I$

For further detailed discussions on unsteady-state diffusion in composite media the reader is referred to the excellent texts by Jost (J1) and Crank (C1).

5.9B Two Slabs in Series, Interface Resistance Present

In some cases there may be a resistance at the interface between two phases in series. If one expresses the contact resistance as a mass transfer coefficient k_c, then at the interface for $K = 1.0$,

$$\frac{-D_{II}\partial c_{II}}{\partial x} = k_c(c_I - c_{II}) \tag{5.9-15}$$

This equation replaces Eq. (5.9-4) at the interface. Solution of these equations (C1) shows that there is a discontinuity at the interface, which narrows as t increases. See Fig. 5.9-3, which briefly sketches the case

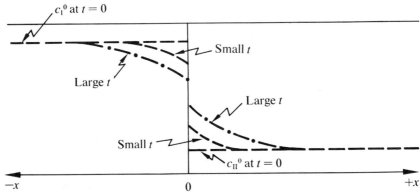

FIGURE 5.9-3 Concentration plot for an interface resistance present between phases I and II; $K = 1.0$

where $D_I = D_{II}$ and $K = 1.0$. At large times the two concentrations at the interface should be equal to each other and also equal to $(c_I^0 + c_{II}^0)/2$. In Section 5.6B a numerical method is given for a similar case.

5.9C Diffusion and Phase Change

Many practical cases of unsteady-state diffusion in solids occur when a solute diffuses into a homogeneous solid and a second phase develops at the surface. This is sketched in Fig. 5.9-4. Originally phase I has an initial uniform concentration of c_I^0 at $t = 0$. At time $t = 0$, the concentration at the surface is suddenly increased to c_0^s and held there, as shown in Fig. 5.9-4a.

In Fig. 5.9-4b at a concentration in phase I of $c_{I,II}$, phase II develops. The concentrations $c_{II,I}$ in phase II and $c_{I,II}$ in phase I are at equilibrium and related by the distribution coefficient K. The point δ moves to the right as diffusion proceeds. Jost (J1, p. 71) gives a solution for this case and also discusses other cases.

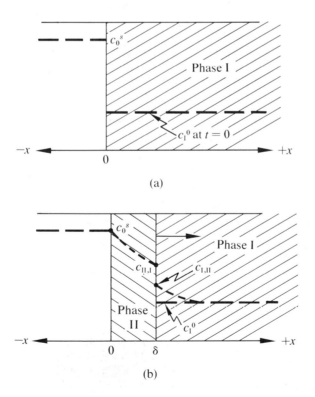

FIGURE 5.9-4 Diffusion into phase I with phase II developing at the surface: (a) concentration at $t = 0$, (b) concentration at $t > 0$

PROBLEMS

5-1 Unsteady-State Diffusion of Gases by Fourier Series

Pure N_2 gas is put in a lower rectangular container of length $L/2$ cm at 1 atm and 25°C. Pure He gas at the same conditions is put in an identical upper chamber. The chambers are separated by a partition. At $t = 0$, the partition is removed and the two gases allowed to interdiffuse. Derive the unsteady-state equation relating the mole fraction y of N_2 with t and position x and solve it using Fourier series. Call $x = 0$ at the bottom end of the N_2 chamber and $x = L$ at the top end of the He chamber. The diffusion is, of course, equimolar counterdiffusion. Calculate and plot the actual concentration versus x for $t = 60$ sec and $L = 20$ cm.

$$\text{Ans.: } y(x,t) = \frac{1}{2} + \frac{2}{\pi}\sum_{n=1}^{\infty}\frac{1}{n}\sin\left(\frac{n\pi}{2}\right)\cos\left(\frac{n\pi x}{L}\right)\exp\left(\frac{-n^2\pi^2 Dt}{L^2}\right)$$

5-2 Unsteady-State Diffusion to a Slab by Fourier Series

Write the original partial differential equation and boundary and initial conditions for Eq. (5.2-39). Solve this by Fourier series.

5-3 Unsteady-State Diffusion in a Cylinder

A wet cylinder of agar gel at 5°C contains a uniform concentration of 1.0×10^{-4} g mole urea/cm^3. The cylinder is 1.20 in. in diameter and is only 1.50 in. long with flat parallel ends. Calculate the concentration at the midpoint of the cylinder after 100 hours for the following cases if the cylinder is suddenly immersed in pure running water. See Example 5.2-2 for diffusivity.

(a) The flat parallel ends are insulated and diffusion occurs only radially.
(b) Diffusion occurs radially and axially.

5-4 Unsteady-State Diffusion and Solution by Laplace Transform

Two salt solutions are separated by a semipermeable membrane M offering no resistance to diffusion of salt from I to II. Region I is kept agitated so its concentration is always constant at c_0. The concentration in II at $t = 0$ is $c_1 x + c_2$ and is shown in Fig. P5-4. Following are the conditions given:

(1) $\partial c/\partial t = D\, \partial^2 c/\partial x^2$.
(2) $c(x, 0) = c_1 x + c_2$ when $t = 0$.
(3) $c(0, t) = c_0$ when $x = 0$.
(4) $c(x, t) = c_1 x + c_2$ as $x \to \infty$.

(a) Find the equation relating c to x and t using the Laplace transform method.
(b) $c_0 = 6 \times 10^{-4}$, $c_1 = 0.1 \times 10^{-4}$, $c_2 = 1 \times 10^{-4}$, $D = 10^{-5}$. From the equation derived in (a), calculate c for x values of 0, 1, 2, and 5 when $t = 10^5$. Do the same for $t = 0$ and $t = \infty$. Plot these values on a graph.

$$\text{Ans.: (a) } c(x, t) = (c_0 - c_2)\,\text{erfc}\left[\frac{x}{2\sqrt{Dt}}\right] + c_1 x + c_2$$

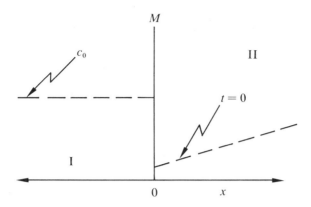

FIGURE P5-4 Concentration profile for Problem 5-4

5-5 Unsteady-State Diffusion in a Slab

A solid slab made of agar gel contains a uniform concentration of 8.33×10^{-4} g mole urea/cm^3 of solid. The slab is 0.4 in. thick. Diffusion is only through the two flat surfaces 0.4 in. apart at 5°C. The ends and sides are impermeable.

(a) Calculate the concentration of urea after 10 hr at the midpoint of the slab (0.2 in. deep) and at 0.05 in. and 0.10 in. from the surface. The slab is immersed in running fresh water. The diffusivity of urea in the solid is 4.70×10^{-6} cm^2/sec at 5°C. Plot c_A on a graph. Use the unsteady-state diffusion charts for calculations.

(b) What is the average concentration after 10 hr? [*Note*: Assume there is no resistance to diffusion at the surfaces.]

5-6 Drying Using Numerical Solution for Unsteady-State

A 1-in.-thick slab of gel is wet with alcohol and is to be dried by the passing of dry air over the top surface. All surfaces of the slab are insulated except the top. The diffusivity of alcohol in gel is 0.00096 ft^2/hr. The initial uniform concentration is 1×10^{-3} g mole alcohol/cm^3. Determine the concentration profile 0·868 hr after exposure to the air. Choose Δx as 0.20 in. and $M = 2.0$. Tabulate final concentrations and also plot them.

5-7 Drying Using Numerical Solution for Digital Computer

Using the conditions of Problem 5–6, solve the problem by the digital computer. Use a $\Delta x = 0.05$ in. Be sure to use the proper equation for the insulated back (opposite) surface. Write the Fortran IV program and plot the final concentrations. Use the explicit method, $M = 2$.

5-8 Unsteady-State Using Digital Computer

For the conditions of Example 5.5-1 in the text, use a $\Delta x = 1.00$ cm. The rear face is not insulated, but at $\theta = 0$ the concentration at the rear face is raised to a constant value of $C = 10 \times 10^{-3}$ g mole A/cm^3 and held there. Using the Crank-Nicolson method, write the Fortran IV program and plot the final concentrations.

5-9 Numerical Solution by Finite Differences

The following partial differential equation holds for steady-state heat transfer in a cylinder of metal:

$$\frac{\partial T}{\partial z} = \alpha\left(\frac{1}{r}\frac{\partial T}{\partial r} + \frac{\partial^2 T}{\partial r^2}\right)$$

where T = temperature,
 r = position (or distance) from center line of cylinder in radial direction,
 z = distance from one end of cylinder in axial direction,
 α = a constant.

Set this equation up so a numerical solution can be obtained by finite differences. Be sure to explain how $1/r$ is obtained and the values that M can have in the explicit method.

5-10 Diffusion through Slabs in Series and Digital Computer Solution

Solute A is diffusing unsteady-state through slabs B and D in series, B being 4.0 cm thick and D being 3.0 cm thick. At time $\theta = 0$, the initial concentration profile of A is linear and is 1.0×10^{-3} g mole A/cm³ at the front face of B and is 2.75×10^{-3} g mole A/cm³ at the rear face of D. The concentration at the front face of phase B is suddenly raised to a value of 8.0×10^{-3} g mole A/cm³ and held there. The rear of phase D is insulated. The diffusivity D_{AB} is 0.5×10^{-4} cm²/sec and $D_{AD} = 0.30 \times 10^{-4}$ cm²/sec. It is desired to calculate the concentration profile in 50,000 sec. Use $M_B = 4.0$ and $\Delta x_B = \Delta x_D = 1.0$ cm. Write the Fortran IV program and compute the results. The distribution coefficient K between the two phases B and D is 1.0.

5-11 Diffusion through Slabs with a Distribution Coefficient

Rederive Eq. (5.6-17), giving the steps in detail, so that $_{\theta+\Delta\theta}C_{iD}$ is solved for in terms of $_\theta C_{iD}$.

5-12 Diffusion through Slabs in Series with a K Value Not 1.0 Using the Digital Computer

Repeat Problem 5-10 but with the following changes. At time $\theta = 0$, the initial concentration in B and D is zero. The distribution coefficient between phases B and D is $C_D = KC_B$, where $K = 0.5$.

5-13 Unsteady-State Diffusion in Composite Media and Laplace Transform

(a) Starting with Eqs. (5.9-1) to (5.9-9), derive Eqs. (5.9-11) to (5.9-14) using the Laplace transform method.
(b) In this case, $c_I^0 = 5 \times 10^{-3}$, $c_{II}^0 = 1 \times 10^{-3}$, $D_I = 1 \times 10^{-5}$, $D_{II} = 5 \times 10^{-5}$, $K = 3.0$. Calculate values of c_I and c_{II} for values of $x = 0, \pm 1, \pm 2,$ and ± 5 for $t = 10^5$. Also do this for $t = \infty$ and $t = 0$. Plot data on a graph and explain the behavior of the curves.

NOTATION

(Boldface symbols are vectors.)

a_n constant in Fourier series
a constant
A cross-sectional area perpendicular to flux, cm^2 or ft^2
A constant
b_n constant in Fourier series
B constant
c_A concentration of A, g mole A/cm^3 or lb mole A/ft^3
c concentration, g mole/cm^3
C_A concentration of A, g mole A/cm^3 or lb mole A/ft^3
c_p heat capacity at constant pressure, cal/g mass-°C or Btu/lb mass-°F
C constant
C concentration, g mole/cm^3
D_A diffusivity of A in solid, cm^2/sec or ft^2/hr
D_{AB} diffusivity, cm^2/sec or ft^2/hr
E total fraction unremoved defined in Eq. (5.2-43)
f function
$f(p)$ Laplace of $f(t)$
h heat transfer coefficient, cal/sec-cm^2-°C or Btu/hr-ft^2-°F
h increment in x, cm or ft
H distance, cm
K distribution coefficient defined by Eq. (5.4-13), dimensionless
k_c mass transfer coefficient, cm/sec or ft/hr
k time increment $= \Delta\theta$, sec or hr
k thermal conductivity, cal-cm/sec-cm^2-°C or Btu-ft/hr-ft^2-°F
L distance, cm or ft
m modulus $= D_{AB}/(k_c x_1)$, dimensionless
M modulus $= (\Delta x)^2/(D_{AB} \Delta\theta)$, dimensionless
N modulus $= k_c \Delta x/D_{AB}$, dimensionless
N_A flux of A relative to a fixed point, g mole A/sec-cm^2 or lb mole A/hr-ft^2
n modulus $= x/x_1$, dimensionless
R_A rate of generation, g mole A/sec-cm^3
R modulus $= D_{AD} \Delta x_B/(D_{AB} \Delta x_D)$, dimensionless
$T(t)$ variable dependent only on t
t time, sec or hr
u variable defined by $x\sqrt{D/D_x}$
$\mathbf{v}$ mass average velocity of stream, cm/sec or ft/hr
v variable defined by $y\sqrt{D/D_y}$
w variable defined by $z\sqrt{D/D_z}$
X modulus $= D\theta/x_1^2$, dimensionless

248 Unsteady-State Diffusion

x distance in x direction, cm or ft
$X(x)$ variable dependent only on x
Y fraction unaccomplished change defined in Eq. (5.2-39)
Y_n/Y_0 value of Y at $n = n$ over value of Y at $n = 0$
$Y(y)$ variable dependent only on y
y distance in y direction, cm or ft
y_A concentration, g mole A/cm^3 or lb mole A/ft^3
z distance in z direction, cm or ft

Greek Letters

α thermal diffusivity, $k/\rho c_p$, cm^2/sec or ft^2/hr
β modulus $= 1/M$, dimensionless
δ equivalent film thickness, cm
θ time, sec or hr
ρ density, g mass/cm^3 or lb mass/ft^3

Subscripts

A component A
a position outside surface in fluid
ave   average
B component B or phase B
D component D or phase D
f rear face position
i interface
i index of position x
j index of time θ
L liquid
n position 0, 1, 2, 3, ...
0   value at $\theta = 0$
x in x direction
y in y direction
z in z direction
1   beginning of diffusion path
2   end of diffusion path
θ time θ
I   phase I
II   phase II

REFERENCES

(C1)   J. Crank, *Mathematics of Diffusion*. Oxford: Clarendon Press, 1956.
(C2)   H. S. Carslaw and J. C. Jaeger, *Conduction of Heat in Solids*, 2d ed. Oxford: Clarendon Press, 1959.
(G1)   H. P. Gurney and J. Lurie, *Ind. Eng. Chem.* **15**, 1170 (1923).

References 249

(H1) M. P. Heisler, *Trans. A.S.M.E.* **69**, 227 (1947).
(J1) W. Jost, *Diffusion in Solids, Liquids, Gases*. New York: Academic Press, Inc., 1960.
(L1) L. Lapidus, *Digital Computation for Chemical Engineers*. New York: McGraw-Hill, Inc., 1962.
(M1) H. S. Mickley, T. K. Sherwood, and C. E. Reed, *Applied Mathematics in Chemical Engineering*, 2d ed. New York: McGraw-Hill, Inc., 1957.
(N1) A. B. Newman, *Ind. Eng. Chem.* **28**, 545 (1936).
(P1) J. H. Perry, *Chemical Engineers' Handbook*, 4th ed. New York: McGraw-Hill, Inc., 1963.
(S1) P. J. Schneider, *Conduction Heat Transfer*. Cambridge, Mass.: Addison-Wesley Publishing Co., Inc., 1955.
(T1) R. E. Treybal, *Mass Transfer Operations*, 2d ed. New York: McGraw-Hill, Inc., 1968.

6 Mass Transfer Coefficients in Laminar and Turbulent Flow

6.1 INTRODUCTION TO TURBULENT TRANSPORT

In the discussion of diffusion in earlier chapters the emphasis has been on molecular transport in fluids that were stagnant or in laminar flow. However, in many cases these diffusion processes may be too slow, and more rapid diffusion or transport is necessary. To speed up this diffusion the fluid velocity is increased so that turbulent transport occurs.

Generally our concern has been with molecular transport in one phase alone. However, to have a fluid in turbulent motion generally requires this fluid to be flowing past another immiscible fluid or (usually) by a solid surface. A good example of this is a fluid flowing in a pipe, where part of the pipe wall is made of a slightly dissolving solid material, such as benzoic acid, which is transported perpendicular to the main stream from the solid wall.

As discussed previously for molecular transport, similar equations can be written for mass, heat, and momentum transport. For molecular mass transport we write

$$J_A^* = \frac{D_{AB}}{z}(c_{A1} - c_{A2}) \qquad (6.1\text{-}1)$$

6.1 Introduction to Turbulent Transport

For turbulent mass transport we can write an equation in a similar fashion as

$$J_A^* = \epsilon(c_{A1} - \bar{c}_A) \qquad (6.1\text{-}2)$$

where ϵ is a convective mass transport coefficient and $\bar{c}_A$ is the average concentration of species A in the bulk fluid. The ϵ is a complex function of the flow and properties of the fluid and the geometry of the system. In the following sections some of the theories and mechanisms on turbulent mass transfer will be reviewed.

It should be pointed out that turbulent mass transport is more complicated than momentum or heat transport, because we are dealing with mixtures and not with a single component. In the diffusing mixture the fluxes of each species are different and interact, so that the fluxes are interdependent of each other. Also, the definitions of velocity get complicated, since molar average, mass average, and volume average velocities can differ.

When a fluid flows past a surface under such conditions that the fluid is in turbulent flow, then the actual velocity of small particles of fluid cannot be described clearly as in laminar flow. In laminar flow the fluid flows in smooth streamlines, and its behavior can usually be described mathematically. In turbulent motion there are no orderly streamlines or equations to describe the behavior. There are large eddies or "chunks" of fluid, which move rapidly in seemingly random fashion. When solute is dissolving from the solid surface and being transferred to the fluid, there is a high concentration of this solute at the surface, and its concentration decreases as the distance from the wall increases. However, if one were able to sample and analyze very minute portions of fluid, he would find that adjacent samples would not always have concentrations close to each other. This occurs because the eddies, which have dissolved solute in them, can move rapidly from one part of the fluid to another, taking with them or transferring relatively large amounts of solute, which contributes to the overall mass transfer process from the surface to the main body of the fluid. This eddy transfer, or turbulent diffusion as it is often called, is very fast in comparison to the relatively slow process of molecular diffusion, where each solute molecule must move by random motion through the fluid.

Adjacent to the surface, a thin laminar-type sublayer film is present. In this region the mass transfer occurs by molecular diffusion, since little or no eddies are present. Since this is a slow process, a large concentration change or drop in concentration occurs across this laminar film.

Adjacent to this is the transition or buffer region. Here some eddies occur, and the transfer is by the sum of molecular diffusion and by turbulent diffusion. In this region there is a gradual and not abrupt transition

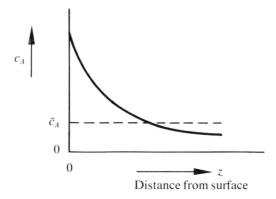

FIGURE 6.1-1 Concentration profile in turbulent mass transport

from the total transfer occurring by almost pure molecular diffusion at the one end to mainly turbulent at the other end. The concentration decrease is much less in this latter region.

In the turbulent region adjacent to the buffer region most of the transfer is by turbulent or eddy diffusion. Molecular diffusion still occurs, but it contributes little to the overall transfer in this region. The concentration decrease is very small here since the rapid eddy movement evens out any gradients tending to exist.

A plot of the concentration profile for the evaporation of water vapor from a surface to turbulent air is given in Fig. 6.1-1 and shows that the concentration drops off very rapidly. This curve is very similar to the shapes also found for heat and for momentum transfer. The average concentration $\bar{c}_A$ is shown, and it is slightly greater than the minimum concentration.

6.2 DERIVATION OF GENERAL EQUATION FOR MASS TRANSFER COEFFICIENTS

Our understanding of laminar flow is quite good, and we can often mathematically describe molecular diffusion in this case. However, at present our knowledge and basic understanding of the movement and sizes of eddies is still incomplete. Since this understanding for turbulent flow is incomplete, we attempt to write the equations for turbulent diffusion in a manner similar to that for molecular diffusion. Hence, we write for turbulent transport,

$$J_A^* = -(D_{AB} + \epsilon_M)\frac{dc_A}{dz} \qquad (6.2\text{-}1)$$

where D_{AB} is the molecular diffusivity and ϵ_M is the mass eddy diffusivity. To integrate this equation, ϵ_M is a variable and is zero at the surface and

6.2 General Equation for Mass Transfer Coefficients

increases as the distance from the wall increases. This relationship is not generally known, so an average $\bar{\epsilon}_M$ is used on integration.

$$J_{A1}^* = \frac{D_{AB} + \bar{\epsilon}_M}{z}(c_{A1} - c_{A2}) \tag{6.2-2}$$

The flux J_{A1}^* is based on the surface area A_1, since the cross-sectional area may vary in diffusing from the interface. Also the value of z, the distance of the path, is usually not known. Hence, Eq. (6.2-2) is written as

$$J_{A1}^* = k_c'(c_{A1} - c_{A2}) \tag{6.2-3}$$

where J_{A1}^* is the flux of A from the surface 1 relative to the whole bulk phase, k_c' is $(D_{AB} + \bar{\epsilon}_M)/z$ and is an experimental mass transfer coefficient in g mole $A/\text{cm}^2\text{-sec-(g mole/cm}^3)$, and c_{A2} is the concentration at point 2. Usually the concentration c_{A2} in a turbulent fluid is not known, so c_{A2} is taken as $\bar{c}_{A2}$, the average bulk concentration.

Since we usually are interested in N_A, the total flux of A relative to stationary coordinates, we should start with the following equation, which is similar to that for molecular diffusion:

$$N_A = -(D_{AB} + \epsilon_M)\frac{dc_A}{dz} + \frac{c_A}{c}(N_A + N_B) \tag{6.2-4}$$

Integrating as before, and assuming $\bar{\epsilon}_M$ to be constant,

$$N_A = \frac{N_A}{N_A + N_B}\frac{(D_{AB} + \bar{\epsilon}_M)c}{z} \ln\left[\frac{\frac{N_A}{N_A + N_B} - \frac{c_{A2}}{c}}{\frac{N_A}{N_A + N_B} - \frac{c_{A1}}{c}}\right] \tag{6.2-5}$$

Now substituting $k_c' = (D_{AB} + \bar{\epsilon}_M)/z$ into Eq. (6.2-5) and $x_A = c_A/c$,

$$N_A = \frac{N_A}{N_A + N_B}k_c'c \ln\left[\frac{\frac{N_A}{N_A + N_B} - x_{A2}}{\frac{N_A}{N_A + N_B} - x_{A1}}\right] \tag{6.2-6}$$

The k_c' is the mass transfer coefficient in Eq. (6.2-3) for flux relative to the bulk phase and has units of cm/sec or ft/hr.

The k_c' is often termed the Colburn-Drew (B1) mass transfer coefficient. Other mass transfer coefficients are sometimes used in the scientific and engineering literature, but this one is generally preferred. This convective mass transfer coefficient k_c' is used when the transport of mass occurs between a surface or fluid and a moving fluid.

The bulk flow correction in Eq. (6.2-4) of $(c_A/c)(N_A + N_B)$ was used in molecular diffusion through a true laminar or stagnant path. However, in turbulent diffusion this correction for bulk flow in a turbulent diffusion path is less certain. This bulk flow term is either from the interface or to the interface in a direction perpendicular to the flowing stream. It is assumed that the fluxes from or to the interface are small enough that the velocity and concentration profiles are unaltered. This idealized situation gives reasonably close approximations to many practical situations where the mass transfer rates or fluxes are small and the bulk flow is only important in calculating the fluxes across the interface.

The final equations (6.2-5) and (6.2-6) are the result of assuming a stagnant or equivalent film of thickness z where diffusion occurs; this is termed the film theory. This theory assumes that the film thickness is unaffected by mass transfer and bulk flow. Other more complicated methods to correct for bulk flow and large fluxes are to use the penetration theory (H1) or boundary layer theory (S1). The correction method to use depends on the flow and geometry conditions. The penetration theory is preferred for liquids at a gas-liquid interface, such as in a packed gas-liquid absorption tower. The boundary layer theory is preferred for laminar flow of a fluid along a stationary plane surface, especially in the case of evaporation of a pure liquid with a large vapor pressure and for pipe entrances. For the gas phase in a liquid-gas absorption column the interface or boundary between the gas and liquid is moving, and there are no adequate theories for this.

In many physical geometries with flow none of the three theories adequately describes the situation. Bird *et al.* (B2) summarize these corrections for bulk flow and high fluxes and show that for bulk flow and low and moderate fluxes the results for the film, penetration, and boundary layer theories are very similar. Hence, because of its simplicity, the film theory is generally preferred under these conditions.

6.3 DEFINITION OF VARIOUS MASS TRANSFER COEFFICIENTS

6.3A Introduction to Theory

It is desirable to be able to write Eq. (6.2-6) in a more abbreviated form, which often is used in the literature (B1):

$$N_A = \frac{k'_c}{\varphi_N} c(x_{A1} - x_{A2}) = \frac{k'_c}{\varphi_N}(c_{A1} - c_{A2}) \qquad (6.3\text{-}1)$$

where φ_N is the bulk flow correction factor, accounting for the fact that there may be a flow of the bulk phase away from the surface or toward it. To solve for the value of φ_N, equate Eq. (6.2-6) to (6.3-1) and solve for φ_N.

$$\varphi_N = \frac{x_{A1} - x_{A2}}{\dfrac{N_A}{N_A+N_B}\ln\left(\dfrac{\dfrac{N_A}{N_A+N_B} - x_{A2}}{\dfrac{N_A}{N_A+N_B} - x_{A1}}\right)} = \frac{\left(\dfrac{N_A}{N_A+N_B} - x_{A2}\right) - \left(\dfrac{N_A}{N_A+N_B} - x_{A1}\right)}{\dfrac{N_A}{N_A+N_B}\ln\left(\dfrac{\dfrac{N_A}{N_A+N_B} - x_{A2}}{\dfrac{N_A}{N_A+N_B} - x_{A1}}\right)}$$

$$(6.3\text{-}2)$$

6.3 Definition of Various Mass Transfer Coefficients

Calling N_R, the flux ratio, equal to $N_A/(N_A+N_B)$,

$$\varphi_N = \frac{(N_R - x_A)_{LM}}{N_R} \quad (6.3\text{-}3)$$

where $(N_R - x_A)_{LM}$ is the ln mean of $(N_R - x_{A2})$ and $(N_R - x_{A1})$. The ratio $N_A/(N_A+N_B)$ is determined as before by nondiffusional considerations such as material balances, total pressure balances, and so on.

The value of φ_N can be determined for various special cases as follows below. For dilute solutions φ_N is close to 1.0.

6.3B Equimolar Counterdiffusion

In this case N_R is infinite, and Eq. (6.3-2) will not easily yield a solution. Starting with Eq. (6.2-4) and setting $N_A = -N_B$, the final equation is the same as Eq. (6.2-3), as expected, and $\varphi_N = 1.0$

$$N_A = k'_c(c_{A1} - c_{A2}) = k'_c c(x_{A1} - x_{A2}) \quad (6.3\text{-}4)$$

Hence, we can see that k'_c is also the mass transfer coefficient for the case of no bulk or convective flow. Equating Eq. (6.3-1) to (6.3-4) and solving for φ_N, $\varphi_N = 1.0$. Since we can define the concentration in several ways, we can define the coefficient in several ways. It is customary to use y_A as mole fraction for a gas phase and x_A as mole fraction for a liquid phase. Hence,

For gases:

$$N_A = k'_G(p_{A1} - p_{A2}) = k'_y(y_{A1} - y_{A2}) = k'_c(c_{A1} - c_{A2}) \quad (6.3\text{-}5)$$

Substituting c_A/c for y_A, and p_A/RT for c_A, the following relationships hold:

$$k'_G P = k'_y = k'_c c \quad (6.3\text{-}6)$$

For liquids:

$$N_A = k'_x(x_{A1} - x_{A2}) = k'_L(c_{A1} - c_{A2}) \quad (6.3\text{-}7)$$

$$k'_x = k'_c c = k'_L c = k'_L \frac{\rho}{M} \quad (6.3\text{-}8)$$

where ρ is the density of the liquid and M the molecular weight.

6.3C Diffusion of A through Stagnant B

Setting $N_B = 0$ in Eq. (6.3-2), $N_R = 1.0$, and then

$$\varphi_N = (1 - x_A)_{LM} = x_{BM} = \frac{x_{B2} - x_{B1}}{\ln(x_{B2}/x_{B1})} \quad (6.3\text{-}9)$$

Hence,

$$N_A = k'_c \frac{c(x_{A1} - x_{A2})}{x_{BM}} \quad (6.3\text{-}10)$$

256   Mass Transfer Coefficients in Laminar and Turbulent Flow

For gases:

$$N_A = k_G(p_{A1} - p_{A2}) = k_y(y_{A1} - y_{A2}) = k_c(c_{A1} - c_{A2}) \quad (6.3\text{-}11)$$

$$k_G P = k_y = k_c c \quad (6.3\text{-}12)$$

For liquids:

$$N_A = k_x(x_{A1} - x_{A2}) = k_L(c_{A1} - c_{A2}) \quad (6.3\text{-}13)$$

Comparing Eqs. (6.3-10) and (6.3-11), then

$$k_y \frac{p_{BM}}{P} = k'_y = k'_c c = k'_G P = k_G p_{BM} \quad (6.3\text{-}14)$$

Table 6.3-1 lists the various relations among the mass transfer coefficients for the several cases. Often experimental values of mass transfer coefficients will be determined for cases where φ_N is not 1.0. To use this for other values of $N_A/(N_A + N_B)$, it is desirable to convert the coefficient first to k'_c and then to correct it to the desired values of $N_A/(N_A + N_B)$. If the mass transfer coefficient was determined experimentally for a mixture of very dilute A in B, then, regardless of the ratio of $N_A/(N_A + N_B)$ used, the experimental value of $k_{c_{\exp}}$ at the experimental conditions is $k_{c_{\exp}} = k_c = k'_c$. This occurs because the bulk flow term in Eq. (6.2-4) is negligible, since $c_A/c \cong 0$.

Example 6.3-1   Vaporization of A from a Solid Surface

Pure gas B at 1.0 atm pressure is flowing over a solid surface (a blotting paper) from which pure A is vaporizing. The liquid A completely wets the surface of the blotting paper. Hence, the partial pressure at the surface is the vapor pressure of pure A at 25°C, which is 0.10 atm. The mass transfer coefficient k_y has been estimated as 0.05 lb mole/hr-ft^2 (mole fraction). Calculate the rate of vaporization of A.

Solution

This is a case of A diffusing through stagnant B. Then $y_{A1} = 0.10/1.00 = 0.10$ at the surface. $y_{A2} = 0$ in the pure gas B. Using Eq. (6.3-11),

$$N_A = k_y(y_{A1} - y_{A2}) = 0.05(0.10 - 0) = 5.0 \times 10^{-3} \text{ lb mole } A \text{ diffusing/hr-ft}^2$$

Example 6.3-2   Effect of Flux Ratio on Mass Transfer Coefficient

In an experiment to be done on equimolar counterdiffusion of A and B the value of k_G was previously determined to be 0.88 lb mole/hr-ft^2-atm pressure difference. For the same flow and also concentrations in this apparatus it is desired to predict the flux or vaporization of A during

Table 6.3-1 Mass Transfer Coefficients and Flux Equations

General Equation

$$N_A = \frac{k'_c}{\varphi_N}(c_{A1} - c_{A2}) = \frac{k'_c}{\varphi_N}c(x_{A1} - x_{A2}) = \frac{k'_y}{\varphi_N}(y_{A1} - y_{A2})$$

where

$$\varphi_N = \frac{(N_R - x_{A2}) - (N_R - x_{A1})}{N_R \ln[(N_R - x_{A2})/(N_R - x_{A1})]}, \qquad N_R = \frac{N_A}{N_A + N_B}$$

Flux Equation for Special Cases

$\varphi_N = 1.0$ Equimolar Counterdiffusion	$\varphi_N = x_{BM}$ Diffusion of A through Stagnant B	Units of Coefficient
GASES		
$N_A = k'_G(p_{A1} - p_{A2})$	$N_A = k_G(p_{A1} - p_{A2})$	$\dfrac{\text{moles transferred}}{\text{time(area)(pressure)}}$
$N_A = k'_y(y_{A1} - y_{A2})$	$N_A = k_y(y_{A1} - y_{A2})$	$\dfrac{\text{moles transferred}}{\text{time(area)(mole fraction)}}$
$N_A = k'_c(c_{A1} - c_{A2})$	$N_A = k_c(c_{A1} - c_{A2})$	$\dfrac{\text{moles transferred}}{\text{time(area)(moles/vol)}}$
	$N_A M_A = k_Y(Y_{A1} - Y_{A2})$	$\dfrac{\text{mass transferred}}{\text{time(area)(mass }A/\text{mass }B)}$
LIQUIDS		
$N_A = k'_x(x_{A1} - x_{A2})$	$N_A = k_x(x_{A1} - x_{A2})$	$\dfrac{\text{moles transferred}}{\text{time(area)(mole fraction)}}$
$N_A = k'_L(c_{A1} - c_{A2})$	$N_A = k_L(c_{A1} - c_{A2})$	$\dfrac{\text{moles transferred}}{\text{time(area)(moles/vol)}}$

Conversions between Coefficients

GASES

$$k'_c c = k'_c \frac{P}{RT} = k_c \frac{p_{BM}}{RT} = k'_G P = k_G p_{BM} = k_y \frac{p_{BM}}{P} = k'_y$$

LIQUIDS

$$k'_L c = k_L x_{BM} c = k'_L \frac{\rho}{M} = k'_x = k_x x_{BM}$$

equimolar counterdiffusion from this surface to a gas mixture containing a mixture of A and B, where $p_{A2} = 0.05$ atm and $p_{B2} = 0.95$ atm. At the surface the partial pressure of A is $p_{A1} = 0.20$ atm and $P = 1.0$ atm.

Solution

The value of k_G for A through stagnant B must be converted to k'_G. Then, $p_{B2} = 0.95$, $p_{B1} = 1.00 - 0.20 = 0.80$. To calculate p_{BM},

$$p_{BM} = \frac{p_{B2} - p_{B1}}{\ln(p_{B2}/p_{B1})} = \frac{0.95 - 0.80}{\ln(0.95/0.80)} = 0.870 \text{ atm}$$

Then from Table 6.3-1, $k'_G = k_G p_{BM}/P = 0.88(0.87)/1.0 = 0.765$. Then, $N_A = k'_G(p_{A1} - p_{A2}) = 0.765(0.20 - 0.05) = 0.115$ lb mole A/hr-ft^2.

Example 6.3-3 Absorption in a Wetted-Wall Tower

In a wetted-wall column, CO_2 is being absorbed from the air to water at 2.0 atm total pressure and 25°C. The mass transfer coefficient k'_c has been predicted to be 22.5 lb mole/hr-ft^2 (lb mole/ft^3). At a given point in the tower the mole fraction of CO_2 in the liquid at the interface is 1.30×10^{-5} and the partial pressure of CO_2 in the air is 0.15 atm. Calculate the rate of absorption of CO_2 in the water.

Solution

The partial pressure of CO_2 in the gas phase at the interface is the equilibrium vapor pressure of CO_2 over the liquid. Henry's law equation from Table A.4-1 in the appendix gives

$$p_{A1}(\text{atm}) = Hx_{A1} = 1640(1.30 \times 10^{-5}) = 0.0213 \text{ atm}$$

The value of $y_{A1} = 0.0213/2.0 = 0.01065$. In the gas phase, $y_{A2} = 0.15/2.0 = 0.075$.

The air (B) is very insoluble in water, so the equation for A diffusing through stagnant B in the gas phase will be used. $T = 298(1.8)°R$, $R = 0.730$ atm-ft^3/lb mole-°R.

$$N_A = k_y(y_{A1} - y_{A2}) = \frac{k'_c P(y_{A1} - y_{A2})}{RT y_{BM}}$$

$$x_{BM} = y_{BM} = \frac{y_{B2} - y_{B1}}{\ln(y_{B2}/y_{B1})} = \frac{(1 - 0.075) - (1 - 0.01065)}{\ln[(1 - 0.075)/(1 - 0.01065)]} = 0.957$$

$$N_A = \frac{22.5(2.0)(0.01065 - 0.075)}{0.730(298)(1.8)(0.957)} = -7.75 \times 10^{-3} \text{ lb mole/hr-ft}^2$$

The flux is negative, and this indicates the direction of diffusion is the opposite of point 1 to 2. This means the diffusion is from the gas phase to the surface of the liquid.

6.3D Multicomponent Turbulent Transport

The prediction of mass transfer rates for turbulent diffusion in multicomponent mixtures is quite complex and is the subject of current research. Some solutions are available for termary systems, such as for A and B diffusing through stagnant C (B2), but in general approximate methods must be used. The general equation used is similar to Eq. (6.2-6), with $N_A + N_B$ replaced by $\sum_{i=A}^{n} N_i$:

$$N_A = \frac{N_A}{\sum_{i=A}^{n} N_i} k'_{cA} c \ln \left[\frac{\frac{N_A}{\sum_{i=A}^{n} N_i} - x_{A2}}{\frac{N_A}{\sum_{i=A}^{n} N_i} - x_{A1}} \right] \quad (6.3\text{-}15)$$

where n = number of components and k'_{cA} is the mass transfer coefficient for component A.

Assuming the mass transfer coefficient k'_{cA} is available from an empirical correlation such as for J_D (which will be discussed later), then the k'_{cB} is related to k'_{cA} as follows:

$$k'_{cB} = k'_{cA} \left(\frac{D_{Bm}}{D_{Am}} \right)^{2/3} \quad (6.3\text{-}16)$$

where D_{Am} is the effective multicomponent diffusivity of A in the mixture as defined by Eqs. (1.10-12) and (1.10-20) previously.

$$\frac{1}{D_{Am}} = \frac{\sum_{j=1}^{n} \frac{1}{D_{Aj}} (x_j N_A - x_A N_j)}{N_A - x_A \sum_{j=1}^{n} N_j} \quad (6.3\text{-}17)$$

The method for calculating the average D_{Am} or D_{im} was discussed in detail in Section 1.10, whereby

$$E_A = \frac{\sum_{j=1}^{n} N_j}{N_A} \quad (6.3\text{-}18)$$

This is also defined in Eq. (1.10-18). Then, using the average values of x_{A0} and $x_{A\delta}$ and also averages for x_B, x_P, x_Q, and x_I, an average value of D_{Am} was calculated by Eq. (1.10-20) or (6.3-17). Finally, N_A was calculated from Eq. (1.10-22) for molecular transport. However, for turbulent transport we can modify Eq. (1.10-22), and in place of D_{Am}/δ we will use

k'_{cA} as follows:

$$N_A = \frac{ck'_{cA}}{E_A} \ln\left(\frac{1-E_A x_{A\delta}}{1-E_A x_{A0}}\right) \qquad (6.3\text{-}19)$$

This assumes the relations between the fluxes are known by stoichiometry. See Example 1.10-3 for the general method.

If the relation between the fluxes is not known, it is trial and error. Then each flux is first estimated by a binary type equation. This method is given in detail in Example 1.10-4 for molecular transport, the only difference again being that Eq. (6.3-19) is used for the flux calculation for N_A. For N_B,

$$N_B = \frac{ck'_{cB}}{E_B} \ln\left(\frac{1-E_B x_{B\delta}}{1-E_B x_{B0}}\right) \qquad (6.3\text{-}20)$$

Similar equations for N_P and N_Q are written, which are similar to Eqs. (1.10-23) to (1.10-32).

In many cases, as in diffusion to catalyst surfaces, the concentrations are known in the bulk stream at point 0 and it is desired to calculate them at point δ on the catalyst surface. The relation between the fluxes is given by stoichiometry, and the absolute value of one flux is set or known. The problem is trial and error, since the concentrations at point δ are not known. First the values of D_{im} are calculated for the first trial, using only the concentrations at point 0. Then Example 1.10-3 is followed to calculate E_i and D_{im} values. Finally, using Eq. (6.3-19), $x_{A\delta}$ is calculated, $x_{B\delta}$, and so on. Using these calculated values, average values of x_i are obtained and new values of D_{im} calculated. The calculation is repeated until the values of $x_{i\delta}$ are constant.

Example 6.3-4 Multicomponent Turbulent Diffusion

Bird et al. (B2, p. 678) give an example of this case for hydrogenating benzene to cyclohexane, where

$$\text{benzene }(A) + 3H_2\,(B) \longrightarrow \text{cyclohexane }(C)$$

It is desired to calculate the sum of the fluxes if the flow conditions and values of x_i at point 0 are given in the bulk gas phase.

Solution

This means $N_B = 3N_A$, $N_C = -N_A$, and

$$\sum_{i=A}^{n} N_i = N_A + N_B + N_C = N_A + 3N_A - N_A = 3N_A$$

If the value of N_A is known and the flow conditions are also set, then the values of k'_{ci} can be predicted (Section 6.5) from a J_D correlation if the values of x_{i0} in the bulk gas phase are used to calculate D_{im}. Later, when a preliminary value of $x_{i\delta}$ is calculated, the average value of D_{im} can be used. Bird *et al.* work this problem out using an average value of x_{i0} in the bulk phase only. Then, of course, the solution is not trial and error.

6.4 MASS TRANSFER COEFFICIENTS IN LAMINAR FLOW

6.4A Introduction

When mass transfer occurs in a fluid flowing in laminar flow, the principal equations are the same or very similar to those for heat transfer by conduction in laminar flow, as discussed previously in Chapter 2. The Schmidt number is substituted for the Prandtl number, the concentration c_A is substituted for the temperature T, and the molecular diffusivity is substituted for the thermal diffusivity.

The phenomena of heat and mass transfer are not always completely analogous, because mass transfer often involves transfer of more than one component, and some simplifications must be made to make them analogous. Also, the flux of mass perpendicular to the direction of the flowing stream must be small so as not to distort the normal laminar velocity profile.

Theoretically it is not necessary to have mass transfer coefficients for laminar flow, since the equations for momentum transfer and the diffusion equation can be solved. However, in many actual situations it is difficult to describe mathematically the conditions of laminar flow for complex geometries, such as flow past a cylinder or through a packed bed. Often the equations become so complex that many simplifications must be made in the solution. Hence, mass transfer coefficients for fluids in laminar flow past various geometries are often measured experimentally and correlated. A detailed discussion of the complex methods to solve mathematically some of the situations for mass transfer in laminar flow is given elsewhere (B2, B4, S2). In the sections that follow, only a simplified discussion will be given of two situations in laminar flow. It will be shown that in mass transfer in laminar flow, where the interface is between a gas and liquid, the equations do not adequately describe the results, since the interface has ripples at the surface and the velocity profile of the liquid may be altered by the flowing gas.

6.4B Mass Transport from a Tube Wall in Laminar Flow

Mass transfer from a tube wall or pipe to fluid flowing inside it in laminar flow can occur, for example, if the wall is made of solid benzoic acid and it is dissolving into flowing water. This is similar to heat transfer from a wall to a flowing fluid, where natural convection is neglected or small. The equation for the parabolic velocity profile when the velocity profile is fully developed has been derived previously [Eq. (2.3-18)]:

$$v_x = v_{max}\left[1 - \left(\frac{r}{r_i}\right)^2\right] \quad (6.4\text{-}1)$$

where v_x is the velocity in the x direction in which the fluid is flowing at a position r from the center. For a pipe, $v_{max} = 2v_{ave}$. The differential mass balance at steady-state in a cylinder is

$$v_x \frac{\partial c_A}{\partial x} = D_{AB}\left(\frac{1}{r}\frac{\partial c_A}{\partial r} + \frac{\partial^2 c_A}{\partial r^2} + \frac{\partial^2 c_A}{\partial x^2}\right) \quad (6.4\text{-}2)$$

If it is assumed that diffusion in the x direction is negligible compared with the bulk flow, $\partial^2 c_A/\partial x^2 = 0$.

Combining Eq. (6.4-1) and (6.4-2), the boundary conditions hold as follows:

B.C. 1: At $x = 0$ (inlet), $c_A = c_{A0}$ at any r. This means the inlet concentration of c_{A0} was uniform.

B.C. 2: At $r = r_i$ (wall), $c_A = c_{Ai}$ at any x. This means the wall concentration is constant at c_{Ai}.

The final solution, given by Sherwood and Pigford (S2), is a complex series similar to the classical Graetz solution for heat transfer and a parabolic velocity profile.

If it is assumed that there is no velocity gradient—that is, the velocity profile is flat as in rodlike flow, so that $v_x = v_{ave}$—the solution is more easily obtained. This final equation, given elsewhere (S2), is similar to the classical Leveque equation for heat conduction in rodlike flow. Another solution is also available for the case where there is a linear velocity profile near the wall and the solute diffuses only a short distance into the fluid from the wall (S2). This is approximately the same as for the parabolic velocity profile at high values of $W/(D_{AB}\rho L)$ and is often called the approximate Leveque equation for parabolic flow. In this approximation the solute has penetrated only a short distance into the fluid. The flow rate is W lb mass/hr and the length is L ft.

Experimental data for evaporation of liquids into flowing gases in a tube do not agree with the solution for parabolic velocity distribution but tend to agree more closely with the solution for rodlike flow. This is attributed to the fact that the parabolic velocity profile is distorted by density gradients. Experimental data for mass transfer to liquids from walls where the Schmidt number is very high agree with the parabolic velocity distribution (B4).

6.4C Mass Transport from a Gas into a Falling Liquid Film in Laminar Flow

The case of mass transfer of a solute from a gas into a falling liquid film in laminar flow is important in wetted-wall columns and in developing theories to explain turbulent mass transfer and mass transfer in stagnant pockets of fluids.

As shown in Fig. 6.4-1, liquid B is a laminar falling film flowing down a vertical wall, and component A in the gas stream is being absorbed at the interface between the liquid and gas. The diffusion of A in the liquid

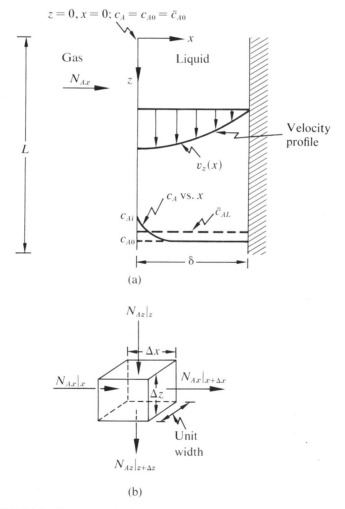

FIGURE 6.4-1 Mass transfer into a falling film: (a) velocity and concentration profiles, (b) mass balance of A

is slow enough so that it has not penetrated the whole distance of $x = \delta$ at the wall. The concentration of A in the inlet liquid is uniform at c_{A0}, the concentration of A at the surface of the liquid is in equilibrium with the concentration in the gas phase and is constant throughout at c_{Ai}, and the distance or time of fall of the fluid is short, so that A has not penetrated all the way to the wall. It is desired to calculate the concentration profile at steady-state at a distance z from the inlet and the mass transfer coefficient k_L to calculate the total amount of A absorbed up to distance z.

It will be assumed that the rate of diffusion of A in the z direction is small compared to the bulk flow of A by the movement of fluid. A mass balance of A will be made on an element Δz by Δx thick, as shown in Fig. 6.4-1, with unit width in the y direction into the paper. For steady-state, rate of input = rate of output.

$$N_{Ax}|_x(1\,\Delta z) + N_{Az}|_z(1\,\Delta x) = N_{Ax}|_{x+\Delta x}(1\,\Delta z) + N_{Az}|_{z+\Delta z}(1\,\Delta x) \quad (6.4\text{-}3)$$

Next we divide through by $\Delta x\,\Delta z$ and let Δx and Δz approach zero. Hence,

$$\frac{\partial N_{Ax}}{\partial x} + \frac{\partial N_{Az}}{\partial z} = 0 \quad (6.4\text{-}4)$$

For the diffusion plus convection flux of A in the x direction,

$$N_{Ax} = -D_{AB}\frac{\partial c_A}{\partial x} + \text{zero convection} = -D_{AB}\frac{\partial c_A}{\partial x} \quad (6.4\text{-}5)$$

For the z direction the diffusion is negligible:

$$N_{Az} = 0 + c_A v_z \quad (6.4\text{-}6)$$

Substituting Eqs. (6.4-5) and (6.4-6) into (6.4-4),

$$v_z \frac{\partial c_A}{\partial z} = D_{AB}\frac{\partial^2 c_A}{\partial x^2} \quad (6.4\text{-}7)$$

This same equation can be derived from the general equation for constant ρ and D_{AB}.

$$\frac{\partial c_A}{\partial t} + (\mathbf{v}\cdot\nabla c_A) - D_{AB}\nabla^2 c_A = R_A \quad (2.6\text{-}12)$$

For diffusion only in the x direction, flow only in the z direction, and steady-state, Eq. (2.6-12) becomes (6.4-7).

The parabolic velocity profile is given by Eq. (6.4-8), which is the same as Eq. (6.4-1):

$$v_z(x) = v_{\max}\left[1 - \left(\frac{x}{\delta}\right)^2\right] \quad (6.4\text{-}8)$$

In a falling film $v_{\max} = (3/2)v_{\text{ave}}$. However, if the solute has penetrated only a short distance into the fluid—that is, there are only short contact

6.4 Mass Transfer Coefficients in Laminar Flow

times t sec of z/v_{max}, then the A species that has diffused has been carried along at a velocity of v_{max}. Hence, v_z in Eq. (6.4-7) will be replaced by v_{max}.

$$\frac{\partial c_A}{\partial (z/v_{max})} = D_{AB}\frac{\partial^2 c_A}{\partial x^2} \quad (6.4\text{-}9)$$

The following are the boundary conditions:

B.C. 1: At $z = 0$, $c_A = c_{A0}$
B.C. 2: At $x = 0$, $c_A = c_{Ai}$
B.C. 3: At $x = \infty$, $c_A = c_{A0}$

Equation (6.4-9) is similar to (5.3-16), which was solved by the Laplace transform. The solution is the same except that z/v_{max} replaces t.

$$\frac{c_A - c_{A0}}{c_{Ai} - c_{A0}} = \text{erfc}\left(\frac{x}{\sqrt{4D_{AB}z/v_{max}}}\right) \quad (6.4\text{-}10)$$

where erfc is the error function, which is an infinite series; values are tabulated in Table 5.3-2 in Chapter 5. To calculate the total moles absorbed and flux at $z = L$ by material balance, the volume flow rate, $v_{max}\delta(1)$ cm³/sec, is multiplied by the difference in average outlet and inlet concentrations, $(\bar{c}_{AL} - \bar{c}_{A0})$, and divided by the cross-sectional area for diffusion of L(1) cm², where $c_{A0} = \bar{c}_{A0}$.

$$N_A = \frac{v_{max}\delta(1)(\bar{c}_{AL} - \bar{c}_{A0})}{L(1)} \quad (6.4\text{-}11)$$

The value of $(\bar{c}_{AL} - \bar{c}_{A0})$ can be determined from Eq. (6.4-10) by obtaining the area under the curve of $(c_A - \bar{c}_{A0})$ vs. x and dividing by δ.

$$\bar{c}_{AL} - \bar{c}_{A0} = \frac{1}{\delta}\int_0^\infty (c_A - \bar{c}_{A0})\,dx = \frac{1}{\delta}\int_0^\infty \text{erfc}\frac{x}{\sqrt{4D_{AB}z/v_{max}}}(c_{Ai} - \bar{c}_{A0})\,dx$$

$$= \frac{1}{\delta}\sqrt{\frac{4D_{AB}L/v_{max}}{\pi}}(c_{Ai} - \bar{c}_{A0}) \quad (6.4\text{-}12)$$

It is customary to define k_L based on the ln mean concentration difference during the flow through the entire length as

$$N_A = \frac{k_L[(c_{Ai} - \bar{c}_{AL}) - (c_{Ai} - \bar{c}_{A0})]}{\ln[(c_{Ai} - \bar{c}_{AL})/(c_{Ai} - \bar{c}_{A0})]} = \frac{k_L(\bar{c}_{A0} - \bar{c}_{AL})}{(\bar{c}_{A0} - \bar{c}_{AL})/(c_{Ai} - \bar{c}_{A0})} \quad (6.4\text{-}13)$$

The last expression in the denominator is good for short contact times, since $\ln(1+\epsilon) \cong \epsilon$ when ϵ is small. Substituting Eq. (6.4-12) for the denominator of Eq. (6.4-13) and equating the results to Eq. (6.4-11),

$$\frac{k_L(\bar{c}_{A0} - \bar{c}_{AL})}{\dfrac{1}{\delta}\sqrt{\dfrac{4D_{AB}}{\pi}L/v_{max}}} = \frac{v_{max}\delta}{L}(\bar{c}_{AL} - \bar{c}_{A0}) \quad (6.4\text{-}14)$$

Hence,

$$k_L = \sqrt{\frac{4D_{AB}v_{\max}}{\pi L}} = \sqrt{\frac{4D_{AB}}{\pi t_L}} \quad (6.4\text{-}15)$$

where t_L is the time of contact, sec, calculated from $v_{\max} = \frac{3}{2}(v_{\text{ave}})$ and L.

This shows that the liquid mass transfer coefficient is proportional to $D_{AB}^{0.5}$ for short contact times. This is the basis for the penetration theory, which will be discussed and also will be compared with the other theories later in this chapter. For wetted-wall towers Eq. (6.4-15) tends to predict values of k_L considerably lower (by up to 50 percent) than experimental values. This is due to several factors. There is a tendency for ripples to form in a falling laminar flow film (T1). The velocity profile of the liquid may be altered by the flowing gas. Also, rapid absorption at the gas-liquid interface may cause spontaneous interfacial turbulance to occur. This is discussed by Sternling and Scriven (S3), who attribute this turbulence to hydrodynamic instability caused by fluctuations in interfacial tension associated with the transfer of solute through the interface.

To determine whether the flow of liquid is laminar, the Reynolds number is defined as

$$N_{\text{Re}} = \frac{4\Gamma}{\mu} \quad (6.4\text{-}16)$$

where Γ is lb mass of liquid flowing per hour per foot of width of wetted tower. Laminar flow occurs for $N_{\text{Re}} < 1200$ (T1). However, Eq. (6.4-15) holds for short contact times or Reynolds numbers above about 100 (S2).

The thickness δ of the laminar film can be calculated from Eq. (6.4-17), which was derived from a momentum balance using Eq. (6.4-8):

$$\delta = \left(\frac{3\mu\Gamma}{\rho^2 g}\right)^{1/3} \quad (6.4\text{-}17)$$

If the length of contact time is so long that the solute penetrates the liquid layer completely, then Eq. (6.4-8) must be used in integrating Eq. (6.4-7), and for $N_{\text{Re}} < 100$,

$$k_L = \frac{3.41 D_{AB}}{\delta} \quad (6.4\text{-}18)$$

Equations (6.4-15) and (6.4-18) hold quite well when rippling is absent. At short tower lengths less than 4 in. or for a Reynolds number less than 25 rippling is absent (P1).

Example 6.4-1 Absorption in a Wetted-Wall Tower

Carbon dioxide is being absorbed by a pure water film flowing down a wetted-wall tower 3.0 ft long and at a rate of 300 lb mass water/hr per foot width at 25°C. The gas is pure CO_2 entering at 1.0 atm. Calculate the amount of CO_2 absorbed.

Solution

The solubility of CO_2 in water at 25°C and 1 atm is $c_{Ai} = 0.0021$ lb mole CO_2/ft^3 solution. $D_{AB} = 2.0 \times 10^{-5}(3.87) = 7.76 \times 10^{-5}$ ft²/hr, density $\rho = 62.3$ lb mass/ft³, viscosity of dilute solution of water = 0.894 cp = $0.894(2.42) = 2.16$ lb mass/ft-hr, $\Gamma = 300$ lb mass/hr-ft, $L = 3.0$ ft.

$$N_{\text{Re}} = \frac{4\Gamma}{\mu} = \frac{4(300)}{2.16} = 556$$

Since this is over 100, Eq. (6.4-15) can be used.

It should be noted that since the gas is pure CO_2, there is no gas film resistance. First the film thickness δ must be calculated by Eq. (6.4-17):

$$\delta = \left(\frac{3\mu\Gamma}{\rho^2 g}\right)^{1/3} = \left[\frac{3(2.16)(300)}{(62.3)^2(4.17 \times 10^8)}\right]^{1/3} = 0.001063 \text{ ft}$$

The average velocity is

$$v_{\text{ave}} = \frac{\Gamma}{\rho\delta} = \frac{300}{62.3(0.001063)} = 4530 \text{ ft/hr}$$

The $v_{\text{max}} = \frac{3}{2}v_{\text{ave}} = \frac{3}{2}(4530) = 6800$ ft/hr. Substituting into Eq. (6.4-15),

$$k_L = \sqrt{\frac{4D_{AB}v_{\text{max}}}{\pi L}} = \sqrt{\frac{4(7.76 \times 10^{-5})(6800)}{\pi(3.0)}} = \frac{0.474 \text{ lb mole}}{\text{hr-ft}^2 \text{ (lb mole/ft}^3)}$$

To calculate the moles absorbed, the outlet concentration $\bar{c}_{AL}$ must be calculated by equating Eq. (6.4-13) to (6.4-11), where v_{ave} is used for the average flow rate.

$$\frac{k_L(\bar{c}_{A0} - \bar{c}_{AL})}{\ln[(c_{Ai} - \bar{c}_{AL})/(c_{Ai} - \bar{c}_{A0})]} = \frac{v_{\text{ave}}\delta(\bar{c}_{AL} - \bar{c}_{A0})}{L}$$

Substituting,

$$\frac{0.474(0 - \bar{c}_{AL})}{\ln[(0.0021 - \bar{c}_{AL})/(0.0021 - 0)]} = \frac{4530(0.001063)(\bar{c}_{AL} - 0)}{3.0}$$

Hence, $\bar{c}_{AL} = 0.000537$ lb mole/ft³. The total moles absorbed are $4530(0.001063)(0.000537) = 0.00258$ lb mole/hr-ft width. Sherwood and Pigford (S2, p. 267) give graphs showing that the experimental value of k_L would be about 1.5 times larger than the predicted value at these conditions.

Equation (6.4-15) can be used to estimate the rate at which rising gas bubbles of pure A are absorbed by liquid B as the bubbles rise at their terminal velocity v_t in a quiescent liquid (B2). For gas bubbles of about 0.3 to 0.5 cm diameter, Hammerton and Garner (H2) show that Eq. (6.4-15) holds, provided the liquid is clean and has no surface active agents present. A rising gas bubble undergoes an internal toroidal circulation. As the bubble rises, it encounters fresh liquid B at the top of the bubble, and this liquid moves relative to the bubble to the bottom of the bubble. Hence, fresh liquid "slides" from the top to the bottom of the bubble. The time of contact is approximately $t_{\exp} \cong D/v_t$, where D is the bubble diameter. Then, using Eq. (6.4-15), the average mass transfer rate for fresh liquid is

$$N_A = k_L(c_{Ai} - 0) = \sqrt{\frac{4D_{AB}}{\pi t_{\exp}}}(c_{Ai} - 0) \tag{6.4-19}$$

where c_{Ai} is the solubility of pure A in B.

6.5 MASS TRANSFER COEFFICIENTS IN TURBULENT FLOW

6.5A Introduction and Dimensionless Numbers Used to Correlate Data

1. *Introduction.* As discussed in Sections 6.4A to 6.4C, even in laminar flow in well-known and defined geometries, such as flow in a pipe or as a film, the predicted mass transfer rates deviated from the experimental values. In turbulent flow, where the eddy diffusivity for mass transfer cannot be predicted, it is necessary to use experimental data to predict the mass transfer coefficients. In geometries such as for flow past blunt objects, such as spheres, attempts to predict mass transfer in turbulent flow are of little practical value.

The experimental data for different kinds of fluids, at different velocities, and in different geometries are correlated in terms of dimensionless numbers, which are very similar to those used in momentum and heat transfer and were discussed in Section 2.9 in the dimensional analysis of the equations of change.

2. *Dimensionless Numbers.* One of the important dimensionless numbers used to indicate the degree of turbulence is the Reynolds number of Eq. (2.9-18):

$$N_{Re} = \frac{Lv\rho}{\mu} \tag{6.5-1}$$

where L is a characteristic length. For spheres, $L = D_p$ of sphere; for pipes, L is the diameter; for flat plates, L usually is the length of plate. The velocity v can be defined in a number of ways. In an empty pipe, v is the bulk or mass average velocity. For flow past a flat plate it can be defined as the velocity approaching the plate or as several other kinds of velocities, depending on the geometry of the enclosure for the plate. In a packed bed the superficial velocity v in the empty cross section of the pipe with no packing is often used. Also used often is the actual interstitial velocity $v' = v/\epsilon$, where ϵ is the void fraction. In summary, the reader is cautioned to be very careful in using the empirical correla-

6.5 Mass Transfer Coefficients in Turbulent Flow

tions in the literature and to be sure that the true meaning of the symbols is clear. Often the author of an article will be quite vague in stating which type of velocity was used.

The Schmidt number is defined as Eq. (2.9-20) and is

$$N_{Sc} = \frac{\mu}{\rho D_{AB}} \quad (6.5\text{-}2)$$

The viscosity is the viscosity of the total average mixture of the fluid phase under consideration. The density is that for the mixture. Hence, for a mixture of 1.0 percent NH_3 in air at 1.0 atm, the density will be that of the bulk mixture. If the mixture is dilute, the properties of the pure fluid can be used. In the Reynolds number the velocity, density, and viscosity of the mixture should also be used. For heat transfer the analogous dimensionless number to the Schmidt number is the Prandtl number. It is especially important with liquids to use the physical properties of the flowing mixture, since viscosity, density, and diffusivity can vary appreciably with concentration.

The dimensionless number containing the mass transfer coefficient is the Sherwood number:

$$N_{Sh} = k'_c \frac{L}{D_{AB}} \quad (6.5\text{-}3)$$

where k'_c is the mass transfer coefficient for equimolar counterdiffusion or no bulk flow. For the case where there is bulk flow also, the following can be substituted for k'_c:

$$k_N \varphi_N = k'_c \quad (6.5\text{-}4)$$

where k_N is the mass transfer coefficient for bulk flow and φ_N is defined by Eq. (6.3-9). Making the substitutions for k'_c from Table 6.3-1,

$$N_{Sh} = \frac{k'_c L}{D_{AB}} = k'_G RT \frac{L}{D_{AB}} = k'_y \frac{RT}{P} \frac{L}{D_{AB}}$$

$$= k_c \frac{p_{BM}}{P} \frac{L}{D_{AB}} = \frac{k_G p_{BM} RT}{P} \frac{L}{D_{AB}}$$

$$= k_y \frac{p_{BM} RT}{(P)(P)} \frac{L}{D_{AB}} = k_N \varphi_N \frac{L}{D_{AB}} \quad (6.5\text{-}5)$$

The dimensionless number in heat transfer corresponding to the Sherwood number is the Nusselt number, $N_{Nu} = hL/k$.

The Peclet number is defined as

$$N_{Pe} = N_{Re} N_{Sc} = \frac{Lv}{D_{AB}} \quad (6.5\text{-}6)$$

270   Mass Transfer Coefficients in Laminar and Turbulent Flow

The Stanton number for mass transfer occurs very often and is as follows for various types of bulk flow:

$$N_{St} = \frac{N_{Sh}}{N_{Re}N_{Sc}} = \frac{k'_c}{v} = \frac{k'_y}{G_M}$$

$$= \frac{k'_G P}{G_M} = \frac{k_c p_{BM}}{G_M RT} = \frac{k'_c c}{G_M} = \frac{k_G p_{BM}}{G_M} = \frac{k_y p_{BM}}{G_M P} = \frac{k_N \varphi_N}{v} \quad (6.5\text{-}7)$$

where $G_M = v\rho/M_{ave} = vc$. The M_{ave} is the average molecular weight of the mixture. The Stanton number for heat transfer will be $N_{Nu}/N_{Re}N_{Pr}$.

The Grashof number in mass transfer is used in cases where there is a density difference $\Delta\rho = (\rho_1 - \rho_2)$ in a given phase and natural convection will arise:

$$N_{Gr} = \frac{gL^3 \Delta\rho}{\rho}\left(\frac{\rho}{\mu}\right)^2 \quad (6.5\text{-}8)$$

Often the mass transfer coefficient is represented as a J_D factor defined as

$$J_D = N_{St}(N_{Sc})^{2/3} = \frac{k'_c}{v}(N_{Sc})^{2/3} = \frac{k'_G P}{G_M}(N_{Sc})^{2/3} = \cdots \quad (6.5\text{-}9)$$

In some cases it is desirable to convert from J_D to the Sherwood number:

$$N_{Sh} = \frac{J_D N_{Re} N_{Sc}}{N_{Sc}^{2/3}} \quad (6.5\text{-}10)$$

$$N_{Sh} = J_D N_{Re} N_{Sc}^{1/3} \quad (6.5\text{-}11)$$

Often experimental data will be reported in the literature for diffusion of A through stagnant B, and the coefficient will be designated as k'_G or k'_c. This usually implies that the experimental solutions were very dilute and $\varphi_N = x_{BM} \cong 1.0$. In fact, for all types of experiments where bulk flow is present, the $\varphi_N = 1.0$ for dilute solutions.

Table 6.5-1 summarizes the dimensionless groups that are important in correlating mass transfer coefficients.

6.5B   Analogies between Mass, Heat, and Momentum Transport

1. *Introduction.* In laminar flow there are many similarities between molecular mass, heat, and momentum transport, which were pointed out in Chapters 1 and 2. There also are similarities in turbulent transport, but these are not nearly as well defined mathematically or physically. Much effort has been devoted to developing analogies between these three transport processes for turbulent flow so as to allow prediction of one from the others and vice versa. Many new theories are constantly being proposed, and these theories are becoming increasingly complex.

6.5 Mass Transfer Coefficients in Turbulent Flow

Table 6.5-1 Dimensionless Numbers for Correlating Mass Transfer Coefficients

Reynolds number	$N_{Re} = \dfrac{Lv\rho}{\mu}$	
Schmidt number	$N_{Sc} = \dfrac{\mu}{\rho D_{AB}}$	
Sherwood number		
General case	$N_{Sh} = k_N \varphi_N \dfrac{L}{D_{AB}}$	
Equimolar counterdiffusion	$N_{Sh} = k_c' \dfrac{L}{D_{AB}} = k_G' RT \dfrac{L}{D_{AB}} = k_y' \dfrac{RT}{P} \dfrac{L}{D_{AB}}$	Gases
	$N_{Sh} = k_L' \dfrac{L}{D_{AB}} = \dfrac{k_x'}{c} \dfrac{L}{D_{AB}}$	Liquids
Diffusion of A through stagnant B	$N_{Sh} = \dfrac{k_c p_{BM}}{P} \dfrac{L}{D_{AB}} = \dfrac{k_G p_{BM} RT}{P} \dfrac{L}{D_{AB}}$	Gases
	$= \dfrac{k_y p_{BM} RT}{P(P)} \dfrac{L}{D_{AB}}$	
	$N_{Sh} = k_L x_{BM} \dfrac{L}{D_{AB}} = k_x \dfrac{x_{BM}}{c} \dfrac{L}{D_{AB}}$	Liquids
Peclet number	$N_{Pe} = N_{Re} N_{Sc} = \dfrac{Lv}{D_{AB}}$	
Stanton number	$N_{St} = N_{Sh}/(N_{Re} N_{Sc})$	
General case	$N_{St} = \dfrac{k_N \varphi_N}{v}$	
Equimolar counterdiffusion ($G_M = vc = v\rho/M_{ave}$)	$N_{St} = \dfrac{k_c'}{v} = \dfrac{k_G' P}{G_M} = \dfrac{k_y'}{G_M}$	Gases
	$N_{St} = \dfrac{k_L'}{v} = \dfrac{k_x'}{G_M}$	Liquids
Diffusion of A through stagnant B	$N_{St} = \dfrac{k_c p_{BM}}{G_M RT} = \dfrac{k_G p_{BM}}{G_M} = \dfrac{k_y p_{BM}}{G_M P}$	Gases
	$N_{St} = \dfrac{k_L x_{BM}}{v} = \dfrac{k_x x_{BM}}{G_M}$	Liquids
J Factor	$J_D = N_{St}(N_{Sc})^{2/3} = \dfrac{k_c'}{v}(N_{Sc})^{2/3} = \cdots$	
	(See above for N_{St}.)	
Conversion from J_D to N_{Sh}	$N_{Sh} = J_D N_{Re} N_{Sc}/(N_{Sc})^{2/3}$	
Grashof number	$N_{Gr} = g \dfrac{L^3 \Delta\rho}{\rho} \left(\dfrac{\rho}{\mu}\right)^2$	

NOTE: For heat transfer substitute $N_{Nu} = hL/k$ for N_{Sh}; $N_{Pr} = c_p \mu/k$ for N_{Sc}; $N_{St} = N_{Nu}/N_{Re} N_{Pr}$ for $N_{Sh}/N_{Re} N_{Sc}$; and J_H for J_D when the analogies hold.

We will first discuss the two earliest theories and then that most commonly used today. Because of our still inexact knowledge of the

272   Mass Transfer Coefficients in Laminar and Turbulent Flow

actual nature of turbulence, many of the analogies still require empirical data to be useful.

2. *Reynolds Analogy.* In 1874 Reynolds noticed a similarity of heat and momentum transport in his research. His early analysis was the start of the present analogies used today. Reynolds (R1) analogy can be restated today for a fluid flowing in turbulent flow in a pipe and being heated by the pipe walls. It is assumed that the heat transfer does not affect the velocity distribution or momentum transfer. This analogy then states (B4):

$$\frac{\text{(a) (loss in momentum by friction)}}{\text{(b) (total or maximum momentum of stream)}}$$

$$= \frac{\text{(c) (heat transfer to fluid)}}{\text{(d)} \begin{pmatrix} \text{maximum heat that could be supplied to fluid} \\ \text{if brought to pipe wall temperature} \end{pmatrix}} \quad (6.5\text{-}12)$$

This analogy can also be written for the case of mass transfer from the walls of the pipe to the fluid.

$$\frac{\text{(e) (mass transfer to fluid)}}{\text{(f)} \begin{pmatrix} \text{maximum mass that could be transferred} \\ \text{and reach equilibrium with the wall} \end{pmatrix}} \quad (6.5\text{-}13)$$

We can equate Eq. (6.5-12) to (6.5-13) by first writing the mathematical expressions for the following.

(a) The loss in momentum is τg_c (lb mass-ft/hr)/hr-ft^2.
(b) The total momentum is $w v_{\text{ave}}$ (lb mass/hr)(ft/hr).
(c) The heat flux is $h(t-t_i)$ Btu/hr-ft^2.
(d) The total heat available is $w c_p(t-t_i)$ Btu/hr.
(e) The mass flux is $k'_y(y_A - y_{Ai})$ lb mole/hr-ft^2.
(f) The total mass available for transport is

$$(w/M_{\text{ave}})(y_A - y_{Ai}) \text{ lb mole/hr}$$

Equating Eq. (6.5-12) to (6.5-13) and substituting the definition of the friction factor f as

$$\tau g_c = \tfrac{1}{2} f v_{\text{ave}}^2 \rho \quad (6.5\text{-}14)$$

$$\frac{f v_{\text{ave}}^2 \rho / 2}{w v_{\text{ave}}} = \frac{h(t-t_i)}{w c_p (t-t_i)} = \frac{k'_y (y_A - y_{Ai})}{(w/M_{\text{ave}})(y_A - y_{Ai})} \quad (6.5\text{-}15)$$

or

$$\frac{f}{2} = \frac{h}{c_p G} = \frac{k'_y}{G_M} \quad (6.5\text{-}16)$$

where $G = v_{\text{ave}} \rho$ and $G_M = v_{\text{ave}} \rho / M_{\text{ave}}$.

Experimental data show that for gases the above analogy holds fairly well to about 35 percent, where the Prandtl and Schmidt numbers are near 1.0. For liquids the analogy is quite inaccurate and should not be used at all. Later in this section the reasons for this discrepancy will be discussed.

3. *Taylor and Prandtl Analogy.* Taylor and Prandtl modified the Reynolds analogy, dividing the transport of mass, heat, or momentum in a pipe into two zones. They reasoned that in the turbulent and buffer or transition zone the transport mechanisms were similar and followed the Reynolds analogy. In the laminar zone next to the wall they used the molecular transport equations. Then they combined the two equations to obtain their analogy.

To derive this we first write the equations for the case of mass transfer from the pipe wall to the turbulent fluid stream. We assume as before that the mass transfer does not affect the momentum transfer. For the turbulent zone we pick arbitrarily two points 1 and 2 close to each other. An interchange of w lb mass/hr-ft² brings about a net transfer of component A of

$$N_A = \frac{w \Delta p_A}{M_{ave} P} \tag{6.5-17}$$

where $\Delta p_A = p_{AF} - p_{AG}$ for the whole turbulent core region. The p_{AG} is the bulk gas phase composition and p_{AF} is the composition at the end of the turbulent core where the molecular diffusion region starts. For momentum transfer,

$$\tau g_c = w(\Delta v) \tag{6.5-18}$$

Substituting w from Eq. (6.5-18) into (6.5-17),

$$\frac{\tau g_c}{\Delta v} = \frac{N_A M_{ave} P}{\Delta p_A} \tag{6.5-19}$$

For the laminar film we can write for molecular mass transfer,

$$N_A = \frac{D_{AB}}{RT z_F}(p_{Ai} - p_{AF}) \tag{6.5-20}$$

where z_F is the thickness of the laminar sublayer film.

For momentum transfer,

$$\tau_i g_c = \frac{\mu}{z_F}(v_F - 0) \tag{6.5-21}$$

where v_F is the velocity at the point z_F from the wall. Substituting z_F from Eq. (6.5-21) into (6.5-20) and rearranging,

$$\frac{\tau_i g_c}{v_F - 0} = \frac{N_A \mu RT}{D_{AB}(p_{Ai} - p_{AF})} \tag{6.5-22}$$

274   Mass Transfer Coefficients in Laminar and Turbulent Flow

Now for the Reynolds analogy to hold in both regions, Eqs. (6.5-19) and (6.5-22) must be the same. To allow this, the following must hold:

$$M_{ave}P = \frac{\mu RT}{D_{AB}} \tag{6.5-23}$$

It can be shown that $M_{ave}P/(RT) = \rho$. Hence,

$$1 = \frac{\mu}{\rho D_{AB}} \tag{6.5-24}$$

This means that for the Reynolds analogy to hold, the N_{Sc} must be 1.0. This explains why the Reynolds analogy breaks down for liquids where the Schmidt number is large.

Continuing with the Taylor and Prandtl analogy, the left-hand side of Eq. (6.5-19) can be written as follows, assuming $\tau_i = \tau$ at $z = z_F$, since the film is very thin:

$$\frac{\tau_i g_c}{v_{ave} - v_F} = \frac{N_A M_{ave} P}{p_{AF} - p_{AG}} \tag{6.5-25}$$

Also,

$$N_A = k'_G(p_{Ai} - p_{AG}) \tag{6.5-26}$$

Combining Eqs. (6.5-14), (6.5-22), (6.5-25), and (6.5-26),

$$\frac{f}{2} = \frac{k'_G P}{G_M}\left[1 - \frac{v_F}{v_{ave}} + \left(\frac{v_F}{v_{ave}}\right)N_{Sc}\right] \tag{6.5-27}$$

Equation (6.5-27) was also shown to give poor results for fluids where the Schmidt number differed from unity.

4. *Chilton and Colburn J Factor Analogy.* Many analogies have been derived since the Taylor and Prandtl analogy (P1). Probably the most successful and most widely used is the Chilton and Colburn J factor. They correlated data and showed empirically that a modified form of Eq. (6.5-27) could be used to represent some cases if the $N_{Sc}^{2/3}$ was substituted for the term in the brackets in Eq. (6.5-27). Also, a similar equation for heat transfer was used as follows:

$$\frac{f}{2} = J_H = \frac{h}{c_p G}N_{Pr}^{2/3} = J_D = \frac{k'_G P}{G_M}N_{Sc}^{2/3} \tag{6.5-28}$$

This J factor correlation has been a useful method to predict mass transfer from experimental momentum or heat transfer data. Bird *et al.* (B2, p. 618) give a theoretical justification for the inclusion of $N_{Sc}^{2/3}$ in the correlations for certain flow systems and this is discussed in Section 6.8E of this chapter.

5. *Use of Analogies.* Here we discuss the use of the J factor analogies

6.5 Mass Transfer Coefficients in Turbulent Flow

and how one obtains mass transfer data when none are available. Also, the general range of validity of the J factor analogies is discussed. In Section 6.7 on the actual correlations of mass transfer more details will be given.

In momentum transport the friction factor f is usually obtained experimentally for the total drag loss, which includes form drag or momentum losses due to flow separation because of the presence of blunt objects and also skin friction.

In heat and mass transfer there is no transfer equivalent to the form drag. Hence, when form drag is present the value of $f/2$ will be greater than J_H or J_D. For example, for flow past a sphere, the $f/2$ will include losses due to impact and flow separation as well as skin friction.

For flow past a flat plate or in a pipe, $f/2 \cong J_H \cong J_D$. For other cases of flow past spheres, in packed beds, and so on, a reasonable assumption is often that $J_H \cong J_D$. This means that if mass transfer experiments are difficult to perform, the heat transfer experiments can be done and the J_D estimated from the J_H data or vice versa. The reader should be cautioned that sometimes experimental correlations of J_D obtained for gases may differ from those for liquids, as will be seen later in this chapter. Hence, extrapolations from gases to liquids and vice versa should be performed with caution.

6.5C Dimensional Analysis and Buckingham Pi Theorem

In some cases it is desirable to ascertain what dimensionless numbers are or could be important in correlating mass transfer coefficients. The celebrated Buckingham pi theorem is often used for this. It states that the functional relationship among q quantities whose units may be given in terms of u fundamental units may be written as $(q-u)$ dimensionless groups (often called π's).

Let us consider as an example the dimensionless groups for a mass transfer coefficient k'_c, where mass transfer is occurring from a wall to a fluid flowing in a vertical pipe (B4). The variables present are

$$k'_c = f(D, \rho, \mu, \Delta\rho, g, v_{\text{ave}}, D_{AB}, L) \tag{6.5-29}$$

where D is the pipe diameter, $\Delta\rho$ the driving force in g mass/cm³, and L the length of pipe. The number of fundamental dimensions (L, M, θ) is $u = 3$. The number of variables $q = 9$. Hence, the number of π's or dimensionless groups is $9 - 3$ or 6. Then

$$\pi_1 = f(\pi_2, \pi_3, \pi_4, \pi_5, \pi_6) \tag{6.5-30}$$

We will choose the first three variables of D, ρ, and μ to be common to all dimensionless groups. Using those three and k'_c,

$$\pi_1 = D^a \rho^b \mu^c (k'_c)^d \tag{6.5-31}$$

or

$$1 = (L)^a \left(\frac{M}{L^3}\right)^b \left(\frac{M}{L\theta}\right)^c \left(\frac{L}{\theta}\right)^d \qquad (6.5\text{-}32)$$

Summing each exponent,

$$\begin{aligned}(L) & \quad 0 = a - 3b - c + d \\ (M) & \quad 0 = b + c \\ (\theta) & \quad 0 = c - d\end{aligned} \qquad (6.5\text{-}33)$$

Solving the equations simultaneously,

$$c = -a = -b = -d \qquad (6.5\text{-}34)$$

Hence,

$$\pi_1 = \left(\frac{\mu}{D\rho k_c'}\right)^c \qquad (6.5\text{-}35)$$

The exponent c can be assumed arbitrarily as unity.

Forming π_2 by relating the first three with $\Delta\rho$, and solving in like manner,

$$\pi_2 = \left(\frac{\Delta\rho}{\rho}\right)^d \qquad (6.5\text{-}36)$$

Continuing in like manner,

$$\pi_3 = \frac{D^3 \rho^2 g}{\mu^2} \qquad (6.5\text{-}37)$$

$$\pi_4 = \frac{D v_{ave} \rho}{\mu} \qquad (6.5\text{-}38)$$

$$\pi_5 = \frac{\mu}{\rho D_{AB}} \qquad (6.5\text{-}39)$$

$$\pi_6 = \frac{L}{D} \qquad (6.5\text{-}40)$$

We can divide π_5 by π_1 and obtain

$$\frac{\pi_5}{\pi_1} = \frac{k_c' D}{D_{AB}} = N_{Sh} \qquad (6.5\text{-}41)$$

Also,

$$\pi_2 \pi_3 \pi_6^3 = N_{Gr} = gL^3 \frac{\Delta\rho}{\rho}\left(\frac{\rho}{\mu}\right)^2 \qquad (6.5\text{-}42)$$

Substituting into Eq. (6.5-30),

$$N_{Sh} = K N_{Gr}^\alpha N_{Sc}^\beta N_{Re}^\gamma \qquad (6.5\text{-}43)$$

When $\gamma = 0$, this is the familiar equation for natural convection. For forced flow π_2 and π_3 would have approximately zero exponents and disappear. Also, π_6 would not enter in the correlation for long pipes and Eq. (6.5-43) becomes the familiar equation

$$N_{Sh} = K N_{Re}^{\gamma} N_{Sc}^{\beta} \tag{6.5-44}$$

The type of analysis above is useful in empirical correlations of data where it is difficult to write differential equations to describe the process. It does not, however, tell us the importance of each dimensionless group discussed previously in Section 2.9.

6.6 CONCENTRATION DRIVING FORCES TO USE WITH MASS TRANSFER COEFFICIENTS

6.6A Ln-Mean Driving Force

In using the mass transfer coefficients in turbulent transport to predict the total amount of transfer when the concentration driving force is varying, we must first set up a differential material and rate balance. Figure 6.6-1 is a sketch of a fluid flowing past a solid surface of A ft² total surface area. The solid is dissolving, so the concentration at the interface is constant at c_{Ai}. The inlet bulk concentration of the fluid is c_{A1} and the outlet is c_{A2}. Component A is diffusing through stagnant component B. The fluid flow is constant, entering at V ft³/hr. The mass transfer coefficient has been estimated from correlations as k_L.

First a material balance is set up for a differential area dA ft². In this

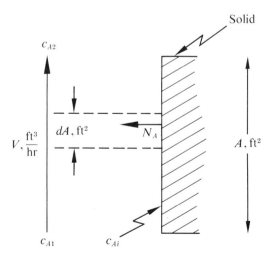

FIGURE 6.6-1 Concentration driving forces for diffusion from a dissolving solid to a liquid

278 Mass Transfer Coefficients in Laminar and Turbulent Flow

element, the mass transfer rate in at steady-state must equal the amount added per unit time by material balance:

$$N_A \, dA = k_L(c_{Ai} - c_A) \, dA = V \, dc_A \tag{6.6-1}$$

Integrating this between 0 and A and c_{A1} and c_{A2}, and substituting k'_L/x_{BM} for k_L to correct for bulk flow,

$$A = \int_0^A dA = -V \int_{c_{A1}}^{c_{A2}} \frac{dc_A}{\dfrac{k'_L}{x_{BM}}(c_A - c_{Ai})} \tag{6.6-2}$$

If the solution is dilute so an average value of k'_L/x_{BM} can be used,

$$A = \frac{V}{k_{L\text{ave}}} \ln\left(\frac{c_{Ai} - c_{A1}}{c_{Ai} - c_{A2}}\right) \tag{6.6-3}$$

The average value of $k_{L\text{ave}}$ can be used as the linear average at the inlet and outlet, where

$$k_{L\text{ave}} = \frac{\dfrac{k'_L}{x_{BM1}} + \dfrac{k'_L}{x_{BM2}}}{2} \tag{6.6-4}$$

$$x_{BM1} = \frac{x_{B1} - x_{Bi}}{\ln\left(\dfrac{x_{B1}}{x_{Bi}}\right)}, \qquad x_{BM2} = \frac{x_{B2} - x_{Bi}}{\ln\left(\dfrac{x_{B2}}{x_{Bi}}\right)} \tag{6.6-5}$$

To get x_B, $x_A + x_B = 1.0$.

Often it is desired to use a mean driving force as follows:

$$N_A = k_{L\text{ave}}(c_{Ai} - c_A)_{\text{mean}} = \frac{V}{A}(c_{A2} - c_{A1}) \tag{6.6-6}$$

Solving for $k_{L\text{ave}}$ in Eq. (6.6-3) and substituting it into Eq. (6.6-6),

$$\frac{V}{A} \ln\left(\frac{c_{Ai} - c_{A1}}{c_{Ai} - c_{A2}}\right)(c_{Ai} - c_A)_{\text{mean}} = \frac{V}{A}(c_{A2} - c_{A1}) \tag{6.6-7}$$

Solving for $(c_{Ai} - c_A)_{\text{mean}}$,

$$(c_{Ai} - c_A)_{\text{mean}} = (c_{Ai} - c_A)_{LM} = \frac{c_{A2} - c_{A1}}{\ln\left(\dfrac{c_{Ai} - c_{A1}}{c_{Ai} - c_{A2}}\right)} = \frac{(c_{Ai} - c_{A1}) - (c_{Ai} - c_{A2})}{\ln\left(\dfrac{c_{Ai} - c_{A1}}{c_{Ai} - c_{A2}}\right)} \tag{6.6-8}$$

The ln-mean driving force defines the mean concentration difference to use during flow through the entire length of path from point 1 to 2. This is entirely analogous to the case of cold water being heated as it flows by a hot surface of uniform temperature. If the physical properties and the heat transfer coefficient are constant, a similar equation is obtained for the ln-mean temperature difference.

6.6B Graphical Integration

If k_L or x_{BM} varies considerably throughout the path, then a graphical integration must be performed that includes x_{BM}. Equation (6.6-5) can be rearranged as follows:

$$x_{BM} = \frac{x_B - x_{Bi}}{\ln\left(\dfrac{x_B}{x_{Bi}}\right)} = \frac{\left(1 - \dfrac{c_A}{c}\right) - \left(1 - \dfrac{c_{Ai}}{c}\right)}{\ln\left(\dfrac{c - c_A}{c - c_{Ai}}\right)} = \frac{1}{c}\frac{c_{Ai} - c_A}{\ln\left(\dfrac{c - c_A}{c - c_{Ai}}\right)} \quad (6.6\text{-}9)$$

Substituting this Eq. (6.6-9) into (6.6-2),

$$A = \frac{V}{k_L'}\int_{c_{A1}}^{c_{A2}} \frac{dc_A}{(c_A - c_{Ai})\dfrac{c}{(c_A - c_{Ai})/\ln\left(\dfrac{c - c_A}{c - c_{Ai}}\right)}}$$

$$= \frac{V}{k_L'}\int_{c_{A1}}^{c_{A2}} \frac{dc_A}{c \ln\left(\dfrac{c - c_A}{c - c_{Ai}}\right)} \quad (6.6\text{-}10)$$

This can be converted into mole fractions as follows:

$$A = \frac{V}{k_L'}\int_{x_{A1}}^{x_{A2}} \frac{dx_A}{\ln\left(\dfrac{1 - x_A}{1 - x_{Ai}}\right)} \quad (6.6\text{-}11)$$

To graphically integrate Eq. (6.6-11) the integral of $f(x)\,dx$ must be evaluated between x_{A1} and x_{A2}. This means a plot of $f(x)$ versus x must be made where $f(x) = 1/\ln[(1 - x_A)/(1 - x_{Ai})]$. The area under the curve is the value of the integral of Eq. (6.6-11). Multiplying this by V/k_L', we obtain the value of A.

Example 6.6-1 Graphical Integration

The following experimental data of $y = f(x)$ were determined.

x	$f(x)$
0	100
0.1	75
0.2	60.5
0.3	53.5
0.4	53
0.5	60
0.6	72.5

It is desired to determine the integral.

$$A = \int_{x=0}^{x=0.6} f(x)\,dx$$

280 Mass Transfer Coefficients in Laminar and Turbulent Flow

Do this by a graphical procedure.

Solution

The data are plotted and a smooth curve is drawn through the points. Taking the area from $x = 0$ to $x = 0.6$, the area is the sum of six rectangles drawn. The height of each rectangle is the average height of each curved interval. The area is the sum of these rectangles and equals

$$(0.1-0)(86.5) + (0.2-0.1)(67.5) + (0.3-0.2)(56.5)$$
$$+ (0.4-0.3)(53) + (0.5-0.4)(56.0) + (0.6-0.5)(66) = 38.55$$

Example 6.6-2 Mass Transfer in a Packed Bed

In an experiment (run 1) by Wilson and Geankoplis (W1) pure water at 26.1°C flowed slowly through a packed bed of 0.251-in. diameter benzoic acid spheres having a total surface area of 0.129 ft². The water flow was 0.0701 ft³/hr. The experimental mass transfer coefficient k_L was 0.0551 ft/hour. The saturation solubility of benzoic acid in water is 0.00184 lb mole benzoic acid/ft³ solution. Calculate the outlet concentration c_{A2} of benzoic acid in the water.

Solution

The $c_{A1} = 0$, $c_{Ai} = 0.00184$ lb mole benzoic acid/ft³ solution, $V = 0.0701$ ft³/hour, $A = 0.129$ ft² area. Equation (6.6-6) can be used, because the solution will be very dilute. The maximum value of any concentration is 0.00184 lb mole/ft³ solution, which is an x_A value of 0.000532 mole fraction. Substituting into Eq. (6.6-6),

$$\frac{k_L(c_{A2}-c_{A1})}{\ln\left(\frac{c_{Ai}-c_{A1}}{c_{Ai}-c_{A2}}\right)} = \frac{0.0551(c_{A2}-0)}{\ln\left(\frac{0.00184-0}{0.00184-c_{A2}}\right)} = \frac{V}{A}(c_{A2}-c_{A1}) = \frac{0.0701}{0.129}(c_{A2}-0)$$

Solving, $c_{A2} = 0.000180$ lb mole benzoic acid/ft³ solution.

6.6C Analog Computer Solution for Integration

If a case arises where a graphical integration is needed, this can often be done using the analog computer. Taking Eq. (6.6-1) and converting to mole fraction units, or differentiating Eq. (6.6-11),

$$dA = \frac{V}{k'_L} \frac{dx_A}{\ln\left(\frac{1-x_A}{1-x_{Ai}}\right)} \tag{6.6-12}$$

6.6 Use of Concentration Driving Forces

Solving for the derivative,

$$\frac{dx_A}{dA} = \frac{k'_L}{V} \ln\left(\frac{1-x_A}{1-x_{Ai}}\right) \quad (6.6\text{-}13)$$

The analog computer diagram to solve Eq. (6.6-13) is easily drawn and contains a log generator. Note that A represents time τ. At the point where the voltage e equals x_{A2}, $\tau = A$.

6.6D Numerical Integration and Simpson's Rule

Often it is desired or necessary to perform a numerical integration by computing the value of a definite integral from a set of numerical values of the integrand $f(x)$. This, of course, can be done graphically, but if data are available in large quantities, numerical methods suitable for the digital computer are desired.

The integral to be evaluated is as follows:

$$\int_{x=a}^{x=b} f(x)\, dx \quad (6.6\text{-}14)$$

where the interval is $b - a$. The most generally used numerical method is the parabolic rule often called Simpson's rule. This method divides the total interval $b - a$ into an even number of subintervals m, where

$$m = \frac{b-a}{h} \quad (6.6\text{-}15)$$

The value of h, a constant, is the spacing in x used. Then, approximating $f(x)$ by a parabola on each subinterval, Simpson's rule is

$$\int_{x=a}^{x=b} f(x)\, dx = \frac{h}{3}[f_0 + 4(f_1+f_3+f_5+\cdots+f_{m-1})$$

$$+ 2(f_2+f_4+f_6+\cdots+f_{m-2}) + f_m] \quad (6.6\text{-}16)$$

where f_0 is the value of $f(x)$ at $x = a$, f_1 the value of $f(x)$ at $x = x_1, \ldots, f_m$ the value of $f(x)$ at $x = b$. The reader should note that m must be an even number and the increments evenly spaced. For other integration formulas, see Lapidus (L4, pp. 47–56) and Mickley et al. (M1, pp. 35–37). This method is well suited for digital computation.

Example 6.6-3 Numerical Integration by Simpson's Method

Using the data from Example 6.6-1, perform the integration numerically by Simpson's method using six intervals.

Solution

The value of m by Eq. (6.6-15) is as follows for $h = 0.1$:

$$m = \frac{b-a}{h} = \frac{0.6-0}{0.1} = 6$$

The value of f_0 is 100, of f_1 is 75, and so on. Substituting into Eq. (6.6-16),

$$\int_{x=0}^{x=0.6} f(x)\, dx = \frac{0.1}{3}[100 + 4(75 + 53.5 + 60) + 2(60.5 + 53) + 72.5]$$

$$= 38.45$$

This compares favorably with the value of 38.55 obtained graphically in Example 6.6-1.

6.6E Digital Computer Solution for Numerical Integration

Equation (6.6-11) can easily be solved by integration using numerical methods on the digital computer. Assume k'_L varies as $k'_L = K_1 + K_2 x_A$, where K_1 and K_2 are constants. The new equation to solve is

$$A = V \int_{x_{A1}}^{x_{A2}} \frac{dx_A}{(K_1 + K_2 x_A)\ln\left(\frac{1-x_A}{1-x_{Ai}}\right)} = V \int_{x_{A1}}^{x_{A2}} f(x_A)\, dx_A \qquad (6.6\text{-}17)$$

Example 6.6-4 *Numerical Integration by Digital Computer*

Using the data of Example 6.6-2 but $k'_L = K_1 + K_2 x_A$, solve Eq. (6.6-17) with the digital computer. Write the Fortran IV program and tabulate the results. For this case, $K_1 = 0.0551$ ft/hr and $K_2 = 185.5$ ft/hr. $A = 0.129$ ft^2, $V = 0.0701$ ft^3/hr, $x_{Ai} = 0.000532$ mole frac, $x_{A1} = 0$, and x_{A2} is the unknown. Use Simpson's method for the integration as discussed in the text.

Solution

In this example the method of Simpson to determine the area must be combined with a modified trial-and-error procedure. In the usual case the values of the integration limits x_{A1} and x_{A2} are known and the area or value of the definite integral is to be determined numerically. In this case the area is known, which is $A = 0.129$ ft^2, but the one integration limit x_{A2} must be determined by trial and error. This is done by first guessing a value of x_{A2} and computing the area by Simpson's method. If the area is incorrect, a new x_{A2} is guessed and the procedure repeated. Each new estimate of x_{A2} is made by comparing the area calculated to

6.6 Use of Concentration Driving Forces

the one previous and modifying it according to some set pattern that will converge to the answer.

The Fortran IV program for this is given in Table 6.6-1 and the results of the various iterations in Table 6.6-2. The detailed steps in the program are as follows.

Table 6.6-1 Fortran Program for Example 6.6-4

```
C       EXAMPLE PROBLEM ON NUMERICAL INTEGRATION USING SIMPSON'S RULE
C
        REAL K1,K2
        F(Z)=V/((K1+K2*Z)*ALOG((1.-Z)/(1.-XAI)))
        DIMENSION FX(101),Z(101)
2       FORMAT(41H1EXAMPLE PROBLEM ON NUMERICAL INTEGRATION)
3       FORMAT(11H1END OF JOB)
4       FORMAT(68H0MASS TRANSFER COEFFICIENT IS GIVEN BY K1 + K2*XA, WHERE
       1 K1,  K2 ARE ,//,1X,2E18.8)
5       FORMAT(1H0/6H0TRIAL,4X,10HXA2(GUESS),4X,12HSIMPSON AREA,4X,14HPER
       1CENT DEVIA)
6       FORMAT(1H0,I6,F17.8,F14.8,F17.5)
7       FORMAT(1H0/21H0CONVERGENCE OBTAINED)
8       FORMAT(1H1/(I10,E18.8))
        IO=6
        XAI=0.000532
        ANS=0.129
        V=0.0701
C       INITIAL TRIAL WILL BE 0.05 OF XAI.   100 INTERVALS TO BE USED AT
C           EACH TRIAL
        GUESS=0.05
        NINT=100
        MINT=NINT+1
        K1=0.0551
        EPS=0.0001
        XA1=0.0
        DO 100 KK=1,2
        GO TO(20,22),KK
20      K2=0.
        FX(1)=F(XA1)
        GO TO 23
22      K2=185.5
C       FIND XA2B, A GUESS BELOW ANSWER(ANS)
23      XA2B=GUESS*XAI
        WRITE(IO,2)
        IT=0
        WRITE(IO,4)K1,K2
        WRITE(IO,5)
24      H=(XA2B-XA1)/FLOAT(NINT)
        XA2G=XA2B
        X=XA1
        DO 30 I=1,NINT
        X=X+H
30      FX(I+1)=F(X)
        IT=IT+1
        CALL QSF(H,FX,Z,MINT)
        AREA=Z(MINT)
        DEV=(AREA-ANS)/ANS*100.
        WRITE(IO,6)IT,XA2G,AREA,DEV
        IF(AREA-ANS)36,61,34
34      XA2B=XA2B*0.5
        GO TO 24
36      XA2G=XA2G*2.
C       FIND XA2G, A GUESS GREATER THAN ANS
        H=(XA2G-XA1)/FLOAT(NINT)
        X=XA1
        DO 39 I=1,NINT
        X=X+H
```

Table 6.6-1 (*continued*)

```
  39    FX(I+1)=F(X)
        IT=IT+1
        CALL QSF(H,FX,Z,MINT)
        AREA=Z(MINT)
        DEV=(AREA-ANS)/ANS*100.
        WRITE(IO,6)IT,XA2G,AREA,DEV
        IF(AREA-ANS)36,61,45
  C     USE TRIAL AND ERROR SEARCH TO FIND ANSWER TO WITHIN EPSILON (EPS)
  45    XA2F=(XA2G+XA2B)*0.5
        H=(XA2F-XA1)/FLOAT(NINT)
        X=XA1
        DO 50 I=1,NINT
        X=X+H
  50    FX(I+1)=F(X)
        IT=IT+1
        CALL QSF(H,FX,Z,MINT)
        AREA=Z(MINT)
        DEV=(AREA-ANS)/ANS*100.
        WRITE(IO,6)IT,XA2F,AREA,DEV
        IF(ABS(ANS-AREA)-EPS)61,56,56
  56    IF(ANS-AREA)57,61,59
  57    XA2G=XA2F
        GO TO 45
  59    XA2B=XA2F
        GO TO 45
  61    WRITE(IO,7)
        WRITE(IO,8)(I,FX(I),I=1,MINT)
 100    CONTINUE
        WRITE(IO,3)
        STOP
        END
```

Table 6.6-2 Results for Example 6.6-4

EXAMPLE PROBLEM ON NUMERICAL INTEGRATION

MASS TRANSFER COEFFICIENT IS GIVEN BY K1 + K2*XA, WHERE K1, K2 ARE

0.55099998E-01   0.18550000E 03

TRIAL	XA2(GUESS)	SIMPSON AREA	PER CENT DEVIA
1	0.00002660	0.06251651	-51.53757
2	0.00005320	0.12320513	-4.49211
3	0.00010640	0.24144608	87.16757
4	0.00006650	0.15305519	18.64748
5	0.00004655	0.10817313	-16.14482
6	0.00005652	0.13069284	1.31232
7	0.00005154	0.11945647	-7.39804
8	0.00005403	0.12507927	-3.03928
9	0.00005528	0.12789005	-0.86039
10	0.00005590	0.12929249	0.22677
11	0.00005559	0.12859160	-0.31655
12	0.00005575	0.12894058	-0.04602

CONVERGENCE OBTAINED

6.6 Use of Concentration Driving Forces

Table 6.6-2 (continued)

EXAMPLE PROBLEM ON NUMERICAL INTEGRATION
MASS TRANSFER COEFFICIENT IS GIVEN BY K1 + K2*XA, WHERE K1, K2 ARE
    0.55099998E-01   0.0

TRIAL	XA2(GUESS)	SIMPSON AREA	PER CENT DEVIA
1	0.00002660	0.06529963	-49.38010
2	0.00005320	0.13412887	3.97591
3	0.00003990	0.09924984	-23.06210
4	0.00004655	0.11657274	-9.63350
5	0.00004987	0.12532067	-2.85215
6	0.00005154	0.12971944	0.55774
7	0.00005071	0.12751776	-1.14898
8	0.00005112	0.12861300	-0.29996
9	0.00005133	0.12917298	0.13413
10	0.00005123	0.12889099	-0.08446
11	0.00005128	0.12902856	0.02218

CONVERGENCE OBTAINED

The number of even increments or intervals $m = (b-a)/h$ is fixed at $100 = \text{NINT}$, where NINT is the symbol in the program used for m. Then for x_A going from x_{A1} to x_{A2}

$$H = \frac{x_{A2} - x_{A1}}{\text{NINT}}$$

where H is the spacing, h, of the interval and x_{A2} is a guess of x_{A2}. The function $f(x_A)$ is

$$f(x_A) = F(X) = \frac{V}{(K_1 + K_2 x_A) \ln\left(\frac{1-x_A}{1-x_{Ai}}\right)}$$

The value of x_{Ai} is XAI and is a constant, as is the value of x_{A1}, which is XA1 = 0.

Because of the physical situation, $x_{A2} < x_{Ai}$, and for the first trial a value of XA2B—that is x_{A2}—of 5 percent of x_{Ai} will be selected. This is then used to calculate H and F(X) at the 100 spaces. Then, using Simpson's rule, the area of the integral times V is calculated. The program for Simpson's rule can be written in detail or a standard subroutine used for the particular machine employed, which is available in most Fortran subroutine libraries. Where such subroutines are available it is preferable to use them rather than develop such a subroutine oneself. In Table 6.6-1 the subroutine CALL QSF(H,FX,Z,MINT) is used, where FX is the input value of the function at each step, Z is the area of the integral up to the point being considered, and MINT is the number of values of FX, which is 101. If this subroutine is not available, the reader is referred to Conte (C1, pp. 136–137), where a Fortran program is available.

If the area calculated in the first trial is not less than $A = 0.129 \text{ ft}^2$, then the first guess is halved. This is repeated until the area is below 0.129 ft^2.

Next in this trial-and-error search pattern a value of x_{A2} is used, called XA2G, which gives an area greater than 0.129 ft^2. This is obtained by repeatedly doubling XA2B until the desired value of XA2G is obtained. Then a value of XA2F, which is the average between XA2B and XA2G, is used to calculate a new area. If this area is greater than that of the ANS(0.129), then XA2F replaces XA2G. Otherwise it replaces XA2B.

The computation is stopped according to the following criterion, which is arbitrary and depends on the accuracy desired:

$$|0.129 - \text{AREA}| < 0.0001 \text{ (EPS)}$$

In Table 6.6-2 the results show that 12 trials were required and that $XA2 = 0.00005575$. The case where $K_2 = 0$, which is Example 6.6-2, was also solved in this program, and $XA2 = 0.00005128$, which compares closely with the value of 0.0000520 calculated analytically in Example 6.6-2.

6.7 EXPERIMENTAL MASS TRANSFER COEFFICIENTS

6.7A Experimental Methods to Determine Mass Transfer Coefficients

Many different experimental methods are used to determine mass transfer coefficients in different systems with varying geometries. The choice of method depends primarily on the complexity of the geometry, the need to eliminate or minimize any resistances that might be present other than the one sought, and the ease of analysis of the concentrations.

For wetted-wall towers a number of methods can be used. In one case, pure gas A is made to flow upward with a film of liquid B flowing downward. The inlet and outlet concentrations of A in B are measured. Then, knowing the flow rates, a Reynolds number, k_L, and J_D are calculated. Since there is no resistance in the gas phase, because pure A is diffusing through A, the only resistance is in the liquid film. If A is diffusing through an inert gas C to get to the liquid interface, then a gas film resistance might exist, which must be corrected for in solving for k_L. To determine the gas film k_G, a pure liquid A flowing can be vaporized into an inert gas

B, such as air. There is no liquid resistance in this case. It might be expected that this gas coefficient k_G would be the same as for mass transfer to the walls of a pipe with gas flowing inside. However, these values for these two cases are not the same, since ripples on a falling film surface greatly affect the results.

To determine the mass transfer coefficient to a sphere, the technique of Steele and Geankoplis (S4) will be described briefly. Figure 6.7-1 shows the basic part of the apparatus. A cast or compressed sphere of benzoic acid $\frac{1}{2}$ inch in diameter is held rigidly by a spindle support at the rear. The rear support was used so that the wake and flow patterns would have minimum disturbance. Before the run, the sphere was weighed. Then it was inserted into the pipe and flow started and continued for a timed interval. Then the flow was stopped and the sphere quickly removed and dried. The sphere was again weighed. From the weight loss the flux N_A was calculated. Less than 0.8 percent of the volume of the sphere was dissolved, so that the sphere did not change shape. Knowing the flux N_A and the driving force $(c_{As} - 0)$, where the c_{As} is the solubility and the water contained no benzoic acid, the k_L was calculated.

This technique of using dissolving solids has been employed to determine mass transfer from pipe walls by coating the walls with dissolving solutes. It has also been used for determining the mass transfer coefficients from solids cast in other varying shapes. This procedure can be used for gases or liquids with solids.

Another useful method is to blow gases over various shapes wet with evaporating liquids. For mass transfer from a flat plate a porous blotter soaked with the liquid serves as the plate. In packed beds porous solids soaked with water or other liquids have been used.

Generally the effects of different gases or liquids are correlated by the Schmidt number. The Schmidt numbers of gases range from 0.6 for air-water vapor to 3 for air-large organic molecules. For liquids the range is very large—from about 300 to 100,000. The J_D has been found to be proportional to $N_{Sc}^{2/3}$ in many cases. However, the reader should be careful in using these correlations outside the experimental range of Schmidt numbers. In some cases the data have been obtained only for gases or only for liquids, and the exponent may be different.

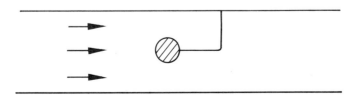

FIGURE 6.7-1 Experimental method to determine mass transfer coefficients to a sphere

288 Mass Transfer Coefficients in Laminar and Turbulent Flow

The fluid properties of viscosity μ and density ρ in the correlations are generally evaluated at the mean film conditions, which means taking the linear average of the property at the phase surface (interface) and the bulk fluid (T4). In some cases the mass velocity G is evaluated as the bulk weight flow rate per ft². However, some investigators do not report the conditions used or often use only bulk fluid properties. For dilute solutions the bulk fluid properties differ very little from the mean film properties.

6.7B Correlations of Mass Transfer Coefficients

The correlations of mass transfer coefficients for the important cases often encountered will be given in the sections that follow. The correlations for gas-liquid flow over tower packings will be given in Chapter 8, Section 8.4. Both laminar and turbulent flow will be covered.

1. *Wetted-Wall Towers and Flow in Pipes.* For mass transfer to a gas or liquid flowing inside a pipe and for mass transfer to a gas flowing inside a wetted-wall tower the same correlations apply. Of course different equations are used for laminar and for turbulent flow of the gas or liquid flowing inside the central core.

(a) *Laminar flow.* When the Reynolds number $Dv\rho/\mu$ is less than 2100, the flow is laminar, where D is inside diameter and v is velocity relative to the pipe wall. This was discussed in Section 6.4B. It is assumed these equations are for fully developed laminar flow. Experimental data were obtained by Gilliland and Sherwood (L1, G1) for gases with vaporization of liquids in a wetted-wall tower. These data are for values of $W/(D_{AB}\rho L)$ less than about 70. These data are plotted in Fig. 6.7-2 as $(c_A - c_{A0})/(c_{Ai} - c_{A0})$ vs. $W/(D_{AB}\rho L)$, where c_A is the exit concentration, c_{A0} inlet concentration, c_{Ai} concentration at the interface, which is the solubility, W is flow in g mass/sec, and L is length of mass transfer section in the direction of flow. We see that these data follow the equation derived for rodlike flow instead of the parabolic flow equation. This has been attributed to natural convection effects. Hence, the empirical equation to use for gases is the rodlike flow plot.

For liquids that have very small values of D_{AB}, data were obtained by Linton and Sherwood (L1) for solids dissolving from a pipe wall to the flowing liquid. In Fig. 6.7-2 these data are shown to follow the approximate Leveque equation for parabolic flow, which is as follows for values of $W/(D_{AB}\rho L)$ over 400:

$$\frac{c_A - c_{A0}}{c_{Ai} - c_{A0}} = 5.5 \left(\frac{W}{D_{AB}\rho L}\right)^{-2/3} \qquad (6.7\text{-}1)$$

The correlation in Fig. 6.7-2, as stated before, is for the fluid flowing inside a pipe or inside a wetted-wall tower. For the mass transfer coefficient from a gas into the liquid film flowing down the walls of a wetted-wall tower, a theoretical discussion was given in Section 6.4C.

6.7 Experimental Mass Transfer Coefficients 289

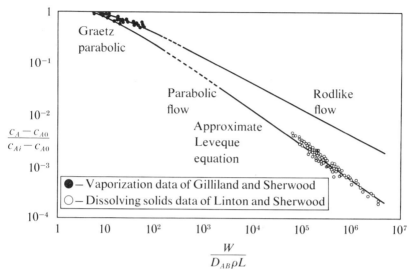

FIGURE 6.7-2 Data and theory for diffusion in streamline flow for a fluid flowing inside a pipe or inside a wetted-wall tower, Ref. (L1). [From W. H. Linton and T. K. Sherwood, *Chem. Eng. Progr.* **46**, 258 (1950). With permission.]

The equations in Section 6.4C should be used and the theoretically predicted values multiplied by about 1.5 for actual values. For Reynolds numbers $4\Gamma/\mu$ greater than 1200, which is the approximate limit for laminar flow with films, Eq. (6.4-15) can be used as an approximation for k_L up to Reynolds numbers of about 3000 and should be multiplied by a factor of 1.5 to 2.5 for actual values (S2).

(b) *Turbulent flow.* For Reynolds numbers $Dv\rho/\mu$ above 2100 and for a gas or a liquid flowing inside a pipe, the data for gases and liquids are correlated by

$$N_{\text{Sh}} = \frac{k_c p_{BM}}{P} \frac{D}{D_{AB}} = 0.023 \left(\frac{Dv\rho}{\mu}\right)^{0.83} \left(\frac{\mu}{\rho D_{AB}}\right)^{0.33} \quad (6.7\text{-}2)$$

or, as J_D,

$$J_D = 0.023 N_{\text{Re}}^{-0.17} \quad (6.7\text{-}3)$$

Note that the Eq. (6.7-2) is written for A diffusing through stagnant B. For equimolar counterdiffusion, from Table 6.3-1, $k_c p_{BM}/P$ is replaced by k_c', and so on. These equations hold for N_{Re} above 2100 and N_{Sc} of 0.6 to 3000 for liquids and gases and are based upon data by Gilliland and Sherwood (G1) and Linton and Sherwood (L1).

2. *Flow Parallel to Flat Plates.* For flow of a fluid past a flat plate in a free stream in an open space the boundary layer is not fully developed. Experimental data have been obtained for this case where mass transfer occurs from the plate to the surrounding fluid. The vaporization of

liquids from a flat surface to a flowing stream is of interest in the evaporation of solvents from paints, for plates in wind tunnels, and in flow channels in chemical process equipment.

For mass transfer to gases or evaporation of liquids Sherwood and Pigford (S2) give experimental data. For the laminar region of Reynolds numbers $N_{Re,L} = Lv\rho/\mu$ less than 15,000 the data can be represented within ±25 percent by the theoretical $J_H = J_D = f/2$ line as follows:

$$J_D = 0.036 N_{Re,L}^{-0.2} \qquad (6.7\text{-}4)$$

where L is the length of the plate in the direction of flow. The actual data cover only a range of Reynolds numbers of 8000 to 15,000 but the equation can be used as an approximation below this.

For gases and $N_{Re,L}$ above 15,000 up to 300,000 the experimental data can be represented within ±30 percent by $J_D = J_H = f/2$ as follows:

$$J_D = 0.036 N_{Re,L}^{-0.2} \qquad (6.7\text{-}5)$$

Data for liquids have been obtained by Litt and Friedlander (L2) for dissolving flat plates of benzoic and cinnamic acids. For the $N_{Re,L}$ of 600 to 50,000 the experimental data are correlated within about ±40 percent by the following equation:

$$J_D = 0.99 N_{Re,L}^{-0.5} \qquad (6.7\text{-}6)$$

This equation is 50 percent higher than Eq. (6.7-4) for gases. Linton and Sherwood (L1) obtained data for dissolving plates in liquids, and their data scatter about the same amount from Eq. (6.7-6) in the $N_{Re,L}$ range of 3000 to 50,000. However, their data are not exactly comparable because their plate stretched completely across the duct. They also obtained data for $N_{Re,L}$ up to 200,000. For the range ot 50,000 to 200,000 their data are about 70 to 100 percent above Eq. (6.7-5) for gases. However, since their data are not exactly comparable to that for flat plates in a free stream, predictions above $N_{Re,L}$ of 50,000 are not accurate.

3. *Flow Past Single Spheres.* For flow past single spheres Steinberger and Treybal (S5) have made a thorough study of the data for gases and liquids and correlated these data. For very low $N_{Re} = D_p v\rho/\mu$, where v is mass average velocity in the empty test section before the sphere, the Sherwood number should approach a value of 2.0. This value can be derived theoretically for a sphere diffusing in a large volume of stagnant fluid. However, they show that natural convection effects can increase this markedly at low flows.

At high flow rates the problem becomes increasingly complex, as flow separation and formation of wakes on the rear of the sphere occur. At very high Reynolds numbers of about 70,000 Steele and Geankoplis (S4) give data that show an abrupt increase and peak in J_D and then an abrupt drop. This abrupt change is a result of transition of the boundary layer, which causes the point of separation of the boundary layer that was formerly adjacent to the sphere to move downstream.

For a Schmidt number range of 0.6 to 3200 and a Reynolds number range of 1 to 16,900 for gases and liquids the data are correlated by Steinberger and Treybal (S5) by

$$N_{Sh} = N_{Sh0} + 0.347(N_{Re}N_{Sc}^{0.5})^{0.62} \tag{6.7-7}$$

where

$$N_{Sh0} = 2.0 + 0.569(N_{Gr}N_{Sc})^{0.25} \tag{6.7-8}$$

The average deviation is ±12.7 percent. Equation (6.7-8) includes the Grashof number, which corrects for natural convection. The equations above also correlate the heat transfer data, the Nusselt number being substituted for the Sherwood, and the Prandtl for the Schmidt.

Often the data to calculate the Grashof number are not available for heat or (especially) mass transfer, so that Eqs. (6.7-7) and (6.7-8) cannot be used. In such cases the following equations can be used. For liquids, the equation of Garner and Suckling (G2, G7) was obtained for data covering a Reynolds number of 2 to 840 but can be used to 2000. The data cover a Schmidt number range of 788 to 1680. This equation, which follows, can be used for heat and mass transfer:

$$N_{Sh} = 2 + 0.95 N_{Re}^{0.50} N_{Sc}^{1/3} \tag{6.7-9}$$

For gases, Steinberger and Treybal (S5) plotted the data of many investigators for a Schmidt number range of 0.6 to 2.7 and a Reynolds number range of about 1 to 48,000. The well-known Froessling equation correlates the heat and mass transfer data well at lower Reynolds numbers but gives low values of N_{Sh} at high Reynolds numbers. Adjusting the exponent of the Reynolds number from 0.50 to 0.53 correlates the data better over this range. This modified equation is

$$N_{Sh} = 2 + 0.552 N_{Re}^{0.53} N_{Sc}^{1/3} \tag{6.7-10}$$

This equation also holds for heat transfer, with the Nusselt number and Prandtl number replacing the Sherwood and Schmidt numbers. Recent data of Evnochides and Thodos (E1) for mass and heat transfer for gases and for a Reynolds number range of 1500 to 12,000 and a Schmidt number range of 0.6 to 1.85 check the equations above.

Since the natural convection correction becomes small at high Reynolds numbers, for a Reynolds number range of 2000 to 16,900 and for liquids the following equation can be used with errors of less than 15 percent to replace Eqs. (6.7-7) and (6.7-8):

$$N_{Sh} = 0.347 N_{Re}^{0.62} N_{Sc}^{1/3} \tag{6.7-11}$$

or

$$J_D = 0.347 N_{Re}^{-0.38} \quad (6.7\text{-}12)$$

For Reynolds numbers above 16,900 and for aqueous solutions the data do not follow any one equation. Figure 6.7-3 plots the data of Steele and Geankoplis (S4) at very high Reynolds numbers — up to 140,000 for organic solutes in water — and the data of Traylor, Burris, and Geankoplis (T2) up to 400,000 for solid uranium in liquid cadmium. It is recommended that for cinnamic acid in water curve 3 be used, for 2-naphthol in water curve 2, and for benzoic acid in water curve 4 for Reynolds numbers above 16,900. Curves 3 and 4 check Eq. (6.7-11) within about 25 percent over the range of N_{Re} of 2000 to 16,900. In this Reynolds number range most of the data for Eq. (6.7-11) were obtained on benzoic acid and are plotted as line 8. No explanation is evident for the lower line for 2-naphthol.

For an approximation for liquid metals line 1 can be used.

4. *Flow Past Single Cylinders.* Bedingfield and Drew (B1) experimentally studied the mass transfer from cylinders when the flow is transverse or perpendicular to the cylinder. Mass transfer to the ends of the cylinder is not considered. Their equation for mass transfer of gases

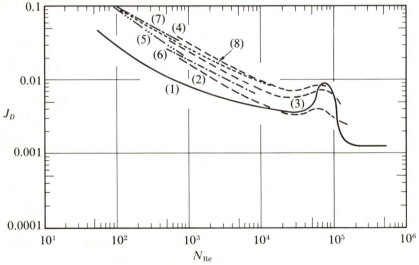

1. Uranium-cadmium (T2)
2. 2-Naphthol (S4)
3. Cinnamic acid (S4)
4. Benzoic acid (S4)
5. Benzoic acid (G7)
6. Benzoic acid (L3)
7. Benzoic, adipic acid (G2)
8. Benzoic acid (S5)

FIGURE 6.7-3 Comparison of J_D for uranium-cadmium and organic-water systems, Ref. (T2). [From E. D. Traylor, Leslie Burris, and C. J. Geankoplis, *Ind. Eng. Chem. Fund.* **4**, 119 (1965). With permission.]

in the Reynolds number $D v \rho / \mu$ range of 400 to 25,000 and Schmidt number range of 0.6 to 2.6 is

$$N_{Sh} = 0.281 N_{Re}^{0.50} N_{Sc}^{0.44} \qquad (6.7\text{-}13)$$

where D is the diameter of the cylinder and v is the average velocity in the duct ahead of the cylinder. This equation also holds for heat transfer with gases where the Nusselt and Prandtl numbers are substituted for the Sherwood and Schmidt numbers. For heat transfer Eq. (6.7-13) can be used up to a Reynolds number of 100,000 with little error.

For liquids little data are available. Linton and Sherwood (L1) obtained data for a Schmidt number range of about 1000 to 3000 and a Reynolds number range of about 750 to 12,000. Their data scatter about Eq. (6.7-14) with a maximum deviation of ± 30 percent.

$$J_D = 0.281 N_{Re}^{-0.4} \qquad (6.7\text{-}14)$$

Hence, for liquids of any Schmidt number up to 3000 it is recommended that Eq. (6.7-14) be used for mass transfer for the Reynolds number range of 400 to 25,000.

5. *Flow through Packed Beds.* The mass transfer of fluids to packed beds of solid particles is very important in the study of the performance of beds of catalyst particles. Using a packed bed a large amount of mass transfer area can be contained in a relatively small volume. Also, the closeness of the packing and the irregular flow channels provide a large amount of turbulence, which enhances the mass transfer coefficient.

(a) *Surface area and void volume in a packed bed.* The void fraction in a packed bed is designated as ϵ, ft³ volume void space between the solid particles divided by the ft³ total volume of void space plus solid particles. These values range from about 0.30 to 0.50 in most beds.

The surface to volume ratio is designated as a, ft² surface area of the solid particles divided by the ft³ volume of total volume. This can be calculated for a bed of spheres, cylinders having equal length and diameter, and cubes by

$$a = \frac{6(1-\epsilon)}{D_p} \qquad (6.7\text{-}15)$$

where D_p is the diameter in feet of the sphere, the length and diameter of a cylinder, or the side of a cube.

Example 6.7-1 Derivation of Surface/Volume Ratio for Cylinders

Derive Eq. (6.7-15) for cylinders in a packed bed where the length and diameter are equal. Also calculate the number, n, of cylinders in a bed with ϵ void fraction.

Solution

Take 1 ft³ as total volume of bed (void plus solid particles). The area of a cylinder is $\pi D_p(h) + 2(\pi D_p^2/4)$, where h is the same as D_p. Hence, area $= \frac{3}{2}\pi D_p^2$. The volume is $\pi D_p^2(D_p)/4$ or $\pi D_p^3/4$. Let n = number of particles in 1 ft³. Then the solid volume of the bed is $(1-\epsilon)$ 1 ft³. Hence,

$$(1-\epsilon)(1) = \frac{n\pi D_p^3}{4}$$

or

$$n = \frac{4(1-\epsilon)}{\pi D_p^3}$$

The definition of a is ft² surface area/ft³ total volume, or

$$a = \frac{n(\frac{3}{2}\pi D_p^2)}{1}$$

Substituting the value of n into the above and solving,

$$a = \frac{6(1-\epsilon)}{D_p}$$

(b) *Mass transfer of gases in a packed bed.* A large quantity of data has been obtained for mass transfer coefficients of gases in packed beds. Before 1963 most of the data were obtained in very short bed heights of 1 to 5 layers of spheres in the bed. Also, the spheres were supported on screens with a large void space before the screen. As a consequence there was considerable flow channeling and the first few layers of spheres were not contacted fully. Gupta and Thodos (G3) showed experimentally that having several layers of inert spheres preceding the active bed to promote good fluid distribution gave mass transfer coefficients considerably higher by factors of up to 2/1 at low Reynolds numbers of 100.

The recommended correlation (G3, G4) for mass and for heat transfer of gases in beds of spheres is

$$\epsilon J_D = \epsilon J_H = 2.06 N_{\text{Re}}^{-0.575} \tag{6.7-16}$$

for a Reynolds number range of 90 to 4000. The Reynolds number is defined as $N_{\text{Re}} = D_p v'\rho/\mu$, where D_p is the diameter of the spheres and v' is the superficial mass average velocity in the empty tube without packing.

For a Reynolds number of 4000 to 10,000 the data show a transitional behavior (G4), and Fig. 6.7-4 should be used for ϵJ_D. For heat transfer, the following can be used over the same range (G4): $\epsilon J_H = 1.05 \epsilon J_D$.

Gupta and Thodos (G5) correlated the data prior to 1962 for packed beds of solids other than spheres and developed correction factors to use for other shapes of solids. The data are low, as discussed above, but these geometry correction factors can be used as approximations with the

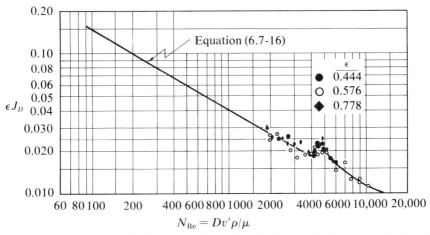

FIGURE 6.7-4 Relation between ϵJ_D and N_{Re} for flow of air through packed and distended beds of spheres (G3, G4). [From A. S. Gupta and George Thodos, *Ind. Eng. Chem. Fund.* **3**, 218 (1964). With permission.]

correlation in Fig. 6.7-4, for gases. These average geometry correction factors were obtained from the plots of Gupta and Thodos (G5) and are tabulated in Table 6.7-1 for a Reynolds number range for spheres of 100 to 4000. The geometry correction factors are used as follows. For example, it is desired to obtain ϵJ_D for 1.0-in. cylinders in a packed bed at a given velocity of fluid. The Reynolds number $D_p v'\rho/\mu$ is calculated for a 1.0-in. sphere. Next the ϵJ_D of the sphere is obtained from Fig. 6.7-4. Then, ϵJ_D of 1-in. cylinder = (0.79) times the ϵJ_D of the 1-in. sphere. Next k_c is obtained from the J_D. The flux of a cylinder in the bed is then

$$N_A(A_{cyl}) = k_c(A_{cyl})(\Delta c_A) \tag{6.7-17}$$

where A_{cyl} is the total area of a cylinder or $\frac{3}{2}\pi D_p^2$.

For other shapes, such as Raschig rings and Berl saddles, reference should be made to the original article (G5). Since the surface area of a 1-in. cylinder is 1.5 times that of a 1-in. sphere, the relative total flux

Table 6.7-1 Approximate Geometry Correction Factors for Mass Transfer Coefficients

(where diameter of sphere = diameter of cylinder = side of cube)

Geometry	Geometry Correction Factor, ϵJ_D geom./ϵJ_D sphere
Sphere	1.0
Cylinder (length = diameter)	0.79
Cube	0.71

$N_A(A)$ of a cylinder is 1.5(0.79) or 1.19, compared to 1.0(1.0) for a sphere. This is 19 percent greater than for a 1-in. sphere.

(c) *Mass transfer of liquids in a packed bed.* For mass transfer of liquids in packed beds of spheres the recent correlation of Wilson and Geankoplis (W1) should be used. The equations are

$$\epsilon J_D = 1.09 N_{\text{Re}}^{-2/3} \qquad (6.7\text{-}18)$$

where $N_{\text{Re}} = D_p v' \rho / \mu$, the Reynolds number range is 0.0016 to 55, and the Schmidt number range is 165 to 70,600. These data are a good confirmation of the $\frac{2}{3}$ exponent on the Schmidt number of J_D.

For a Reynolds number range of 55 to 1500 and a Schmidt number range of 165 to 10,690,

$$\epsilon J_D = 0.250 N_{\text{Re}}^{-0.31} \qquad (6.7\text{-}19)$$

As a preliminary estimate with liquids in packed beds of different geometries other than spheres, Table 6.7-1 can be used.

(d) *Mass transfer in fluidized beds.* For fluidized beds of spheres the correlation of Gupta and Thodos (G6) can be used for a Reynolds number range of 20 to 3000 and gases and liquids:

$$J_D = 0.010 + \frac{0.863}{N_{\text{Re}}^{0.58} - 0.483} \qquad (6.7\text{-}20)$$

An alternate correlation of Gupta and Thodos (G5) can also be used.

Example 6.7-2 Mass Transfer Coefficient for Different Shapes of Solids

It is desired to estimate the value of the mass transfer coefficient k_G of water vapor in a flowing air stream in a duct being transfered to different solid shapes. The average velocity in the empty duct before the solids is 12.0 ft/sec at 150°F and 1.0 atm. It is assumed that the water vapor concentration in the air is small, so the physical properties of the gas stream are that of pure air. Do this for the following shapes and cases: (a) for the air parallel to a flat plate 1.0 in. long, (b) a single 1-in. by 1-in. cylinder normal to the air stream, (c) a single 1-in. sphere, (d) a packed bed of 1-in. spheres with a void fraction ϵ of 0.35. (e) For case (c) the partial pressure of water in the air is 0.150 atm and the solid is a drying agent. Calculate the total moles of water absorbed in 10 sec by the single sphere.

Solution

For pure air at 150°F and 1 atm, $\mu = (0.0195 \text{ cp}) 2.42 = 0.0473$ lb mass/ft-hr. Density $\rho = (29/359)[492/(460+150)] = 0.0652$ lb mass/ft³. Diffus-

6.7 Experimental Mass Transfer Coefficients

ivity of H_2O in air at 42°C or 108°F = 0.288 cm²/sec. Correcting this to 150°F,

$$D_{AB} = 0.288 \left(\frac{460+150}{460+108}\right)^{1.75} = 0.330 \text{ cm}^2/\text{sec}$$

or $0.330(3.87) = 1.28$ ft²/hr. Hence,

$$N_{Sc} = \frac{\mu}{\rho D_{AB}} = \frac{0.0473}{0.0652(1.28)} = 0.569$$

$$G_M = \frac{v\rho}{M_{ave}} = \frac{(12)(3600)(0.0652)}{29} = 97.0 \text{ lb mole/hr-ft}^2$$

For a flat plate,

$$N_{Re,L} = \frac{Lv\rho}{\mu} = \frac{1}{12}\frac{(12)(3600)(0.0652)}{0.0473} = 4970$$

Hence, Eq. (6.7-4) can be used.

$$J_D = 0.664 N_{Re,L}^{-0.5} = 0.664(4970)^{-0.5} = 0.00940$$

From Table 6.5-1,

$$J_D = \frac{k_G p_{BM}}{G_M} N_{Sc}^{2/3}$$

Since the air is dilute, $p_A = 0$ and $p_{BM} = 1.0$. Substituting into the J_D equation,

$$0.00940 = \frac{k_G(1.0)(0.569)^{2/3}}{97.0}$$

Hence $k_G = 1.33$ lb mole/hr-ft²-atm   [Ans. to (a)].

For a 1-in. cylinder,

$$N_{Re} = \frac{D_p v\rho}{\mu} = \frac{1}{12}\frac{(12)(3600)(0.0652)}{0.0473} = 4970$$

Use Eq. (6.7-13):

$$N_{Sh} = 0.281 N_{Re}^{0.60} N_{Sc}^{0.44} = 0.281(4970)^{0.60}(0.569)^{0.44} = 36.3$$

From Table 6.5-1,

$$N_{Sh} = \frac{k_G p_{BM} RT D_p}{P D_{AB}} = \frac{k_G(1.0)(0.730)(610)(\frac{1}{12})}{(1.0)(1.28)} = 36.3$$

Solving for k_G, $k_G = 1.25$ lb mole/hr-ft²-atm   [Ans. to (b)].

For a 1-in. sphere, $N_{Re} = 4970$. Use Eq. (6.7-10), since Grashof data are not available for Eq. (6.7-7).

$$N_{Sh} = 2 + 0.552 N_{Re}^{0.53} N_{Sc}^{1/3} = 2 + 0.552(4970)^{0.53}(0.569)^{1/3} = 44.0$$

298 Mass Transfer Coefficients in Laminar and Turbulent Flow

Making the same substitution for the Sherwood number as for the cylinder, $k_G = 1.52$ lb mole/hr-ft^2-atm   [Ans. to (c)].

For a packed bed of spheres, $N_{Re} = 4970$. Use Fig. 6.7-4 and $\epsilon J_D = 0.020$. Hence, $J_D = 0.020/0.35 = 0.0571$. Then,

$$J_D = k_G \frac{p_{BM}}{G_M} N_{Sc}^{2/3}$$

Solving, $k_G = 8.08$ lb mole/hr-ft^2-atm   [Ans. to (d)].

For the case where the $p_A = 0.150$ atm of water vapor, the film properties used must be the average of the interface and bulk physical properties. The film viscosity is calculated as the mole fraction average of the water and air. The film average $p_A = (0.150 + 0)/2 = 0.075$ atm or mole fraction. The film $\mu_{ave} = 0.075(0.0115) + 0.925(0.0195) = (0.0186$ cp$)(2.42) = 0.0449$ lb mass/ft-hr. For more accurate methods for viscosity of gas mixtures, see Reid and Sherwood (R2). The average film molecular weight is $0.075(18) + 0.925(29) = 28.2$. The film density $\rho = (28.2/359)[492/(460 + 150)] = 0.0628$ lb mass/ft^3. The film Schmidt number $= 0.0449/[0.0628(1.28)] = 0.560$.

$$G_M = \frac{(12)(3600)(0.0628)}{28.2} = 95.8 \text{ lb mole/hr-ft}^2$$

$$N_{Re} = \frac{1}{12} \frac{(12)(3600)(0.0628)}{0.0449} = 5020$$

Using Eq. (6.7-10) again, $N_{Sh} = 44.2$.

The p_{A1} at the surface $= 0$, $p_{A2} = 0.15$ atm, $p_{B1} = 1.0$, $p_{B2} = 0.85$,

$$p_{BM} = \frac{1.0 - 0.85}{\ln\left(\frac{1.0}{0.85}\right)} = 0.925 \text{ atm}$$

$$N_{Sh} = 44.2 = \frac{k_G p_{BM}}{P} RT \frac{D_p}{D_{AB}} = k_G \frac{(0.925)(0.730)(610)(\frac{1}{12})}{(1.00)(1.28)}$$

Hence, $k_G = 1.65$ lb mole/hr-ft^2-atm. To calculate the flux, $N_A = k_G(p_{A1} - p_{A2}) = 1.65(0 - 0.15) = -0.248$ lb mole/hr-ft^2. The surface area of the sphere $\pi D^2 = \pi(\frac{1}{12})^2$ ft^2. Hence,

moles absorbed in 10 sec $= (0.248)(\pi)(\frac{1}{12})^2(\frac{10}{3600}) = 1.50 \times 10^{-5}$ lb mole

[Ans. to (e)]

6. *Liquid Metals Mass Transfer.* In recent years liquid metals have been used increasingly for heat and mass transfer. Only a few correlations for mass transfer coefficients are available for different geometries. In Table 6.7-2 the diffusivities and Schmidt numbers for a few liquid metal systems are given.

Table 6.7-2 Diffusion Coefficients of Liquid Metals

Solute	Solvent	Temp., °C	Solute Conc., Wt %	Diffusivity, $(cm^2/sec)(10^5)$	Schmidt Number, $\mu/\rho D$	Ref.
Tin	Mercury	20	0	1.63	71.4	(D1)
Tin	Mercury	30	0	1.73	66.0	(D1)
Pb	Mercury	20	0	1.13	103.4	(D1)
Pb	Mercury	30	0	1.21	93.3	(D1)
Zn	Mercury	20	0	1.63	71.8	(D1)
Zn	Mercury	30	0	1.72	65.3	(D1)
Cd	Mercury	20	0	1.46	79.6	(D1)
Cd	Mercury	30	0	1.60	70.3	(D1)
U	Cd	600	0	2.07	96.3	(T2)

Dunn et al. (D1) studied the mass transfer of cylinders of tin, cadmium, zinc, and lead in mercury in natural convection and found that the Sherwood numbers obtained agreed with the Nusselt numbers for heat transfer of nonmetals. Data for turbulent mass transfer in zinc tubes or beds of lead spheres in mercury also agreed with nonmetal correlations of J_D versus Reynolds number. It was necessary to prewet the metals first. Bennett and Lewis (B3) found approximately the same mass transfer correlation for rotating cylinders of tin and lead in mercury and benzoic acid in water. Traylor, Burris, and Geankoplis (T2) found reasonably similar correlations for mass transfer of uranium spheres to cadmium and for organic spheres to water, as shown in Fig. 6.7-3. Dunn et al. (D1) conclude that with moderate safety factors the correlations for other fluids in mass transfer may be used for liquid metals. Care must be taken in wetting the materials. Also, if the solid is an alloy, there may exist a substantial resistance to diffusion in the solid phase.

6.8 MASS TRANSFER COEFFICIENTS AND TURBULENCE THEORIES AND MODELS

Mass transfer coefficients have been used for many years to correlate experimental data and in the design of process equipment for transfer processes. As has been shown earlier in the present chapter in discussions of momentum, heat, and mass transfer, the boundary layer theory, eddy diffusivity theory, film theory, and penetration theory have been used as models to explain turbulent mass transfer. In the following sections the analogies and turbulence theories and models will be briefly presented again, emphasizing how they pertain specifically to mass transfer and how they can be used to extend the empirical correlations.

6.8A Film Theory

The most elementary theory is the film theory, which states that all the resistance to mass transfer is in a stagnant film next to the boundary of the fluid. Figure 6.8-1 shows a plot of the concentration profile for mass transfer from a fluid flowing by in turbulent motion to the wall. The distance z_1 is the thickness of the laminar sublayer where molecular diffusion is occurring. The distance z_f is the thickness of the equivalent film for molecular diffusion, which has the same resistance to mass transfer as the actual process occurring in the laminar, transition, and turbulent core. It is assumed that molecular diffusion is occurring only in z_f, and the equation for equimolar mass transport is

$$N_A = k'_c(c_{A1} - c_{A2}) = \frac{D_{AB}}{z_f}(c_{A1} - c_{A2}) \quad (6.8\text{-}1)$$

where c_{A1} is the mean concentration $\bar{c}_A$. Hence,

$$k'_c = \frac{D_{AB}}{z_f} \quad (6.8\text{-}2)$$

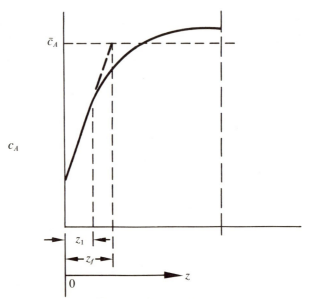

z_1 = laminar sublayer thickness
z_f = thickness of equivalent fictitious film

FIGURE 6.8-1 Concentration profile for film theory

6.8 Mass Transfer Coefficients and Turbulence Theories

The mass transfer coefficient is proportional to $D_{AB}^{1.0}$. However, we have seen that for the majority of the mass transfer coefficients the J_D type equation correlates the data well, where

$$J_D = \frac{k_c'}{v}(N_{Sc})^{2/3} = \frac{k_c'}{v}\left(\frac{\mu}{\rho D_{AB}}\right)^{2/3} \tag{6.8-3}$$

Hence, for constant flow k_c' is proportional to $D_{AB}^{2/3}$. Obviously the exponent on the diffusivity is incorrect, and the film theory is of little help to us in understanding and predicting turbulent mass transfer.

The great advantage of this film theory model is its simplicity. It is generally not used except in describing complex situations and in cases where diffusion and chemical reaction occur simultaneously.

6.8B Penetration Theory

The penetration theory originally derived by Higbie (H1) was derived in this chapter for mass transfer or penetration into a laminar falling film for short contact times in Eq. (6.4-15):

$$k_L = \sqrt{\frac{4D_{AB}}{\pi t_L}} \tag{6.4-15}$$

The mass transfer in turbulent flow is pictured as molecular diffusion in unsteady-state at the wall or boundary of the fluid, where t_L is the time of contact. Danckwerts (D2) extended this and postulated that a fluid eddy or "chunk" of fluid that has a uniform concentration in the turbulent core is swept to the surface, where it stays for a random amount of time and undergoes unsteady-state molecular diffusion. Then it is swept away to the eddy core, where it is mixed and the amount absorbed is distributed in the eddy core. Other eddies then are swept to the surface. Danckwerts defines a mean surface renewal factor, s, which is $1/t_{ave}$, where t_{ave} is the average time of residence at the surface.

$$k_L = \sqrt{s D_{AB}} \tag{6.8-4}$$

Hence, in the surface renewal theory the k_L is proportional to $D_{AB}^{0.5}$. In some systems, such as mass transfer coefficients of liquids flowing over a packing where some semistagnant pockets occur and where the surface is continually renewed, the mass transfer coefficient is approximately proportional to $D_{AB}^{0.5}$.

At present the value of s cannot be predicted and must be determined experimentally. Others (D3, H3, T3) have derived a combination film-surface renewal theory, which predicts a gradual change of the exponent of D_{AB} from 0.5 to 1.0 depending on turbulence conditions. Experimental data are still needed for the parameters.

6.8C Eddy Diffusivity Theory

In the eddy diffusivity theory the equation for mass transfer in turbulent flow is

$$J_{A1}^* = \frac{D_{AB} + \bar{\epsilon}_M}{z}(c_{A1} - c_{A2}) \qquad (6.2\text{-}2)$$

The $\bar{\epsilon}_M$ is the average eddy diffusivity over the whole diffusion path. For complex geometries the measurement or prediction of this eddy diffusivity at each point in the path z and the obtaining of the average value is almost impossible. Hence, this theory is not very useful at present, and its use awaits a more sophisticated understanding of the theory of turbulence.

6.8D J Factor Empirical Model

As shown in the correlations for mass transfer coefficients, the J factor model correlates most data quite well. However, it is not a model as such, since it really does not explain turbulent mass transfer but does allow us to correlate the data and to some extent to predict new data. If an experimental J_D value and k_c' value have been obtained for component A diffusing in B in a given geometry at a given Reynolds number, then the k_c' for another solute E diffusing in F in the same geometry and at the same Reynolds number can be predicted by

$$k_{cE}' = k_{cA}' \left[\left(\frac{\mu_{Bm}}{\rho_{Bm} D_{AB}} \right) \bigg/ \left(\frac{\mu_{Fm}}{\rho_{Fm} D_{EF}} \right) \right]^{2/3} \qquad (6.8\text{-}5)$$

where μ_{Bm} is the viscosity of the solution A in B, ρ_{Bm} the density of the solution A in B, and μ_{Fm} and ρ_{Fm} are for solution E in F. For mass transfer of gas and liquid flows over packings the exponent in Eq. (6.8-5) may be lower and closer to 0.5.

6.8E Boundary Layer Theory

Boundary layer theory can be used to evaluate or explain mass transfer coefficients. As an example, an exact analysis can be made for a fluid in laminar flow with a uniform velocity past a flat plate of v_∞ in the x direction. Mass transfer in the y direction is occurring from the fluid having a concentration of $c_{A\infty}$ to the plate with a concentration of c_{AS}. The velocity $v_x = 0$ at the surface and increases as one goes further from the plate to v_∞. This region is called the velocity boundary layer and a similar region exists for the concentration boundary layer as shown in Figure 6.8-2. As the distance x from the forward edge of the plate increases, the thickness of the velocity boundary layer δ increases. The thickness of the boundary layer is often arbitrarily taken as the distance

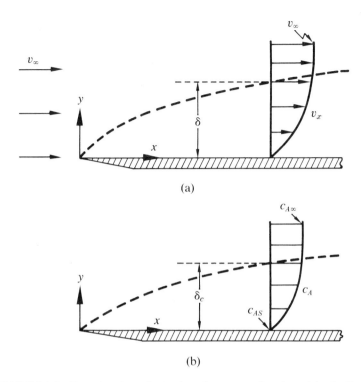

FIGURE 6.8-2 Boundary layer for laminar flow past a flat plate: (a) velocity boundary layer, (b) concentration boundary layer

away from the surface where the velocity is 99 percent of the free-stream velocity v_∞.

The fluid is assumed to have a constant density and the flow is steady-state in two dimensions x and y. The continuity Eq. (2.2-8) becomes as follows:

$$\frac{\partial v_x}{\partial x}+\frac{\partial v_y}{\partial y}=0 \tag{6.8-6}$$

The equation of motion in the x direction is Eq. (2.3-9) and it becomes as follows:

$$v_x\frac{\partial v_x}{\partial x}+v_y\frac{\partial v_x}{\partial y}=\frac{\mu}{\rho}\frac{\partial^2 v_x}{\partial y^2} \tag{6.8-7}$$

The dimensionless boundary conditions are $v_x/v_\infty = 0$ at $y = 0$ and $v_x/v_\infty = 1$ at $y = \infty$. Using Eq. (2.6-12) we obtain the following for negligible diffusion in the x direction.

$$v_x\frac{\partial c_A}{\partial x}+v_y\frac{\partial c_A}{\partial y}=D_{AB}\frac{\partial^2 c_A}{\partial y^2} \tag{6.8-8}$$

Choosing c_{AS} as greater than $c_{A\infty}$ for convenience, the boundary conditions are in dimensionless form $(c_{AS} - c_A)/(c_{AS} - c_{A\infty}) = 0$ at $y = 0$ and $(c_{AS} - c_A)/(c_{AS} - c_{A\infty}) = 1$ at $y = \infty$.

Equations (6.8-6) and (6.8-7) are used after transformation of the variables to obtain a single ordinary nonlinear differential equation. This was solved using a series solution. The numerical results give the velocity distribution in the laminar boundary region. This solution can be extended to include the diffusion of A for the same geometry and laminar flow. If D_{AB} in Eq. (6.8-8) is set equal to μ/ρ in Eq. (6.8-7) or $N_{Sc} = 1.0$ and the dependent variable c_A is changed to the dimensionless variable $(c_{AS} - c_A)/(c_{AS} - c_{A\infty})$, the similarity between the two equations is obvious. The solutions are identical and hence the dimensionless concentration and velocity profiles are the same and the velocity and concentration boundary layers are of equal thickness.

Using the method of Pohlhausen, the above solution was extended to cover fluids with a Schmidt number not equal to 1.0 (B4) and concentration profiles at various values of Schmidt numbers were obtained. The following relation between the velocity and concentration boundary layers holds in laminar flow.

$$\frac{\delta}{\delta_c} = N_{Sc}^{1/3} \tag{6.8-9}$$

Using this the final result was expressed as the Sherwood number and holds for a Schmidt number above 0.6 and for a low mass flux at the surface that does not alter the velocity profile.

$$N_{Sh} = 0.664 N_{Re,L}^{0.5} N_{Sc}^{1/3} \tag{6.8-10}$$

This checks the empirical correlation of Eq. (6.7-4) and recent experimental data (C2) and also justifies the J_D relation where k'_c is proportional to $D_{AB}^{2/3}$.

When the configuration is complicated or the flow is turbulent, an approximate integral analysis of the boundary layer developed by von Kármán can be made where the forms of the velocity and concentration profiles are assumed. Using a power series expression for the profiles, the resultant expression for laminar flow past a flat plate is quite similar to Eq. (6.8-10) and indicates that this method may be used with confidence in cases where an exact solution is unknown (B4). A similar method can be used for turbulent flow.

PROBLEMS

6-1 Bulk Flow Correction Factor

Component A in the flowing gas phase at a total pressure of P is diffusing to a catalyst surface, where it reacts instantaneously as follows: $A \rightarrow 2B$. Gas B diffuses back into the flowing gas phase. The partial pressure of A in the gas phase is p_{A1} atm. The experimental value of k_G was determined for A diffusing through a very dilute mixture of A in B with B not diffusing. Derive the equation to calculate k_{GR} for the above case of $A \to 2B$ from the value of k_G. The equation for k_{GR} is

$$N_A = k_{GR}(p_{A1} - p_{A2}) = \frac{k'_c}{\varphi_N}(c_{A1} - c_{A2})$$

6-2 Conversion of Mass Transfer Coefficients

Prove or show the following relationships: (a) conversion of k_G to k_y and k_c, (b) conversion of k_L to k_x and k'_x, (c) conversion of k'_c to k_y and k_G.

6-3 Distillation and Flux Ratios

A mixture of A and B at 1.0 atm total pressure and 100°C is being distilled from a liquid to the vapor state. The vapor composed of A and B is blown over the vaporizing surface at a velocity so that the value of k_c is predicted as 10.0 lb mole/hr-ft^2 (lb mole/ft^3) for a mixture of very dilute A in B at 1.0 atm total pressure. Since the mixture is dilute, $k_c p_{BM} = k'_c P$ and $p_{BM} = P$. So $k'_c = k_c$. At the interface, $p_{A1} = 0.80$ atm and $p_{B1} = 0.20$ atm, and in the gas phase $p_{A2} = 0.60$ atm and $p_{B2} = 0.40$ atm. $D_{AB} = 1.00$ cm^2/sec at 100°C and 1.0 atm.

(a) Calculate the flux N_A if the latent heats of A and B are equal. [*Hint*: $N_A + N_B = 0$ in this case.]

(b) Calculate the flux N_A if the latent heat of A is twice that of B. [*Hint*: How are N_A and N_B related?]

6-4 Dilute Mass Transfer Coefficients

Prove, using Eq. (6.2-4) for a very dilute concentration of A in B, that for any ratio of $N_A/(N_A + N_B)$ the value of φ_N is approximately 1.0 and $k'_G = k_G$.

6-5 Mass Transfer in a Wetted-Wall Tower

In a wetted-wall tower, H_2S is being absorbed from the air to the water at 1.50 atm total pressure and 30°C. The gas phase mass transfer coefficient k'_c has been predicted to be 11.3 lb mole/hr-ft^2 (lb mole/ft^3). At a given point in the tower the mole fraction of H_2S in the liquid at the interface is $2.0(10^{-5})$ and the partial pressure of H_2S in the air is 0.05 atm. Calculate the rate of absorption of H_2S in the water. The Henry's law equation is p_A (atm) $= 609 x_A$ (mole fraction).

6-6 Mass Transfer in Liquid and Solid Phases

In the drawing in Fig. P6-6 a fluid is flowing past the face (1) of a cylinder of diameter D, and the sides of the cylinder are insulated. At the rear face (2) the concentration is kept constant in the solid at the known value of c_{As2} g moles of A per cm^3.

The concentration in the bulk liquid is constant at c_{AL0}. At the surface at (1) the mass transfer coefficient is constant at k_L, where $N_A = k_L \Delta c_{AL}$. The solute A diffuses from the liquid, to the solid, and then through the solid according to Fick's law. The solute A in the liquid is dilute.

At point (1), the equilibrium relation is $c_{AL} = mc_{As}$, where m is a constant. The diffusivity of A in the solid is a variable and is $D_s = K_1 + K_2 c_{As}$, where K_1 and K_2 are constants.

Set up the basic differential equations at steady-state to calculate the flux N_A

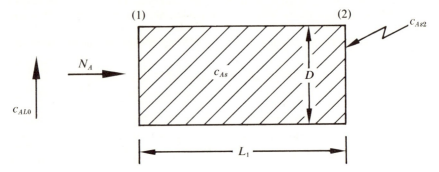

FIGURE P6-6 Mass transfer to a solid for Problem 6-6

in the liquid phase and in the solid. Integrate these equations. Number the final equations. State which are the unknowns and how to solve for them.

6-7 Mass Transfer, Interphase Transfer, Analog Computers

In Fig. P6-7 fluids are flowing past the front face and the rear face of a cylinder that is insulated at the sides but not at the circular ends, having an area of A_1 cm² and a radius of a_1. At steady-state, the concentration of solute A in the front in the liquid is c_{A0} g mole A/cm³ and is constant. At the rear the constant concentration is c_{A3} in the liquid. At the surface at (1) the mass transfer coefficient is k_L, where $N_A = k_L \Delta c_A$. The solute A diffuses into the porous solids I and II according to Fick's law. At the surface at (3) the mass transfer coefficient is infinite. D_I and D_II are given in cm²/sec. At the surface at (1) the distribution coefficient is 1.0.

At point (2), which is at the interface, $c_{A\mathrm{I}} = Kc_{A\mathrm{II}}$, where K is a constant.

(a) Set up the basic differential equations at steady-state and integrate them. Number the final integrated equations.

(b) Calculate the concentration at point (2) in I and II. $a_1 = 1.0$ cm, $D_\mathrm{I} = 4 \times 10^{-5}$ cm²/sec, $D_\mathrm{II} = 2 \times 10^{-5}$, $c_{A0} = 1 \times 10^{-4}$ g mole A/cm³, $c_{A3} = 0$, $k_L = 0.030$, $L_1 = 2$ cm, $L_2 = 20$ cm, $K = 2.5$.

(c) Set up the analog computer diagram to calculate the concentration in phase I as a function of distance. Put in initial conditions. Use only letters and not actual numbers.

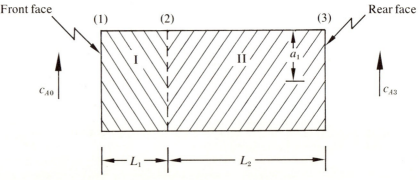

FIGURE P6-7 Diffusion through solids in series for Problem 6-7

(d) In parts (a), (b), and (c) above it is assumed that at point (2) c_{AI} and c_{AII} are in equilibrium. If it is assumed that they are not in equilibrium but that there is a resistance at this interface at point (2) of R, where $N_A = \Delta c_A / R$, set up the basic equations at steady-state and integrate them. Number the final equations. Use only letters here. In this case $K = 1.0$ at point (2).

6-8 Binary Turbulent Diffusion and Analog Solution

Using $N_A = -KN_B$ in Eq. (6.2-4), $(D_{AB} + \bar{\varepsilon}_M)$ is given as 10.0 cm²/sec, $x_{A1} = 0.80$, $x_{A2} = 0.20$, total pressure = 1.0 atm at 25°C, $K = 2.0$, $L = z_2 - z_1 = 0.10$ cm. Rewrite Eq. (6.2-4) in terms of x_A and draw the scaled analog computer diagram. A slow recorder is available so that the problem should take 20 sec. The voltage range available is -10 to $+10$ volts. Be sure to scale the initial conditions too.

6-9 Mass Transfer to a Falling Film

Carbon monoxide is being absorbed by a pure water film flowing down a wetted-wall tower 2.0 ft long and at a rate of 250 lb mass water/hr per ft width at 20°C. The gas is pure CO at 1.0 atm, and the solubility of CO in water at 1.0 atm and 20°C is $c_{Ai} = 0.647(10^{-4})$ lb mole CO/ft³ solution. Calculate the amount of CO absorbed. The diffusivity of CO must be estimated.

6-10 Mass Transfer to a Rising Bubble

A bubble of pure chlorine gas (A) at 1.0 atm and 0.5 cm in diameter is rising at a velocity of 20 cm/sec in pure water (B) at 16°C. The diffusivity $D_{AB} = 1.26 (10^{-5})$ cm²/sec. The saturation solubility of chlorine in water is 0.823 g mass Cl_2 per 100 g mass water. What is the average rate of absorption of the bubble, N_A, when the bubble is 0.5 cm in diameter?

6-11 Mass Transfer to a Packed Bed and J_D Factors

Pure water is flowing slowly through a packed bed of 0.251-in. diameter benzoic acid spheres at a rate of $V = 0.0701$ ft³/hr at 26.1°C. The mass transfer coefficient k_L is 0.0551 ft/hr and the solubility of benzoic acid in water is 0.00184 lb mole/ft³ solution.

(a) Calculate the area of bed needed in A ft² of surface of the spheres for the outlet concentration in the water to reach 0.00050 lb mole benzoic acid/ft³ solution.

(b) Assuming the mass transfer J_D factor is proportional to $N_{Re}^{-2/3}$, predict the outlet concentration if the flow rate is doubled.

6-12 Analog Computer Solution of Mass Transfer

For the conditions in Eq. (6.6-13), draw the analog computer solution to solve this. Assume $k_L' = K_1 + K_2 x_A$. Also scale the problem using $\tau = \alpha A$ and $e = \beta x_A$.

6-13 Digital Computer Solution

Write the Fortran IV program to solve Eq. (6.6-11). The k_L' varies as follows: $k_L' = (K_1 + K_2 x_A)^{0.5}$, where $K_1 = 0.00303$ and $K_2 = 0.020$. Use Simpson's method for the integration with 100 increments. The unknown is the area A. Knowns are $x_{A1} = 0$, $x_{A2} = 0.000200$, $x_{Ai} = 0.000532$, $V = 0.0701$ ft³/hr.

6-14 Surface Area in a Reactor

A reactor is packed with spheres having a diameter of D_p feet. The void volume is ϵ.

(a) Derive Eq. (6.7-15) for a, ft²/ft³ volume, and for n, the number of spheres/ft³ total volume.

(b) If $D_p = 0.10$ ft and $\epsilon = 0.38$, calculate n and a.

6-15 Mass Transfer to Definite Shapes

Estimate the value of the mass transfer coefficient of naphthalene vapor in a stream of flowing air at 52.6°C in a duct for the following shapes. The average velocity in the duct is 5.0 ft/sec at 52.6°C and 2.0 atm. The gas stream is essentially pure air. The diffusivity of naphthalene in air is 0.200 ft²/hr at 0°C and 1.0 atm.

(a) For air parallel to a flat plate 6.0 in. long.

(b) A single sphere 0.5 in. in diameter.

(c) A bed of spheres of 0.5 in. diameter and a void fraction of 0.40.

(d) A bed of cylinders of 0.5 in. diameter and 0.5 in. long and a void fraction of 0.40.

(e) Calculate the flux N_A and $(N_A A)$ for one cylinder in the bed in case (d) if the partial pressure of naphthalene in the air is zero and the bed is composed of naphthalene cylinders. The vapor pressure of naphthalene is 1.0 mm at 52.6°C.

6-16 Estimation of Mass Transfer Coefficients

For water flowing over a packed bed of 0.50-in. benzoic acid spheres an experimental value of ϵJ_D at 25°C was determined to be 0.13 at a Reynolds number of 10.0.

(a) Calculate the value of k_c.

(b) Estimate the value of k_c for mass transfer of dilute ammonia in water to a bed of spheres of the same size at the same temperature and Reynolds number.

6-17 Transfer from Naphthalene and Film Theory

A 1.0-in. diameter sphere of solid naphthalene was placed in an air stream at 215°F. The vapor pressure of naphthalene is 20 mm Hg at 215°F. The mass transfer coefficient k_G was determined experimentally to be 0.84 lb mole/hr-ft²-atm. The $D_{AB} = 0.346$ ft²/hr at 215°F. Calculate the flux, N_A, and the effective film thickness at 1.0 atm if the film theory is used.

6-18 Liquid Metals Mass Transfer

Predict the J_D and the mass transfer coefficient k_c for mercury at 26.5°C flowing over a bed of lead spheres of 0.0825-in. diameter with a void fraction of 0.499. The superficial velocity is 0.0721 ft/sec. The solubility of lead in mercury is 1.721 wt percent. The Schmidt number for the concentration range is 124.1. The viscosity of the solution is 1.06×10^{-3} lb mass/ft-sec and the density 13.53 g mass/cm³. Compare the predicted J_D with the experimental value of 0.076 (corrected to the J_D of Table 6.5-1) by Dunn et al. (D1) at these conditions in run 1 of their data. Use Eq. (6.7-19) to predict the J_D value.

Ans.: $J_D = 0.078$.

NOTATION

a	surface/volume ratio, ft^2/ft^3
A	cross-sectional area perpendicular to flux, cm^2 or ft^2
A	surface area, cm^2 or ft^2
c	total concentration, g mole/cm^3 or lb mole/ft^3
c_A	concentration of A, g mole A/cm^3 or lb mole A/ft^3
$\bar{c}_A$	mean concentration of A, g mole A/cm^3 or lb mole A/ft^3
c_p	heat capacity at constant pressure, cal/g mass-°C or Btu/lb mass-°F
D_{AB}	diffusivity, cm^2/sec or ft^2/hr
D_{Am}	mean or effective diffusivity, cm^2/sec or ft^2/hr [Eq. (1.10-12)]
D_p	diameter of particle, cm or ft
E_A	flux ratio defined by Eq. (6.3-18)
e	dependent variable on analog computer, volts
f	Fanning friction factor, dimensionless
g_c	980 g mass-cm/g force-sec^2 or 4.17×10^8 lb mass-ft/lb force-hr^2
g	acceleration due to gravity, cm/sec^2 or ft/hr^2
G	mass velocity $= v\rho$, g mass/sec-cm^2 or lb mass/hr-ft^2
G_M	molar mass velocity $= v\rho/M$, g mole/sec-cm^2 or lb mole/hr-ft^2
h	heat transfer coefficient, Btu/hr-ft^2-°F
J_A^*	flux of A relative to the molar average velocity, g mole A/sec-cm^2 or lb mole A/hr-ft^2
J_D	mass transfer J factor defined by Eq. (6.5-9), dimensionless
J_H	heat transfer J factor defined in Table 6.5-1, dimensionless
k	thermal conductivity, cal-cm/sec-cm^2-°C or Btu-ft/hr-ft^2-°F
$k_c, k_c', k_G,$ $k_L, k_x, \ldots$	mass transfer coefficients, g mole/sec-cm^2-concentration difference or lb mole/hr-ft^2-concentration difference (see Table 6.3-1)
K, K_1, K_2	constants
L	linear dimension, cm or ft
L	distance, cm or ft
L	length of plate in direction of flow, cm or ft
L	length of wetted-wall tower, cm or ft
M	molecular weight, g mass/g mole or lb mass/lb mole
M	mass, g mass or lb mass
n	number of particles
N_A	flux of A relative to a fixed point, g mole A/sec-cm^2 or lb mole A/hr-ft^2
N_{Gr}	Grashof number, dimensionless; see Table 6.5-1
N_{Nu}	Nusselt number, dimensionless; see Table 6.5-1
N_{Pe}	Peclet number $= N_{Re}N_{Sc}$, dimensionless; see Table 6.5-1
N_{Pr}	Prandtl number $= c_p\mu/k$, dimensionless
N_{Re}	Reynolds number $= Lv\rho/\mu$, dimensionless
N_R	flux ratio, $N_A/(N_A+N_B)$, dimensionless
N_{Sc}	Schmidt number $= \mu/\rho D_{AB}$, dimensionless
N_{St}	Stanton number $= N_{Sh}/(N_{Re}N_{Sc})$ dimensionless; see Table 6.5-1
N_{Sh}	Sherwood number $= k_c'L/D_{AB}$, dimensionless; see Table 6.5-1

310   Mass Transfer Coefficients in Laminar and Turbulent Flow

P total pressure, atm abs
p_A partial pressure of A, atm
p_{BM} ln mean of p_{B1} and p_{B2}, atm
r radius, cm or ft
R universal gas law constant, 82.06 cm³-atm/g mole-°K or 0.730 ft³-atm/lb mole-°R
s mean surface renewal factor, sec⁻¹
t time, sec or hr
t_L time of contact, sec or hr; see Eq. (6.4-15)
T absolute temperature, °K or °R
v mass average velocity relative to stationary coordinates, cm/sec or ft/hr
v' superficial mass average velocity in empty tube, cm/sec or ft/hr
V volume flow rate, cm³/sec or ft³/hr
W weight flow rate, g mass/sec or lb mass/hr
w weight flow rate, lb mass/hr or lb mass/hr-ft²
x_A mole fraction of A
x distance in x direction, cm or ft
x_{BM} ln mean of x_{B1} and x_{B2}; see Eq. (6.3-9) or (6.6-9)
y_A mole fraction of A
y distance in y direction, cm or ft
Y_A weight ratio, g mass A/g mass B or lb mass A/lb mass B
z distance in z direction, cm or ft

Greek Letters

α analog computer time scale factor in $\tau = \alpha t$
α constant, dimensionless
β analog computer amplitude scale factor in $e = \beta x$
β constant, dimensionless
γ constant, dimensionless
Γ flow rate, g mass/sec-cm width or lb mass/hr-ft width
δ thickness of boundary layer or laminar film, cm
δ distance of diffusion path, cm or ft
ϵ void fraction in bed, dimensionless
ϵ turbulent transfer coefficient, cm/sec or ft/hr
ϵ_M mass transfer turbulent diffusivity, cm²/sec or ft²/hr
$\bar{\epsilon}_M$ mean mass transfer turbulent diffusivity, cm²/sec or ft²/hr
θ time, sec or hr
μ viscosity, g mass/cm-sec or lb mass/ft-hr
π dimensionless group
ρ density, g mass/cm³ or lb mass/ft³
τ shear stress, lb force/ft²
τ analog computer time, sec
φ_N bulk flow correction factor; see Eq. (6.3-3)

Subscripts

A	component A
ave	average
B	component B
c	concentration
exp	experimental
f	film
F	film
G	bulk gas phase
i	interface or wall
i	component i
max	maximum
m	effective or average
M	mass transfer
N	bulk flow
n	last of n components
LM	ln mean
S	surface
x, y, z	in the $x, y,$ or z direction
1	beginning of diffusion path
2	end of diffusion path
∞	infinity

REFERENCES

(B1) C. H. Bedingfield Jr. and T. B. Drew, *Ind. Eng. Chem.* **42**, 1164 (1950).
(B2) R. B. Bird, W. E. Stewart, and E. N. Lightfoot, *Transport Phenomena*. New York: John Wiley & Sons, Inc., 1960.
(B3) J. A. R. Bennett and J. B. Lewis, *AIChEJ* **4**, 418 (1958).
(B4) C. O. Bennett and J. E. Myers, *Momentum, Heat and Mass Transfer*. New York: McGraw-Hill, Inc., 1962.
(C1) S. D. Conte, *Elementary Numerical Analysis*. New York: McGraw-Hill, Inc., 1965.
(C2) W. J. Christian and S. P. Kezios, *AIChEJ* **5**, 61 (1959).
(D1) W. E. Dunn, C. F. Bonilla, C. Ferstenberg, and B. Gross, *AIChEJ* **2**, 184 (1956).
(D2) P. V. Danckwerts, *Ind. Eng. Chem.* **43**, 1460 (1951).
(D3) W. E. Dobbins, in pt. 2-1, McCabe and Eckenfelder, eds., *Biological Treatment of Sewage and Industrial Wastes*. New York: Van Nostrand-Reinhold Company, 1956.
(E1) Spyros Evnochides and George Thodos, *AIChEJ* **7**, 78 (1961).
(G1) E. R. Gilliland and T. K. Sherwood, *Ind. Eng. Chem.* **26**, 516 (1934).
(G2) F. H. Garner and R. D. Suckling, *AIChEJ* **4**, 114 (1958).
(G3) A. S. Gupta and George Thodos, *AIChEJ* **9**, 751 (1963).
(G4) ———, *Ind. Eng. Chem. Fund.* **3**, 218 (1964).

(G5) _____, *Chem. Eng. Progr.* **58** (No. 7), 58 (1962).
(G6) _____, *AIChEJ* **8**, 609 (1962).
(G7) F. H. Garner and R. W. Grafton, *Proc. Roy. Soc. (London)* **A224**, 64 (1954).
(H1) Ralph Higbie, *Trans. AIChE* **31**, 365 (1935).
(H2) D. Hammerton and F. H. Garner, *Trans. Inst. Chem. Engrs. (London)* **32**, S18 (1954).
(H3) Peter Harriott, *Chem. E. Sci.* **17**, 149 (1962).
(L1) W. H. Linton Jr. and T. K. Sherwood, *Chem. Eng. Progr.* **46**, 258 (1950).
(L2) Michael Litt and S. K. Friedlander, *AIChEJ* **5**, 483 (1959).
(L3) M. Linton and L. L. Sutherland, *Chem. Eng. Sci.* **12**, 214 (1960).
(L4) L. Lapidus, *Digital Computation for Chemical Engineers.* New York: McGraw-Hill, Inc., 1962.
(M1) H. S. Mickley, T. K. Sherwood, and C. E. Reed, *Applied Mathematics in Chemical Engineering*, 2d ed. New York: McGraw-Hill, Inc., 1957.
(P1) J. H. Perry, *Chemical Engineers' Handbook*, 3d ed. New York: McGraw-Hill, Inc., 1950.
(R1) O. Reynolds, *Proc. Manchester Lit. Phil. Soc.* **14**, 7 (1874).
(R2) R. C. Reid and T. K. Sherwood, *The Properties of Gases and Liquids*, 2d ed. New York: McGraw-Hill, Inc., 1966.
(S1) W. E. Stewart, Sc.D. thesis, Massachusettes Institute of Technology, Cambridge, Mass. (1950).
(S2) T. K. Sherwood and R. L. Pigford, *Absorption and Extraction*, 2d ed. New York: McGraw-Hill, Inc., 1952.
(S3) C. V. Sternling and L. E. Scriven, *AIChEJ* **5**, 514 (1959).
(S4) L. R. Steele and C. J. Geankoplis, *AIChEJ* **5**, 178 (1959).
(S5) R. L. Steinberger and R. E. Treybal, *AIChEJ* **6**, 227 (1960).
(T1) W. J. Thomas and S. Portalski, *Ind. Eng. Chem.* **50**, 1081 (1958).
(T2) E. D. Traylor, Leslie Burris, and C. J. Geankoplis, *Ind. Eng. Chem. Fund.* **4**, 119 (1965).
(T3) H. L. Toor and J. M. Marshello, *AIChEJ* **4**, 97 (1958).
(T4) R. E. Treybal, *Mass-Transfer Operations*, 2d ed. New York: McGraw-Hill, Inc., 1968.
(W1) E. J. Wilson and C. J. Geankoplis, *Ind. Eng. Chem. Fund.* **5**, 9 (1966).

7 Interphase Mass Transport

7.1 INTRODUCTION TO INTERPHASE MASS TRANSPORT

In many industrial processes mass, heat, or momentum transfer occurs from one phase to another. In previous chapters on mass transfer we have considered primarily the transfer of a solute in one phase. For example, we considered the dissolving of benzoic acid from the surface of a sphere and its diffusing through one fluid phase. In mass transfer through two phases, the solute may diffuse through a gas phase and then diffuse through and be absorbed in an adjacent and immiscible liquid phase. In a heat exchanger for two different fluids heat is being transferred from a hot fluid phase through a tube wall to a cold fluid phase.

In heat transfer between two phases the contact area between the phases is usually a solid wall and the area is known. Also, the two phases are kept separated. In mass transfer the two phases are usually in direct contact with each other in a packed, tray, or spray type tower, and the interfacial area is less well defined. In two-phase mass transfer there will exist a concentration gradient in each phase, causing the diffusion to occur. If there is no concentration gradient, diffusion will stop and the two phases will be in equilibrium. Hence, in mass transfer equilibrium relations between the two phases are also important and more complex than the equilibrium relations in heat transfer.

7.2 EQUILIBRIUM RELATIONS BETWEEN PHASES

7.2A Experimental Determination

As an example of the experimental determination of the equilibrium distribution between two phases, the system SO_2, air, and water will be considered. A given amount of gaseous SO_2, gaseous air, and water are put in a closed container and shaken repeatedly until equilibrium is reached at a given temperature. The partial pressure p_A or mole fraction y_A of SO_2 (A) in the gas and the concentration or mole fraction x_A of SO_2 in the liquid are determined by sampling each bulk phase. More SO_2 is added and the process repeated until a complete equilibrium plot of p_A vs. x_A is determined. Figure 7.2-1 shows a plot of the partial pressure of SO_2, p_A, in atm in the vapor in equilibrium with the mole fraction of SO_2, x_A, in the liquid at 20°C.

7.2B Use of Phase Rule and Equilibrium Relations

1. *Phase Rule.* The equilibrium between two phases is restricted as follows by the phase rule:

$$P + F = C + 2 \qquad (7.2\text{-}1)$$

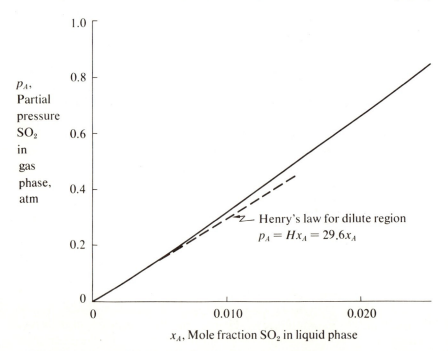

FIGURE 7.2-1 Equilibrium data for SO_2-water system at 20°C

7.2 Equilibrium Relations between Phases

where P is the number of phases at equilibrium, C is the number of components in the two phases when no chemical reactions are occurring, and F is the number of variants or degrees of freedom of the system. For example, for the SO_2-air-water system there are two phases and three components (considering air as one inert component); hence, $F = 3$. This means that if the total pressure and temperature are set, one variable is left that can be arbitrarily set. Hence, if the composition of x_A in the liquid is set, y_A or p_A is automatically determined.

2. *Equilibrium between Phases.* The phase rule does not tell us the partial pressure p_A in equilibrium with the selected x_A. This must be determined experimentally. Each system will have a different relation between the vapor and liquid compositions at equilibrium. Also, the two phases do not have to be a gas and a liquid. For example, the equilibrium distribution of acetic acid between two solvents, water and isopropyl ether, has been determined for extraction processes. Where we discussed the diffusion of gases in solids, an equilibrium partial pressure of hydrogen existed between the gas phase and the concentration of hydrogen in the solid rubber phase. In liquid-solid systems, such as used in leaching processes, an equilibrium will exist between the concentration of solute A in the liquid phase and the concentration of A in the solid phase. Even in solid-solid systems an equilibrium exists of the concentration of solute A in one solid phase and the concentration of solute A in a second solid phase that is immiscible with the first solid phase.

3. *Henry's Law.* In many cases the equilibrium relation between the partial pressure of the solute A in the gas phase and the concentration in the liquid phase can be expressed by a straight-line equation called Henry's law at low concentrations:

$$p_A = H x_A \qquad (7.2\text{-}2)$$

where p_A is the partial pressure in mm or atm and H is Henry's law constant in mm/mole fraction or atm/mole fraction for the given system. This relationship generally holds at low partial pressures and low values of x_A. Equation (7.2-2) can be converted to y_A by dividing p_A by P; then H' is the Henry's law constant as mole fraction in vapor/mole fraction in liquid:

$$y_A = H' x_A \qquad (7.2\text{-}3)$$

Care should be exercised in using H', because it depends on total pressure P while H does not.

In Fig. 7.2-1 the data follow Henry's law up to a low concentration of x_A of about 0.005, giving an H value of 29.6 atm/mole fraction or H of 22,500 mm/mole fraction.

When the concentrations are low, most data can be represented by

Henry's law, with the values of H depending on the system and the temperature. In general, up to a total pressure P of about 5 atm the value of p_A in equilibrium with x_A in Eq. (7.2-2) is independent of P. If Henry's law holds for a system, then it usually holds for values of p_A up to about 1 atm (H1, P1).

In certain systems that are ideal the mixture will follow Raoult's law, where H in Eq. (7.2-2) is replaced by the vapor pressure of pure A at the given temperature. This is discussed more fully in Section 9.2A.

Example 7.2-1 Henry's Law Constant and Equilibrium Relations

The Henry's law constant for CO_2 in air-water at 30°C is $H = 0.186 \times 10^4$ atm/mole fraction. If the partial pressure of CO_2 in air is 100 mm and the total pressure is 850 mm, what is the value of x_A in equilibrium and y_A?

Solution

$p_A = 100$ mm/760 $= 0.1315$ atm. Then, using Eq. (7.2-2),

$$p_A = Hx_A \quad \text{or} \quad 0.1315 = (0.186 \times 10^4)x_A$$

Hence, $x_A = 7.06 \times 10^{-5}$. The value of $y_A = 100/850 = 0.1163$.

Example 7.2-2 Concentration of Dissolved Oxygen

The partial pressure of oxygen in air is 0.21 atm. What will be the concentration of oxygen dissolved in water at 25°C? Henry's law constant is 4.38×10^4 atm/mole fraction.

Solution

Using Eq. (7.2-2), $p_A = Hx_A$ or $0.21 = 4.38 \times 10^4 x_A$. Then $x_A = 4.80 \times 10^{-6}$. This means 4.80×10^{-6} moles O_2 are dissolved in 1.00 moles water plus oxygen or 0.000853 parts O_2 per 100 parts water.

4. *Distribution in Liquid-Liquid Systems.* When a solute is distributed between two essentially immiscible phases, the distribution coefficient K is given as

$$K = \frac{c_{A(B)}}{c_{A(C)}} \tag{7.2-4}$$

where $c_{A(B)}$ is the concentration of A in g mole A per cm^3 of solution of B and $c_{A(C)}$ is the concentration of A in solution C. The constant K is dependent on concentrations, but at dilute concentrations it is relatively constant.

7.2 Equilibrium Relations between Phases

Example 7.2-3 Equilibrium Concentration in Extraction

The distribution coefficient $K = $ g mole acetic acid per cm³ in benzene phase/g mole acetic acid per cm³ in water phase is 0.023 at 25°C. Benzene (B) and water phases (C) are essentially immiscible. If 1 liter of benzene containing 0.10 g mole acetic acid is shaken up with 2 liters of water, what are the final equilibrium concentrations?

Solution

Let $c_{A(B)}$ be concentration of A g mole A/cm^3 in B phase and $c_{A(C)}$ in C phase at equilibrium. Then $K = c_{A(B)}/c_{A(C)}$. At equilibrium, the total number of moles of $A = 0.10$. A material balance gives

$$0.10 = c_{A(B)}(1000) + c_{A(C)}(2000)$$

Also, $K = 0.023 = c_{A(B)}/c_{A(C)}$. Solving the two equations for c_A, $c_{A(B)} = 0.00000114$ g mole A/cm^3, $c_{A(C)} = 0.0000494$ g mole A/cm^3.

5. *Distribution and Activity Coefficients.* When there is an equilibrium distribution of component A between two phases, such as a gas and liquid, liquid and liquid, or gas and solid, the chemical potentials of A in both phases are equal. The new symbol $a_{A(B)}$ is the activity of A in phase B, and $a_{A(C)}$ is the activity of A in phase C. The activity coefficient $\gamma_{A(B)}$ times $x_{A(B)}$ is equal to $a_{A(B)}$. For a gas-liquid two-phase system and using $x_{A(B)}$ as y_A and $x_{A(C)}$ as x_A in Eq. (7.2-3),

$$x_{A(B)} = H' x_{A(C)} \quad (7.2\text{-}5)$$

or

$$H' = \frac{x_{A(B)}}{x_{A(C)}}$$

Since $a_{A(B)} = x_{A(B)}\gamma_{A(B)}$, $a_{A(C)} = x_{A(C)}\gamma_{A(C)}$, then

$$H' = \frac{\gamma_{A(C)}}{\gamma_{A(B)}}\left(\frac{a_{A(B)}}{a_{A(C)}}\right) \quad (7.2\text{-}6)$$

It can be shown thermodynamically that $a_{A(B)}/a_{A(C)}$ is constant at constant T. Hence, H' is constant only if the γ_A values are constant. In heat transfer with two immiscible phases B and C, which can be gas-solid, liquid-liquid, gas-liquid, solid-solid, or liquid-solid, the distribution coefficients between temperatures T_B and T_C have been shown thermodynamically to be 1.0.

$$\frac{T_B}{T_C} = 1.0 \quad (7.2\text{-}7)$$

7.3 MASS TRANSPORT BETWEEN TWO PHASES

7.3A Concentration Profiles

Present in the majority of mass transfer systems are usually two phases, which are essentially immiscible in each other, and an interface between the two phases. Suppose a solute A is in the bulk gas phase G and is diffusing into the liquid phase L. Since the solute A is being transferred from one phase, across the interface, and to another phase, a concentration gradient must exist to cause diffusion through the resistances in each phase. This is shown in Fig. 7.3-1, where the average or bulk concentration of A in the gas phase in mole fraction units is y_{AG}, and x_{AL} in the liquid phase in mole fraction units.

The concentration in the bulk gas phase is y_{AG}, and it decreases to y_{Ai} at the interface. In the liquid phase the concentration starts at x_{Ai} and falls to x_{AL}. The bulk phase concentrations y_{AG} and x_{AL} are certainly not at equilibrium; otherwise there would be no driving force and diffusion would not occur.

At the interface, since it appears that there would be no resistance to transfer across the interface, the concentrations y_{Ai} and x_{Ai} are in equilibrium and are related by the equilibrium distribution relation

$$y_{Ai} = f(x_{Ai}) \tag{7.3-1}$$

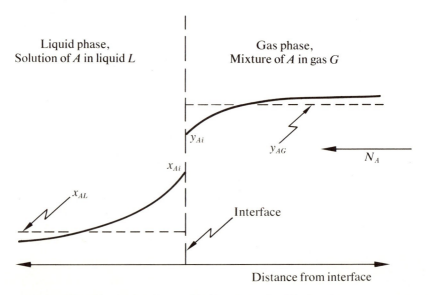

FIGURE 7.3-1 Concentration profile of solute A diffusing from one phase to another

where y_{Ai} is a function of x_{Ai}, the two being related by an equilibrium plot such as that in Fig. 7.2-1, where y_A has been converted to p_A by $p_A = y_A P$. If the system follows Henry's law, then $y_A P$ or p_A and x_A are related by Eq. (7.2-2).

The resistance at the interface has been shown to be essentially negligible for most cases of mass transfer, such as absorption of many common gases from air to aqueous solutions and extraction of organic solutes from one phase to another, where chemical reactions do not occur. There are, however, a number of notable exceptions. Certain substances or surface active compounds may concentrate at the interface of a liquid and cause an "interfacial" resistance that slows down the rate of diffusion of solute molecules A across this boundary. Cetyl alcohol is such a substance, concentrating on the surface of water and slowing down the diffusion or evaporation of water molecules to the air. In mass transfer of a metal solute A from a liquid metal to a solid metal surface, the solute A must deposit at the surface and then form a crystalline structure. This finite time required to form regular crystals causes an "interfacial resistance." Theories to predict when interfacial resistance may occur are still obscure and very uncertain.

7.3B Interface Compositions and Film Mass Transfer Coefficients

1. *Equimolar Counterdiffusion.* For the case of equimolar counter-diffusion the concentration conditions shown in Fig. 7.3-1 can be plotted graphically as in Fig. 7.3-2. In this figure point P represents the bulk phase compositions y_{AG} and x_{AL}, and point M represents the concentrations y_{Ai} and x_{Ai} at the interface. The equations (L1) for the flux of A when A is diffusing from a gas to a liquid and there is equimolar counter-diffusion of B from the liquid to the gas are given by

$$N_A = k'_y(y_{AG} - y_{Ai}) = k'_x(x_{Ai} - x_{AL}) \qquad (7.3\text{-}2)$$

The values $(y_{AG} - y_{Ai})$ and $(x_{Ai} - x_{AL})$ are the differences in concentration or driving forces in each phase. Rearranging Eq. (7.3-2),

$$\frac{-k'_x}{k'_y} = \frac{y_{AG} - y_{Ai}}{x_{AL} - x_{Ai}} \qquad (7.3\text{-}3)$$

Hence, the slope of the line PM is $-k'_x/k'_y$. This means that if the two film mass transfer coefficients are known, then the interface compositions can be determined by line PM with slope $-k'_x/k'_y$. Or, if the equilibrium line can be represented by an equation of

$$y_{Ai} = f(x_{Ai}) \qquad (7.3\text{-}1)$$

then the compositions y_{Ai} and x_{Ai} can be solved for directly by substituting Eq. (7.3-1) into (7.3-3).

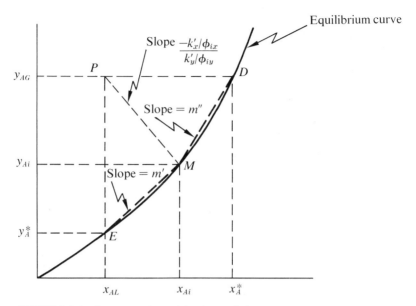

FIGURE 7.3-2 Concentration driving forces in interphase mass transfer

It should be emphasized that only the bulk concentrations y_{AG} and y_{AL} can be determined experimentally by sampling the mixed bulk liquid phase and gas phase and analyzing the samples. The concentrations at the interface cannot be determined experimentally because of the impossibility of sampling right at the interface. These concentrations must be determined by Eq. (7.3-3).

2. *Diffusion of A through Stagnant B.* For the case of A diffusing through stagnant B the concentration conditions can also be plotted as in Fig. 7.3-2, where P represents the bulk phase compositions and M the interface. The equations for A diffusing through a stagnant gas and through a stagnant liquid are given by

$$N_A = k_y(y_{AG} - y_{Ai}) = k_x(x_{Ai} - x_{AL}) \tag{7.3-4}$$

But $k_y = k'_y/(1-y_A)_{iM}$ and $k_x = k'_x/(1-x_A)_{iM}$, where

$$(1-y_A)_{iM} = \frac{(1-y_{Ai}) - (1-y_{AG})}{\ln\left(\dfrac{1-y_{Ai}}{1-y_{AG}}\right)} \tag{7.3-5}$$

$$(1-x_A)_{iM} = \frac{(1-x_{AL}) - (1-x_{Ai})}{\ln\left(\dfrac{1-x_{AL}}{1-x_{Ai}}\right)} \tag{7.3-6}$$

Then,

$$N_A = \frac{k'_y}{(1-y_A)_{iM}}(y_{AG} - y_{Ai}) = \frac{k'_x}{(1-x_A)_{iM}}(x_{Ai} - x_{AL}) \quad (7.3\text{-}7)$$

Solving as before,

$$\frac{-k'_x/(1-x_A)_{iM}}{k'_y/(1-y_A)_{iM}} = \frac{y_{AG} - y_{Ai}}{x_{AL} - x_{Ai}} \quad (7.3\text{-}8)$$

Hence, the slope of the line *PM* for the case of *A* diffusing through stagnant *B* is given by the left side of Eq. (7.3-8).

It should be noted that the slope of Eq. (7.3-8) differs from that of Eq. (7.3-3) for equimolar counterdiffusion by the bulk flow correction terms $(1-y_A)_{iM}$ and $(1-x_A)_{iM}$. When *A* is diffusing through stagnant *B* and the solutions are dilute, the bulk flow correction terms are approximately 1.0, and Eq. (7.3-3) can also be used instead of Eq. (7.3-8). Many texts do not specifically point out the differences between Eq. (7.3-3) and (7.3-8) and use Eq. (7.3-3).

The use of Eq. (7.3-8) to get the slope is by necessity trial and error, because the left-hand side contains the interface concentrations y_{Ai} and x_{Ai} that are being sought. This is easily handled by using Eq. (7.3-3) as the first trial. Then, using those preliminary values of y_{Ai} and x_{Ai}, a value of the left-hand side of Eq. (7.3-8) is computed and a new slope drawn to get new values of y_{Ai} and x_{Ai}. The second trial is repeated until the values of y_{Ai} and x_{Ai} with successive trials do not change. Three trials are usually sufficient.

3. *General Case for Diffusion of A and B.* For the general case of both *A* and *B* diffusing, the concentration can again be represented in Fig. 7.3-2, where *P* represents the bulk phase and *M* the interface compositions (K1, T1). The general equations for both *A* and *B* diffusing as given in Table 6.3-1 are

$$N_A = \frac{k'_y}{\varphi_{iy}}(y_{AG} - y_{Ai}) = \frac{k'_x}{\varphi_{ix}}(x_{Ai} - x_{AL}) \quad (7.3\text{-}9)$$

where

$$\varphi_{iy} = \frac{[N_A/(N_A+N_B) - y_{Ai}] - [N_A/(N_A+N_B) - y_{AG}]}{\frac{N_A}{N_A+N_B} \ln\left[\frac{N_A/(N_A+N_B) - y_{Ai}}{N_A/(N_A+N_B) - y_{AG}}\right]} \quad (7.3\text{-}10)$$

$$\varphi_{ix} = \frac{[N_A/(N_A+N_B) - x_{AL}] - [N_A/(N_A+N_B) - x_{Ai}]}{\frac{N_A}{N_A+N_B} \ln\left[\frac{N_A/(N_A+N_B) - x_{AL}}{N_A/(N_A+N_B) - x_{Ai}}\right]} \quad (7.3\text{-}11)$$

Solving as before,

$$\frac{-k'_x/\varphi_{ix}}{k'_y/\varphi_{iy}} = \frac{y_{AG} - y_{Ai}}{x_{AL} - x_{Ai}} \qquad (7.3\text{-}12)$$

The slope of the line PM in Fig. 7.3-2 is given by the left side of Eq. (7.3-12). As before, the equation is trial and error, where the interface compositions y_{Ai} and x_{Ai} must be assumed for the first trial. The ratio of the fluxes $N_A/(N_A + N_B)$ must be known or set. For equimolar counterdiffusion or very dilute solutions Eq. (7.3-12) reduces to Eq. (7.3-3), and for A diffusing through stagnant B it reduces to Eq. (7.3-8).

Example 7.3-1 Effect of Flux Ratio on Interface Compositions

Component A is being separated from a gas mixture of A and B in a wetted-wall absorption tower with the liquid mixture flowing downward along the wall. At a certain point in the tower the bulk gas concentration y_A is 0.70 mole fraction and the bulk liquid x_A is 0.10 mole fraction. At the temperature 30°C and 1.0 atm at which the tower is operating, the equilibrium data are given below as mole fractions.

x_A	y_A
0	0
0.10	0.155
0.20	0.340
0.25	0.465
0.30	0.650
0.34	0.905

From the correlations for wetted-wall towers and for dilute solutions, the film mass transfer coefficient for A is $k_x = 1.55$ lb mole/hr-ft^2-mole fraction and $k_y = 1.03$ lb mole/hr-ft^2-mole fraction. Calculate the interface compositions y_{Ai} and x_{Ai} and the local mass transfer flux N_A for the following cases: (a) Equimolar counterdiffusion of A and B through the gas and liquid films. The liquid is also composed of A and B. (b) Diffusion of A through stagnant B, where the liquid is a nondiffusing solvent. (c) Diffusion of A and B where the liquid is also composed of A and B. The fluxes are related by $N_B = -\tfrac{1}{2}N_A$.

Solution

In part (a), the mass transfer coefficients k_x and k_y are for diffusion of A through stagnant B. Since the solutions are dilute, x_{BM} and y_{BM} are approximately 1.0 and the coefficients are then the same as k'_y and k'_x. The equation to use to calculate the slope of the line PM in Fig. 7.3-3 is

7.3 Mass Transport between Two Phases

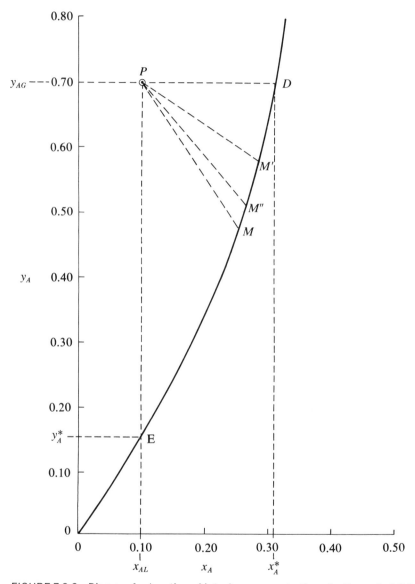

FIGURE 7.3-3 Diagram for location of interface concentrations for Example 7.3-1

Eq. (7.3-3) when $y_{AG} = 0.70$ and $x_{AL} = 0.10$.

$$\frac{-k'_x}{k'_y} = -\frac{1.55}{1.03} = -1.506 = \frac{y_{AG} - y_{Ai}}{x_{AL} - x_{Ai}} = \frac{0.70 - y_{Ai}}{0.10 - x_{Ai}}$$

A line through point P with a slope of -1.506 is plotted in Fig. 7.3-3.
At the intersection of this line with the equilibrium line at M, $y_{Ai} =$

324 Interphase Mass Transport

0.470 and $x_{Ai} = 0.253$. To calculate the flux, Eq. (7.3-2) is used.

$$N_A = k'_y(y_{AG} - y_{Ai}) = 1.03(0.700 - 0.470) = 0.237 \text{ lb mole/hr-ft}^2$$

$$N_A = k'_x(x_{Ai} - x_{AL}) = 1.55(0.253 - 0.100) = 0.237 \text{ lb mole/hr-ft}^2$$

[Ans. to (a)]

For part (b), Eqs. (7.3-5), (7.3-6), and (7.3-8) must be used. To calculate $(1-y_A)_{iM}$ and $(1-x_A)_{iM}$ a preliminary estimate of y_{Ai} and x_{Ai} of 0.470 and 0.253, respectively, will be used. Then

$$(1-y_A)_{iM} = \frac{(1-y_{Ai}) - (1-y_{AG})}{\ln\left(\dfrac{1-y_{Ai}}{1-y_{AG}}\right)} = \frac{(1-0.470) - (1-0.700)}{\ln\left(\dfrac{1-0.470}{1-0.700}\right)} = 0.407$$

$$(1-x_A)_{iM} = \frac{(1-x_{AL}) - (1-x_{Ai})}{\ln\left(\dfrac{1-x_{AL}}{1-x_{Ai}}\right)} = \frac{(1-0.100) - (1-0.253)}{\ln\left(\dfrac{1-0.100}{1-0.253}\right)} = 0.820$$

Substituting the above values into Eq. (7.3-8),

$$\frac{-k'_x/(1-x_A)_{iM}}{k'_y/(1-y_A)_{iM}} = -\frac{1.55/0.820}{1.03/0.407} = -0.745$$

A line through P with a slope of -0.745 is plotted in Fig. 7.3-3 and the values of $y_{Ai} = 0.564$ and $x_{Ai} = 0.281$ are read off the intersection of this line with the equilibrium line. Using these values of y_{Ai} and x_{Ai}, a new value of the slope of -0.683 is plotted in Fig. 7.3-3. The new values of $y_{Ai} = 0.573$ and $x_{Ai} = 0.284$ are read off the plot. Using these new values again, a slope of -0.673 is calculated and plotted. The final values of $y_{Ai} = 0.575$ and $x_{Ai} = 0.285$ are determined. This final line is plotted as PM' in Fig. 7.3-3. To calculate the flux, Eq. (7.3-7) is used.

$$N_A = \frac{k'_y}{(1-y_A)_{iM}}(y_{AG} - y_{Ai}) = \frac{1.03}{0.360}(0.700 - 0.575)$$

$$= 0.358 \text{ lb mole/hr-ft}^2$$

$$N_A = \frac{k'_x}{(1-x_A)_{iM}}(x_{Ai} - x_{AL}) = \frac{1.55}{0.801}(0.285 - 0.100)$$

$$= 0.358 \text{ lb mole/hr-ft}^2 \quad \text{[Ans. to (b)]}$$

For part (c), Eqs. (7.3-9), (7.3-10), (7.3-11), and (7.3-12) must be used. $N_B = -\frac{1}{2}N_A$. Hence,

$$\frac{N_A}{N_A + N_B} = \frac{N_A}{N_A - \frac{1}{2}N_A} = 2.0$$

For the first trial assume $y_{Ai} = 0.470$ and $x_{Ai} = 0.253$ as for part (a).

7.3 Mass Transport between Two Phases

Then for φ_{iy},

$$\varphi_{iy} = \frac{(2.0-0.470)-(2.0-0.700)}{2.0 \ln\left(\frac{2.0-0.470}{2.0-0.700}\right)} = 0.700$$

$$\varphi_{ix} = \frac{(2.0-0.100)-(2.0-0.253)}{2.0 \ln\left(\frac{2.0-0.100}{2.0-0.253}\right)} = 0.910$$

$$\frac{-k'_x/\varphi_{ix}}{k'_y/\varphi_{iy}} = -\frac{1.55/0.910}{1.03/0.700} = -1.16 = \frac{y_{AG}-y_{Ai}}{x_{AL}-x_{Ai}}$$

A line with this slope is plotted through P and it intersects the equilibrium line at $y_{Ai} = 0.510$ and $x_{Ai} = 0.265$. Using these new values, a new slope of -1.157 was calculated. The final line PM'' is drawn in Fig. 7.3-3, and the final values of $y_{Ai} = 0.510$ and $x_{Ai} = 0.265$ are obtained. Then,

$$N_A = \frac{k'_y}{\varphi_{iy}}(y_{AG}-y_{Ai}) = \frac{1.03}{0.695}(0.700-0.510)$$

$$= 0.282 \text{ lb mole/hr-ft}^2$$

$$N_A = \frac{k'_x}{\varphi_{ix}}(x_{Ai}-x_{AG}) = \frac{1.55}{0.905}(0.265-0.100)$$

$$= 0.282 \text{ lb mole/hr-ft}^2 \quad \text{[Ans. to (c)]}$$

7.3C Overall Mass Transfer Coefficients and Driving Forces

1. *Introduction to Overall Mass Transfer Coefficients.* Experimentally the single phase or film mass transfer coefficients k'_y and k'_x or k_y and k_x are often difficult to measure except in special experiments, designed so that the concentration difference across one phase is small and can be neglected. For convenience then the overall mass transfer coefficients K'_y and K'_x are measured based on the gas phase or on the liquid phase. This is analogous to heat transfer, where often film heat transfer coefficients are not determined but an overall heat transfer coefficient is obtained based on the inside or on the outside areas.

The overall mass transfer coefficient K'_y can be defined by

$$N_A = K'_y(y_{AG}-y_A^*) \tag{7.3-13}$$

where K'_y is the overall mass transfer coefficient based on the overall gas phase driving force and y_A^* is the value in equilibrium with x_{AL}. The y_A^* is a measure of the x_{AL} and it is on the same basis as y_{AG}. The value of y_A^* is shown in Fig. 7.3-2 in equilibrium with x_{AL}.

The overall mass transfer coefficient K'_x can be defined as

$$N_A = K'_x(x_A^*-x_{AL}) \tag{7.3-14}$$

where K'_x is the overall mass transfer coefficient based on the overall liquid phase driving force and x_A^*, that is in equilibrium with y_{AG}, is a measure of the y_{AG}, and is on the same basis as x_{AL}. This value of x_A^* is shown in Fig. 7.3-2 in equilibrium with y_{AG}.

2. *Equimolar Counterdiffusion and/or Diffusion in Dilute Solutions.* When equimolar counterdiffusion is occurring, Eq. (7.3-2) holds, or when the solutions are quite dilute, then φ_{iy} and φ_{ix} in Eq. (7.3-9) are 1.0 and Eqs. (7.3-2) and (7.3-9) become identical. This result is

$$N_A = \frac{k'_y}{1.0}(y_{AG} - y_{Ai}) = \frac{k'_x}{1.0}(x_{Ai} - x_{AL}) \qquad (7.3\text{-}15)$$

From the geometry shown in Fig. 7.3-2,

$$y_{AG} - y_A^* = (y_{AG} - y_{Ai}) + (y_{Ai} - y_A^*) \qquad (7.3\text{-}16)$$

The slope m' between points E and M can be given as

$$m' = \frac{y_{Ai} - y_A^*}{x_{Ai} - x_{AL}} \qquad (7.3\text{-}17)$$

Then,

$$m'(x_{Ai} - x_{AL}) = (y_{Ai} - y_A^*) \qquad (7.3\text{-}18)$$

Substituting Eq. (7.3-18) into (7.3-16) for $(y_{Ai} - y_A^*)$,

$$y_{AG} - y_A^* = (y_{AG} - y_{Ai}) + m'(x_{Ai} - x_{AL}) \qquad (7.3\text{-}19)$$

Then, on substituting Eqs. (7.3-13) and (7.3-15) into (7.3-19),

$$\frac{N_A}{K'_y} = \frac{N_A}{k'_y} + \frac{m' N_A}{k'_x} \qquad (7.3\text{-}20)$$

Canceling out N_A, the final equation relating the overall gas phase coefficient to the individual gas film coefficient and the liquid film coefficient is

$$\frac{1}{K'_y} = \frac{1}{k'_y} + \frac{m'}{k'_x} \qquad (7.3\text{-}21)$$

where m' is the slope of the chord between points E and M in Fig. 7.3-2. The left-hand side of Eq. (7.3-21) can be looked at as the total resistance based on the overall gas driving force, which is equal to the sum of the gas film resistance $1/k'_y$ and the liquid film resistance m'/k'_x. This is very similar to the case in heat transfer, where the total resistance is equal to the sum of the individual film resistances.

In a similar manner, from the geometry shown in Fig. 7.3-2,

$$(x_A^* - x_{AL}) = (x_A^* - x_{Ai}) + (x_{Ai} - x_{AL}) \qquad (7.3\text{-}22)$$

The slope between the points M and D is

$$m'' = \frac{y_{AG} - y_{Ai}}{x_A^* - x_{Ai}} \qquad (7.3\text{-}23)$$

Then,

$$\frac{1}{m''}(y_{AG} - y_{Ai}) = x_A^* - x_{Ai} \qquad (7.3\text{-}24)$$

Substituting Eqs. (7.3-14), (7.3-15), and (7.3-24) into (7.3-22) and canceling out N_A,

$$\frac{1}{K_x'} = \frac{1}{m''k_y'} + \frac{1}{k_x'} \qquad (7.3\text{-}25)$$

As before, the left-hand side is the total resistance and it is equal to the sum of the individual resistances.

Some special cases of Eqs. (7.3-21) and (7.3-25) will now be considered. Generally the numerical values of k_x' and k_y' are very roughly similar. Then the value of the slope of the chords in the equilibrium curve is quite important.

If m' is very small, so the equilibrium curve in Fig. 7.3-2 is very flat, then only a very small concentration of y_A in the gas will give a very large value of x_A in equilibrium in the liquid. This means the gas solute A is very soluble in the liquid phase and, hence, the term m'/k_x' in Eq. (7.3-21) becomes very small or negligible. Then

$$\frac{1}{K_y'} \cong \frac{1}{k_y'} \qquad (7.3\text{-}26)$$

and the major resistance is said to be in the gas phase, or the "gas phase is controlling." Also,

$$y_{AG} - y_A^* \cong y_{AG} - y_{Ai} \qquad (7.3\text{-}27)$$

and the point M has moved down so that M and E are at almost the same point in Fig. 7.3-2 when the equilibrium curve is redrawn very flat or almost horizontal.

In a similar manner, when m'' is very great or the solute A is very insoluble in the liquid, then the term $1/(m''k_y')$ becomes very small and

$$\frac{1}{K_x'} \cong \frac{1}{k_x'} \qquad (7.3\text{-}28)$$

The major resistance to mass transfer is then in the liquid, the "liquid phase is controlling," and

$$x_A^* - x_{AL} \cong x_{Ai} - x_{AL} \qquad (7.3\text{-}29)$$

This means the equilibrium curve has a very steep slope and point M in Fig. 7.3-2 is very close to point D. Systems that are similar to Eq. (7.3-28) are those such as the absorption of oxygen or hydrogen from air by water.

Equations (7.3-26) and (7.3-28) are very useful in helping determine how to increase overall mass transfer in two-phase gas-liquid systems.

If the system is similar to the case in Eq. (7.3-26), where the major resistance is in the gas phase, then efforts should be centered on decreasing the gas phase resistance by increasing the gas phase turbulence or using equipment that specifically will have high turbulence in this gas phase. For the case of Eq. (7.3-28) the turbulence in the liquid phase should be increased. Also, increasing the temperature will significantly increase the k'_x and have only a small effect on k'_y.

It should be pointed out that Eqs. (7.3-2) through (7.3-29) also hold for any two-phase system where the y stands for the one phase and the x for the other phase. For example, in the extraction of acetic acid A from water (y phase) by isopropyl ether (x phase), the equilibrium curve will be for the distribution of acetic acid between these two phases and Eqs. (7.3-2) through (7.3-29) will hold.

3. *Diffusion of A through Stagnant B.* For the case of diffusion of A through stagnant B, Eqs. (7.3-7) and (7.3-19) hold.

$$N_A = \frac{k'_y}{(1-y_A)_{iM}} (y_{AG} - y_{Ai}) = \frac{k'_x}{(1-x_A)_{iM}} (x_{Ai} - x_{AL}) \qquad (7.3-7)$$

$$y_{AG} - y_A^* = (y_{AG} - y_{Ai}) + m'(x_{Ai} - x_{AL}) \qquad (7.3-19)$$

However, we must define a new overall mass transfer coefficient as

$$N_A = \frac{K'_y}{(1-y_A)*_M} (y_{AG} - y_A^*) = \frac{K'_x}{(1-x_A)*_M} (x_A^* - x_{AL}) \qquad (7.3-30)$$

where K'_y is the overall gas mass transfer coefficient for equimolar counterdiffusion, and

$$(1-y_A)*_M = \frac{(1-y_A^*) - (1-y_{AG})}{\ln\left(\frac{1-y_A^*}{1-y_{AG}}\right)} \qquad (7.3-31)$$

Substituting Eqs. (7.3-7) and (7.3-30) into (7.3-19) there results (T1)

$$\frac{1}{K'_y/(1-y_A)*_M} = \frac{1}{k'_y/(1-y_A)_{iM}} + \frac{m'}{k'_x/(1-x_A)_{iM}} \qquad (7.3-32)$$

Similarly for the K'_x,

$$\frac{1}{K'_x/(1-x_A)*_M} = \frac{1}{m''k'_y/(1-y_A)_{iM}} + \frac{1}{k'_x/(1-x_A)_{iM}} \qquad (7.3-33)$$

where

$$(1-x_A)*_M = \frac{(1-x_{AL}) - (1-x_A^*)}{\ln\left(\frac{1-x_{AL}}{1-x_A^*}\right)} \qquad (7.3-34)$$

7.3 Mass Transport between Two Phases

Example 7.3-2 Overall Mass Transfer Coefficients and Flux Ratios

Using the same data as in Example 7.3-1, do the following: (a) For equimolar counterdiffusion of A and B calculate the overall mass transfer coefficients, flux, and the percent resistance in the gas and liquid films. (b) Repeat for A diffusing through stagnant B.

Solution

For part (a), from Fig. 7.3-3, $y_A^* = 0.155$, which is in equilibrium with the bulk liquid $x_{AL} = 0.10$. The slope of chord m' is then between E and M and is

$$m' = \frac{y_{Ai} - y_A^*}{x_{Ai} - x_{AL}} = \frac{0.470 - 0.155}{0.253 - 0.100} = 2.06$$

Then for equimolar counterdiffusion, since $k_y' = 1.03$ and $k_x' = 1.55$, Eq. (7.3-21) is

$$\frac{1}{K_y'} = \frac{1}{k_y'} + \frac{m'}{k_x'} = \frac{1}{1.03} + \frac{2.06}{1.55} = 0.970 + 1.330 = 2.30$$

K_y' is equal to 0.435 lb mole/hr-ft²-mole fraction. The percent resistance in the gas film is $(0.97/2.30)100 = 42.1$ percent and 57.9 percent in the liquid film. The flux is given by Eq. (7.3-13):

$$N_A = K_y'(y_{AG} - y_A^*) = 0.435(0.700 - 0.155) = 0.237 \text{ lb mole/hr-ft}^2$$

This is the same value, of course, as calculated in Example 7.3-1 for the film equations $N_A = k_y'(y_{AG} - y_{Ai})$ and $N_A = k_x'(x_{Ai} - y_{AL})$. It should be noted that the overall coefficient cannot be obtained without knowing the film coefficients and the interface values unless the equilibrium line is straight.

To calculate K_x', we first calculate m'' from Eq. (7.3-23), which is the slope between M and D in Fig. 7.3-3, using $x_A^* = 0.310$ from Fig. 7.3-3.

$$m'' = \frac{y_{AG} - y_{Ai}}{x_A^* - x_{Ai}} = \frac{0.700 - 0.470}{0.310 - 0.253} = 4.03$$

$$\frac{1}{K_x'} = \frac{1}{m''k_y'} + \frac{1}{k_x'} = \frac{1}{4.03(1.03)} + \frac{1}{1.55} = 0.240 + 0.645 = 0.885$$

Hence, $K_x' = 1.13$. As before, the percent resistance in the gas film = $(0.240/0.885)100 = 27.1$ percent. This is different than before because the equilibrium line is not straight over the range of concentrations. Then

$$N_A = K_x'(x_A^* - x_{AL}) = 1.13(0.310 - 0.100)$$
$$= 0.237 \text{ lb mole/hr-ft}^2 \quad [\text{Ans. to (a)}]$$

For part (b), point M' in Fig. 7.3-3 represents $y_{Ai} = 0.575$ and $x_{Ai} = 0.285$. For m',

$$m' = \frac{y_{Ai} - y_A^*}{x_{Ai} - x_{AL}} = \frac{0.575 - 0.155}{0.285 - 0.100} = 2.27$$

For diffusion of A through stagnant B, from Example 7.3-1,

$$\frac{k_y'}{(1-y_A)_{iM}} = \frac{1.03}{0.360} \quad \text{and} \quad \frac{k_x'}{(1-x_A)_{iM}} = \frac{1.55}{0.801}$$

Using Eq. (7.3-31),

$$(1-y_A)_{*M} = \frac{(1-y_A^*) - (1-y_{AG})}{\ln\left(\frac{1-y_A^*}{1-y_{AG}}\right)} = \frac{(1-0.155) - (1-0.700)}{\ln\left(\frac{1-0.155}{1-0.700}\right)} = 0.526$$

Then, using Eq. (7.3-32),

$$\frac{1}{K_y'/(1-y_A)_{*M}} = \frac{1}{k_y'/(1-y_A)_{iM}} + \frac{m'}{k_x'/(1-x_A)_{iM}}$$

$$\frac{1}{K_y'/0.526} = \frac{1}{1.03/0.360} + \frac{2.27}{1.55/0.801} = 0.350 + 1.174 = 1.524$$

$K_y' = 0.344$. The percent resistance in the gas film is $(0.350/1.524)100 = 23.0$ percent and 77.0 percent in the liquid film. The flux is

$$N_A = \frac{K_y'}{(1-y_A)_{*M}}(y_{AG} - y_A^*) = \frac{0.344}{0.526}(0.700 - 0.155) = 0.358 \text{ lb mole/hr-ft}^2$$

This is the same value as calculated in Example 7.3-1 for the film equations.

4. *General Case for Diffusion of A and B*. For the case of both A and B diffusing, Eqs. (7.3-9), (7.3-10), (7.3-11), (7.3-19), and (7.3-22) hold, and the final equations are

$$N_A = \frac{K_y'}{\varphi*_y}(y_{AG} - y_A^*) = \frac{K_x'}{\varphi*_x}(x_A^* - x_{AL}) \tag{7.3-35}$$

$$\frac{1}{K_y'/\varphi*_y} = \frac{1}{k_y'/\varphi_{iy}} + \frac{m'}{k_x'/\varphi_{ix}} \tag{7.3-36}$$

$$\frac{1}{K_x'/\varphi*_x} = \frac{1}{m''k_y'/\varphi_{iy}} + \frac{1}{k_x'/\varphi_{ix}} \tag{7.3-37}$$

where $\varphi*_y$ and $\varphi*_x$ are identical to Eqs. (7.3-10) and (7.3-11) but y_A^* is substituted for y_{Ai} and x_A^* for x_{Ai}.

5. *Discussion of Overall Coefficients*. In order to actually do a design for heat transfer equipment, the overall heat transfer coefficient is

7.3 Mass Transport between Two Phases 331

predicted or synthesized from the individual film heat transfer coefficients. The individual film heat transfer coefficients are predicted from correlations for nonflowing fluids or fluids flowing past objects of definite geometries. These are then converted to resistances and added together to give the overall resistance and, hence, the overall heat transfer coefficient. Usually some sort of a barrier or solid wall separates the two fluids between which heat is being transferred.

In mass transfer the same general procedure is used in synthesizing the overall mass transfer coefficient from the individual film mass transfer coefficients. These film coefficients are obtained from the correlations developed in Chapters 6 and also 8. Care must be exercised to be sure that conditions of the physical system being used are the same or very similar to those of the correlations. The various turbulence and hydrodynamic factors and the Reynolds number must be similar. Special care must be taken where two fluids are in direct contact and the film mass transfer coefficient of each fluid phase is estimated from a correlation. Often the correlation was developed from data for mass transfer from this phase to a second pure phase with zero resistance or a second phase with negligible resistance. Hence, when these two fluids are combined in an actual apparatus, there may be an interaction and the two resistances may not be exactly additive. For example, if a solute is being transferred between the two phases, an interfacial resistance could develop owing to concentration of surface active molecules at the interface. Or increased interfacial turbulence could occur because of the solute transfer. Correlations of mass transfer coefficients for gas-liquid flow over tower packings are given in Chapter 8, Section 8.4.

As is evident in Eqs. (7.3-21) and (7.3-25) and Fig. 7.3-3, the slope of the equilibrium curve m' or m'' can vary markedly throughout an apparatus. Hence, the overall mass transfer coefficients K'_y or K'_x can vary markedly. If the equilibrium and operating lines are straight, these coefficients remain constant. As a result, when individual film coefficients are available, it is preferable to use them in predicting the amount absorbed. Unfortunately, often only experimental overall coefficients are available. Then approximations and average values must be used, which will be discussed in the next chapter.

6. *Discussion of Film Coefficients.* As mentioned in the preceding chapter, the coefficients given by Eqs. (7.3-7) and (7.3-9) for the gas and liquid films are corrected for the bulk flow terms. The film theory was used instead of the penetration or boundary layer theories to correct for this effect of bulk flow on the concentration profiles. The penetration theory appears to give better results for liquids at a gas-liquid interface in a packed column. For the gas phase, since the boundary at the gas-liquid interface is moving, no one theory gives an adequate description.

The film theory is generally preferred for both the gas and liquid coefficients in a packed gas-liquid tower because of its simplicity and the fact that uncertainties exist as to which theory is adequate. For bulk flow and low and moderate solute concentrations the results for all three theories are very similar, and at quite low fluxes and concentrations the results are essentially identical. For the liquid phase film, where the mole fraction of the solute is usually less than 10 percent, the differences between the theories are less than 2 or 3 percent. Also, in the gas phase in a countercurrent absorption column even at high mole fractions, y_A, the differences in the theories are less than several percent because the change in y_A or Δy_A across the film is small compared to the absolute value of y_A.

7.4 MASS TRANSPORT AND CHEMICAL REACTION IN TWO PHASES

7.4A Introduction to Two-Phase Reactions

Usually in absorption or extraction of a solute from one fluid phase to another that is relatively immiscible with the first, it is desired to remove as much as possible of the solute. However, if the absorbing second fluid phase becomes too concentrated in solute A, the solute can build up an equilibrium back pressure so that the rate of dissolution of A is slowed considerably. This means that the size of the absorber or extractor used becomes quite large, or else a dilute solution must be used, which may mean excessive amounts of this second phase are needed.

An important method to reduce this equilibrium back pressure is to use a second fluid phase that reacts with the solute A and reduces the concentration of A and its equilibrium back pressure. An example is absorption of CO_2 from a gas phase by a liquid phase containing a dilute solution of NaOH in water. This solution reacts with the CO_2 to give a nonvolatile carbonate, and the normally high equilibrium back pressure of CO_2 is reduced to zero. Hence, the amount of absorbing fluid or the size of the absorber can be reduced. This chemical reaction alters the concentration profile in the liquid and also affects the diffusion rate into the liquid.

Usually it is desired to regenerate the absorbing liquid after it has absorbed the solute A. For example, to absorb CO_2, Na_2CO_3 is used, where

$$Na_2CO_3 + CO_2 + H_2O \rightleftarrows 2NaHCO_3 \quad (7.4\text{-}1)$$

In the absorption at room temperatures the equilibrium is far to the right, so there is little CO_2 equilibrium back pressure. To regenerate the solution, it is heated to the boiling point, driving the reaction to the left and giving almost pure CO_2 after condensation of the vapors and Na_2CO_3.

7.4 Mass Transport and Chemical Reaction in Two Phases

Sometimes in purely physical absorption in water, certain gases will react or ionize. For example, Cl_2 absorbed in water hydrolyzes at an appreciable rate; hence the amount of unreacted chlorine is less, which reduces the equilibrium back pressure.

The occurrence of a chemical reaction of the solute A in the second fluid phase does not necessarily make the resistance due to mass transport negligible. The overall rate of absorption in the fluid phase will depend upon the mass transfer resistance in the phase and the chemical reaction rate. Obviously, the relative magnitude of these two processes can vary markedly. At one extreme the rate of chemical reaction may be very fast, so that only the mass transport rate need be considered. At the other extreme the reaction rate may be very slow, so that it mainly determines the overall rate of the process. Many cases in between these two extremes can exist, and these are usually the more difficult to analyze.

The chemical reaction rate is often difficult to obtain or predict, and this will often limit the accuracy of predictions of the overall rates. Often, however, certain extreme values of this rate can be used to predict limiting cases accurately.

In the following, we will cover only a few of the fundamental cases for simultaneous diffusion and chemical reaction. Many cases are so complex that the equations cannot be solved analytically and must be solved numerically.

The approaches to the setting up of the models are similar to those discussed in Chapter 6 for the models used for the mass transfer coefficients. The simplest is the film theory (L1), where it is assumed that all the resistance to mass transfer is in a thin stagnant film in which steady-state is essentially present. This film is assumed to have a negligible capacity to absorb solute. The second approach is to consider that the mass transfer process consists of the penetration theory, where fresh fluid elements appear at the interface and unsteady-state diffusion occurs in which the solute A can accumulate. The third analysis considers and uses the boundary layer theories. Most of the work has been done using the first two theories.

In the following sections of this chapter various cases for mass transfer and chemical reaction are discussed. A recent monograph by Danckwerts (D2) covers in detail this field of mass transfer and reaction with many applied problems being solved. The reader is referred there for further advanced study.

7.4B Slow First-Order Irreversible Reaction Using the Film Theory

Lightfoot (L2), Bird *et al.* (B1), and Sherwood and Pigford (S1) consider the case where a gas A dissolves in a turbulent pure liquid B or a solution with a high concentration of B and the following slow, first-order irreversible reaction occurs:

$$A + B \rightarrow AB \tag{7.4-2}$$

The solute A is assumed sparingly soluble or dilute in the liquid phase containing B, and the reaction is irreversible and fast enough to insure that the reaction occurs only in the liquid film. The concentration c_{A0} at the interface is assumed constant; also the bulk phase concentration $c_{A\delta}$ shown in Fig. 7.4-1 remains essentially constant. If the liquid is not pure B, but the concentration of B is high compared to A, then the reaction is first-order with respect to A.

Since the solution is quite dilute in A, we can assume Eq. (2.6-12) holds for constant ρ and D_{AB}:

$$\frac{\partial c_A}{\partial t} + (\mathbf{v} \cdot \nabla c_A) - D_{AB} \nabla^2 c_A = R_A \tag{7.4-3}$$

For steady-state, $\partial c_A / \partial t = 0$. Since the solutions are dilute, we will neglect the bulk flow and set $\mathbf{v} = 0$. The generation rate is $R_A = -k'c_A$ for a first-order or pseudo first-order reaction. Then, for diffusion in the z direction,

$$D_{AB} \frac{d^2 c_A}{dz^2} - k' c_A = 0 \tag{7.4-4}$$

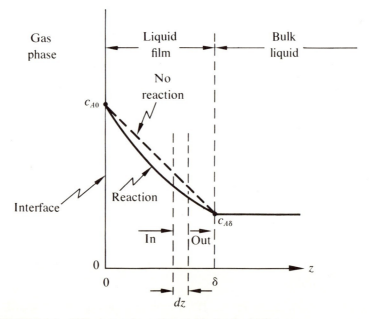

FIGURE 7.4-1 Diffusion and reaction of A in the liquid film

7.4 Mass Transport and Chemical Reaction in Two Phases

which is the same as Eq. (3.3-16). For steady-state the boundary conditions are

$$\text{B.C. 1:} \quad \text{at } z = 0, \quad c_A = c_{A0}$$
$$\text{B.C. 2:} \quad \text{at } z = \delta, \quad c_A = c_{A\delta} \tag{7.4-5}$$

The value c_{A0} is the concentration at the interface, which is in equilibrium with the gas phase at this point. The gas phase resistance is not being considered here.

The solution of Eq. (7.4-4), which is an ordinary second-order differential equation, is

$$\frac{c_A}{c_{A0}} = \frac{(c_{A\delta}/c_{A0}) \sinh bZ + \sinh b(1-Z)}{\sinh b} \tag{7.4-6}$$

where $Z = z/\delta$ and $b = \delta\sqrt{k'/D_{AB}}$. When there is no chemical reaction for steady-state diffusion, $d^2c_A/dz^2 = 0$ from Eq. (7.4-4), and integration gives

$$\frac{c_A}{c_{A0}} = \frac{c_{A\delta}}{c_{A0}} \frac{z}{\delta} + \left(1 - \frac{z}{\delta}\right) \tag{7.4-7}$$

Equations (7.4-6) and (7.4-7) are plotted in Fig. 7.4-1, showing the concentration profiles.

Generally we are interested in getting the mass flux. For chemical reaction, the flux of A to the liquid is N_{AR} at the surface, where $z = 0$,

$$N_{AR(z=0)} = -D_{AB}\left(\frac{dc_A}{dz}\right)_{z=0} = \frac{D_{AB}c_{A0}}{\delta}\left(\frac{b \cosh b - bc_{A\delta}/c_{A0}}{\sinh b}\right) \tag{7.4-8}$$

For no chemical reaction, integration of

$$N_{A(z=0)} = -D_{AB}\frac{dc_A}{dz} \tag{7.4-9}$$

gives

$$N_{A(z=0)} = \frac{D_{AB}}{\delta}(c_{A0} - c_{A\delta}) = \frac{D_{AB}c_{A0}}{\delta}\left(1 - \frac{c_{A\delta}}{c_{A0}}\right) \tag{7.4-10}$$

The ratio of N_{AR} with reaction to N_A with no reaction is Eq. (7.4-8) divided by (7.4-10):

$$\frac{N_{AR}}{N_A} = \frac{c_{A0}\left(\dfrac{b \cosh b - bc_{A\delta}/c_{A0}}{\sinh b}\right)}{c_{A0} - c_{A\delta}} \tag{7.4-11}$$

Often it is desired to write the equation for the flux with chemical reaction in a manner similar to that for no chemical reaction as

$$N_{AR} = k_{LR}(c_{A0} - c_{A\delta}) \tag{7.4-12}$$

where k_{LR} is the "mass transfer coefficient" for chemical reaction using the same driving force as for no reaction. Equating Eq. (7.4-12) to (7.4-8) and solving for k_{LR}/k_L,

$$\frac{k_{LR}}{k_L} = c_{A0}\frac{\left(\dfrac{b\cosh b - bc_{A\delta}/c_{A0}}{\sinh b}\right)}{c_{A0} - c_{A\delta}} \qquad (7.4\text{-}13)$$

where $k_L = D_{AB}/\delta$ by the film theory. Data for diffusion and chemical reaction are often obtained experimentally or predicted as the ratio of k_{LR} for chemical reaction to k_L for no reaction. The value of k_L can, of course, be obtained from the various mass transfer correlations using the Sherwood number and/or the various J factors in Chapter 6.

Sherwood and Pigford (S1, p. 328) derive equations for the same case given above but using the unsteady-state penetration theory. They also give more details on the behavior of the film model and how to combine it with the overall coefficient. Danckwerts (D1) shows that the film and penetration theories give very similar results. Hence, the film theory is often preferred.

In the special case where $c_{A\delta}$ in the main body is low, as is often the case, Eq. (7.4-13) becomes

$$\frac{k_{LR}}{k_L} = \frac{b}{\tanh b} \qquad (7.4\text{-}14)$$

where N_{AR} is calculated from Eq. (7.4-12) as $k_{LR}(c_{A0} - c_{A\delta})$.

Sherwood and Pigford (S1) have derived the case for a slow first-order reaction but reversible, using the chemical equilibrium constant K of the reversible reaction $A \leftrightarrows B$, where K is equal to the forward over the reverse rate constant. When the value of K is ∞, the reaction is irreversible. They give a graph of k_{LR}/k_L as a function of time, reaction velocity constant, and K (S1, p. 332). They used the unsteady-state theory. Danckwerts (D1) again showed that the film and penetration theories give results within about 10 percent of each other.

7.4C Very Fast Second-Order Irreversible Reaction and Film Theory

Here we are considering the very fast irreversible second-order chemical reaction where solute gas A is absorbed from the gas stream by a solution containing B and reacts as follows:

$$A + B \rightarrow AB$$

The reaction occurs in the liquid film. At first the solute B, which is initially at a uniform concentration in the liquid, is present at the interface and B and A react. As B is depleted, it must diffuse from the main liquid body through the liquid film to meet A. Also, A must diffuse through the liquid film to meet B. Steady-state is established very quickly; the profiles are shown in Fig. 7.4-2. Since the reaction is infinitely fast, A and B cannot exist simultaneously at any given point. The derivation below is due primarily to Hatta (H2) and follows the method of Sherwood and Pigford (S1).

For the mass flux in the gas phase we can write, assuming dilute solutions,

$$N_A = k_{AG}(p_A - p_{Ai}) \tag{7.4-15}$$

For the liquid film,

$$N_A = k_{AL}(c_{Ai} - 0)\frac{\delta}{\delta_A} = \frac{D_A}{\delta_A}(c_{Ai} - 0) \tag{7.4-16}$$

where k_{AL} is the mass transfer coefficient given in the literature for straight mass transfer with no chemical reaction and is based on transfer through the whole film of length δ. For component B,

$$-N_B = N_A = k_{BL}(c_B - 0)\frac{\delta}{\delta_B} = \frac{D_B}{\delta_B}(c_B - 0) \tag{7.4-17}$$

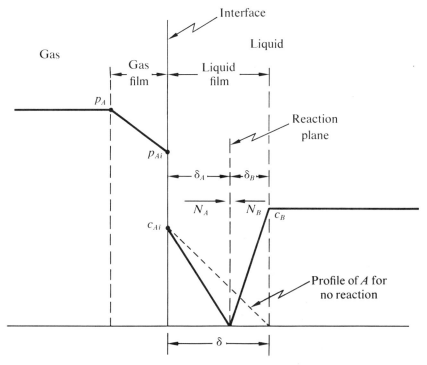

FIGURE 7.4-2 Concentration profiles for very fast irreversible reaction $A + B \rightarrow AB$ for film theory

where it is assumed for the film theory that

$$\frac{k_{AL}}{k_{BL}} = \frac{D_A/\delta}{D_B/\delta} = \left(\frac{D_A}{D_B}\right)^{1.0} \tag{7.4-18}$$

Assuming Henry's law,

$$p_{Ai} = c_{Ai} H_A \tag{7.4-19}$$

Also,

$$\delta = \delta_A + \delta_B \tag{7.4-20}$$

Solving for δ_A in Eq. (7.4-16) and δ_B in (7.4-17) and substituting into Eq. (7.4-20), we obtain after rearrangement

$$N_{AR} = \frac{D_A}{\delta}\left(c_{Ai} + \frac{D_B}{D_A}c_B\right) = k_{AL}\left(c_{Ai} + \frac{D_B}{D_A}c_B\right) \tag{7.4-21}$$

where N_{AR} is N_A for chemical reaction. For no chemical reaction,

$$N_A = k_{AL}(c_{Ai} - 0) \tag{7.4-22}$$

which gives the profile for A in Fig. 7.4-2 for no reaction. Again we define k_{ALR}, the mass transfer coefficient with chemical reaction, by

$$N_{AR} = k_{ALR}(c_{Ai} - 0) \tag{7.4-23}$$

Equating Eq. (7.4-23) to (7.4-21) and solving,

$$\frac{k_{ALR}}{k_{AL}} = \left(1 + \frac{D_B}{D_A}\frac{c_B}{c_{Ai}}\right) \tag{7.4-24}$$

This gives the ratio for the liquid film only. Hence, the rate in the liquid is increased by a factor greater than 1, depending on the ratios of diffusivities and concentrations.

Next we will combine the gas and liquid film equations to obtain an overall rate through the two films in series. Substituting Eq. (7.4-19) into (7.4-15) and using N_{AR} for N_A, we obtain

$$c_{Ai} = -\frac{N_{AR}}{k_{AG} H_A} + \frac{p_A}{H_A} \tag{7.4-25}$$

Substituting Eq. (7.4-25) into Eq. (7.4-21) to eliminate c_{Ai} and solving for N_{AR},

$$N_{AR} = \frac{p_A/H_A + (D_B/D_A)c_B}{1/k_{AL} + 1/(H_A k_{AG})} \tag{7.4-26}$$

This equation can be used directly to calculate the flux N_{AR}. For no chemical reaction and zero bulk concentration of A in the liquid, we obtain the usual two-film equation:

$$N_A = \frac{(p_A/H_A - 0)}{1/k_{AL} + 1/(H_A k_{AG})} \tag{7.4-27}$$

7.4 Mass Transport and Chemical Reaction in Two Phases

Certain restrictions and special conditions must, however, be observed in using the equations above. Suppose c_B is increased until the reaction zone moves to the left to the interface, as in Fig. 7.4-3a. Then p_{Ai} will be 0. For this case we equate Eqs. (7.4-26) and (7.4-15).

$$N_{AR} = \frac{p_A/H_A + (D_B/D_A)c_B}{1/k_{AL} + 1/(H_A k_{AG})} = k_{AG}(p_A - p_{Ai}) \qquad (7.4\text{-}28)$$

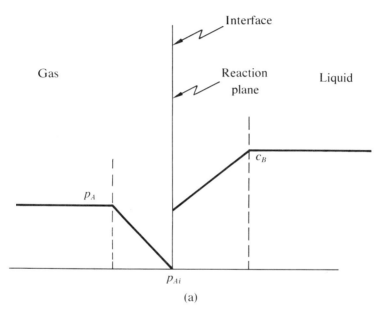

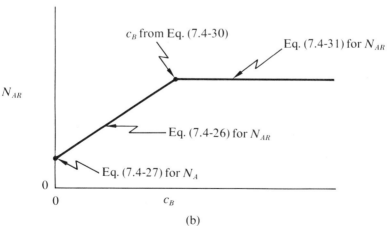

FIGURE 7.4-3 Effect of c_B on concentration profile and flux for chemical reaction: (a) concentration profile for reaction plane at interface, (b) flux N_{AR} for chemical reaction as a function of c_B

Solving for p_{Ai},

$$p_{Ai} = \frac{-\dfrac{p_A}{k_{AG}H_A} - \dfrac{D_B}{D_A}\dfrac{c_B}{k_{AG}}}{1/k_{AL} + 1/(H_A k_{AG})} + p_A \geq 0 \qquad (7.4\text{-}29)$$

Or, solving for c_B, the critical value of c_B is

$$c_B \leq \frac{k_{AG}}{k_{BL}} p_A \qquad (7.4\text{-}30)$$

Hence for Eq. (7.4-26) to be valid the reaction zone must not be at the interface, which means c_B must be equal to or less than the value given in Eq. (7.4-30).

If c_B is greater than the value in Eq. (7.4-30), then the reaction zone is at the interface and Eq. (7.4-15) is used with $p_{Ai} = 0$, or the gas film controls and N_{AR} is calculated by

$$N_{AR} = k_{AG}(p_A - 0) \qquad (7.4\text{-}31)$$

This means in an absorption tower Eq. (7.4-26) could hold in part of the tower when p_A is large and (7.4-31) when p_A gets small. The reverse holds for c_B. Figure 7.4-3b shows the variation of the flux N_{AR} versus c_B. When $c_B = 0$, Eq. (7.4-27) holds for no chemical reaction. As c_B increases, the flux from Eq. (7.4-26) increases linearly. At the critical value of c_B from Eq. (7.4-30), the flux remains constant with further increases in c_B.

7.4D Very Fast Second-Order Irreversible Reaction and Unsteady-State or Penetration Theory

In this section we will consider the same very fast second-order irreversible reaction where solute A is absorbed and reacts with B:

$$A + B \rightarrow AB$$

However, we will analyze this system using unsteady-state penetration or diffusion into the liquid instead of assuming a stagnant steady-state film.

Initially the liquid layer at the surface has a uniform concentration of B. It can be assumed that this liquid was brought from the main body of the liquid by an eddy. Solute A enters the liquid and reacts instantly with B. As the molecules of B are consumed near the interface, they are replenished by diffusion of more B molecules from the bulk stream of the liquid. The plane of reaction, which originally was at the surface, keeps moving back into the liquid. Hence, the rate of mass transfer of A diminishes with time as the length of the diffusion path increases.

This situation, where the reaction zone moves inward with time, is shown in Fig. 7.4-4. The unsteady-state equations describing this are

7.4 Mass Transport and Chemical Reaction in Two Phases

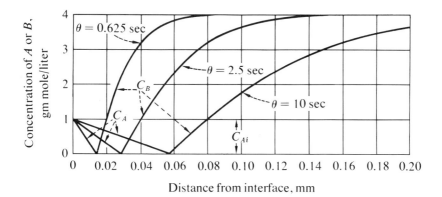

FIGURE 7.4-4 Calculated concentration profiles during rapid second-order reaction in liquid film. Based on $D_A = 3.9 \times 10^{-5}$ ft²/hr, $D_B = 1.95 \times 10^{-5}$ ft²/hr. [From R. H. Perry and R. L. Pigford, *Ind. Eng. Chem.* **45**, 1247 (1953). With permission.]

given below. For components A and B the equations for unsteady-state diffusion in the x direction (D1, S1) are

$$\frac{\partial c_A}{\partial \theta} = D_A \frac{\partial^2 c_A}{\partial x^2}, \quad 0 < x < x_R \quad (7.4\text{-}32)$$

$$\frac{\partial c_B}{\partial \theta} = D_B \frac{\partial^2 c_B}{\partial x^2}, \quad x_R < x < \infty \quad (7.4\text{-}33)$$

where x_R is the distance of the reaction zone from the interface. It can be shown (S1) that the reaction zone moves away from the interface at a rate proportional to $\sqrt{\theta}$. The boundary and initial conditions that apply in the liquid phase are

I.C.: $c_B = c_{B0}$ when $\theta = 0, x > 0$
B.C. 1: $c_A = c_{Ai}$ when $\theta > 0, x = 0$
B.C. 2: $c_B = c_{B0}$ when $\theta \geq 0, x = \infty$ (7.4-34)
B.C. 3: $c_A = c_B = 0$ when $x = x_R$
B.C. 4: $D_A \dfrac{\partial c_A}{\partial x} = -D_B \dfrac{\partial c_B}{\partial x}$ when $x = x_R$

The solutions of Eqs. (7.4-32) to (7.4-34) are given as error functions and must be solved by trial and error (S1). They have been solved for only a narrow range of variables. For the special case where $D_A = D_B$ an exact solution is obtained, giving

$$\frac{k_{LR}}{k_L} = \left(1 + \frac{c_B}{c_{Ai}}\right) \quad (7.4\text{-}35)$$

This is the same result as given by the film theory in Eq. (7.4-24) for $D_A = D_B$.

Danckwerts and Kennedy (D1) and Sherwood and Pigford (S1) give a detailed comparison of the film theory model and the various penetration models for this very fast irreversible second-order reaction and show that the models agree to within about 10 percent for a wide range of variables. For a very fast first-order reversible or irreversible reaction the models also are shown to check reasonably well, as discussed previously.

7.4E Slow Second-Order Reaction

For the case of irreversible second-order reactions, where

$$A + \nu B \rightarrow AB \tag{7.4-36}$$

and where the reaction is not infinitely fast, Perry (P1) gives a plot covering a wide range of variables to determine k_{LR}/k_L, which is given in Fig. 7.4-5, where

B_0 = initial concentration, g mole B/liter
A_i = concentration of A at interface, g mole A/liter
k_{II} = second-order reaction rate constant, liters/g mole-sec
t = time, sec

This plot was derived by using the film theory and modifying it to account for the diffusivity as given by the penetration theory.

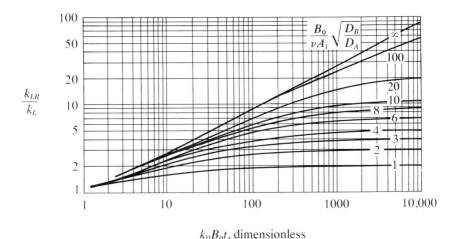

FIGURE 7.4-5 Second-order irreversible chemical reaction and diffusion. (From J. H. Perry, *Chemical Engineers' Handbook,* 4th ed. New York: McGraw-Hill, Inc., 1963. With permission.)

When k_{II} or t is very large, then the lines in Fig. 7.4-5 are horizontal and also give answers for the case where the reaction is infinitely fast. Perry also gives a graph (P1, p. 14–16) to correct for the effect of reversibility.

Example 7.4-1 Effect of Chemical Reaction on Mass Transfer Coefficient

For a second-order irreversible very fast reaction the following data are given: $D_A = 2 \times 10^{-5}$ cm^2/sec, $D_B = 3 \times 10^{-5}$ cm^2/sec, $\nu = 2$, $A_i = 1.0 \times 10^{-2}$ g mole/liter, $B_0 = 2.0 \times 10^{-2}$ g mole/liter. (a) Calculate k_{LR}/k_L using the film theory. (b) Repeat, using Fig. 7.4-5.

Solution

For part (a), Eq. (7.4-24) does not hold, since this is for 1 mole A and 1 mole B. Starting with Eq. (7.4-17) and rewriting it for the reaction above,

$$-N_B = \nu N_A = \frac{D_B}{\delta_B}(c_B - 0) \qquad (7.4\text{-}37)$$

Then, carrying through the derivation as before,

$$\frac{k_{ALR}}{k_{AL}} = \frac{k_{LR}}{k_L} = \left(1 + \frac{D_B}{D_A}\frac{c_B}{\nu c_{Ai}}\right) \qquad (7.4\text{-}38)$$

Substituting the given data into the above,

$$\frac{k_{LR}}{k_L} = \left(1 + \frac{3 \times 10^{-5}(2 \times 10^{-2})}{2 \times 10^{-5}(2)(1 \times 10^{-2})}\right) = \frac{2.5}{1} \qquad \text{[Ans. to (a)]}$$

For part (b), we first calculate

$$\frac{B_0}{\nu A_i}\sqrt{\frac{D_B}{D_A}} = \frac{2 \times 10^{-2}}{(2)(1 \times 10^{-2})}\sqrt{\frac{3 \times 10^{-5}}{2 \times 10^{-5}}} = 1.23$$

From Fig. 7.4-5 for $k_{II}B_0 t$ very large (very fast irreversible),

$$\frac{k_{LR}}{k_L} = 2.2 \qquad \text{[Ans. to (b)]}$$

PROBLEMS

7-1 Henry's Law and Gas Solubility

The concentration of CO_2 in water at 30°C is $1.03(10^{-4})$ g CO_2/g water. The Henry's law constant is $H = 0.186(10^4)$ atm/mole fraction. What partial pressure of CO_2 must be kept in the gas phase to keep the CO_2 from vaporizing from the solution?

7-2 Solution of a Gas in a Liquid

An aqueous solution of oxygen dissolved in water containing $1.0(10^{-3})$ g $O_2/100$ g water is in contact with a large volume of ordinary air at a total pressure of 2.50 atm at 40°C. The Henry's law constant for oxygen is $H = 5.35(10^4)$ atm/mole fraction. Will the solution lose oxygen or gain oxygen? Also, what will the concentration of oxygen be in the final equilibrium solution?

7-3 Equilibrium in Liquid Phases

The distribution coefficient $K = $ g mole benzoic acid per liter of chloroform/g mole benzoic acid per liter of water is 2.40 at 25°C. The water and chloroform phases are essentially immiscible. If 2.5 liters of chloroform containing 0.050 g mole benzoic acid/liter are contacted with 1.0 liter of water, what are the final equilibrium concentrations?

7-4 Film Coefficients, Interface Compositions, and Overall Coefficients

The gas SO_2 is being absorbed in a wetted-wall absorption tower with the liquid water mixture flowing downward along the wall at 30°C and 1 atm total pressure. At a certain point in the tower the bulk gas concentration is $y_A = 0.60$ mole fraction and $x_A = 0.0080$ mole fraction. The equilibrium data at 30°C and 1 atm total pressure are given in the appendix, Table A.4-3.

From the correlations for wetted-wall towers and for dilute solutions, the film mass transfer coefficients for SO_2 are $k_x = 1.10$ lb mole/hr-ft²-mole fraction and $k_y = 1.30$ lb mole/hr-ft²-mole fraction. Assume water vapor does not diffuse.

(a) Calculate the interface compositions y_{Ai} and x_{Ai} and the local mass transfer flux through the gas and through the liquid.

(b) Calculate the overall mass transfer coefficients, the percent resistance in the gas film, and the total flux using both overall mass transfer coefficients.

7-5 Mass Transfer and Reaction of A and Film Theory

For the case where dilute solute A is diffusing into pure B and reacting by a slow first-order irreversible reaction, do the following for the film theory:

(a) Derive Eq. (7.4-7) below for no reaction, starting from $d^2c_A/dz^2 = 0$.

$$\frac{c_A}{c_{A0}} = \frac{c_{A\delta}}{c_{A0}}\frac{z}{\delta} + \left(1 - \frac{z}{\delta}\right) \tag{7.4-7}$$

(b) Do the same but start with Eq. (7.4-6).

$$\frac{c_A}{c_{A0}} = \frac{(c_{A\delta}/c_{A0})\sinh bZ + \sinh b(1-Z)}{\sinh b} \tag{7.4-6}$$

[*Hints*: For no reaction, $b = 0$. This gives 0/0. Use L'Hospital's rule.]

(c) Prove that when $b = 0$, Eq. (7.4-13) below reduces to $k_{LR}/k_L = 1.0$.

$$\frac{k_{LR}}{k_L} = \frac{c_{A0}\left(\dfrac{b\cosh b - bc_{A\delta}/c_{A0}}{\sinh b}\right)}{c_{A0} - c_{A\delta}} \tag{7.4-13}$$

(d) For b having values of 0, 0.2, 1.0, and 2.0 and $c_{A0}/c_{A\delta} = 2$, calculate k_{LR}/k_L. Plot the values versus b. Discuss the shape of the curve.

Ans.: (d) For $b = 1.0$, $k_{LR}/k_L = 1.76/1$

7-6 Fast Irreversible Reaction and Film Theory

For the reaction

$$A + \nu B \to C$$

where ν is an integer, derive the final equation similar to Eq. (7.4-26) for a fast irreversible reaction. The solute A is being absorbed into a liquid, with the solute B dissolved in the liquid. Also solve for the critical value of c_B.

$$\text{Ans.: } N_{AR} = \frac{p_A/H_A + \dfrac{D_B}{D_A}\dfrac{c_B}{\nu}}{1/k_{AL} + 1/(H_A k_{AG})}; \quad c_B \leq \frac{\nu k_{AG}}{k_{BL}} p_A$$

NOTATION

(Boldface symbols are vectors.)

- a_A activity of $A = x_A \gamma_A$
- b dimensionless number $= \delta \sqrt{k'/D_{AB}}$
- C number of components in the phase rule
- c_A concentration of A, g mole A/cm^3 or lb mole A/ft^3
- $c_{A(B)}$ concentration of A in B, g mole A/cm^3
- $c_{A(C)}$ concentration of A in C, g mole A/cm^3
- D_A diffusivity of A in solution, cm^2/sec or ft^2/hr
- D_{AB} diffusivity, cm^2/sec or ft^2/hr
- F number of degrees of freedom
- H Henry's law constant, atm/mole frac, mm/mole frac, or atm/(g mole A/cm^3)
- H' Henry's law constant, mole frac in vapor/mole frac in liquid
- k_{II} homogeneous second-order reaction velocity constant, liters/g mole-sec
- k' homogeneous first-order reaction velocity constant, sec^{-1}
- K equilibrium distribution coefficient $= c_{A(B)}/c_{A(C)}$, dimensionless
- k_G gas film mass transfer coefficient, g mole/sec-cm^2-atm
- k_L liquid film mass transfer coefficient, cm/sec
- k_{LR} liquid film mass transfer coefficient with reaction, cm/sec
- k'_y gas film mass transfer coefficient, g mole/sec-cm^2-mole frac or lb mole/hr-ft^2-mole frac
- K'_y overall gas mass transfer coefficient, g mole/sec-cm^2-mole frac or lb mole/hr-ft^2-mole frac; defined by Eq. (7.3-13)
- k'_x liquid film mass transfer coefficient, g mole/sec-cm^2-mole frac or lb mole/hr-ft^2-mole frac
- K'_x overall liquid mass transfer coefficient, g mole/sec-cm^2-mole frac or lb mole/hr-ft^2-mole frac; defined by Eq. (7.3-14)

346 Interphase Mass Transport

k_y, k_x mass transfer coefficients, g mole/sec-cm^2-mole frac or lb mole/hr-ft^2-mole frac; see Table 6.3-1
m' slope of equilibrium line; defined by Eq. (7.3-17)
m'' slope of equilibrium line; defined by Eq. (7.3-23)
$\mathbf{N}_A$ flux of A relative to a fixed point, g mole A/sec-cm^2 or lb mole A/hr-ft^2
N_{AR} flux of A relative to a fixed point when chemical reaction occurs, g mole A/sec-cm^2
p_A partial pressure of A, atm
P total pressure, atm
P number of phases at equilibrium
R_A rate of generation, g mole A/sec-cm^3
T absolute temperature, °K or °R
T_B temperature in phase B, °K or °R
T_C temperature in phase C, °K or °R
t time, sec
$\mathbf{v}$ mass average velocity vector relative to stationary coordinates, cm/sec
x_A mole fraction of A in liquid
x_A^* mole fraction of A in equilibrium with y_{AG}
x_{AL} mole fraction of A in bulk liquid phase
x distance in x direction, cm
x_R distance of reaction zone from interface, cm
y distance in y direction, cm
y_A mole fraction of A in gas
y_A^* mole fraction of A in equilibrium with x_{AL}
y_{AG} mole fraction of A in bulk gas phase
$(1-x_A)_{iM}$ inert mole fraction defined by Eq. (7.3-6)
$(1-x_A)_{*M}$ inert mole fraction defined by Eq. (7.3-34)
$(1-y_A)_{iM}$ inert mole fraction defined by Eq. (7.3-5)
$(1-y_A)_{*M}$ inert mole fraction defined by Eq. (7.3-31)
z distance in z direction, cm
Z ratio z/δ, dimensionless

Greek Letters

γ_A activity coefficient of A
δ film thickness, cm
θ time, sec
ν integer
ρ density, g mass/cm^3 or lb mass/ft^3
φ_{ix} bulk flow factor defined by Eq. (7.3-11)
φ_{iy} bulk flow factor defined by Eq. (7.3-10)
φ_{*x} bulk flow factor defined by Eq. (7.3-11) with x_A^* substituted for x_{Ai}
φ_{*y} bulk flow factor defined by Eq. (7.3-10) with y_A^* substituted for y_{Ai}

Subscripts

- A component A
- B component B
- C component C
- i interface
- G gas phase
- L liquid phase
- 0   interface
- δ distance δ from interface
- 1   beginning of diffusion path
- 2   end of diffusion path

REFERENCES

(B1)   R. B. Bird, W. E. Stewart and E. N. Lightfoot, *Transport Phenomena*. New York: John Wiley & Sons, Inc., 1960.

(D1)   P. V. Danckwerts and A. M. Kennedy, *Tr. I.Ch.E.* (London) **32**, S49, S53 (1954).

(D2)   P. V. Danckwerts, *Gas-Liquid Reactions*. New York: McGraw-Hill, Inc., 1970.

(H1)   O. A. Hougen, K. M. Watson, and R. A. Ragatz, *Chemical Process Principles*, pt. I, 2d ed. New York: John Wiley & Sons, Inc., 1962.

(H2)   S. Hatta, *Techol. Repts. Tohoku Imp. Univ.* **8**, 1 (1928–1929).

(K1)   E. R. Kent and R. L. Pigford, *AIChEJ* **2**, 363 (1956).

(L1)   W. K. Lewis and W. G. Whitman, *Ind. Eng. Chem.* **16**, 1215 (1924).

(L2)   E. N. Lightfoot, *AIChEJ* **4**, 499 (1958).

(P1)   J. H. Perry, *Chemical Engineers' Handbook*, 4th ed. New York: McGraw-Hill, Inc., 1963.

(S1)   T. K. Sherwood and R. L. Pigford, *Absorption and Extraction*, 2d ed. New York: McGraw-Hill, Inc., 1952.

(T1)   R. E. Treybal, *Mass-Transfer Operations*, 2d ed. New York: McGraw-Hill, Inc., 1968.

8 Continuous Two-Phase Mass Transport Processes

8.1 INTRODUCTION TO CONTINUOUS TWO-PHASE MASS TRANSPORT PROCESSES

8.1A Classification of Two-Phase Continuous Processes

In this chapter we will study continuous two-phase mass transport processes. The two phases are in direct contact with each other and are only somewhat miscible in each other or immiscible. As typical examples, the two-phase pair can be gas-liquid, gas-solid, liquid-liquid, or liquid-solid. All of these two-phase processes have the general requirement that the two phases be in more or less intimate contact so that a solute or solutes can diffuse from one to the other.

Generally the equipment used is designed to provide intimate contact, with large interfacial areas and relatively high turbulence, in one or both of the two phases to increase the film mass transfer coefficients.

When the two contacting fluids are a gas and a liquid, the process is called *absorption*. In absorption a solute A or several solutes are absorbed from the gas phase into a liquid phase. The process can involve diffusion of solute A through a stagnant gas B into a stagnant liquid, where the liquid does not vaporize or diffuse back into the gas stream. An example is absorption of ammonia from air by the solvent water. Generally the exit water-ammonia solution from the absorption tower is distilled to recover relatively pure ammonia. Another example of absorption is the absorbing of hydrocarbons from a gas stream by a light, nonvolatile oil, as in the petroleum industry. A third is removal of SO_2 from flue gases by absorption in alkaline solutions. The reverse of absorption is called *desorption* or *stripping*, and the same basic principles and theories hold. When the gas is pure air and the liquid pure water, the process is called *humidification*.

When the two fluids are a volatile vapor and a liquid that vaporizes, the process is called *distillation*. In this case the process is often equimolar counterdiffusion, where A and B are counterdiffusing. In some cases the process is such that the fluxes of A and of B are not equimolar but are related by heat balances.

When the two fluids are liquids, where a solute or solutes are removed from one liquid phase to another phase, the process is called *extraction*. An example is extraction of acetic acid from a water solution by isopropyl ether.

When the one fluid is air and the other phase is a solid from which water or moisture is being removed, the process is called *drying*. The theory for this operation becomes quite specialized and complex, since the diffusion in the solid depends on physical structure and on chemical attachment of the moisture. This theory is adequately covered in many texts (P1, T1, M1, F1).

If one fluid is a liquid that is being used to extract a solute A from a solid, the process is termed *leaching*. Examples are leaching copper from solid ores by sulfuric acid and leaching vegetable oils from solid soybeans by organic solvents. The theory can become quite complicated, since at times the solute A is primarily leached from the solid surface and at other times the solvent must penetrate the solid and interact physically or chemically to break free the solute A and then allow it to diffuse back out of the solid. Since the theory is very complicated, usually experimental data must be obtained on each physical system.

8.1B Introduction to the Theory of Two-Phase Continuous Processes

The integration of the heat transfer equations for continuous two-phase flow, which usually occurs in heat exchangers, is quite similar to the integration of the mass transfer equations. In the present chapter we will discuss the integration of the differential mass transfer diffusion and material balance equations over the height of a tower in interphase, continuous-contacting equipment.

Note, however, that mass transfer is more complicated than heat transfer for several reasons: the contacting may be by stages, or in packing where different types of diffusion can occur, rather than continuous as in heat exchangers; the two phases may be partially miscible, which causes complications in the material balance equations; and more than one solute can be diffusing.

We point out that all of the rate equations and factors studied in Chapters 1 through 7 hold in this chapter. The complicating factor is that these equations and/or factors can often vary from point to point in a tower, and methods must be developed to handle these variations throughout a column.

The main factors and equations that must be known to design a continuous mass transfer process are as follows:

1. Material balance relations between all streams and components.
2. Equilibrium data as discussed in Chapter 7.
3. Mass transfer coefficients as given in Chapters 6, 7, and 8.
4. The type of column to be used and its physical characteristics, such as allowable flow rates, pressure drops, type of packing, and so on. These latter data are given and well summarized elsewhere (P1, T1).

8.2 MATERIAL BALANCES AND OPERATING LINES

8.2A Countercurrent Processes

In making material balances in countercurrent mass transfer processes, the total moles of each of the two flowing phases may change throughout the column. Hence, the material balance equations (often called operating lines) that relate the two streams may not be straight lines when plotted on mole fraction coordinates y versus x. In the sections that follow, the equations will first be derived in a general form that holds for all cases, and then simplified for the special cases of equimolar flow and flow where A diffuses through stagnant B. The operating lines hold for a differential contact tower, such as a packed tower, where there are no sudden discrete changes in composition, and for a stage contact or plate tower, where there are discrete changes between trays.

1. *General Case.* In this derivation an operating line or material balance equation will be derived that holds for all cases of flow and diffusion.

Figure 8.2-1 shows the flows of phase V, total moles/hr, and phase L, total moles/hr. The phases V and L can be the following pairs: gas-liquid, liquid-liquid, gas-solid, or liquid-solid. The mole fraction of solute A in the V phase is y_A or y and in the L phase it is x_A or x. The solute A is being transferred from phase V to phase L. The following are some examples of cases that can occur for which the general equations to be derived here will hold.

(a) *Diffusion of A through phase V and then diffusion of A through phase L*. The second component of phase V is inert B, which does not diffuse. The second component of phase L is C, which is

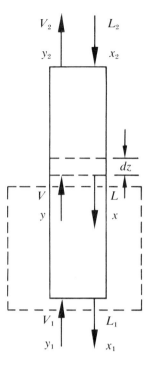

FIGURE 8.2-1 Material balance or operating line of a countercurrent process

nondiffusing. An example of this is ammonia (A) diffusing through an inert gas, air (B), into and through an inert liquid (C), which is water. Hence, we see that in the gas phase we have a binary A–B with A diffusing, and in the liquid phase a binary A–C with only A diffusing. Another example is extraction of acetic acid (A) from water phase (B) in the V stream to the L stream composed of methyl-isobutyl ketone (C). The water B and ketone C are essentially immiscible when the acetic acid is dilute.

(b) *Equimolar counterdiffusion of A and B in phase V and also equimolar counterdiffusion of A and B in phase L.* The V phase is a binary A–B and the L phase is also a binary A–B. An example of this is distillation in a tower of benzene (B) and toluene (A), where the vapor phase is V and the liquid L. Here the latent heats of A and B are equal, so if a net flow of 1 mole of A diffuses into the liquid from the vapor (that is, condenses), one mole of B must counterdiffuse to the vapor or vaporize.

(c) *Both A and B diffusing but not equimolar nor only A diffusing.* An example is in the distillation of benzene A and butane B, which are in the vapor V and the liquid L. The latent heat of benzene is approximately 1.6 times that of butane. Hence, the fluxes will be related by $N_B = -1.6 N_A$.

Referring again to Fig. 8.2-1, let us derive the material balance or operating line equation. First a total material balance is made over the whole tower, where the total moles/hr into the tower equal the total out:

$$L_2 + V_1 = L_1 + V_2 \quad (8.2\text{-}1)$$

Making a material balance only on component A over the whole tower,

$$L_2 x_2 + V_1 y_1 = L_1 x_1 + V_2 y_2 \quad (8.2\text{-}2)$$

Equation (8.2-1) and/or Eq. (8.2-2) can be used if one of the streams or compositions is unknown. For example, suppose the inlet and outlet flows and compositions of V are set and also the inlet L_2 and x_2 are set. Then by Eqs. (8.2-1) and (8.2-2) the outlet L_1 and x_1 can be determined.

Drawing a box shown as the dotted line in Fig. 8.2-1, we obtain the following total material balance and balance on component A:

$$L + V_1 = L_1 + V \quad (8.2\text{-}3)$$

$$Lx + V_1 y_1 = L_1 x_1 + Vy \quad (8.2\text{-}4)$$

In Eqs. (8.2-3) and (8.2-4) the total flow rates V or L may vary in the tower. If A is being removed from V, then V will decrease going up the tower and L will increase going down the tower. Equation (8.2-4) relates the composition y of stream V and x of L at any point in the tower. Equation (8.2-4) is a material balance equation but is usually called an operating line.

A plot of Eq. (8.2-4) is made in Fig. 8.2-2, where y and x are mole fraction units. This operating line is curved. Referring to Eq. (8.2-4), L and V vary throughout the tower, so the equation when plotted as y versus x is not straight. If Eq. (8.2-4) is solved for y,

$$y = \frac{L}{V} x + \frac{V_1 y_1 - L_1 x_1}{V} \quad (8.2\text{-}5)$$

If the solutions are relatively dilute, the operating line is nearly straight, and then average values of L and V can be used to obtain the slope of the operating line in Fig. 8.2-2, which is approximately $L_{\text{ave}}/V_{\text{ave}}$.

The equations given above are the same whether the tower is a countercurrent differential contact type as a packed tower or a countercurrent stage contact tower as a plate column. For a plate tower there exist only discrete values of y and x for each plate, which will be discussed later in this chapter.

Equation (8.2-4) can also be given in terms of partial pressures p_A or p and total pressure P, which will be assumed constant.

$$Lx + V_1 \frac{p_1}{P} = L_1 x_1 + V \frac{p}{P} \quad (8.2\text{-}6)$$

8.2 Material Balances and Operating Lines

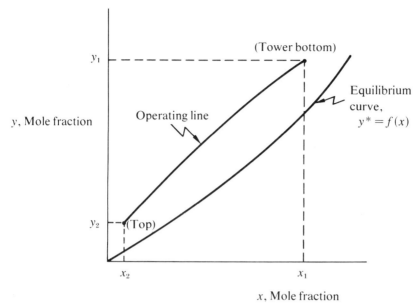

FIGURE 8.2-2 Operating line using mole-fraction units

Hence, a plot of p versus x can be used for the operating and equilibrium lines.

In Fig. 8.2-2 the operating line is above the equilibrium line, which means the solute A is being removed from stream V and absorbed into stream L. When the operating line is below the equilibrium line, the solute A is being removed or stripped from stream L and enters stream V. These two cases are shown in Fig. 8.2-3.

Equations (8.2-1) to (8.2-6) cannot be solved unless the flux ratio N_A/N_B is specified first. For example, it must be specified if the process is A diffusing through stagnant B ($N_B = 0$), equimolar counterdiffusion ($N_A = -N_B$), or some other relation between N_A and N_B [$N_B = f(N_B)$] that is determined by heat balances or other considerations. In the following sections these main special cases will be considered. In the following derivations for mass transfer during the countercurrent flow of the two phases, it is assumed that there is no axial dispersion or mixing axially of each phase with itself, that is, all elements in a phase travel at the same rate through the apparatus. However, such axial dispersion sometimes occurs and decreases the mass transfer rates. For further details see Levenspiel (L1) for the theory and others (E1, S4) for mathematical models. Danckwerts (D1) concludes that in a packed tower with a height many times the size of the packing, the axial dispersion effect is negligible. However, actual channeling can increase it.

354 Continuous Two-Phase Mass Transport Processes

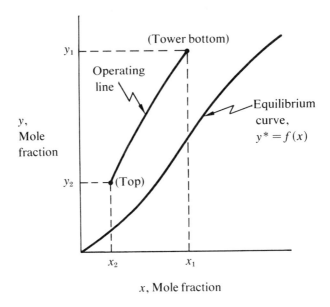

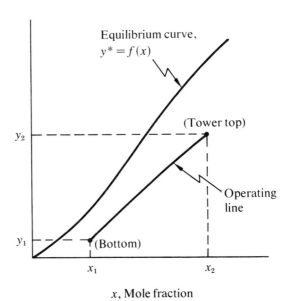

FIGURE 8.2-3 Location of operating lines for absorption and for stripping: (a) absorption or transfer of solute A from V to L stream, (b) stripping or transfer of solute A from L to V stream

8.2 Material Balances and Operating Lines

2. *Diffusion of A through Stagnant B.* In the case of diffusion of A through stagnant B, as in the absorption of ammonia in air by water, the moles of inert air and inert water remain constant throughout the tower. If the rates are V' mole inert air/hr and L' mole inert solvent water/hr, a material balance on component A (Fig. 8.2-1) is

$$L'\left(\frac{x_2}{1-x_2}\right) + V'\left(\frac{y_1}{1-y_1}\right) = L'\left(\frac{x_1}{1-x_1}\right) + V'\left(\frac{y_2}{1-y_2}\right) \quad (8.2\text{-}7)$$

A balance around the dotted-line box in Fig. 8.2-1 gives

$$L'\left(\frac{x}{1-x}\right) + V'\left(\frac{y_1}{1-y_1}\right) = L'\left(\frac{x_1}{1-x_1}\right) + V'\left(\frac{y}{1-y}\right) \quad (8.2\text{-}8)$$

Or, using partial pressures, p, of component A in place of y, Eq. (8.2-8) becomes

$$L'\left(\frac{x}{1-x}\right) + V'\left(\frac{p_1}{P-p_1}\right) = L'\left(\frac{x_1}{1-x_1}\right) + V'\left(\frac{p}{P-p}\right) \quad (8.2\text{-}9)$$

The operating line equation (8.2-8), when plotted as in Fig. 8.2-2, will be curved. If x and y are very dilute, then $(1-x)$ and $(1-y)$ can be neglected and taken as 1.0, and Eq. (8.2-8) becomes

$$L'x + V'y_1 \cong L'x_1 + V'y \quad (8.2\text{-}10)$$

This has a slope of L'/V', and the operating line is approximately straight.

Example 8.2-1 *Absorption of A from Stagnant B in a Tower*

Component A in an inert air stream B is being absorbed in water at 25°C. The inlet liquid stream $L_2 = 100$ lb mole/hr and $x_2 = 0$ mole fraction. The inlet gas stream $V_1 = 100$ lb mole/hr and $y_1 = 0.50$ mole fraction. It is desired to reduce the y at the exit so $y_2 = 0.10$. Derive the operating line equation and plot on a graph.

Solution

The process flow diagram is given in Fig. 8.2-1. The knowns are $L_2 = 100$, $x_2 = 0$, $V_1 = 100$, $y_1 = 0.50$, and $y_2 = 0.10$. A material balance on component A as in Eq. (8.2-7) is

$$L'\left(\frac{x_2}{1-x_2}\right) + V'\left(\frac{y_1}{1-y_1}\right) = L'\left(\frac{x_1}{1-x_1}\right) + V'\left(\frac{y_2}{1-y_2}\right)$$

Now $L' = 100(1-x_2) = 100(1-0) = 100$ lb mole inert water/hr. Also, $V' = 100(1-y_1) = 100(1-0.50) = 50$ lb mole inert air/hr. Substituting into Eq. (8.2-7),

$$100\left(\frac{0}{1-0}\right) + 50\left(\frac{0.50}{1-0.50}\right) = 100\left(\frac{x_1}{1-x_1}\right) + 50\left(\frac{0.10}{1-0.10}\right)$$

solving, this gives $x_1 = 0.307$. Then,

$$L_1(1-x_1) = L'$$
$$L_1(1-0.307) = 100$$
$$L_1 = 144.45$$
$$V_2(1-y_2) = V'$$
$$V_2(1-0.10) = 50$$
$$V_2 = 55.55$$

To calculate intermediate points on the operating line, Eq. (8.2-8) is used.

$$L'\left(\frac{x}{1-x}\right) + V'\left(\frac{y_1}{1-y_1}\right) = L'\left(\frac{x_1}{1-x_1}\right) + V'\left(\frac{y}{1-y}\right)$$

Selecting $x = 0.196$, y is calculated as follows:

$$100\left(\frac{0.196}{1-0.196}\right) + 50\left(\frac{0.50}{1-0.50}\right) = 100\left(\frac{0.307}{1-0.307}\right) + 50\left(\frac{y}{1-y}\right)$$

Hence, $y = 0.375$. To calculate another intermediate point, set $x = 0.10$; y is calculated as 0.250. These four sets of points are plotted in Fig. 8.2-4. The operating line is significantly curved.

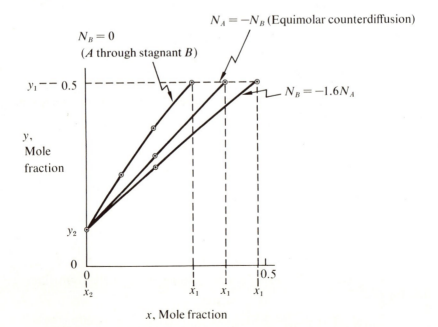

FIGURE 8.2-4 Effect of flux ratios on operating lines

8.2 Material Balances and Operating Lines

3. *Equimolar Counterdiffusion of A and B.* In the case of equimolar counterdiffusion of A and B, $N_A = -N_B$. Then Eqs. (8.2-2) and (8.2-4) hold, and L and V are constant and not variables, as in the case of A diffusing through stagnant B.

$$L_2 x_2 + V_1 y_1 = L_1 x_1 + V_2 y_2 \quad (8.2\text{-}2)$$

$$Lx + V_1 y_1 = L_1 x_1 + Vy \quad (8.2\text{-}4)$$

The operating line equation, Eq. (8.2-4), is straight and has a slope of L/V.

Example 8.2-2 Equimolar Absorption in a Tower

The component benzene (B) and the component toluene (A) in the vapor stream are being contacted with a liquid stream of benzene and toluene (distillation). Since the latent heats are the same, equimolar counterdiffusion occurs. The inlet liquid stream $L_2 = 100$ lb mole/hr and $x_2 = 0$. The inlet vapor stream $V_1 = 100$ lb mole/hr and $y_1 = 0.50$. It is desired to reduce the y at the outlet to $y_2 = 0.10$. Derive and plot the operating line equation.

Solution

The process flow diagram is as in Fig. 8.2-1. The flows are $L_2 = L_1 = 100$ and $V_1 = V_2 = 100$, since it is equimolar. Then, by Eq. (8.2-2),

$$L_2 x_2 + V_1 y_1 = L_1 x_1 + V_2 y_2$$

$$100(0) + 100(0.50) = 100(x_1) + 100(0.10)$$

Solving, $x_1 = 0.40$. Since the equation is a straight line, it is not necessary to calculate an intermediate point. Then, substituting into Eq. (8.2-4), the operating line equation is

$$Lx + V_1 y_1 = L_1 x_1 + Vy$$

$$100(x) + 100(0.50) = 100(0.40) + 100(y)$$

This is plotted again in Fig. 8.2-4.

Example 8.2-3 Absorption with Flux Ratio Set by Latent Heats

In the distillation of benzene (A) and butane (B), which are in the vapor and the liquid, the latent heat of benzene is approximately 1.6 times that of butane. Hence, the fluxes will be related by $N_B = -1.6 N_A$. The inlet liquid stream to the tower is $L_2 = 100$ lb mole/hr and $x_2 = 0$. The inlet vapor stream $V_1 = 100$ lb mole/hr and $y_1 = 0.50$. It is desired to reduce y to $y_2 = 0.10$. Derive and plot the operating line equation.

Solution

The flow diagram is the same as in Fig. 8.2-1. An overall material balance and a material balance for component A cannot be solved unless the relation between the fluxes N_A and N_B are included. Using A_1 as the interfacial area ft^2, $N_A(A_1) = $ lb mole A/hr. To relate this flux to the V stream, the lb mole A/hr entering in the V_1 stream minus the lb mole A/hr leaving in the V_2 stream are equal to the lb mole A/hr transferred:

$$V_1(y_1) - V_2(y_2) = N_A(A_1)$$

Substituting, $100(0.50) - V_2(0.10) = N_A(A_1)$. Writing the same type of material balance equation for component B, the flux is N_B or $-1.6N_A$.

$$V_1(1-y_1) - V_2(1-y_2) = -1.6N_A(A_1)$$
$$100(1-0.50) - V_2(1-0.10) = -1.6N_A(A_1)$$

Now with two equations and the two unknowns, V_2 and $N_A(A_1)$, they can be solved, giving $V_2 = 122.7$ lb mole/hr and $N_A(A_1) = 37.7$ lb mole A/hr.

Making an overall balance, Eq. (8.2-1),

$$L_2 + V_1 = L_1 + V_2 \quad \text{or} \quad 100 + 100 = L_1 + 122.7$$

Hence, $L_1 = 77.30$ lb mole/hr. Using Eq. (8.2-2),

$$L_2 x_2 + V_1 y_1 = L_1 x_1 + V_2 y_2 \quad \text{or} \quad 100(0) + 100(0.50) = 77.3(x_1) + 122.7(0.10)$$

This gives $x_1 = 0.488$.

To calculate an intermediate point on the operating line, pick $N_A(A_1) = 20.0$. Then for a balance on A as before,

$$V_1(y_1) - V(y) = N_A(A_1) \quad \text{or} \quad 100(0.50) - Vy = 20$$

For a balance on B,

$$V_1(1-y_1) - V(1-y) = -1.6N_A(A_1)$$

or

$$100(1-0.50) - V(1-y) = -1.6(20)$$

Solving simultaneously for V and y, $V = 112.0$ lb mole/hr and $y = 0.268$. Making a total material balance,

$$L + V_1 = L_1 + V \quad \text{or} \quad L + 100 = 77.3 + 112$$

This gives $L = 89.3$ lb mole/hr. Making a balance on A again,

$$Lx + V_1 y_1 = L_1 x_1 + Vy \quad \text{or} \quad 89.3(x) + 100(0.50) = 77.3(0.488) + 112(0.268)$$

Hence, $x = 0.199$.

The two end points, x_1, y_1, x_2, y_2, and the intermediate point are plotted in Fig. 8.2-4, together with the operating lines for the three

8.2 Material Balances and Operating Lines

different flux ratios for comparison. Only the equimolar counterdiffusion flux operating line is straight.

Example 8.2-4 Extraction of Nicotine in a Tower

The solute nicotine (A) in a water solution L_{W2} lb mass total/hr is to be stripped with a light petroleum oil V_{W1} lb mass total/hr at 68°F. The inlet water solution contains $x_{W2} = 0.015$ wt fraction nicotine in water with $L_{W2} = 100$, and the inlet oil is $V_{W1} = 200$ lb mass total/hr with $y_{W1} = 0.0005$ wt fraction. It is desired to remove 90 percent of the nicotine. Derive and plot the operating line. The water and light oil solvents are essentially insoluble. The equilibrium data are as follows (C1).

x_w, Wt Fraction Nicotine in Water	y_w, Wt Fraction Nicotine in Light Oil
0.001010	0.000806
0.00246	0.001959
0.00500	0.00454
0.00746	0.00682
0.00988	0.00904
0.0202	0.0185

Solution

The process flow diagram is given in Fig. 8.2-5 and the equilibrium data in Fig. 8.2-6. Since 90 percent is to be removed from the L_W stream, first calculate L'_W lb mass inert water/hr.

$$L'_W = L_{W2}(1 - x_{W2}) = 100(1 - 0.015) = 98.50 \text{ lb mass/hr}$$

Also for V'_W,

$$V'_W = V_{W1}(1 - y_{W1}) = 200(1 - 0.0005) = 199.9 \text{ lb mass/hr}$$

The pounds mass per hr of nicotine entering in L_{W2} are $L_{W2}(x_{W2}) = 100(0.015) = 1.50$ lb mass/hr. Then $0.10(1.50) = 0.150$ lb mass nicotine/hr in L_{W1}. Then,

$$x_{W1} = \frac{0.150}{98.5 + 0.150} = 0.00152 \quad \text{and} \quad L_{W1} = 98.5 + 0.15 = 98.65.$$

An overall material balance using pounds mass and weight fractions gives $L_{W2} + V_{W1} = L_{W1} + V_{W2}$ or $100 + 200 = 98.65 + V_{W2}$. Then $V_{W2} = 201.35$. A balance on component A gives

$$L_{W2}x_{W2} + V_{W1}y_{W1} = L_{W1}x_{W1} + V_{W2}y_{W2}$$

$$100(0.015) + 200(0.0005) = 98.65(0.00152) + 201.35(y_{W2})$$

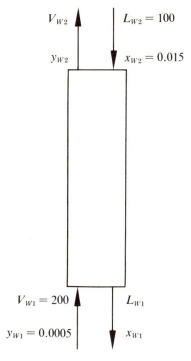

FIGURE 8.2-5 Flow diagram for extraction process of Example 8.2-4

Then $y_{W2} = 0.00720$. These two points on the operating line are plotted in Fig. 8.2-6.

An intermediate point will be calculated at $x_W = 0.0080$ to see whether the operating line is curved appreciably. Using Eq. (8.2-8) in terms of pounds mass and weight fractions,

$$L'_W\left(\frac{x_W}{1-x_W}\right) + V'_W\left(\frac{y_{W1}}{1-y_{W1}}\right) = L'_W\left(\frac{x_{W1}}{1-x_{W1}}\right) + V'_W\left(\frac{y_W}{1-y_W}\right)$$

$$98.50\left(\frac{0.008}{1-0.008}\right) + 199.9\left(\frac{0.0005}{1-0.0005}\right) = 98.50\left(\frac{0.00152}{1-0.00152}\right) + 199.9\left(\frac{y_W}{1-y_W}\right)$$

Solving, $y_W = 0.00371$. This point is plotted in Fig. 8.2-6; the operating line is quite straight, as expected.

4. *Limiting L'/V' Ratios.* In the designing of a countercurrent tower with phases V and L, the operating line may be above or below the equilibrium curve. When a solute is being removed from an inlet V stream by the L stream, we generally call it extraction if V and L are liquid streams and absorption if V is a gas and L a liquid.

In the absorption process the inlet gas flow V_1 and its composition y_1 are generally set. The exit concentration y_2 is also usually set by the

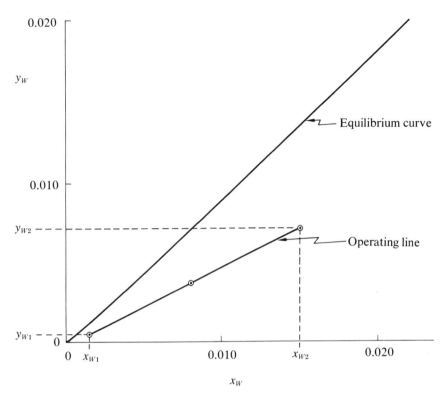

FIGURE 8.2-6 Operating and equilibrium lines for extraction process of Example 8.2-4

designer. Also, the concentration of the entering liquid x_2 is usually fixed by process requirements. The amount of the liquid flow L_2 is open to a choice by the designer.

In Fig. 8.2-7a the concentrations y_2, x_2, and y_1 are set. When the liquid flow is a minimum, the operating line has a minimum slope and touches the equilibrium line at y_1 and $x_{1_{\max}}$, which is point P. At this pinch point P the driving forces $y - y^*$, $y - y_i$, $x^* - x$, or $x_i - x$ are all zero and no mass transfer can occur. Hence, the minimum value of L is the one that causes the driving force to be zero at any point in the tower. The value of x_1 is a maximum when L is a minimum. To solve for the L' minimum the values of y_1 and $x_{1_{\max}}$ are substituted into the operating line equation.

In some cases, as in Fig. 8.2-7b, the minimum value of L is reached by the operating line becoming tangent to the equilibrium line instead of intersecting it as in Fig. 8.2-7a. In this case it is more or less trial and error to get the minimum value of L', because the operating line is curved. To minimize the trial-and-error process, the operating line is first

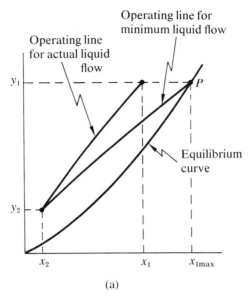

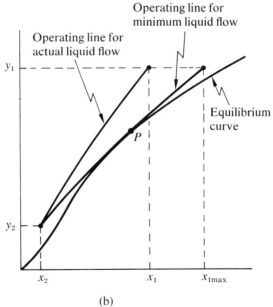

FIGURE 8.2-7 Minimum liquid/gas ratio for absorption

assumed straight, and this preliminary tangent point P is used, the y and x values at this point being substituted into the operating line equation. Then a value of L' is obtained, and using it the curved operating line is plotted. This line will not touch the equilibrium line at the same point P. However, a new estimate of P is made by observing where this line would touch again if rotated slightly counterclockwise or clockwise about y_2 and x_2 as a pivot. This second estimate of P is used again to calculate L' minimum. Two trials are usually sufficient.

When the solute (A) is being transferred from a liquid phase (C) to a gas phase (B), the process is called desorption or stripping. The operating line touches the equilibrium line at point P.
The flow of the liquid L' and x_2 and x_1 are specified. The inlet composition y_1 of the V phase is also set. The minimum gas flow occurs when the operating line touches the equilibrium line at point P.

The choice of the optimum L'/V' ratio to use depends on economics. For an absorption tower, too high a value requires a large liquid flow and, hence, a large-diameter tower. Also, the cost of recovering the solute from the liquid will be high. To small a liquid flow results in a high tower, which is costly. The optimum liquid flow for absorption is obtained by using a value of about 1.5 for the ratio of the average slope of the operating line to that of the equilibrium line. However, this factor can vary depending on the value of the solute and the type of tower used.

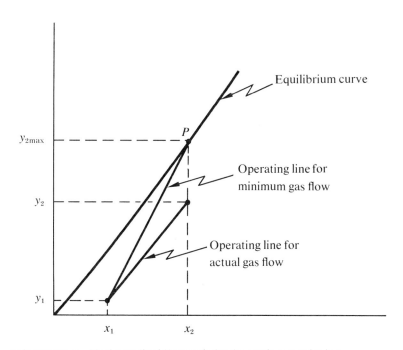

FIGURE 8.2-8 Maximum liquid/gas ratio for desorption or stripping

Example 8.2-5 Minimum Liquid Flow in a Tower

An ammonia gas stream from a reactor contains 25 percent ammonia and the rest inert gases. The total flow is 1000 lb mole/hr to an absorption tower at 86°F and 1 atm, where water containing 0.0050 mole fraction ammonia is the scrubbing liquid. It is desired to reduce the gas to an outlet concentration of 2.0 percent ammonia. What is the minimum liquid flow? Using 1.5 times the minimum, plot the operating line. The mole fraction equilibrium data at 1.0 atm are given in the appendix, Table A.4-2.

Solution

We have $y_1 = 0.25$, $V_1 = 1000$, $y_2 = 0.020$, $x_2 = 0.0050$. These values are plotted in Fig. 8.2-9. The value of $x_{1_{max}}$ is 0.143. The value of $V' = 1000(1 - y_1) = 1000(1 - 0.25) = 750$. Using Eq. (8.2-7),

$$L'\left(\frac{x_2}{1-x_2}\right) + V'\left(\frac{y_1}{1-y_1}\right) = L'\left(\frac{x_1}{1-x_1}\right) + V'\left(\frac{y_2}{1-y_2}\right)$$

$$L'\left(\frac{0.0050}{1-0.0050}\right) + 750\left(\frac{0.25}{1-0.25}\right) = L'\left(\frac{0.143}{1-0.143}\right) + 750\left(\frac{0.020}{1-0.020}\right)$$

Solving, $L'_{min} = 1448$ lb mole/hr. Using Eq. (8.2-8) and picking an intermediate value of x on this operating line as $x = 0.08$, we solve for a value of y.

$$L'\left(\frac{x}{1-x}\right) + V'\left(\frac{y_1}{1-y_1}\right) = L'\left(\frac{x_1}{1-x_1}\right) + V'\left(\frac{y}{1-y}\right)$$

$$1448\left(\frac{0.08}{1-0.08}\right) + 750\left(\frac{0.25}{1-0.25}\right) = 1448\left(\frac{0.143}{1-0.143}\right) + 750\left(\frac{y}{1-y}\right)$$

Hence, $y = 0.152$. This point is plotted to give the operating line at minimum L' rate in Fig. 8.2-9. The line is curved slightly.

Using $1.5L'_{min}$, $L' = 1.5(1448) = 2172$. Substituting this value into Eq. (8.2-7) to obtain x_1,

$$2172\left(\frac{0.0050}{1-0.0050}\right) + 750\left(\frac{0.25}{1-0.25}\right) = 2172\left(\frac{x_1}{1-x_1}\right) + 750\left(\frac{0.020}{1-0.020}\right)$$

Then $x_1 = 0.1017$ at the operating value of $L' = 2172$. Picking an intermediate value of $x = 0.06$, and using Eq. (8.2-8), we obtain a value of $y = 0.160$. These values are plotted in Fig. 8.2-9. To completely establish the final operating line curvature, several more intermediate points should be calculated.

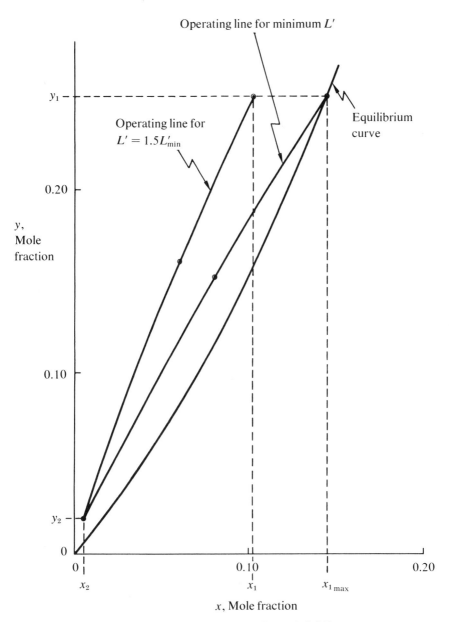

FIGURE 8.2-9 Minimum liquid-to-gas ratio for Example 8.2-5

5. *Multistage Countercurrent Processes.* In packed towers the two phases undergo continuous differential contact between these two phases. In tray towers and similar types of equipment the two phases

undergo stepwise contact of the L and the V phases, often called multistage cascades. For example, in a tray tower the two L and V streams enter a tray, are contacted intimately, and then are separated so that the exit L and V streams are considered in equilibrium. Such a tray or cascade is called one theoretical tray or stage. In this section only the multistage countercurrent tower will be considered. Chapter 9 will give a detailed treatment of multistage processes.

To determine the number of ideal stages required to bring about a given separation, the calculation is most conveniently done graphically. Figure 8.2-10 shows an absorber and the compositions on each tray. Starting at tray 1, y_0 and x_1 are on the operating line. The vapor y_1 leaving is in equilibrium with the leaving x_1. Then x_2 and y_1 are on the operating line and y_2 is in equilibrium with x_2, and so on. The further apart the operating and equilibrium lines the fewer the number of stages that are required. It should be pointed out that any set of consistent units, such as mole ratios or partial pressures and liquid mole fractions, can be used for the operating and equilibrium lines.

If the operating and equilibrium lines are nearly straight, mathematical expressions can be used to determine the number of ideal stages. Sherwood and Pigford (S1, p. 146) and Treybal (T1, p. 117) give convenient graphs to determine the number of ideal stages. Equations for this case are also derived in Chapter 9 of this text.

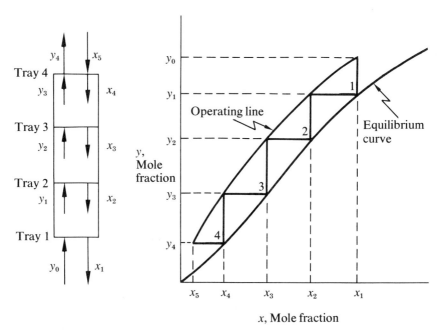

FIGURE 8.2-10 Determination of the number of ideal trays or stages in an absorber

Example 8.2-6 Number of Trays for Stream Stripping

A nonvolatile oil contains 4.00 mole percent propane and it is being stripped by direct superheated steam in a stripping tower to reduce the propane content to 0.20 percent. The total pressure is 2.0 atm abs, and the temperature is held constant at 300°F by internal heating in the tower. For 100 lb mole of oil plus propane feed, 3.808 lb mole of direct steam are used. At 2.0 atm and 300°F the vapor-liquid equilibria may be represented by $y = 25x$, where y is the mole fraction of propane in the steam and x the mole fraction of propane in the oil. Steam will not condense at the conditions used.

Plot the operating and equilibrium lines and determine the number of theoretical trays needed.

Solution

This is a case of propane (A) diffusing through stagnant oil (C) and stagnant steam (B). Here $L_2 = 100$ lb mole liquid/hr, $x_2 = 0.040$ mole fraction, $x_1 = 0.0020$, $y_1 = 0$, and $V_1 = V' = 3.808$ lb mole steam/hr.

Then $L' = L_2(1-x_2) = 100.0(1-0.04) = 96.0$. Making a material balance using Eq. (8.2-7),

$$L'\left(\frac{x_2}{1-x_2}\right) + V'\left(\frac{y_1}{1-y_1}\right) = L'\left(\frac{x_1}{1-x_1}\right) + V'\left(\frac{y_2}{1-y_2}\right)$$

$$96.0\left(\frac{0.040}{1-0.040}\right) + 3.808\left(\frac{0}{1-0}\right) = 96.0\left(\frac{0.002}{1-0.002}\right) + 3.808\left(\frac{y_2}{1-y_2}\right)$$

Solving, $y_2 = 0.50$. Setting $x = 0.010$, and using Eq. (8.2-8),

$$L'\left(\frac{x}{1-x}\right) + V'\left(\frac{y_1}{1-y_1}\right) = L'\left(\frac{x_1}{1-x_1}\right) + V'\left(\frac{y}{1-y}\right)$$

$$96.0\left(\frac{0.010}{1-0.010}\right) + 3.808\left(\frac{0}{1-0}\right) = 96.0\left(\frac{0.002}{1-0.002}\right) + 3.808\left(\frac{y}{1-y}\right)$$

Solving, $y = 0.170$. Again setting $x = 0.022$, $y = 0.340$. These points on the operating line are plotted in Fig. 8.2-11. Then, starting at the top, 5.5 theoretical trays are required.

8.2B Cocurrent Processes

In most processes countercurrent flow has many advantages over cocurrent flow of phases L and V. In countercurrent flow very low exit concentrations can often be achieved and still allow relatively large driving forces for reasonable tower heights. However, cocurrent flow is used in some instances. Some towers are operated in cocurrent flow in order to obtain high velocities without flooding, giving high mass transfer

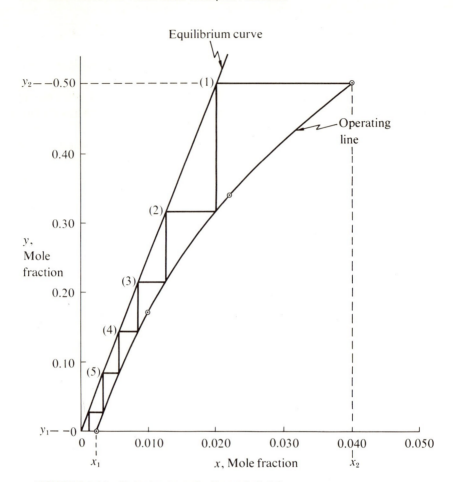

FIGURE 8.2-11 Stripping tower for Example 8.2-6

coefficients. Cocurrent flow also can be used if the solute A to be removed from V is absorbed by a liquid composed of pure A. Then there is no advantage to be obtained by countercurrent operation. The equilibrium relation is a horizontal line and the equilibrium concentration of y in equilibrium with the liquid is constant and unaffected by the amount of absorption.

Making an overall balance in Fig. 8.2-12,

$$V_1 + L_1 = V_2 + L_2 \qquad (8.2\text{-}11)$$

Making a balance on component A,

$$V'\left(\frac{y_1}{1-y_1}\right) + L'\left(\frac{x_1}{1-x_1}\right) = V'\left(\frac{y_2}{1-y_2}\right) + L'\left(\frac{x_2}{1-x_2}\right) \qquad (8.2\text{-}12)$$

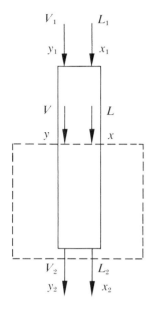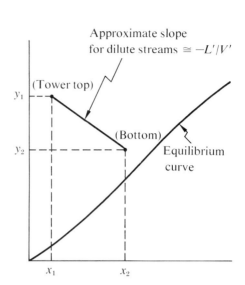

FIGURE 8.2-12 Cocurrent tower

Using the dotted-line box and making a balance on A, we obtain the operating line equation:

$$V'\left(\frac{y}{1-y}\right)+L'\left(\frac{x}{1-x}\right)=V'\left(\frac{y_2}{1-y_2}\right)+L'\left(\frac{x_2}{1-x_2}\right) \quad (8.2\text{-}13)$$

If the solutions are dilute, the slope $dy/dx \cong -L'/V'$, which is the negative slope for the countercurrent processes. A typical cocurrent operating line is shown in Fig. 8.2-12.

8.3 DESIGN METHODS FOR CONTINUOUS COUNTERCURRENT PROCESSES

8.3A Method Using Film Mass Transfer Coefficients

1. *Mass Transfer Coefficients on a Volume Basis in Packed Towers.* In packed towers the interfacial area A ft^2 between phases L and V is generally unknown. It is very difficult to measure this A experimentally and also the film coefficients k'_x and k'_y experimentally. Generally in a tower the measurements yield only a mass transfer coefficient that combines both the interfacial area and the film mass transfer coefficient.

Figure 8.3-1 shows a packed tower, where dz is the differential tower height in ft and S is the cross-sectional area of the tower, ft^2, without the

370 Continuous Two-Phase Mass Transport Processes

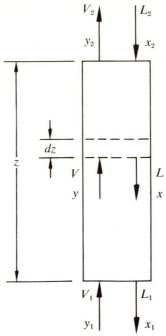

FIGURE 8.3-1   Packed tower and flows

packing. If a is defined as the (unknown) interfacial area in ft^2 per ft^3 volume of packed section, then the volume of packing in a height dz ft is $S\,dz$, and

$$dA = aS\,dz \tag{8.3-1}$$

where a is ft^2 interfacial area/ft^3 packing volume.

Since a is unknown, it is combined with the gas and liquid mass transfer coefficients as follows:

$$k'_y a = \frac{\text{lb mole}}{\text{hr-ft}^3 \text{ packing-(mole frac)}} \tag{8.3-2}$$

$$k'_x a = \frac{\text{lb mole}}{\text{hr-ft}^3 \text{ packing-(mole frac)}} \tag{8.3-3}$$

The values of $k'_y a$ and $k'_x a$ are usually determined experimentally in towers rather than a, k'_y, and k'_x separately.

2. *Design Method for A Diffusing through Stagnant B in Packed Towers.* For the design of packed towers the material balance equations (8.2-1) through (8.2-10) as derived previously still hold. Referring to Fig. 8.3-1 again and using the case for A diffusing through stagnant B, the operating line equation as given before is

8.3 Design Methods for Countercurrent Processes

$$L'\left(\frac{x}{1-x}\right) + V'\left(\frac{y_1}{1-y_1}\right) = L'\left(\frac{x_1}{1-x_1}\right) + V'\left(\frac{y}{1-y}\right) \quad (8.2\text{-}8)$$

The balance on A is, as before,

$$L'\left(\frac{x_2}{1-x_2}\right) + V'\left(\frac{y_1}{1-y_1}\right) = L'\left(\frac{x_1}{1-x_1}\right) + V'\left(\frac{y_2}{1-y_2}\right) \quad (8.2\text{-}7)$$

Making a balance on the differential height of tower, dz ft, the change in total moles in the V stream must equal the change in the liquid stream:

$$dV = dL \quad (8.3\text{-}4)$$

Also, for component A the amount of A leaving V must equal the amount entering the L stream:

$$d(Vy) = d(Lx) \quad (8.3\text{-}5)$$

where $d(Vy) = d(Lx) =$ lb mole A transferred/hr in dz tower height. Now the lb mole A transferred/hr as calculated from the material balance equation (8.3-5) must be equal to the lb mole A transferred/hr as calculated from the rate equation for N_A. First Eq. (7.3-7) is given as

$$N_A = \frac{k'_y}{(1-y_A)_{iM}}(y_{AG} - y_{Ai}) = \frac{k'_x}{(1-x_A)_{iM}}(x_{Ai} - x_{AL}) \quad (7.3\text{-}7)$$

where

$$(1-y_A)_{iM} = \frac{(1-y_{Ai}) - (1-y_{AG})}{\ln\left[\frac{(1-y_{Ai})}{(1-y_{AG})}\right]} \quad (7.3\text{-}5)$$

$$(1-x_A)_{iM} = \frac{(1-x_{AL}) - (1-x_{Ai})}{\ln\left[\frac{(1-x_{AL})}{(1-x_{Ai})}\right]} \quad (7.3\text{-}6)$$

Now, multiplying the left-hand side N_A of Eq. (7.3-7) by dA and the two right-side terms by $aS\,dz$, which is the equivalent of dA as given by Eq. (8.3-1),

$$N_A\,dA = \frac{k'_y a}{(1-y_A)_{iM}}(y_{AG} - y_{Ai})S\,dz = \frac{k'_x a}{(1-x_A)_{iM}}(x_{Ai} - x_{AL})S\,dz \quad (8.3\text{-}6)$$

where $N_A\,dA =$ lb mole A transferred/hr in height dz feet. Equating Eq. (8.3-5) to (8.3-6) and using y_{AG} as the mole fraction in the bulk gas phase and x_{AL} for the bulk liquid phase,

$$d(Vy_{AG}) = \frac{k'_y a}{(1-y_A)_{iM}}(y_{AG} - y_{Ai})S\,dz = d(Lx_{AL}) \quad (8.3\text{-}7)$$

$$d(Lx_{AL}) = \frac{k'_x a}{(1-x_A)_{iM}}(x_{Ai} - x_{AL})S\,dz \quad (8.3\text{-}8)$$

For the diffusion of A through stagnant B, $V = V'/(1 - y_{AG})$. Then, rewriting $d(Vy_{AG})$ as

$$d(Vy_{AG}) = d\left(\frac{V'}{1-y_{AG}} y_{AG}\right) = V' d\left(\frac{y_{AG}}{1-y_{AG}}\right) = \frac{V' dy_{AG}}{(1-y_{AG})^2} \quad (8.3\text{-}9)$$

But $V'/(1 - y_{AG})$ is equal to V and can be substituted in Eq. (8.3-9):

$$d(Vy_{AG}) = \frac{V \, dy_{AG}}{1 - y_{AG}} \quad (8.3\text{-}10)$$

Equating Eq. (8.3-10) to (8.3-7),

$$\frac{V \, dy_{AG}}{1 - y_{AG}} = \frac{k'_y a}{(1 - y_A)_{iM}} (y_{AG} - y_{Ai}) S \, dz \quad (8.3\text{-}11)$$

Doing the same for Eq. (8.3-8), since $L = L'/(1 - x_{AL})$,

$$\frac{L \, dx_{AL}}{1 - x_{AL}} = \frac{k'_x a}{(1 - x_A)_{iM}} (x_{Ai} - x_{AL}) S \, dz \quad (8.3\text{-}12)$$

Rearranging Eqs. (8.3-11) and (8.3-12) and integrating while dropping the subscripts A, G, and L for brevity,

$$\int_0^z dz = z = \int_{y_2}^{y_1} \frac{V \, dy}{\dfrac{k'_y a S}{(1-y)_{iM}} (1-y)(y-y_i)} \quad (8.3\text{-}13)$$

$$\int_0^z dz = z = \int_{x_2}^{x_1} \frac{L \, dx}{\dfrac{k'_x a S}{(1-x)_{iM}} (1-x)(x_i-x)} \quad (8.3\text{-}14)$$

Equations (8.3-13) and (8.3-14) must be integrated graphically in the general case, because the equilibrium and operating lines are generally curved and $k'_x a$ and $k'_y a$ may vary with total flows. The general steps to follow are given below and shown in Fig. 8.3-2.

1. The operating line Eq. (8.2-8) is plotted as in Fig. 8.3-2.
2. The values of $k'_y a$ and $k'_x a$ are obtained from empirical correlations available. They usually are functions of G_y^n and G_x^m, where G_y and G_x are total flow rates in lb mass/hr-ft² column cross-sectional area for gas and liquid, respectively, and n and m are exponents in the range of 0.2 to about 0.8. Using the material balance equation, values of G_y and G_x and, hence, $k'_y a$ and $k'_x a$ must be solved for at different values of y and x in the tower. Of course, if the tower is dilute, G_y and G_x will not vary much, and average values of them and $k'_x a$ and $k'_y a$ can be used.
3. Starting with the bottom of the tower at point P_1 (Fig. 8.3-2), the interface compositions at point M_1 are obtained. To do this, $(1-y)_{iM}$ and $(1-x)_{iM}$ must first be calculated from Eqs. (7.3-5) and (7.3-6) by trial and error. For the first trial assume y_i and y are the same and x_i and x are the same.

8.3 Design Methods for Countercurrent Processes

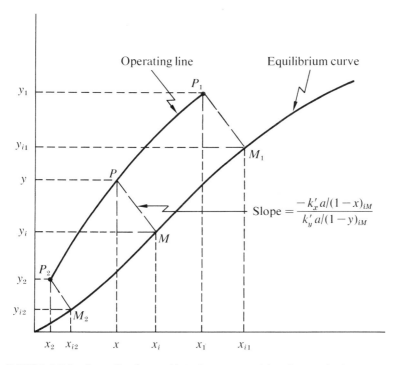

FIGURE 8.3-2 Operating line and interface compositions in a packed tower

Calculate the slope of line $P_1 M_1$ from Eq. (8.3-15), which is similar to Eq. (7.3-8) only with the "a" values included.

$$\text{slope} = \frac{-k'_x a/(1-x)_{iM}}{k'_y a/(1-y)_{iM}} \qquad (8.3\text{-}15)$$

Be sure to use the $k'_x a$ and $k'_y a$ values calculated in step 2 for the bottom of the tower, since G_y and G_x vary. Plot line $P_1 M_1$ and determine y_{i1} and x_{i1}. Using these values of y_{i1} and x_{i1} for the second trial, recalculate the slope in Eq. (8.3-15) and determine y_{i1} and x_{i1} the second time. If the values are close, no third trial is necessary.

Select another point in the tower at P and repeat the above. Do this for several intermediate points and also at point P_2 at the top of the tower. If the liquid solution is dilute where the values of x and x_i are less than 0.01 or 0.02 mole fraction, then $(1-x)_{iM}$ can be assumed as 1.0 which simplifies Eq. (8.3-15). Also, if the gas phase is dilute, $(1-y)_{iM}$ can be assumed as 1.0.

4. Using the values of y_i and x_i determined in step 3, graphically integrate Eq. (8.3-13) or (8.3-14) to obtain the tower height by plotting

$$\frac{V}{\dfrac{k'_y aS}{(1-y)_{iM}}(1-y)(y-y_i)}$$

versus y between y_2 and y_1 or

$$\frac{L}{\dfrac{k'_x aS}{(1-x)_{iM}}(1-x)(x_i-x)}$$

versus x between x_2 and x_1. Again, if x and x_i are small in dilute solutions, $(1-x)_{iM}$ can be assumed as 1.0 in Eq. (8.3-14). Similarly, $(1-y)_{iM}$ can be assumed as 1.0 if the gas phase is dilute.

Example 8.3-1 Design of an Absorption Tower

A column packed with 1-in. ceramic rings is to be designed to absorb SO_2 from an air stream by using pure water as the liquid at 68°F. The gas entering contains 40 percent SO_2 ($y_1 = 0.40$) and the leaving gas is to contain $y_2 = 0.02$ mole fraction. The total pressure is constant at 1.0 atm absolute. The rate of water used will be two times the minimum flow. The total inert air flow rate is 150 lb mass air/hr-ft² column cross-sectional area. The column cross section $S = 1.0$ ft².

The data of R. P. Whitney and J. E. Vivian (W1) are to be used to estimate the film mass transfer coefficients. Equilibrium data are available at 68°F and 1 atm in the appendix, Table A.4-3.

The empirical correlations for the film mass transfer coefficients for SO_2 at 68°F are as follows for 1-in. rings:

$$k_y a = 0.028 G_y^{0.7} G_x^{0.25}, \qquad k_x a = 0.152 G_x^{0.82}$$

where G_x and G_y are lb mass total liquid or gas respectively per hour per ft² of tower cross section.

Solution

The equilibrium data in mole fraction units are plotted in Fig. 8.3-3. Also $y_2 = 0.020$, $x_2 = 0$, $y_1 = 0.40$, and x_1 is undetermined as yet. For minimum liquid rate, the operating line will intersect the equilibrium line at $y_1 = 0.40$ and x_1 (from the curve) = 0.0125. The operating line is curved concavely when facing the equilibrium line, so it will not be tangent to it before this point. Using these values and substituting into Eq. (8.2-7),

$$L'\left(\frac{x_2}{1-x_2}\right) + V'\left(\frac{y_1}{1-y_1}\right) = L'\left(\frac{x_1}{1-x_1}\right) + V'\left(\frac{y_2}{1-y_2}\right)$$

$$L'\left(\frac{0}{1-0}\right) + V'\left(\frac{0.40}{1-0.40}\right) = L'\left(\frac{0.0125}{1-0.0125}\right) + V'\left(\frac{0.020}{1-0.020}\right)$$

Solving, $L'_{\min} = 51.2 V'$. Using two times $L'_{\min}$, $L' = 2(51.2)V' = 102.4 V'$ lb mole inert water/hr.

8.3 Design Methods for Countercurrent Processes

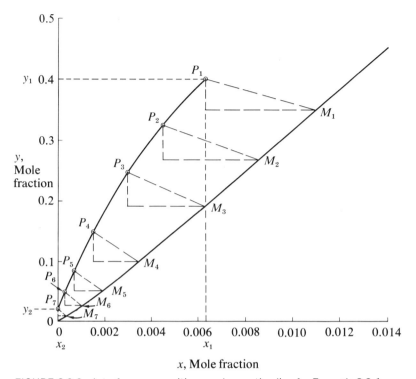

FIGURE 8.3-3 Interface compositions and operating line for Example 8.3-1

The total inert air flow = 150 lb mass air/hr-ft². Converting this to lb mole/hr-ft², $V'/S = 150/(29) = 5.18$ lb mole air/hr-ft². Hence, $L'/S = (102.4 V')/S = 102.4(5.18) = 530$ lb mole water/hr-ft². Since $S = 1.0$ ft², $V' = 5.18$ lb mole air/hr, and $L' = 530$ lb mole water/hr.

The final overall balance operating line equation is Eq. (8.2-7):

$$L'\left(\frac{x_2}{1-x_2}\right) + V'\left(\frac{y_1}{1-y_1}\right) = L'\left(\frac{x_1}{1-x_1}\right) + V'\left(\frac{y_2}{1-y_2}\right)$$

Substituting,

$$530\left(\frac{0}{1-0}\right) + 5.18\left(\frac{0.40}{1-0.40}\right) = 530\left(\frac{x_1}{1-x_1}\right) + 5.18\left(\frac{0.020}{1-0.020}\right)$$

Solving, $x_1 = 0.00628$.

The final operating line is

$$530\left(\frac{x}{1-x}\right) + 5.18\left(\frac{0.40}{1-0.40}\right) = 530\left(\frac{0.00628}{1-0.00628}\right) + 5.18\left(\frac{y}{1-y}\right)$$

Setting $x = 0.0045$ in the equation above and solving for y,

$$530\left(\frac{0.0045}{1-0.0045}\right) + 5.18\left(\frac{0.40}{1-0.40}\right) = 530\left(\frac{0.00628}{1-0.00628}\right) + 5.18\left(\frac{y}{1-y}\right)$$

$y = 0.325$. Selecting other values of x and solving for y, the following points on the operating line were calculated and plotted in Fig. 8.3-3.

x	y	G_y	G_x	$k'_x a$	$k'_y a$	x_i	y_i
0	0.0200	156.7	9550	279	8.58	0.00036	0.0082
0.00031	0.0500	167.5	9561	279	9.00	0.00103	0.025
0.00070	0.0845	180.7	9574	279	9.46	0.00195	0.050
0.00150	0.1485	207.8	9601	280	10.47	0.00350	0.100
0.00300	0.247	258.7	9652	281	12.17	0.00630	0.190
0.00450	0.325	310	9703	282	13.90	0.00855	0.266
0.00628	0.400	371	9764	283	15.80	0.01095	0.348

The mass velocities G_x and G_y must be calculated to estimate the mass transfer coefficients. The lb mole air/hr-ft^2 = 5.18. Lb mole SO$_2$/hr-ft^2 = $5.18[y/(1-y)]$. Total lb mole air + SO$_2$/hr-ft^2 = $5.18 + 5.18[y/(1-y)]$. Total lb mass/hr-ft^2 equals

$$G_y = 5.18(29) + 5.18\left(\frac{y}{1-y}\right)64.1 = 150 + 332\left(\frac{y}{1-y}\right)$$

Setting $y = 0.020$, $G_y = 150 + 332(0.02)/(0.98) = 156.7$ lb mass total gas/hr-ft^2. In like manner G_y is calculated for all points and tabulated.

For the liquid flow, lb mole inert liquid/hr-ft^2 = 530. Total lb mole liquid/hr-ft^2 = $530 + 530[x/(1-x)]$. Total lb mass/hr-ft^2 equals

$$G_x = 530(18) + 530\left(\frac{x}{1-x}\right)(64.1) = 9550 + 33{,}900\left(\frac{x}{1-x}\right)$$

Setting $x = 0.00150$,

$$G_x = 9550 + 33{,}900\frac{(0.00150)}{(0.9985)} = 9601 \text{ lb mass total liquid/hr-ft}^2$$

In like manner G_x is calculated for all points.

The experimental data of Whitney and Vivian for the liquid film were reported as $k_x a$. Since their solutions were very dilute, $(1-x)_{iM} \cong 1.0$ and $k'_x a = k_x a(1-x)_{iM} = k_x a(1.0)$. They determined $k_x a$ from experimental values of $K'_x a/(1-x)*_M$, where x was very small.

For calculation of $k'_x a$, $k'_x a = 0.152 G_x^{0.82}$. For $x = 0$, $G_x = 9550$ and $k'_x a = 0.152(9550)^{0.82} = 279$. In a similar manner the other values of $k'_x a$ were calculated. Since G_x does not vary much, an average value of $k'_x a$ could have been used.

For $k_y a$ as given, the data were obtained for an average $(1-y)_{iM}$ of

8.3 Design Methods for Countercurrent Processes 377

about 0.90. Hence an approximate correction will be made. For this correction, $k'_y a = k_y a(0.90) = 0.028(0.90) G_y^{0.7} G_x^{0.25} = 0.0252 G_y^{0.7} G_x^{0.25}$. Usually most experimenters obtain $k_y a$ using dilute solutions, so that the $k_y a$ reported is the same as $k'_y a$ or is already corrected for $(1-y)_{iM}$. For $x = 0$, $k'_y a = 0.0252(156.7)^{0.7}(9550)^{0.25} = 8.58$. The rest of the values are calculated and tabulated. Note that the values of $k'_y a$ vary greatly in the tower and an average value cannot be used.

The interface compositions must next be determined. For the point $y = 0.400$, $x = 0.00628$ we must make a preliminary estimate of $(1-y)_{iM}$ and $(1-x)_{iM}$. For y, assume $y_i = y = 0.40$.

$$(1-y)_{iM} = \frac{(1-y_i)-(1-y)}{\ln\left(\frac{1-y_i}{1-y}\right)} = 0.60$$

For x, assume $x_i = x = 0.00628$.

$$(1-x)_{iM} = \frac{(1-x)-(1-x_i)}{\ln\left(\frac{1-x}{1-x_i}\right)} = 0.993$$

The slope of the line $P_1 M_1$ by Eq. (8.3-15) is

$$\frac{-k'_x a/(1-x)_{iM}}{k'_y a/(1-y)_{iM}} = \frac{-283/0.993}{15.80/0.60} = -10.82$$

Plotting this on Fig. 8.3-3, $y_i = 0.355$, $x_i = 0.0112$. Using these for a second trial,

$$(1-y)_{iM} = \frac{(1-0.355)-(1-0.40)}{\ln\left(\frac{1-0.355}{1-0.40}\right)} = 0.622$$

$$(1-x)_{iM} = \frac{(1-0.00628)-(1-0.0112)}{\ln\left(\frac{1-0.00628}{1-0.0112}\right)} = 0.991$$

The slope by Eq. (8.3-15) is then $(-283/0.991)/(15.80/0.622) = -11.22$. Plotting this, $y_i = 0.348$, $x_i = 0.01095$. This final value of point M_1 is shown in Fig. 8.3-3.

For the point $y = 0.325$, $x = 0.00450$, assume $y_i = y = 0.325$ for the first trial. $(1-y)_{iM} = 0.675$. For x, assume $x_i = x = 0.00450$. $(1-x)_{iM} = 0.995$. Slope of $P_2 M_2 = -(282/0.995)/(13.90/0.675) = -13.80$. Plotting this, a second estimate of $y_i = 0.270$ and $x_i = 0.0086$. Then

$$(1-y)_{iM} = \frac{(1-0.270)-(1-0.325)}{\ln\left(\frac{1-0.270}{1-0.325}\right)} = 0.701$$

Also,

$$(1-x)_{iM} = \frac{(1-0.00450) - (1-0.0086)}{\ln\left(\frac{1-0.00450}{1-0.0086}\right)} = 0.993$$

The slope is then $(-282/0.993)/(13.90/0.701) = -14.3$. Plotting this, the final values are $y_i = 0.266$, $x_i = 0.00855$. In a similar manner the slopes of P_3M_3, P_4M_4, P_5M_5, P_6M_6, and P_7M_7 are calculated. It can be seen that the slopes are increasing markedly in going up the tower. The values of y_i and x_i are tabulated above. The reader should note that in using Eq. (8.3-15) to obtain the interface compositions, $(1-x)_{iM}$ could have been assumed as 1.0 with an error of 1 percent or less in the slope since the liquid solution is dilute. This assumption is often made for dilute solutions and greatly simplifies the calculations.

Equation (8.3-13) will be graphically integrated.

$$\int_0^z dz = z = \int_{y_2}^{y_1} \frac{V\,dy}{\dfrac{k'_y a}{(1-y)_{iM}} S(1-y)(y-y_i)}$$

In the equation above, values of total V are needed. At $y = 0.40$,

$$V = 5.18 + 5.18\left(\frac{0.40}{1-0.40}\right) = 8.63 \text{ lb mole/hr}$$

Values of V for other values of y are calculated and tabulated in the accompanying table. For $y = 0.020$,

$$\frac{V}{\dfrac{k'_y a}{(1-y)_{iM}} S(1-y)(y-y_i)} = \frac{5.29}{\dfrac{8.58}{0.986}(1.0)(0.980)(0.0118)} = 52.6$$

Similarly, the other values are calculated and tabulated. Figure 8.3-4 gives the graphical integration.

y	y_i	$(1-y)$	$(y-y_i)$	$(1-y)_{iM}$	V	$\dfrac{V}{\dfrac{k'_y a}{(1-y)_{iM}} S(1-y)(y-y_i)}$
0.0200	0.0082	0.980	0.0118	0.986	5.29	52.6
0.0500	0.0250	0.950	0.0250	0.962	5.44	24.5
0.0845	0.0500	0.9155	0.0345	0.932	5.66	17.65
0.1485	0.100	0.8515	0.0485	0.876	6.09	12.40
0.247	0.190	0.753	0.057	0.781	6.88	10.28
0.325	0.266	0.675	0.059	0.704	7.63	9.73
0.400	0.348	0.600	0.052	0.625	8.63	10.95

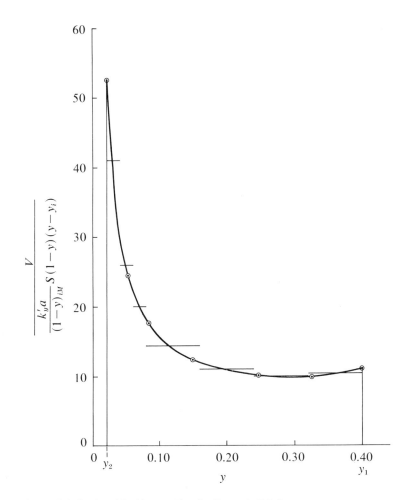

FIGURE 8.3-4   Graphical integration for Example 8.3-1

The total area in Fig. 8.3-4 is the sum of the seven rectangles:

$$0.820 + 0.520 + 0.400 + 1.152 + 0.880 + 0.800 + 0.832 = 5.404.$$

Hence, column height $z = 5.40$ ft.

3. *Design Method for Equimolar Counterdiffusion of A and B in Packed Towers.* For the case of equimolar counterdiffusion the material balance equations (8.2-2) and (8.2-4) hold and V and L are constant.

$$L_2 x_2 + V_1 y_1 = L_1 x_1 + V_2 y_2 \tag{8.2-2}$$

$$Lx + V_1 y_1 = L_1 x_1 + Vy \tag{8.2-4}$$

The operating line is straight. Making a balance on a differential height of tower as before,

$$d(Vy) = V\,dy = d(Lx) = L\,dx \tag{8.3-16}$$

The flux Eq. (7.3-2) is

$$N_A = k'_y(y - y_i) = k'_x(x_i - x) \tag{7.3-2}$$

Multiplying the left side N_A by dA and the right-side terms by $aS\,dz$,

$$N_A\,dA = k'_y a(y - y_i) S\,dz = k'_x a(x_i - x) S\,dz \tag{8.3-17}$$

Equating Eq. (8.3-16) to (8.3-17),

$$V\,dy = k'_y a(y - y_i) S\,dz \tag{8.3-18}$$

$$L\,dx = k'_x a(x_i - x) S\,dz \tag{8.3-19}$$

Now, rearranging Eqs. (8.3-18) and (8.3-19) and integrating,

$$\int_0^z dz = z = \int_{y_2}^{y_1} \frac{V\,dy}{k'_y a(y - y_i) S} = \frac{V}{S}\int_{y_2}^{y_1} \frac{dy}{k'_y a(y - y_i)} \tag{8.3-20}$$

$$\int_0^z dz = z = \int_{x_2}^{x_1} \frac{L\,dx}{k'_x a(x_i - x) S} = \frac{L}{S}\int_{x_2}^{x_1} \frac{dx}{k'_x a(x_i - x)} \tag{8.3-21}$$

Since V and L are considered constant, they can be removed from the integral. The mass transfer coefficients will be constant if the mass flow rates G_x and G_y are constant. The molar flow rates V and L are constant, but their compositions will vary somewhat throughout the tower. Hence, assuming changes in physical properties in the tower are small, G_x and G_y are approximately constant,

$$z = \frac{V}{Sk'_y a}\int_{y_2}^{y_1} \frac{dy}{y - y_i} \tag{8.3-22}$$

$$z = \frac{L}{Sk'_x a}\int_{x_2}^{x_1} \frac{dx}{x_i - x} \tag{8.3-23}$$

The quantities inside the integral must be determined by graphical integration if the equilibrium line is curved. To get the interface compositions the slope is, from Eq. (7.3-3),

$$\frac{-k'_x a}{k'_y a} = \frac{y - y_i}{x - x_i} \tag{8.3-24}$$

This slope should be constant throughout the column.

8.3B Method Using Overall Mass Transfer Coefficients

1. *Design Method for A Diffusing through Stagnant B in Packed Towers.* For A diffusing through stagnant B, Eq. (7.3-30) can be written

8.3 Design Methods for Countercurrent Processes

as follows, using Eq. (8.3-1):

$$N_A \frac{dA}{S\,dz} = \frac{K'_y a}{(1-y)*_M}(y - y^*) = \frac{K'_x a}{(1-x)*_M}(x^* - x) \quad (8.3\text{-}25)$$

Also, multiplying Eqs. (7.3-32) and (7.3-33) by $1/a$,

$$\frac{1}{K'_y a/(1-y)*_M} = \frac{1}{k'_y a/(1-y)_{iM}} + \frac{m'}{k'_x a/(1-x)_{iM}} \quad (7.3\text{-}32)$$

$$\frac{1}{K'_x a/(1-x)*_M} = \frac{1}{m''k'_y a/(1-y)_{iM}} + \frac{1}{k'_x a/(1-x)_{iM}} \quad (7.3\text{-}33)$$

where

$$(1-y)*_M = \frac{(1-y^*) - (1-y)}{\ln\left[\frac{(1-y^*)}{(1-y)}\right]} \quad (7.3\text{-}31)$$

$$(1-x)*_M = \frac{(1-x) - (1-x^*)}{\ln\left[\frac{(1-x)}{(1-x^*)}\right]} \quad (7.3\text{-}34)$$

Using Eq. (8.3-25) and rearranging as before,

$$N_A\,dA = \frac{K'_y a}{(1-y)*_M}(y-y^*)S\,dz = \frac{K'_x a}{(1-x)*_M}(x^*-x)S\,dz \quad (8.3\text{-}26)$$

After equating Eq. (8.3-26) to $d(Vy)$ from Eq. (8.3-10) and $d(Lx)$,

$$z = \int_{y_2}^{y_1} \frac{V\,dy}{\dfrac{K'_y a}{(1-y)*_M} S(1-y)(y-y^*)} \quad (8.3\text{-}27)$$

$$z = \int_{x_2}^{x_1} \frac{L\,dx}{\dfrac{K'_x a}{(1-x)*_M} S(1-x)(x^*-x)} \quad (8.3\text{-}28)$$

Often the average values can be taken outside the integral.

$$z = \left(\frac{V}{K'_y a S}\right) \int_{y_2}^{y_1} \frac{(1-y)*_M\,dy}{(1-y)(y-y^*)} \quad (8.3\text{-}29)$$

$$z = \left(\frac{L}{K'_x a S}\right) \int_{x_2}^{x_1} \frac{(1-x)*_M\,dx}{(1-x)(x^*-x)} \quad (8.3\text{-}30)$$

Equations (8.3-27) to (8.3-30) can be used in many different modifications. Some of these are discussed below.

1. If the $K'_y a$ or $K'_x a$ does not vary much from one end of the tower to the other, then an average value can be used outside of the integral. Often

382 Continuous Two-Phase Mass Transport Processes

experimental data are only given as an average $K'_y a$ or $K'_x a$ and no film coefficients are available. This average, then, must be used in the design. Care must be taken that the coefficient is used for design over a concentration range and V and L range similar to the experimental range. This is especially important when the equilibrium line is markedly curved.

2. If the film coefficients $k'_x a$ and $k'_y a$ are known, it is preferable to use either of them in the design rather than predicting an overall coefficient and using it for design. Also, if the m or slope of the equilibrium line varies markedly, an average $K'_y a$ or $K'_x a$ can be in error.

3. If the solutions are not very concentrated, the ratio $(1-y)*_M/(1-y)$ or $(1-x)*_M/(1-x)$ may be close to 1.0 at both ends of the tower and an average value used outside of the integral. Even in concentrated solutions this approximation can often be used with little error.

Example 8.3-2   Tower Height Using Overall Mass Transfer Coefficient

For Example 8.3-1 recalculate the data using the overall mass transfer coefficient $K'_y a$.

Solution

To calculate $K'_y a$, Eq. (7.3-32) is as follows:

$$\frac{1}{K'_y a/(1-y)*_M} = \frac{1}{k'_y a/(1-y)_{iM}} + \frac{m'}{k'_x a/(1-x)_{iM}}$$

The values will be calculated at the top and the bottom of the tower. For the bottom, the slope of the chord m' is given by Eq. (7.3-17):

$$m' = \frac{y_i - y^*}{x_i - x} = \frac{0.348 - 0.190}{0.01095 - 0.00628} = 33.7$$

Then,

$$(1-y)*_M = \frac{(1-y^*) - (1-y)}{\ln\left(\frac{1-y^*}{1-y}\right)} = \frac{(1-0.190)-(1-0.40)}{\ln\left(\frac{1-0.190}{1-0.40}\right)} = 0.700$$

$$(1-y)_{iM} = \frac{(1-y_i) - (1-y)}{\ln\left(\frac{1-y_i}{1-y}\right)} = \frac{(1-0.348)-(1-0.40)}{\ln\left(\frac{1-0.348}{1-0.40}\right)} = 0.625$$

$$(1-x)_{iM} = \frac{(1-x)-(1-x_i)}{\ln\left(\frac{1-x}{1-x_i}\right)} = \frac{(1-0.00628)-(1-0.01095)}{\ln\left(\frac{1-0.00628}{1-0.01095}\right)} = 0.991$$

$$\frac{1}{K'_y a/0.700} = \frac{1}{15.80/0.625} + \frac{33.7}{283/0.991} = 0.0395 + 0.118 = 0.1575$$

$$K'_y a = 4.43$$

8.3 Design Methods for Countercurrent Processes

Then,

$$\frac{V}{K'_y aS} = \frac{8.63}{4.43(1.0)} = 1.95$$

For the top of the tower, repeating the calculations,

$$m' = \frac{0.0082 - 0}{0.00036 - 0} = 22.8$$

$$(1-y)*_M = \frac{(1-0)-(1-0.020)}{\ln\left(\frac{1-0}{1-0.020}\right)} = 0.990$$

$$(1-y)_{iM} = \frac{(1-0.0082)-(1-0.020)}{\ln\left(\frac{1-0.0082}{1-0.020}\right)} = 0.986$$

$$(1-x)_{iM} = \frac{(1-0)-(1-0.00036)}{\ln\left(\frac{1-0}{1-0.00036}\right)} = 0.999$$

$$\frac{1}{K'_y a/0.990} = \frac{1}{8.58/0.986} + \frac{22.8}{279/0.999}$$

$$= 0.1147 + 0.0818 = 0.1965$$

$$K'_y a = 5.03$$

The average $K'_y a = (5.03 + 4.43)/2 = 4.73$. This does not vary appreciably. Then,

$$\frac{V}{K'_y aS} = \frac{5.29}{5.03(1.0)} = 1.05$$

The average value of $V/K'_y aS = (1.95 + 1.05)/2 = 1.50$. It is evident that this overall value varies appreciably throughout the tower. Hence, the V will be kept inside the integral and an average $K'_y a$ outside. To graphically integrate Eq. (8.3-27), the following values are obtained and plotted in Fig. 8.3-5, with the average $K'_y a$ kept outside of the integral.

y	y^*	$(1-y)$	$(y-y^*)$	$(1-y^*)$	$(1-y)*_M$	V	$\dfrac{V(1-y)*_M}{(1-y)(y-y^*)}$
0.0200	0	0.980	0.0200	1.00	0.990	5.29	267
0.0500	0.0070	0.950	0.0430	0.993	0.971	5.44	129.5
0.0845	0.0165	0.9155	0.0680	0.9835	0.949	5.66	86.2
0.1485	0.038	0.8515	0.1105	0.962	0.906	6.09	58.8
0.247	0.084	0.753	0.163	0.916	0.833	6.88	46.8
0.325	0.132	0.675	0.193	0.868	0.771	7.63	45.2
0.400	0.190	0.600	0.210	0.810	0.700	8.63	48.1

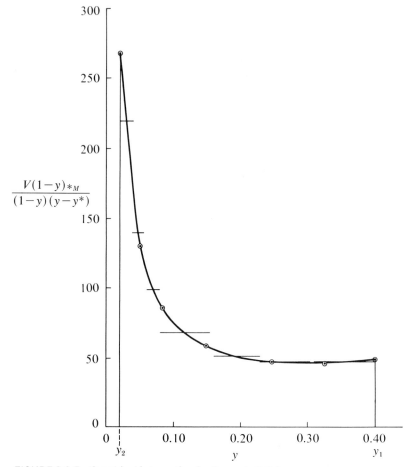

FIGURE 8.3-5 Graphical integration for Example 8.3-2

The total area in Fig. 8.3-5 is the sum of the seven rectangles:

$$4.40 + 2.80 + 1.98 + 5.48 + 4.10 + 3.76 + 3.76 = 26.28$$

Then,

$$z = \frac{1}{K'_y aS}(\text{area}) = \frac{1}{4.73(1)}(26.28) = 5.55 \text{ ft}$$

This value is reasonably close to the value of 5.40 ft obtained in Example 8.3-1 using $k'_y a$. However, it was still necessary to have the film coefficients to obtain y_i and x_i.

The ratio $(1-y)_{*M}/(1-y)$ at the tower top is $(0.99)/(0.98)$ or 1.01 and $0.700/0.600$ or 1.16 at the bottom. The average is $(1.16 + 1.01)/2 = 1.085$. Plotting only $V/(y-y^*)$ versus y, the area is 24.92. Then the tower height is

$$z = \frac{(1.085)}{(4.73)(1)}(24.92) = 5.71 \text{ ft}$$

This value is reasonably close to 5.40 ft. Hence, the method of using an average value of $(1-y)_{*M}/(1-y)$ outside the integral works reasonably well even for a concentrated mixture.

2. *Design Method for Equimolar Counterdiffusion of A and B in Packed Towers.* Using the same methods used to derive Eqs. (8.3-22) and (8.3-23), we can derive the following equations for equimolar counterdiffusion with overall mass transfer coefficients:

$$z = \frac{V}{S}\int_{y_2}^{y_1} \frac{dy}{K'_y a(y - y^*)} \tag{8.3-31}$$

$$z = \frac{L}{S}\int_{x_2}^{x_1} \frac{dx}{K'_x a(x^* - x)} \tag{8.3-32}$$

Since V and L are constant, they are removed outside of the integrals. Also, if the equilibrium lines are reasonably straight over the range of concentrations used, $K'_y a$ and $K'_x a$ can be removed from the integrals and average values used. This can be done provided the physical properties remain reasonably constant through the tower.

8.3C Method Using Transfer Units

1. *Design Method for A Diffusing through Stagnant B in Packed Towers.* A newer and in some ways more useful method for design of packed towers is the use of the transfer unit concept. For the diffusion of A through stagnant B, Eqs. (8.3-13), (8.3-14), (8.3-29), and (8.3-30) can be written as

$$z = \int_{y_2}^{y_1} \left(\frac{V}{k'_y aS}\right) \frac{(1-y)_{iM}\, dy}{(1-y)(y-y_i)} = H_G \int_{y_2}^{y_1} \frac{(1-y)_{iM}\, dy}{(1-y)(y-y_i)} \tag{8.3-33}$$

$$z = \int_{x_2}^{x_1} \left(\frac{L}{k'_x aS}\right) \frac{(1-x)_{iM}\, dx}{(1-x)(x_i - x)} = H_L \int_{x_2}^{x_1} \frac{(1-x)_{iM}\, dx}{(1-x)(x_i - x)} \tag{8.3-34}$$

$$z = \int_{y_2}^{y_1} \left(\frac{V}{K'_y aS}\right) \frac{(1-y)_{*M}\, dy}{(1-y)(y-y^*)} = H_{OG} \int_{y_2}^{y_1} \frac{(1-y)_{*M}\, dy}{(1-y)(y-y^*)} \tag{8.3-35}$$

$$z = \int_{x_2}^{x_1} \left(\frac{L}{K'_x aS}\right) \frac{(1-x)_{*M}\, dx}{(1-x)(x^* - x)} = H_{OL} \int_{x_2}^{x_1} \frac{(1-x)_{*M}\, dx}{(1-x)(x^* - x)} \tag{8.3-36}$$

where

$$H_G = \frac{V}{k'_y aS} = \frac{V}{k_y a(1-y)_{iM} S} = \frac{V}{k_G aP(1-y)_{iM} S} \quad (8.3\text{-}37)$$

$$H_L = \frac{L}{k'_x aS} = \frac{L}{k_x a(1-x)_{iM} S} = \frac{L}{k_L ac(1-x)_{iM} S} \quad (8.3\text{-}38)$$

$$H_{OG} = \frac{V}{K'_y aS} = \frac{V}{K_y aS(1-y)*_M S} = \frac{V}{K_G aP(1-y)*_M S} \quad (8.3\text{-}39)$$

$$H_{OL} = \frac{L}{K'_x aS} = \frac{L}{K_x a(1-x)*_M S} = \frac{L}{K_L ac(1-x)*_M S} \quad (8.3\text{-}40)$$

The units of each H are in feet. The values of the heights of transfer units H are generally more constant than the mass transfer coefficients themselves. For example, since $k'_y a$ is often proportional to $V^{0.8}$, then the $H_G \propto V^{1.0}/V^{0.8} \propto V^{0.2}$. Also, since $k'_x a$ is often proportional to $L^{0.8}$, then $H_L \propto L^{1.0}/L^{0.8} \propto L^{0.2}$. These heights of transfer units are removed from the integral and average values used. The average value of $(1-y)_{iM}$, $(1-y)*_M$, $(1-x)_{iM}$, and $(1-x)*_M$ must be used.

The number of transfer units N is defined as the integral on the right side of Eqs. (8.3-33) to (8.3-36). Hence,

$$N_G = \int_{y_2}^{y_1} \frac{(1-y)_{iM}\, dy}{(1-y)(y-y_i)} \quad (8.3\text{-}41)$$

$$N_L = \int_{x_2}^{x_1} \frac{(1-x)_{iM}\, dx}{(1-x)(x_i-x)} \quad (8.3\text{-}42)$$

$$N_{OG} = \int_{y_2}^{y_1} \frac{(1-y)*_M\, dy}{(1-y)(y-y^*)} \quad (8.3\text{-}43)$$

$$N_{OL} = \int_{x_2}^{x_1} \frac{(1-x)*_M\, dx}{(1-x)(x^*-x)} \quad (8.3\text{-}44)$$

The height of the tower z is then

$$z = H_G N_G = H_L N_L = H_{OG} N_{OG} = H_{OL} N_{OL} \quad (8.3\text{-}45)$$

Basically these equations are no different than the equations using mass transfer coefficients, but the height of transfer unit will be more constant throughout a tower, so it is removed from the integral. One still needs $k'_y a$ and $k'_x a$ to evaluate the interface concentrations y_i and x_i.

A physical interpretation of one transfer unit can be given as follows. If in Eq. (8.3-33) we disregard the ratio $(1-y)_{iM}/(1-y)$, which is near 1.0, then the remainder is the number of times the driving force $(y-y_i)$ divides into the change of gas concentration (y_1-y_2). This is really a measure of the difficulty of absorption, since a large (y_1-y_2) gives a large number of transfer units as does a small driving force $(y-y_i)$.

8.3 Design Methods for Countercurrent Processes

The film and overall heights of transfer units can be combined by substituting into Eqs. (7.3-32) and (7.3-33) for $m' = m'' = m$,

$$H_{OG} = H_G \frac{(1-y)_{iM}}{(1-y)*_M} + \frac{mV}{L} H_L \frac{(1-x)_{iM}}{(1-y)*_M} \tag{8.3-46}$$

$$H_{OL} = H_L \frac{(1-x)_{iM}}{(1-x)*_M} + \frac{L}{mV} H_G \frac{(1-y)_{iM}}{(1-x)*_M} \tag{8.3-47}$$

Example 8.3-3 Tower Height Using Number of Transfer Units

For Example 8.3-1 calculate the tower height using the H_G and the number of transfer units N_G.

Solution

From the calculations in Example 8.3-1, at the top of the tower $y_2 = 0.020$, $k'_y a = 8.58$, $V = 5.29$, and then

$$H_G = \frac{V}{k'_y a S} = \frac{5.29}{8.58(1.0)} = 0.615 \text{ ft}$$

At the bottom,

$$H_G = \frac{8.63}{15.80(1.0)} = 0.547 \text{ ft}$$

The average is

$$H_G = \frac{0.615 + 0.547}{2} = 0.581 \text{ ft}$$

To calculate the number of transfer units, Eq. (8.3-41) is

$$N_G = \int_{y_2}^{y_1} \frac{(1-y)_{iM} \, dy}{(1-y)(y-y_i)}$$

The calculated data are available in Example 8.3-1. It remains only to calculate $(1-y)_{iM}/[(1-y)(y-y_i)]$ for each value of y. For $y = 0.0200$,

$$\frac{(1-y)_{iM}}{(1-y)(y-y_i)} = \frac{0.986}{0.980(0.0118)} = 85.0$$

These values are calculated for each y and plotted, and the area by graphical integration is $N_G = 9.30$ transfer units. Column height $z = H_G N_G = 0.581(9.30) = 5.40$ ft. This is the same value as in Example 8.3-1, using $k'_y a$.

2. *Design Method for Equimolar Counterdiffusion of A and B in Packed Towers.* The transfer unit concept can also be applied to the case

of equimolar counterdiffusion of A and B. The equations are simpler than those for A through stagnant B.

$$z = H_G N_G = \left(\frac{V}{k'_y aS}\right) \int_{y_2}^{y_1} \frac{dy}{y - y_i} \tag{8.3-48}$$

$$z = H_L N_L = \left(\frac{L}{k'_x aS}\right) \int_{x_2}^{x_1} \frac{dx}{x_i - x} \tag{8.3-49}$$

$$z = H_{OG} N_{OG} = \left(\frac{V}{K'_y aS}\right) \int_{y_2}^{y_1} \frac{dy}{y - y^*} \tag{8.3-50}$$

$$z = H_{OL} N_{OL} = \left(\frac{L}{K'_x aS}\right) \int_{x_2}^{x_1} \frac{dx}{x^* - x} \tag{8.3-51}$$

8.3D Simplified Design Methods for Dilute Gases and A Diffusing through Stagnant B

When the solutions are reasonably dilute, various simplifications are possible of the mass transfer coefficient equations (8.3-13), (8.3-14), (8.3-29), and (8.3-30) and of the transfer unit equations (8.3-33) to (8.3-44). The concentrations can be considered dilute for engineering design purposes when all of the mole fractions of y and x in the two streams are less than about 10 percent. Then the flows will vary by less than 10 percent and the mass transfer coefficients by less than this. Hence, average values of the mass transfer coefficients and flows V and L can be taken outside of the integral in these equations. Also, the terms $(1-y)_{iM}/(1-y)$, $(1-y)*_M/(1-y)$, $(1-x)*_M/(1-x)$, and $(1-x)_{iM}/(1-x)$ can be taken outside of the integral and average values used. When quite dilute these terms can be dropped out as 1.0. The equations are then

$$z = \left[\frac{V(1-y)_{iM}}{k'_y aS(1-y)}\right]_{ave} \int_{y_2}^{y_1} \frac{dy}{y - y_i} \tag{8.3-52}$$

$$z = \left[\frac{L(1-x)_{iM}}{k'_x aS(1-x)}\right]_{ave} \int_{x_2}^{x_1} \frac{dx}{x_i - x} \tag{8.3-53}$$

$$z = \left[\frac{V(1-y)*_M}{K'_y aS(1-y)}\right]_{ave} \int_{y_2}^{y_1} \frac{dy}{y - y^*} \tag{8.3-54}$$

$$z = \left[\frac{L(1-x)*_M}{K'_x aS(1-x)}\right]_{ave} \int_{x_2}^{x_1} \frac{dx}{x^* - x} \tag{8.3-55}$$

$$z = H_G N_G = H_G \left[\frac{(1-y)_{iM}}{1-y}\right]_{ave} \int_{y_2}^{y_1} \frac{dy}{y - y_i} \tag{8.3-56}$$

$$z = H_L N_L = H_L \left[\frac{(1-x)_{iM}}{1-x}\right]_{ave} \int_{x_2}^{x_1} \frac{dx}{x_i - x} \tag{8.3-57}$$

8.3 Design Methods for Countercurrent Processes

$$z = H_{OG}N_{OG} = H_{OG}\left[\frac{(1-y)*_M}{1-y}\right]_{ave}\int_{y_2}^{y_1}\frac{dy}{y-y^*} \quad (8.3\text{-}58)$$

$$z = H_{OL}N_{OL} = H_{OL}\left[\frac{(1-x)*_M}{1-x}\right]_{ave}\int_{x_2}^{x_1}\frac{dx}{x^*-x} \quad (8.3\text{-}59)$$

The equations above hold whether the equilibrium line is curved or straight. The graphical integration of the integral takes this into account.

A further simplification can be obtained if the solution is dilute and the equilibrium line is straight over the range of concentrations used. Then $(y - y_i)$ varies linearly with y and with x.

$$(y - y_i) = ky + b \quad (8.3\text{-}60)$$

where k and b are constants. The operating line will be reasonably straight and the slopes in Eq. (8.3-15) will be constant. Then the integral in Eq. (8.3-52) can be integrated.

$$\int_{y_2}^{y_1}\frac{dy}{y-y_i} = \frac{y_1 - y_2}{(y-y_i)_M} \quad (8.3\text{-}61)$$

where

$$(y - y_i)_M = \frac{(y-y_i)_1 - (y-y_i)_2}{\ln\left[\frac{(y-y_i)_1}{(y-y_i)_2}\right]} \quad (8.3\text{-}62)$$

Equation (8.3-61) is the amount absorbed $(y_1 - y_2)$ divided by the ln-mean driving force given in Eq. (8.3-62). Each of the integrals in Eqs. (8.3-52) to (8.3-59) can be written as above. If $(1-y)_{iM}/(1-y)$ is considered 1.0, then Eq. (8.3-61) is the number of transfer units, N_G. Equations (8.3-52) and (8.3-61) can be written then as follows:

$$\frac{V}{S}(y_1 - y_2) = k'_y az(y - y_i)_M = k'_G azP(y - y_i)_M \quad (8.3\text{-}63)$$

where the left-hand side is lb mole absorbed/hr-ft² and the right-hand side is the rate equation. Equations similar to Eq. (8.3-63) can be written for $K'_y a$ using $(y - y^*)_M$, for $k'_x a$ using $(x_i - x)_M$, and so on. Also, Eq. (8.3-61) can be used in a similar manner with Eq. (8.3-48) for equimolar counter-diffusion.

Example 8.3-4 Absorption Using Number of Transfer Units N_G

Ammonia is being absorbed from a reactor gas stream, which can be considered to be inert air at 68°F and 1 atm, by pure water. The packed tower is operated with a water flow of 100 lb mole/hr and a total inlet gas flow of 85 lb mole air and ammonia/hr. The tower diameter is 2.0 ft.

390 Continuous Two-Phase Mass Transport Processes

The inlet gas contains 4.0 mole percent NH_3 and the outlet 0.5 percent.
From correlations for dilute solutions the average H_L has been estimated as 0.85 ft and average H_G as 1.6 ft. The equilibrium data in mole fraction of NH_3 at 1.0 atm in the liquid and vapor are as follows:

x	y
0.0208	0.0158
0.0309	0.0240
0.0503	0.0418

(a) Plot the operating and equilibrium lines. (b) Convert the transfer unit coefficients to $k'_y a$ and $k'_x a$ and find the interface compositions. Then calculate the number of transfer units N_G and the tower height. (c) Repeat part (b) using ln-mean driving forces.

Solution

$L' = 100$, $V' = 85$, $y_1 = 0.040$, $y_2 = 0.005$, $x_2 = 0$. Then $V' = 85(1-y_1) = 85(1-0.04) = 81.6$ lb mole inert air/hr. The operating line equation (8.2-7) is

$$L'\left(\frac{x_2}{1-x_2}\right) + V'\left(\frac{y_1}{1-y_1}\right) = L'\left(\frac{x_1}{1-x_1}\right) + V'\left(\frac{y_2}{1-y_2}\right)$$

$$100\left(\frac{0}{1-0}\right) + 81.6\left(\frac{0.04}{1-0.04}\right) = 100\left(\frac{x_1}{1-x_1}\right) + 81.6\left(\frac{0.005}{1-0.005}\right)$$

Hence, $x_1 = 0.0291$. To calculate an intermediate point on the operating line, set $y = 0.020$; then,

$$100\left(\frac{x}{1-x}\right) + 81.6\left(\frac{0.04}{1-0.04}\right) = 100\left(\frac{0.0291}{1-0.0291}\right) + 81.6\left(\frac{0.020}{1-0.020}\right)$$

Solving, $x = 0.0124$. Setting $y = 0.030$, $x = 0.02064$.

These points are plotted as the operating line in Fig. 8.3-6. No more points will be calculated, since the operating line is very nearly straight. [Ans. to (a)]

To convert the coefficients to $k'_y a$ and $k'_x a$, Eq. (8.3-37) is $H_G = V/k'_y aS$. $S = \pi D^2/4 = \pi(2.0)^2/4 = 3.14$ ft^2. The H_G used is the average H_G. Hence, the average V must be used.

$$V_1 = 85.0, \qquad V_2 = 81.6\left(\frac{1}{1-0.005}\right) = 82.0$$

$$V_{ave} = \frac{85.0 + 82.0}{2} = 83.5$$

Then,

$$k'_y a = \frac{V}{SH_G} = \frac{83.5}{3.14(1.6)} = 16.6$$

8.3 Design Methods for Countercurrent Processes

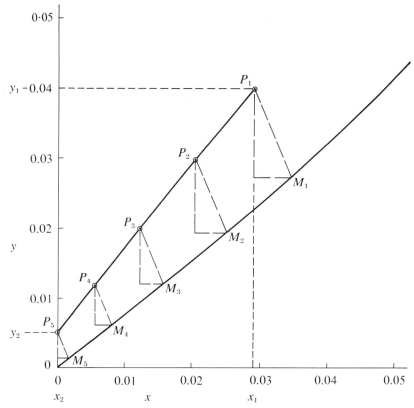

FIGURE 8.3-6 Graphical construction for Example 8.3-4

For $k'_x a$, $H_L = L/k'_x aS$.

$$L_2 = 100, \quad L_1 = 100\left(\frac{1}{1-0.0291}\right) = 103, \quad L_{\text{ave}} = 101.5$$

Then,

$$k'_x a = \frac{L}{SH_L} = \frac{101.5}{3.14(0.85)} = 38.0$$

To determine the interface compositions, Eq. (8.3-15) is used:

$$\frac{-k'_x a/(1-x)_{iM}}{k'_y a/(1-y)_{iM}} = \text{slope of line from } P_1 \text{ to } M_1$$

This is trial and error, since x_i and y_i values must be estimated.

For the point $y = 0.04$, $x = 0.0291$, assume $y_i = 0.03$ and $x_i = 0.04$ as a first estimate. Then,

$$(1-y)_{iM} = \frac{(1-y_i)-(1-y)}{\ln\left(\frac{1-y_i}{1-y}\right)} = \frac{(1-0.03)-(1-0.04)}{\ln\left(\frac{1-0.03}{1-0.04}\right)} = 0.965$$

$$(1-x)_{iM} = \frac{(1-x)-(1-x_i)}{\ln\left(\frac{1-x}{1-x_i}\right)} = \frac{(1-0.0291)-(1-0.04)}{\ln\left(\frac{1-0.0291}{1-0.04}\right)} = 0.965$$

Then the slope is $(-38.0/0.965)/(16.6/0.965) = -2.29$. Plotting this in Fig. 8.3-6, values at M_1 are $y_i = 0.0272$ and $x_i = 0.0346$. Recalculating $(1-y)_{iM}$ and $(1-x)_{iM}$, the values are essentially unchanged.

For other points in the tower the ratio $(1-y)_{iM}/(1-x)_{iM}$ is very close to 1.0; hence, the slope of all lines to the interface is constant at $-k'_x a/k'_y a = -2.29$. We draw a series of parallel lines and tabulate the interface values:

x	y	y_i	x_i	$y-y_i$	$\dfrac{1}{y-y_i}$
0.0291	0.040	0.0272	0.0346	0.0128	78.2
0.02064	0.030	0.0193	0.0250	0.0107	93.3
0.0124	0.020	0.0120	0.0158	0.0080	125.0
0.0056	0.012	0.0062	0.0080	0.0058	172
0	0.005	0.0014	0.0016	0.0036	277

For the graphical integration Eq. (8.3-56) will be used, since $(1-y)_{iM}/(1-y)$ will not vary much. This ratio will be calculated at the top and the bottom and the average used. At the bottom,

$$(1-y)_{iM} = \frac{(1-0.0272)-(1-0.040)}{\ln\left(\frac{1-0.0272}{1-0.040}\right)} = 0.966$$

$1-y = 1.0 - 0.040 = 0.960$. Ratio $= 0.966/0.960 = 1.006$. At the top,

$$(1-y)_{iM} = \frac{(1-0.0014)-(1-0.005)}{\ln\left(\frac{1-0.0014}{1-0.005}\right)} = 0.996$$

$1-y = 1.0 - 0.005 = 0.995$. Ratio $= 0.996/0.995 = 1.001$. Hence, the average of $(1.006+1.001)/2 = 1.003$ will be used, and its removal from the integral was justified.

For the graphical integration, a plot of $1/(y-y_i)$ versus y is made in Fig. 8.3-7. The area $= 0.850 + 1.080 + 1.530 + 1.135 = 4.595$. Then,

$$N_G = \left[\frac{(1-y)_{iM}}{1-y}\right]_{ave} \int_{y_2}^{y_1} \frac{dy}{y-y_i} = 1.003(4.595) = 4.61 \text{ transfer units}$$

The tower height $= H_G N_G = 1.6(4.61) = 7.38$ ft. [Ans. to (b)]
For part (c) the ln-mean driving force, Eq. (8.3-62), is

$$(y-y_i)_M = \frac{(y-y_i)_1 - (y-y_i)_2}{\ln\left[\frac{(y-y_i)_1}{(y-y_i)_2}\right]} = \frac{(0.04-0.0272)-(0.005-0.0014)}{\ln\left(\frac{0.04-0.0272}{0.005-0.0014}\right)}$$
$$= 0.00725$$

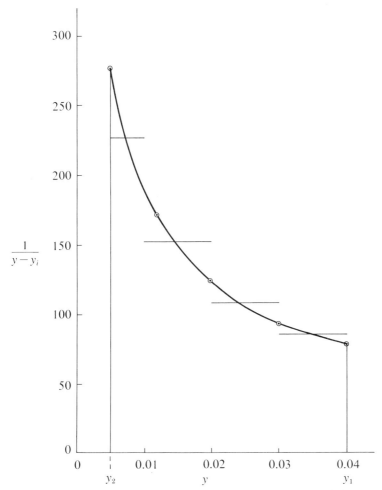

FIGURE 8.3-7 Graphical integration of $\int [dy/(y-y_i)]$ for Example 8.3-4

Equation (8.3-63) is

$$z = \frac{V(y_1-y_2)}{k'_y aS(y-y_i)_M} = \frac{83.5(0.040-0.005)}{16.6(3.14)(0.00725)} = 7.71 \text{ ft}$$

This value is greater than the value by graphical integration, which is 7.38 ft. The latter method is correct if the equilibrium line is straight over the range of concentrations from M_5 to M_1. In Fig. 8.3-6 the line is slightly concave, facing upward. This means the driving force is slightly larger with the true curved equilibrium line that makes the true tower height slightly smaller, as was found. [Ans. to (c)]

All of the design methods in Section 8.3 are also well suited for use with the digital computer. The integration of the equations for the tower height can be performed using the numerical methods given in Chapter 6,

Section 6.6D. When the equilibrium line is curved it can be approximated by a polynomial equation and it can be used along with equations such as Eq. (8.3-15) to solve for the interface compositions.

8.4 ESTIMATION OF MASS TRANSFER COEFFICIENTS IN PACKED TOWERS

8.4A Introduction and Experimental Methods

1. *Introduction.* The mass transfer processes occurring in a packed bed between a gas stream and liquid stream flowing countercurrent to each other are quite complex. In Chapter 6 mass transfer coefficients of fluids flowing past packed beds were presented, and the data are quite accurate. Here the interfacial area is known, which is the surface area of the packing. When two fluids, a gas and a liquid, are flowing, the interfacial area a, ft^2/ft^3 volume, is the area between the gas and the liquid—which is difficult to measure (S2).

The distribution of flow of the gas and liquid over the solids or packing and possibilities of chemical interaction in the liquid further complicate the process. Since two fluids are involved, two mass transfer coefficients or resistances are present. Also, since the interface concentrations cannot be measured directly, only overall mass transfer coefficients can be obtained experimentally. Individual film mass transfer coefficients and interfacial areas cannot be measured directly.

Individual film mass transfer coefficients k'_y and k'_x depend generally upon Schmidt number, Reynolds number, and the size and shape of the packing. Except for diffusivity, these variables as well as interfacial tension between the phases and wetting of the packing affect the interfacial area, a. As a result, the correlations for mass transfer coefficients are highly empirical.

The reliability of these predicted mass transfer coefficients is still far from satisfactory. Deviations up to 25 percent from the correlations are not uncommon. A principal difficulty arises from the fact that an overall resistance is measured experimentally that represents the two film resistances in series. Then, to get the single-phase coefficients, the experimental system is so arranged that the second film resistance is negligible or can be approximately calculated. Other sources of error are the entrance effects often present in short beds and the channeling effects in very long beds.

In some specialized cases an overall coefficient is employed that has been determined under conditions comparable to those of the proposed design. This is especially true for cases in which other interactions can occur in the liquid, such as hydrolysis of Cl_2 being absorbed in water, and the like. A number of references give good compilations of such data—for example, Perry (P1, pp. 18-35 to 18-45, 14-35 to 14-39), Shulman *et al.* (S2), and Cornell *et al.* (C2). However, when no interactions are present, use of film coefficients is preferred.

2. *Liquid Phase Measurements.* For measuring single-phase mass transfer coefficients of the liquid phase, the general approach has been to choose a system in which the gas phase resistance is negligible. These systems are absorption or desorption of very insoluble gases such as O_2, CO_2, or H_2 in water. Then experiments in such systems give $K'_x a$ equal to $k'_x a$, since the resistance in the gas phase is negligible. These techniques give relatively accurate values of $k'_x a$.

3. *Gas Phase Measurements.* For measuring the gas phase film mass transfer coefficients it is desired that the solute be so soluble in the liquid that the liquid phase resistance is negligible. However, most such systems, such as NH_3-air-water, have a liquid phase resistance of about 10 percent or so. By subtracting this known liquid resistance (obtained from correcting the O_2, CO_2, or H_2 absorption data of $k'_x a$ to NH_3 data for $k'_x a$) from the overall measured resistance in Eq. (7.3-32), we obtain the gas phase resistance and coefficient $k'_y a$.

Some investigators have obtained gas phase coefficients by a second method: evaporating a pure liquid into a gas. Here there is no liquid film resistance, since no concentration gradient exists in the pure liquid. However, these mass transfer coefficients $k'_y a$ are larger by almost a factor of 2 when compared to $k'_y a$ values obtained by absorption experiments, such as for NH_3 (S1, p. 286).

A third method is to absorb a gas such as CO_2 from air in a caustic solution where an irreversible fast chemical reaction occurs. Then the equilibrium partial pressure p^* is zero for no liquid film resistance. These values of $k'_y a$ obtained are also about twice those for absorption of NH_3 (S1).

Shulman *et al.* (S2) have recently explained the differences obtained by the three methods, showing that the first method, using absorption, is correct. In vaporization they show that the value of the interfacial area, a, is greater than for absorption. In absorption much of the interfacial area is in the form of semistagnant pools of liquid in the pockets of packing where diffusion in the liquid is slow, and this adds an extra resistance that is not present in vaporization. In vaporization this area is completely effective because there is no concentration gradient in the pure liquid. Also in the third method the stagnant pools are completely effective, because the fast chemical reaction at the surface is not affected by diffusion in the liquid phase.

8.4B Correlations for Film Coefficients

1. *Gas Film Coefficients.* The main data for the gas film coefficients $k'_x a$ or H_G are Fellinger's, published in graphical form (P1, P2, S1 p. 281), for absorption of NH_3 from air by water in different packings. These data are corrected for the liquid film resistance, which is about 10 percent. Over certain ranges of gas and liquid flows the data can be represented by an equation of the form

$$H_G = \alpha G_y^{\beta} G_x^{\gamma} N_{Sc}^{0.5} \qquad (8.4\text{-}1)$$

where G_y is total gas flow lb mass/hr-ft² cross-sectional area, G_x total liquid flow lb mass/hr-ft², and α, β, and γ are constants for a given packing given in Table 8.4-1. For higher flow rates the values of H_G fall rapidly and the actual curves should be consulted (P1, P2, S1). The exponent on the Schmidt number has been assumed to be that for the penetration theory and agrees with that for wetted-wall towers.

The temperature has been shown to have a negligible effect on H_G, since the Schmidt number is relatively insensitive to temperature.

The effect of pressure on H_G can be shown as follows. The $k'_y = k'_c c$ and

Table 8.4-1 Height of a Gas Film Transfer Unit, H_G, in Feet

$$H_G = \alpha G_y^{\beta} G_x^{\gamma} N_{Sc}^{0.5}$$

where G_y = lb mass total gas/hr-ft², G_x = lb mass total liquid/hr-ft², $N_{Sc} = \mu/\rho D$.

Packing Type	α	β	γ	Range of Values G_y	G_x
Raschig rings					
$\tfrac{3}{8}$ in.	2.32	0.45	−0.47	200–500	500–1500
1 in.	6.41	0.32	−0.51	200–600	500–4500
1$\tfrac{1}{2}$ in.	17.3	0.38	−0.66	200–700	500–1500
1$\tfrac{1}{2}$ in.	2.58	0.38	−0.40	200–700	1500–4500
2 in.	3.82	0.41	−0.45	200–800	500–4500
Berl saddles					
$\tfrac{1}{2}$ in.	32.4	0.30	−0.74	200–700	500–1500
$\tfrac{1}{2}$ in.	0.811	0.30	−0.24	200–700	1500–4500
1 in.	1.97	0.36	−0.40	200–800	400–4500
1$\tfrac{1}{2}$ in.	5.05	0.32	−0.45	200–1000	400–4500

SOURCE: Data from Fellinger (P2, S1) as given by (T2) R. E. Treybal, *Mass Transfer Operations*, 1st ed., p. 239 (New York: McGraw-Hill, Inc., 1955). With permission.

$k'_c \propto D_{AB} \propto 1/P$. Also $c \propto P$. Hence, $k'_y a$ is independent of P. Then $H_G \propto 1/k'_y a$, and H_G is independent of P. Also, $k'_y a = k'_G aP$. Hence, $k'_G a \propto 1/P$. See Table 6.3-1 for other relations.

Adding a wetting agent to the liquid phase affects the surface tension but has little affect on H_g. However, the liquid phase resistance can be greatly affected (S1).

Equation (8.4-1) can be used to correct existing data available on a specific packing. Suppose an experimental value for acetone being absorbed in the gas film is available for a certain type of packing. Then to correct this H_G value to that for another solute, say ethanol, the H_G for ethanol will be calculated by

$$H_{G\text{ethanol}} = H_{G\text{acetone}} \left(\frac{N_{\text{Sc}_{\text{ethanol}}}}{N_{\text{Sc}_{\text{acetone}}}}\right)^{0.5} \quad (8.4\text{-}2)$$

These values, of course, must both be at the same G_y and G_x values in the tower.

2. *Liquid Film Coefficients*. The data for the liquid film coefficients are quite reliable. The data show that H_L is independent of gas rate until the "loading" occurs where the value of H_L falls. The H_L increases as L increases. Temperature has an effect on decreasing liquid viscosity so that resistance to diffusion decreases as temperature increases. The data of Sherwood and Holloway (S3) furnish the best basis for the correlation in Eq. (8.4-3):

$$H_L = \theta \left(\frac{G_x}{\mu_L}\right)^\eta N_{\text{Sc}}^{0.5} \quad (8.4\text{-}3)$$

where H_L is in feet, μ_L is viscosity of liquid in lb mass/ft-hr, G_x is lb mass total liquid/hr-ft^2, N_{Sc} is $\mu_L/\rho D$. The values for different packings are given in Table 8.4-2. In using Tables 8.4-1 and 8.4-2, gas and liquid flows must be below loading as given in Table 8.4-3.

Equation (8.4-3) can be used to correct existing data on a given packing that are not in Table 8.4-2. For example, if the same liquid, temperature, and velocities are used but the solute is changed, then the new H_L can be corrected by change in Schmidt numbers to the 0.5 exponent.

Example 8.4-1 Prediction of Mass Transfer Coefficients

It is desired to predict the H_G, H_L, $k'_y a$, $k'_x a$, H_{OG}, and $K'_y a$ for the absorption of NH$_3$ from air by water in dilute solutions in a packed tower containing 1-in. Raschig rings at 1 atm total pressure and 30°C. Compare the predicted value with the experimental value of $H_{OG} = 0.73$ ft by Fellinger (S1). The $G_y = 200$ lb mass/hr-ft^2 and $G_x = 1500$ lb mass/hr-ft^2.

398 Continuous Two-Phase Mass Transport Processes

Table 8.4-2 Height of a Liquid-Film Transfer Unit, H_L, in Feet

$$H_L = \theta \left(\frac{G_x}{\mu_L}\right)^\eta N_{Sc}^{0.5}$$

where G_x = lb mass total liquid/hr-ft², μ_L = viscosity of liquid lb mass/ft-hr, $N_{Sc} = \mu_L/(\rho D)$. G_y is less than loading[a]

Packing Type	θ	η	Range of G_x
Raschig rings			
$\frac{3}{8}$ in.	0.00182	0.46	400–15,000
$\frac{1}{2}$ in.	0.00357	0.35	400–15,000
1 in.	0.0100	0.22	400–15,000
$1\frac{1}{2}$ in.	0.0111	0.22	400–15,000
2 in.	0.0125	0.22	400–15,000
Berl saddles			
$\frac{1}{2}$ in.	0.00666	0.28	400–15,000
1 in.	0.00588	0.28	400–15,000
$1\frac{1}{2}$ in.	0.00625	0.28	400–15,000

[a] For a correlation for loading velocities of G_x and G_y see (P1, p. 18-30; S1, p. 248). Some values at loading are given in Table 8.4-3. The velocities G_y and G_x must be less than these to be below loading.

SOURCE: Based on data by Sherwood and Holloway (S3) as given by (T2) R. E. Treybal, *Mass Transfer Operations*, 1st ed., p. 237 (New York: McGraw-Hill, Inc., 1955). With permission.

Table 8.4-3 Loading Velocities

G_y/φ	$G_x\varphi/G_y$	G_y/φ	$G_x\varphi/G_y$
$\frac{1}{2}$-in. Rings		2-in. Rings	
600	0.5	1300	0.5
420	5	1000	5
200	50	520	50
$\frac{1}{2}$-in. Saddles		$1\frac{1}{2}$-in. Saddles	
820	0.5	1600	0.5
650	5	1200	5
260	50	520	50

$\varphi = (\rho/0.075)^{0.5}$, where ρ = gas density, lb mass/ft³.

Solution

The density of the gas air at 30°C is $\rho = 0.0728$ lb mass/ft³. Hence, $\varphi = (0.0728/0.075)^{0.5} = 0.99$. First the flows will be converted to lb mole/hr-ft²:

$$\frac{V}{S} = \frac{200}{29} = 6.90 \text{ lb mole/hr-ft}^2$$

8.4 Mass Transfer Coefficients in Packed Towers   399

$$\frac{L}{S} = \frac{1500}{18} = 83.3 \text{ lb mole/hr-ft}^2$$

From Table 8.4-3, for $\frac{1}{2}$-in. Raschig rings and a G_y/φ of 200/0.99 or 202, loading is at $G_x\varphi/G_y = 50$. G_x at loading $= 50(202) = 10,100$. Hence, the G_x used of 1500 is below loading. There are no data given for 1-in. rings, but loading velocities increase with size.

For 30°C the equilibrium data are $y = 1.20x$ at 30°C. The Schmidt number is needed for NH_3 in air. The viscosity of air $= 0.0183$ cp $= 0.0183(2.42) = 0.0443$ lb mass/ft-hr. Density of air $= (29/359)(273/303) = 0.0728$ lb mass/ft^3. The diffusivity of NH_3-air at 0°C $= 0.198$ cm^2/sec. Correcting it for temperature to 30°C,

$$D_{AB} = 0.198\left(\frac{303}{273}\right)^{1.75} = 0.239 \text{ cm}^2/\text{sec}$$

$$= 0.239(3.87) = 0.925 \text{ ft}^2/\text{hr}$$

$$N_{Sc} = \frac{\mu}{\rho D} = \frac{0.0443}{0.0728(0.925)} = 0.660$$

For the gas phase Eq. (8.4-1) is as follows for 1-in. rings using Table 8.4-1:

$$H_G = 6.41(G_y)^{0.32}(G_x)^{-0.51}N_{Sc}^{0.5} = 6.41(200)^{0.32}(1500)^{-0.51}(0.660)^{0.5}$$

$$= 0.684 \text{ ft}$$

For the liquid phase, the diffusivity of NH_3 in water at 15°C $= 1.77 \times 10^{-5}$ cm^2/sec. To correct it to 30°C, use Eq. (3.2-9), where $D_{AB} \propto 1/\mu^{1.1}$. The viscosity of water at 15°C $= 1.1404$ cp and 0.8007 cp at 30°C. Hence,

$$D_{AB} = \left(\frac{1.1404}{0.8007}\right)^{1.1}(1.77 \times 10^{-5}) = 2.62 \times 10^{-5} \text{ cm}^2/\text{sec}$$

$$= (2.62 \times 10^{-5})(3.87) = 10.15 \times 10^{-5} \text{ ft}^2/\text{hr}$$

$$N_{Sc} = \frac{\mu}{\rho D} = \frac{0.8007(2.42)}{62.2(10.15 \times 10^{-5})} = 307$$

For the liquid phase, Eq. (8.4-3) is as follows using Table 8.4-2:

$$H_L = 0.010\left(\frac{G_x}{\mu_L}\right)^{0.22}(N_{Sc})^{0.5} = 0.010\left(\frac{1500}{0.8007 \times 2.42}\right)^{0.22}(307)^{0.5} = 0.755 \text{ ft}$$

Equation (8.3-46) is

$$H_{OG} = \frac{H_G(1-y)_{iM}}{(1-y)_{*M}} + \frac{mV}{L}H_L\frac{(1-x)_{iM}}{(1-y)_{*M}}$$

But since the solutions are dilute,

$$H_{OG} = H_G + \frac{mV}{L}H_L$$

Substituting,

$$H_{OG} = 0.684 + 1.20\left(\frac{6.90}{83.3}\right)(0.755) = 0.684 + 0.075 = 0.759 \text{ ft}$$

This value is close to the experimental value given of 0.73 ft.
To convert to $k'_y a$,

$$H_G = \frac{V}{k'_y aS} \quad \text{or} \quad k'_y a = \frac{V}{SH_G} = \frac{6.90}{0.684} = 10.09$$

$$k'_x a = \frac{L}{SH_L} = \frac{83.3}{0.755} = 110.3$$

Then, for dilute solutions,

$$\frac{1}{K'_y a} = \frac{1}{k'_y a} + \frac{m}{k'_x a} = \frac{1}{10.09} + \frac{1.20}{110.3} = 0.0991 + 0.0110 = 0.1101. \quad K'_y a = 9.08$$

The percent resistance in the liquid film is $(0.0110/0.1101)100 = 10$ percent.

Example 8.4-2 Mass Transfer Coefficients for CO_2 Absorption

It is desired to predict the mass transfer coefficients for absorption of CO_2 from air by water in the same packing and using the same flow conditions as Example 8.4-1.

Solution

For 30°C, the equilibrium data are $y = 1.86(10^3)x$. For the gas phase the diffusivity of CO_2 in air at 3.2°C $= 0.142$ cm²/sec. Correcting to 30°C,

$$D_{AB} = 0.142\left(\frac{303}{276.2}\right)^{1.75} = 0.167 \text{ cm}^2/\text{sec} = 0.167(3.87) = 0.649 \text{ ft}^2/\text{hr}.$$

The Schmidt number using the same viscosity and density as Example 8.4-1 is

$$N_{Sc} = \frac{0.0443}{0.0728(0.649)} = 0.940$$

For the gas phase,

$$H_G = 6.41(200)^{0.32}(1500)^{-0.51}(0.940)^{0.5} = 0.811 \text{ ft}$$

For the liquid phase the diffusivity of CO_2 in water is 2.0×10^{-5} cm²/sec at 25°C. The viscosity of water at 25°C $= 0.8937$ cp. Correcting as before,

$$D_{AB} = \left(\frac{0.8937}{0.8007}\right)^{1.1}(2.0 \times 10^{-5}) = 2.26 \times 10^{-5} \text{ cm}^2/\text{sec}$$

$$= (2.26 \times 10^{-5})(3.87) = 8.78 \times 10^{-5} \text{ ft}^2/\text{hr}$$

$$N_{Sc} = \frac{0.8007(2.42)}{62.2(8.78 \times 10^{-5})} = 356$$

Then,

$$H_L = 0.010 \left(\frac{1500}{0.8007 \times 2.42}\right)^{0.22} (356)^{0.5} = 0.815 \text{ ft}$$

By Eq. (8.3-47),

$$H_{OL} = H_L + \frac{L}{mV} H_G = 0.815 + \frac{83.3}{(1.86 \times 10^3)(6.90)}(0.811)$$

$$= 0.815 + 0.00526 = 0.820 \text{ ft}$$

$$k'_y a = \frac{V}{SH_g} = \frac{6.90}{0.811} = 8.50$$

$$k'_x a = \frac{L}{SH_L} = \frac{83.3}{0.815} = 102.2$$

By Eq. (7.3-33),

$$\frac{1}{K'_x a} = \frac{1}{k'_x a} + \frac{1}{m k'_y a} = \frac{1}{102.2} + \frac{1}{(1.86 \times 10^3)(8.50)}$$

$$= 0.00979 + 0.0000633 = 0.009853$$

$$K'_x a = 101.5$$

The percent resistance in the gas film is $0.0000633(100)/0.009853 = 0.7$ percent. This shows that for a very insoluble gas, even though the $k'_y a$ is not very large, the large value of m causes the gas phase resistance to be negligible.

8.5 HEAT EFFECTS IN ABSORPTION COLUMNS

In the design procedures discussed so far it has been assumed that the tower has been operating at a constant temperature. In most absorbers and strippers with dilute gas mixtures and liquids it is usually satisfactory to assume this. When large amounts of solute are absorbed, the heat effects cannot be ignored. Often the tower is provided with intercoolers on the downcomers of each tray or between packed sections, and sometimes with external coolers that keep the tower temperature relatively constant.

If heat removal is not provided and the heat effects are mild—that is, the temperature rise is not large—the main effect of this rise is to change the equilibrium line. The equilibrium solubility of the solute will be reduced, and less absorption will result. Using an approximate heat balance at different points in the tower, we can calculate the temperature at these points. Then the equilibrium line can be corrected for tempera-

PROBLEMS

8-1 Absorption of Ammonia from Reactor Effluent

Ammonia in a reactor gas stream is being removed by scrubbing with water at 30°C and 1 atm total pressure. The mole fraction of ammonia is 0.20 and the inert gas can be considered as air. If 20.0 lb mole/hr-ft² is the total gas inlet stream and it is desired to recover 95 percent of the ammonia by scrubbing with 50.0 lb mole/hr-ft² of pure water, derive the operating line equation and plot it, using mole fraction units for a countercurrent tower.

8-2 Equimolar Counterdiffusion

For Example 8.2-2, involving benzene (B) and toluene (A), the inlet liquid stream contains $L_2 = 200$ lb mole/hr and $x_2 = 0.050$. The inlet vapor stream contains $V_1 = 260$ lb mole/hr and $y_1 = 0.40$. It is desired to reduce y so $y_2 = 0.080$. Derive and plot the operating line equation for a countercurrent tower.

8-3 Diffusion of Both A and B

In a distillation process the binary mixture A and B is being distilled in a stripping tower, with B being the more volatile. Hence, the A is going to the liquid and the B to the vapor stream. The latent heat of component B is two times that of component A. The inlet vapor stream is $V_1 = 100$ lb mole/hr containing $y_1 = 0.35$ (component A) and the outlet $y_2 = 0.05$. The inlet liquid stream is $L_2 = 150$ lb mole/hr and $x_2 = 0.02$. Derive and plot the operating line equation for countercurrent flow.

8-4 Extraction of Nicotine and Minimum Water Rate

The solute nicotine (A) is in a petroleum solvent. This total flow rate is $V_{W1} = 100$ lb mass/hr and $y_{W1} = 0.012$ wt fraction nicotine. This nicotine is to be extracted in a countercurrent tower with pure water. It is desired to remove 90 percent of the nicotine. Use equilibrium data from Example 8.2-4.
  (a) Calculate the minimum water rate to do this.
  (b) Using 50 percent above the minimum, plot the operating line.
  (c) On this graph determine the number of theoretical stages required in a stage tower.

8-5 Absorption of Acetone and Plate Tower

In a chemical process acetone evaporates into the air and it is desired to recover this acetone. The stream contains 10 volume percent acetone in air at 68°F. The total gas to be treated is 50,000 ft³/hr at 68°F and 1.0 atm. It is desired to remove 90 percent of the acetone by scrubbing with pure water in a countercurrent plate tower. If the water rate used is 20 percent above the minimum, determine the number of theoretical stages required. The equilibrium data are given in the appendix, Table A.4-5.

8-6 Absorption, Minimum Liquid/Gas Ratio, and Number of Theoretical Plates

The ammonia gas in Problem 8-1 contains 0.20 mole fraction ammonia with a total inlet gas flow of 20.0 lb mole/hr-ft^2. The scrubbing inlet water contains 0.0020 mole fraction ammonia. It is desired to remove 95 percent of the ammonia. The equilibrium data at 30°C and 1 atm are given in the appendix, Table A.4-2.

(a) Determine the minimum L'/V' to be used.
(b) Using 1.4 times the minimum L'/V', calculate all liquid and gas flows and plot the operating line.
(c) On the graph determine the number of theoretical plates required.

8-7 Stripping and Number of Plates

An aqueous solution of SO_2 containing $x_2 = 0.020$ mole fraction SO_2 is being stripped by an inert gas stream at 20°C and 1 atm to remove the SO_2. The entering gas contains no SO_2. It is desired to remove 90 percent of this SO_2. For 100 moles of total liquid feed, do the following. (The liquid contains a slight amount of dissolved inert solids).

(a) Calculate the minimum gas flow V' necessary.
(b) Using 1.2 times the minimum V', determine the outlet gas concentration and plot the operating line.
(c) Determine the number of theoretical plates required if a plate tower is used.

[*Note*: Since the other dissolved solids in the liquid are inert, it will be assumed that the equilibrium data for the SO_2-air-water system in the appendix, Table A.4-3, hold.]

8-8 Steam Stripping and Number of Plates

An absorption oil contains 4.35 mole percent normal butane and is being regenerated by stripping with direct steam to reduce the butane content to 0.50 mole percent. The tower operates at an average pressure of 1 atm gage, and the temperature is held constant at 300°F by internal heating in the tower. The butane-free oil may be assumed to be nonvolatile and has an average molecular weight of 375. Seven moles of direct steam will be used per 100 moles of butane-free oil.

Estimate the number of theoretical trays necessary to perform the stripping. It is assumed that steam will not condense at the conditions given. In plotting the equilibrium data, use an expanded scale for the x coordinate (abcissa). Step off trays from the top.

The vapor-liquid equilibrium data may be represented by

$$y = 23x$$

where y = mole fraction of butane in vapor (mixture of butane-steam),
       x = mole fraction of butane in liquid (mixture of butane-absorption oil).

8-9 Design of SO_2 Tower and Liquid Film Coefficients

Using the data of Example 8.3-1, calculate the height of the tower using Eq. (8.3-14), which is based on the liquid film mass transfer coefficient $k'_x a$. Note that the interface values of x_i have already been obtained. It will be necessary to calculate the total lb mole/hr, L, for each value of x in order to graphically integrate the equation.

8-10 Design of SO_2 Tower Using Overall Coefficients

Using the data of Examples 8.3-1 and 8.3-2, calculate the height of the tower using Eq. (8.3-28), which is based on the overall $K'_x a$. Note that the values x^* need to be determined, so that the operating and equilibrium lines must be plotted to read off the x^* values. Do this integration two ways: first keeping $(1-x)*_M/(1-x)$ inside the integral, then using an average value outside the integral.

8-11 Design of SO_2 Tower

A tower with 1-in. ceramic rings is to be designed to absorb SO_2 from air by water at 68°F and 1 atm. The overall mass transfer coefficient for this system at the given flows was found experimentally to be $K'_x a = 43.0$ lb mole/hr-ft³-mole fraction. The inert water flow is 1500 lb mass/hr-ft² and the inert air flow is 104 lb mass/hr-ft². Inlet $y_1 = 0.127$ mole fraction SO_2 and outlet $y_2 = 0.076$. Inlet water contains no SO_2.

(a) Use the overall coefficient $K'_x a$. Plot the operating and equilibrium lines (use equilibrium data of Example 8.3-1). By graphical integration determine the tower height needed. Compare this with run 109 of Whitney and Vivian (W1), who used 2.0 ft.

(b) Calculate the overall height of a transfer unit H_{OG} and graphically determine the number of transfer units N_{OG}. Then calculate the tower height and compare with part (a).

8-12 Heights of Transfer Unit

Using the data of Examples 8.3-1 and 8.3-2, calculate the tower height, using (a) the H_{OG} and the number of transfer units N_{OG}, (b) the H_L and the number of transfer units N_L.

8-13 Heights of Transfer Unit

Using the data of Examples 8.3-1 and 8.3-2, calculate the tower height using H_{OL} and N_{OL}.

8-14 Number of Transfer Units

Using the data of Example 8.3-4, do the following.

(a) Plot the operating and equilibrium lines. Calculate the H_{OG} for the whole tower. For this use the average m as the slope between points M_5 and M_1. Then calculate N_{OG} by graphical integration and tower height.

(b) Calculate N_L by graphical integration and tower height.

(c) Repeat (a) and (b) but use a ln-mean driving force.

8-15 Prediction of Film Coefficients

For the absorption of dilute mixtures of methanol from air at 39.9°C and 1 atm total pressure by pure water predict the H_G, H_L, $k'_y a$, $k'_x a$, H_{OG}, $K'_y a$, and $K'_x a$ for a tower with 1-in. Berl saddles. The $G_y = 250$ lb mass/hr-ft² and $G_x = 3000$ lb mass/hr-ft². The Henry's law equation for dilute solutions at 1 atm total pressure is $y = 0.66x$. The diffusivity of methanol in air at 0°C is 0.132 cm²/sec. Also calculate the percent resistance in the liquid film.

8-16 Correction of Experimental Data

Experimental data in a given tower at given gas and liquid flows for absorption of dilute NH_3 in air by water at 68°F give $H_G = 0.45$ ft and $H_L = 0.69$ ft. It is desired to predict H_G, H_L, H_{OG}, $k'_y a$, $k'_x a$, and $K'_y a$ for absorption of acetone from air by water at the same conditions in the same packing. The Henry's law equation for acetone-air-water is $y = 1.18x$, and the diffusivity of acetone in air at 0°C is 0.109 cm²/sec at 1 atm. The flows are $G_y = 275$ lb mass/hr-ft² and $G_x = 2820$ lb mass/hr-ft².

NOTATION

a	interfacial area, ft²/ft³ packed section
A_1, A	interfacial area, ft²
b	constant
c	total concentration, lb mole/ft³
D	diffusivity, cm²/sec or ft²/hr
G_x	total liquid rate, lb mass/hr-ft²
G_y	total gas rate, lb mass/hr-ft²
H_G	height of a gas film transfer unit $= V/k'_y aS$, ft; defined by Eq. (8.3-37)
H_L	height of a liquid film transfer unit $= L/k'_x aS$, ft; defined by Eq. (8.3-38)
H_{OG}	height of an overall gas transfer unit $= V/K'_y aS$, ft; defined by Eq. (8.3-39)
H_{OL}	height of an overall liquid transfer unit $= L/K'_x aS$, ft; defined by Eq. (8.3-40)
k	constant
k_G	gas film mass transfer coefficient, lb mole/hr-ft²-atm
k_L	liquid film mass transfer coefficient, lb mole/hr-ft²-(lb mole/ft³)
$k'_x a$	liquid film mass transfer coefficient, lb mole/hr-ft³-mole frac
$k'_y a$	gas film mass transfer coefficient, lb mole/hr-ft³-mole frac
$K'_x a$	overall liquid mass transfer coefficient, lb mole/hr-ft³-mole frac; defined by Eq. (7.3-33)
$K'_y a$	overall gas mass transfer coefficient, lb mole/hr-ft³ mole frac; defined by Eq. (7.3-32)
k_y, k_x	mass transfer coefficients, lb mole/hr-ft²-mole frac; see Table 6.3-1
L	total liquid rate, lb mole/hr

L' inert liquid rate, lb mole/hr
L_W total liquid rate, lb mass/hr
L'_W inert liquid rate, lb mass/hr
m slope of equilibrium line
m' slope of equilibrium line; defined by Eq. (7.3-17)
m'' slope of equilibrium line, defined by Eq. (7.3-23)
N_A flux of A relative to a fixed point, lb mole A/hr-ft^2
N_G number of gas film transfer units
N_L number of liquid film transfer units
N_{OG} number of overall gas transfer units
N_{OL} number of overall liquid transfer units
N_{Sc} Schmidt number $= \mu/\rho D$, dimensionless
p partial pressure, atm
P total pressure, atm
p^* equilibrium partial pressure, atm
S cross-sectional area of tower, ft^2
V total gas rate, lb mole/hr
V' inert gas rate, lb mole/hr
V_W total gas rate, lb mass/hr
V'_W inert gas rate, lb mass/hr
x mole fraction in bulk liquid
x_i mole fraction in liquid at the interface
x^* mole fraction in the liquid in equilibrium with the bulk gas phase
x_W wt fraction in bulk liquid
y mole fraction in bulk gas
y_i mole fraction in gas at the interface
y^* mole fraction in the gas in equilibrium with the bulk liquid phase
y_W wt fraction in bulk gas
$(1-x)_{iM}$ inert mole fraction defined by Eq. (7.3-6)
$(1-x)_{*M}$ inert mole fraction defined by Eq. (7.3-34)
$(1-y)_{iM}$ inert mole fraction defined by Eq. (7.3-5)
$(1-y)_{*M}$ inert mole fraction defined by Eq. (7.3-31)
z tower height, ft

Greek Letters

α, β, γ constants in Eq. (8.4-1)
θ, η constants in Eq. (8.4-3)
μ viscosity, lb mass/ft-hr
ρ density, lb mass/ft^3
φ $= (\rho/0.075)^{0.5}$, where ρ is gas density, lb mass/ft^3

Subscripts

A component A
B component B

C component C
ave   average
G gas phase
i interface
L liquid phase
N point or tray N
max   maximum
min   minimum
V vapor phase
1   bottom of tower
2   top of tower

REFERENCES

(C1)   J. B. Claffey, C. O. Badgett, C. D. Skalamera, and G. W. Phillips, *Ind. Eng. Chem.* **42**, 166 (1950).

(C2)   D. Cornell, W. G. Knapp, and J. R. Fair, *Chem. Eng. Progr.* **56**, 68 (1960).

(D1)   P. V. Danckwerts, *Gas-Liquid Reactions*. New York: McGraw-Hill, Inc., 1970.

(E1)   N. Epstein, *Can. J. Chem. Eng.* **36**, 210 (1958).

(F1)   A. S. Foust, L. A. Wenzel, C. W. Clump, L. Maus, and L. B. Anderson, *Principles of Unit Operations*. New York: John Wiley & Sons, Inc., 1960.

(L1)   O. Levenspiel, *Chemical Reaction Engineering*. New York: John Wiley & Sons, Inc., 1962.

(M1)   W. L. McCabe and J. C. Smith, *Unit Operations of Chemical Engineering*, 2d ed. New York: McGraw-Hill, Inc., 1967.

(P1)   J. H. Perry, *Chemical Engineers' Handbook*, 4th ed. New York: McGraw-Hill, Inc., 1963.

(P2)   *Ibid.*, 3d ed., 1950.

(S1)   T. K. Sherwood and R. L. Pigford, *Absorption and Extraction*, 2d ed. New York: McGraw-Hill, Inc., 1952.

(S2)   H. L. Shulman and coworkers, *AIChEJ* **1**, 247, 253, 259 (1955); **3**, 157 (1957); **5**, 290 (1959); **6**, 175, 469 (1960); **9**, 479 (1963).

(S3)   T. K. Sherwood and F. A. L. Holloway, *TAIChE* **30**, 39 (1940).

(S4)   C. A. Sleicher, Jr., *AIChEJ* **5**, 145 (1959).

(T1)   R. E. Treybal, *Mass Transfer Operations*, 2d ed. New York: McGraw-Hill, Inc., 1968.

(T2)   *Ibid.*, 1st ed., 1955.

(W1)   R. P. Whitney and J. E. Vivian, *Chem. Eng. Progr.* **45**, 323 (1949).

9 Mass Transport in Stage Processes

9.1 INTRODUCTION TO STAGE PROCESSES

In Chapters 7 and 8 we were concerned with the rate of mass transport between two phases that were in intimate contact. The solute was transferred from one phase to another in continuous contacting equipment, such as a packed column. In these continuous contacting devices the amount of solute transferred from one phase to another was dependent upon the mass transport rate equations. Increased turbulence and larger interfacial areas enhanced the rate of transport.

Frequently the equipment for transport of a solute from one phase to another consists of a single stage where the two phases are brought into intimate contact and the solute being transferred from one phase to the other. Such a device could consist of a mixer, where the two phases are mixed thoroughly, and a settler, where the two phases are allowed to separate into separate phases. Or it could be a distillation tray where the vapor and liquid phases enter, are intimately contacted, and then are separated. Perry (P1) and Treybal (T1, T2) give excellent physical descriptions of such equipment, and the reader is referred there.

While the two phases are being contacted in a single stage, the solute is being transported from one phase to another and the mass transfer rate equations derived in Chapter 8 hold. These equations state:

$$\begin{pmatrix} \text{amount} \\ \text{transferred} \end{pmatrix} \propto (t) \begin{pmatrix} \text{mass transfer} \\ \text{coefficient} \end{pmatrix} \begin{pmatrix} \text{interfacial} \\ \text{area} \end{pmatrix} \begin{pmatrix} \text{driving} \\ \text{force} \end{pmatrix} \quad (9.1\text{-}1)$$

where t is time. In a stage process, such as a mixer-settler, the time of contact is relatively large, and high turbulence yields a large mass transfer coefficient and a large interfacial area. Hence, the two exit phases tend to be relatively close to equilibrium.

An equilibrium stage or an ideal stage is defined as one in which the two exit phases are in equilibrium with each other. This can be approached by allowing a large contact time and a high degree of turbulence in the mixer part of the stage.

An actual stage does not accomplish as large a change in composition of the entering and leaving phases as an ideal stage. We can define a stage efficiency as the actual composition change in a stage divided by the change in an equilibrium stage. Detailed discussions and data on such efficiencies are given in Perry (P1).

Figure 9.1-1 shows an equilibrium stage. Here the two streams L_0 and V_2 enter the stage and undergo mixing and settling. The two exit streams L_1 and V_1 are in equilibrium with each other.

9.2 PHASE EQUILIBRIUM RELATIONS

9.2A Vapor-Liquid Equilibrium

1. *Phase Rule.* The phase rule given previously in Chapter 7 defines the number of degrees of freedom F between a vapor and a liquid in equilibrium:

$$P + F = C + 2 \tag{9.2-1}$$

where C is the number of components and P the number of phases present.

To show the use of the phase rule for a vapor-liquid system we will use the system ammonia-water for an example. For two phases, liquid and vapor, and two components, F from Eq. (9.2-1) is 2. This means we have two degrees of freedom. The four variables are pressure, temperature, and the composition of NH_3 in the liquid phase and the vapor phase. If the pressure is fixed, then only one more variable can be fixed. If we fix

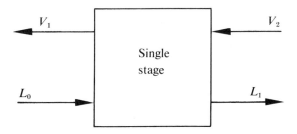

FIGURE 9.1-1 An equilibrium stage

the liquid concentration, then the temperature and composition of the vapor in equilibrium are fixed. Suppose we have only a vapor phase present. Then by Eq. (9.2-1), $F = 3$. This means temperature, pressure, and composition of the vapor phase can be fixed.

2. *Raoult's Law.* In vapor-liquid phases we can define an ideal law, Raoult's law, which is

$$p_A = P_A x_A \qquad (9.2\text{-}2)$$

where p_A is the partial pressure of component A in the vapor, in mm or atm, P_A is the vapor pressure of pure A, in mm or atm, and x_A is the mole fraction of A in the liquid. This law holds only for ideal solutions. Examples are benzene-toluene, pentane-hexane, methyl alcohol-ethyl alcohol, and so on, which are usually members of homologous series. Examples of mixtures not following Raoult's law are ammonia-water, ethyl alcohol-water, and acetone-water. Many of these systems follow Henry's law in dilute solutions, as discussed in Section 7.2, where P_A in Eq. (9.2-2) is replaced by a constant H.

3. *Boiling-Point and Equilibrium Diagrams.* In many cases it is convenient to represent equilibrium relations in a binary mixture of A and B as a boiling-point diagram. Such a diagram for the system benzene (A)-toluene (B) at a total pressure of 1.0 atm is shown in Fig. 9.2-1. Component A is the more volatile and boils at the lower temperature. The upper line is the saturated vapor line and the lower line is the saturated liquid line. The two-phase region is in the region between these two lines.

Suppose that we have a vapor-liquid mixture at equilibrium at temperature T_1 in Fig. 9.2-1. By the phase rule for two phases and the temperature and pressure fixed, we have no degrees of freedom left. This means x_{A1} and y_{A1} are fixed. They are shown in Fig. 9.2-1, where $T_1 = 98°C$, $x_{A1} = 0.318$, and $y_{A1} = 0.532$.

For all points at or above the saturated vapor line, called the dew-point line, the mixture is all vapor. In the region at or below the saturated liquid line, called the bubble-point line, the mixture is all liquid.

If we start with a liquid mixture of $x_{A1} = 0.318$ at 70°C and heat it slowly, it will start to boil at 98.0°C, and the composition of the first amount of vapor will be $y_{A1} = 0.532$. If we continue to boil the mixture, the composition of x_A will move to the left, since y_A is richer in A.

Since benzene and toluene follow Raoult's law, we can calculate the boiling-point diagram from the pure vapor pressure data given in Table 9.2-1. The equations to be used are

$$p_A + p_B = P \qquad (9.2\text{-}3)$$

$$P_A x_A + P_B (1 - x_A) = P \qquad (9.2\text{-}4)$$

$$y_A = \frac{p_A}{P} = \frac{P_A x_A}{P} \qquad (9.2\text{-}5)$$

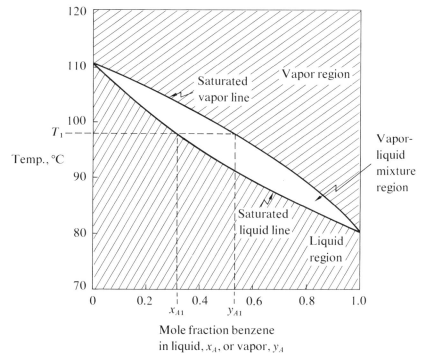

FIGURE 9.2-1 Boiling-point diagram at 1.0 atm total pressure for system benzene (A)-toluene (B)

Table 9.2-1 Data for Benzene-Toluene System

Temp., °C	Vapor Pressure, mm Hg		Mole Fraction Benzene at 1.0 Atm Pressure[a]	
	Benzene	Toluene	x_A	y_A
80.1	760		1.000	1.000
85	877	345	0.780	0.900
90	1016	405	0.581	0.777
95	1168	475	0.411	0.632
100	1344	557	0.258	0.456
105	1532	645	0.130	0.261
110	1748	743	0.017	0.039
110.6	1800	760	0	0

[a]Calculated using Raoult's law.

Example 9.2-1 Boiling-Point Diagram Using Raoult's Law

Using the vapor pressure data of Table 9.2-1 for pure benzene and toluene, calculate the compositions of the vapor and liquid in equilibrium at 100°C and 1.0 atm total pressure.

Solution

From Table 9.2-1 at 100°C, $P_A = 1344$ mm and $P_B = 557$ mm. Substituting into Eq. (9.2-4),

$$1344(x_A) + 557(1 - x_A) = 760$$

Solving, $x_A = 0.258$. $x_B = 1 - x_A = 1 - 0.258 = 0.742$. Substituting into Eq. (9.2-5),

$$y_A = \frac{P_A x_A}{P} = \frac{1344(0.258)}{760} = 0.456, \qquad y_B = 1 - y_A = 0.544$$

The data are often plotted as y_A vs. x_A instead of temperature vs. composition. In Fig. 9.2-2 the equilibrium diagram for the benzene-toluene system is shown. A 45° line is plotted to show that the vapor y_A is richer in component A than x_A. To obtain the temperature for a given composition, the boiling-point diagram is needed.

The boiling-point diagram in Fig. 9.2-1 is typical of an ideal system that follows Raoult's law. Systems that are nonideal have boiling-point curves that differ substantially. In Fig. 9.2-3a the boiling-point and equilibrium diagrams are given for a maximum boiling azeotrope. The

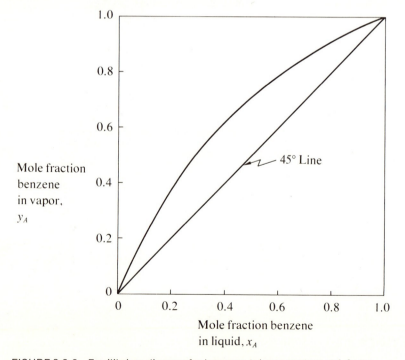

FIGURE 9.2-2 Equilibrium diagram for benzene-toluene system at 1.0 atm

9.2 Phase Equilibrium Relations 413

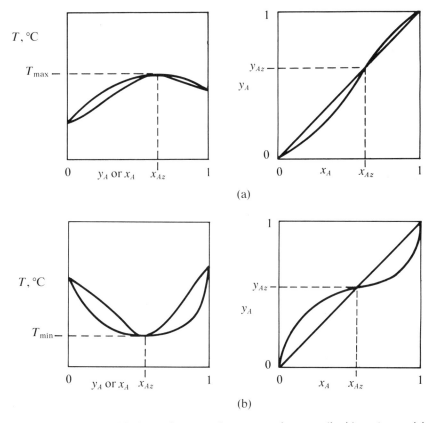

FIGURE 9.2-3. Equilibrium diagrams for azeotropic vapor-liquid systems: (a) boiling-point and equilibrium diagram for maximum-boiling azeotrope, (b) boiling point and equilibrium diagram for minimum-boiling azeotrope

temperature T_{max} that corresponds to a concentration x_{Az} is the highest temperature reached by any mixture. The concentration of the vapor y_{Az} at this point is equal to x_{Az}. Hence, no separation is possible at this point. A typical example of such systems is the acetone-chloroform mixture.

Figure 9.2-3b shows a minimum boiling azeotrope. Again at the azeotrope point the vapor composition is equal to the liquid composition. Typical examples of such systems are ethanol-water, benzene-ethanol, and carbon disulfide-acetone.

4. *Enthalpy-Concentration Diagrams.* Vapor-liquid equilibrium data for binary systems are often plotted on coordinates of enthalpy per unit mass or enthalpy per unit mole versus concentration in mass fraction or mole fraction at constant pressure. The data from the appendix, Table

A.4-8, for ethanol-water at 1 atm total pressure are plotted in Fig. 9.2-4.

The upper line in Fig. 9.2-4 represents the saturated vapor line of enthalpy as H Btu/lb mass of vapor at the dew point versus mass fraction y_A. The lower line represents the enthalpy h Btu/lb mass of liquid at the boiling point versus mass fraction x_A. The region in between represents a mixture of vapor and liquid. The tie line ab at 184.5°F represents the compositions and enthalpies of the two liquid and vapor phases in equilibrium. This tie line was obtained from the vapor-liquid equilibrium diagram below by drawing lines dc, cb, and da. Other lines can be drawn in a similar manner. To obtain the temperature of a tie line a boiling-point diagram must be plotted.

Example 9.2-2 Use of Enthalpy-Concentration Diagram

Use the enthalpy-concentration diagram in Fig. 9.2-4 for ethanol-water. (a) A liquid mixture of ethanol-water of mass fraction $x_A = 0.30$ is heated to the boiling point. What is the enthalpy of the liquid and the vapor in equilibrium with it? (b) If the liquid in part (a) is vaporized completely, what is the latent heat? Assume the vapor is kept in contact during this process. (c) What is the latent heat of pure ethanol where $x_A = 1.0$?

Solution

For part (a), the enthalpy of the liquid at point a on the graph is $h = 135$ Btu/lb mass. The enthalpy of the vapor in equilibrium at point b is, for $y_A = 0.71$, $H = 652$ Btu/lb mass.

For part (b), if the liquid is completely vaporized, then for $y_A = 0.30$, $H = 943$ Btu/lb mass. The latent heat is $H - h = 943 - 135 = 808$ Btu/lb mass. This is simply the vertical distance in Btu/lb mass between point a and the dew-point line.

In part (c), the latent heat is the vertical distance between the vapor and liquid lines at $y_A = x_A = 1.0$. This is $458 - 89 = 369$ Btu/lb mass.

9.2B Liquid-Liquid Equilibrium

1. *Phase Rule.* In a liquid-liquid system we can also apply the phase rule. Suppose we have three components, A, B, and C, and two liquid phases in equilibrium. Then, by Eq. (9.2-1), the number of degrees of freedom is three. The variables are temperature, pressure, and four concentrations. The four concentrations occur because only two of the three mass fraction concentrations can be specified in a phase, since the third must make the total add up to 1.0. If temperature and pressure are set, then setting one concentration in a phase fixes the system. In most cases the phase diagrams are determined for a constant pressure and temperature.

2. *Equilibrium Diagrams.* In Fig. 9.2-5 a phase diagram is shown

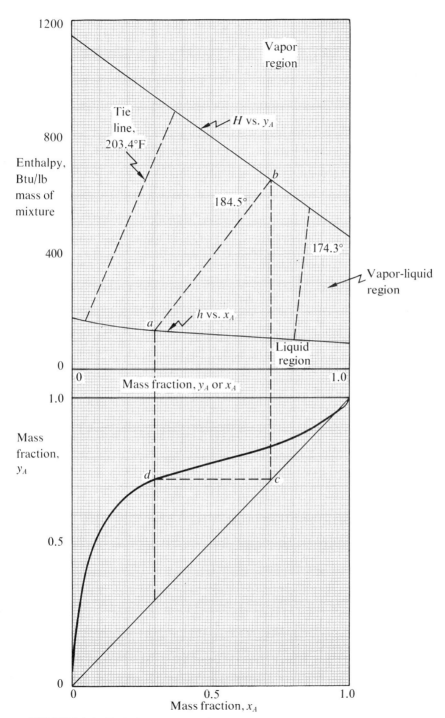

FIGURE 9.2-4 Enthalpy-concentration diagram for ethanol (A)-water (B) system at 1.0 atm

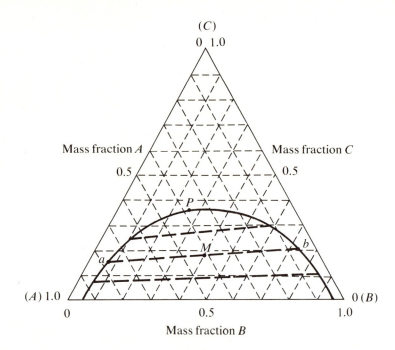

FIGURE 9.2-5 Liquid-liquid phase diagram where A and B are partially miscible

on triangular coordinates for a system where components A and B are partially miscible. The two-phase region is included inside below the curved envelope. A mixture of composition M will separate into two phases, a and b, which are on the tie line through point M. Other tie lines are shown. At point P, the Plait point, the two phases are identical.

A more useful method of plotting the three component data is to use rectangular coordinates. In Fig. 9.2-6 the data from the appendix, Table A.4-6, for the system acetic acid (A), water (B), and isopropyl ether (C) are plotted. In this system the water (B) and ether (C) phases are partially miscible. A tie line, ab, is shown connecting the water-rich layer, called the *raffinate*, and the ether-rich layer, which is the solvent layer and is called the *extract*. To construct a tie line using the y_A-x_A equilibrium plot at the bottom, lines da, dc, and cb are drawn. The two-phase region is inside the envelope and the one-phase region is outside.

Example 9.2-3 Use of Liquid-Liquid Phase Diagram

Using Fig. 9.2-6, a mixture containing a composition of 0.10 mass fraction of A and 0.30 mass fraction of C is allowed to separate after equilibration. What are the compositions of the two phases?

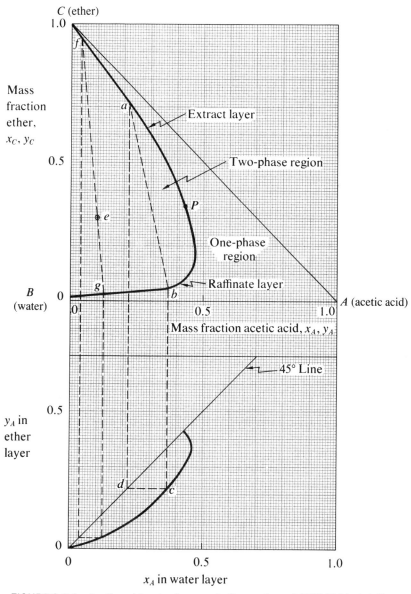

FIGURE 9.2-6 Acetic acid–water–isopropyl ether system at 20°C (Table A.4-6)

Solution

This mixture is plotted as point e on the graph. By trial and error, we draw the tie line fg through this point by using a construction like the one used for tie line *ab*. The composition of the extract or ether layer at f is $y_A = 0.039$, $y_C = 0.943$, and $y_B = 0.018$; and the water layer is $x_A = 0.125$, $x_C = 0.025$, and $x_B = 0.850$.

Figure 9.2-6 is typical of many types of liquid-liquid systems. Another common type is shown in Fig. 9.2-7, where only the pair of liquids *A-B* are soluble instead of two pairs *A-B* and *A-C*, as in Fig. 9.2-6. Often a change of temperature can cause the type of curve in Fig. 9.2-6 to change to the type in Fig. 9.2-7.

3. *Equilibrium Diagrams on a Solvent-Free Basis.* Often it is convenient to convert liquid-liquid data to a solvent-free basis. The system styrene (*A*)-ethylbenzene (*B*)-diethylene glycol (*C*) has a phase diagram similar to Fig. 9.2-7. The data for this system are given in the appendix, Table A.4-7. These data can be converted to a solvent (*C*)-free basis using the following definitions:

$$X_A = \frac{\text{lb mass } A}{\text{lb mass } A + \text{lb mass } B} = \frac{x_A}{x_A + x_B} \qquad (9.2\text{-}6)$$

$$Y_A = \frac{\text{lb mass } A}{\text{lb mass } A + \text{lb mass } B} = \frac{y_A}{y_A + y_B} \qquad (9.2\text{-}7)$$

$$s = \frac{\text{lb mass } C}{\text{lb mass } A + \text{lb mass } B} = \frac{x_C}{x_A + x_B} \qquad (9.2\text{-}8)$$

$$S = \frac{\text{lb mass } C}{\text{lb mass } A + \text{lb mass } B} = \frac{y_C}{y_A + y_B} \qquad (9.2\text{-}9)$$

$$X_A + X_B = 1.0, \qquad Y_A + Y_B = 1.0 \qquad (9.2\text{-}10)$$

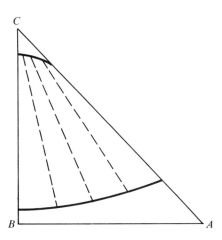

FIGURE 9.2-7 Phase diagram where only liquids *A-B* are soluble

The quantities X_A and Y_A are the mass fractions on a solvent (C)-free basis and are the two phases in equilibrium. Also, S and s are the solvent ratios in each of the two phases in equilibrium. These solvent ratios may have values up to infinity for pure solvent. The layer containing X_A is the raffinate and Y_A the extract layer.

Example 9.2-4 Plot of Data on a Solvent-Free Basis

Data for the system styrene (A)-ethylbenzene (B)-diethylene glycol (C) are given in Table A.4-7 as ordinary mass fractions. Convert these data to a solvent-free basis and plot the data.

Solution

Taking the one data point for the ethylbenzene-rich layer of $x_A = 8.63$ percent, $x_B = 90.56$, $x_C = 0.81$ and for the glycol-rich layer, $y_A = 1.64$ percent, $y_B = 9.85$, $y_C = 88.51$, we substitute into Eqs. (9.2-6) to (9.2-9):

$$X_A = \frac{8.63}{8.63 + 90.56} = 0.087, \qquad Y_A = \frac{1.64}{1.64 + 9.85} = 0.143$$

$$s = \frac{0.81}{8.63 + 90.56} = 0.0082, \qquad S = \frac{88.51}{1.64 + 9.85} = 7.71$$

In a similar manner the data were calculated to be as follows for the two layers in equilibrium:

Ethylbenzene-rich layer		Glycol-rich layer	
X_A	s	Y_A	S
0	0.0068	0	8.62
0.087	0.0082	0.143	7.71
0.189	0.0094	0.273	6.81
0.288	0.0101	0.386	6.04
0.384	0.0110	0.480	5.44
0.464	0.0122	0.565	4.95
0.573	0.0140	0.674	4.37
0.781	0.0183	0.833	3.47
1.00	0.0256	1.00	2.69

The data above are plotted in Fig. 9.2-8. This diagram is very similar to the enthalpy-concentration diagram for ethanol-water in Fig. 9.2-4. One can think of the solvent (C) in Fig. 9.2-8 as the heat in Btu in Fig. 9.2-4. The lower parts of both figures contain the equilibrium diagram on a binary basis. As will be shown later, the same types of equations can be used for the stage processes.

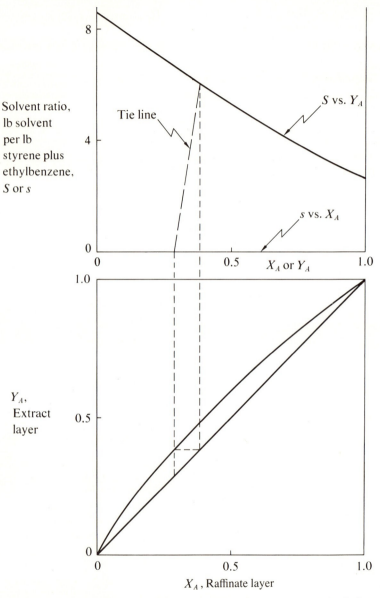

FIGURE 9.2-8 Solvent-concentration diagram for styrene (A)-ethylbenzene (B)-diethylene glycol solvent (C) system, Example 9.2-4

9.2C Gas-Liquid Equilibrium

For gas-liquid equilibrium for a system such as SO_2-air-water, a thorough discussion of the use of the phase rule and of the equilibrium data was given in Section 7.2.

9.3 THEORY FOR ANALYTICAL AND GRAPHICAL CALCULATIONS IN A SINGLE STAGE

9.3A Derivation of Equations

Mass balance calculations on a single stage are easily done analytically, but they can also be done graphically on a rectangular diagram. In Fig. 9.3-1 two streams, L lb mass and V lb mass, containing A, B, and C, are mixed to give a resulting stream Σ lb total mass. The compositions of component A are given on the diagram as mass fraction A.

Writing an overall mass balance and a balance on A,

$$V + L = \Sigma \tag{9.3-1}$$

$$V y_A + L x_A = \Sigma z_A \tag{9.3-2}$$

where z_A is the mass fraction of A in the Σ stream.

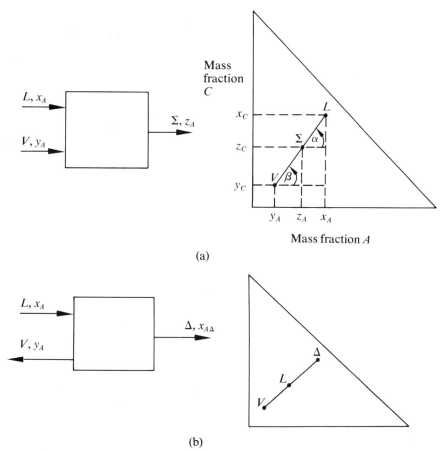

FIGURE 9.3-1 Graphical (a) addition and (b) subtraction

A similar equation can be written for component C:

$$Vy_C + Lx_C = \Sigma z_C \tag{9.3-3}$$

It is not necessary to write a balance on B, since $x_B = 1 - x_A - x_C$ and also $z_B = 1 - z_A - z_C$. Substituting Eq. (9.3-1) into (9.3-2) to eliminate Σ,

$$\frac{L}{V} = \frac{y_A - z_A}{z_A - x_A} \tag{9.3-4}$$

Substituting Eq. (9.3-1) into (9.3-3),

$$\frac{L}{V} = \frac{y_C - z_C}{z_C - x_C} \tag{9.3-5}$$

Equating (9.3-4) to (9.3-5) and rearranging,

$$\frac{x_C - z_C}{x_A - z_A} = \frac{z_C - y_C}{z_A - y_A} \tag{9.3-6}$$

In Fig. 9.3-1a the points L, V, and Σ are plotted on the rectangular diagram. The point Σ is on a straight line between V and L, since in Eq. (9.3-6) the left-hand side is the tan α and the right-hand side is tan β. Thus, since the slope of line $L\Sigma$ is the same as ΣV, Σ must be on a straight line between them.

It also can be shown that if an enthalpy-concentration chart is used, as in Fig. 9.2-4, the resultant of mixing two streams is on a straight line between the two streams. The same holds for the solvent-free diagram in Fig. 9.2-8. Hence, this general method holds on a phase diagram regardless of the units used (such as enthalpy, mass, or moles); however, the units must be consistent. For example, Btu/lb mole and mole fraction can be used.

This method of graphical addition is called the *inverse lever arm rule*. From Eq. (9.3-4),

$$\frac{\text{lb mass } L}{\text{lb mass } V} = \frac{\overline{V\Sigma}}{\overline{L\Sigma}} \tag{9.3-7}$$

where $\overline{V\Sigma}$ is the distance of the line between V and Σ and $\overline{L\Sigma}$ the distance between L and Σ. It can be also shown that

$$\frac{\text{lb mass } L}{\text{lb mass } \Sigma} = \frac{\overline{V\Sigma}}{\overline{LV}} \tag{9.3-8}$$

For the subtraction of two streams, as in Fig. 9.3-1b,

$$L - V = \Delta \tag{9.3-9}$$

$$Lx_A - Vy_A = \Delta x_{A\Delta} \tag{9.3-10}$$

Rearranging Eq. (9.3-10), we can show it to be equivalent to adding Δ to V as follows:

9.3 Theory for Calculations in a Single Stage

$$\Delta x_{A\Delta} + V y_A = L x_A \qquad (9.3\text{-}11)$$

Hence, the inverse lever arm rule holds; the subtraction is shown in the graph in Fig. 9.3-1b.

Example 9.3-1 Amounts of Phases in Extraction

In Example 9.2-3 a mixture contained $x_A = 0.10$, $x_C = 0.30$, and $x_B = 0.60$. It separated into two layers, the extract being $y_A = 0.039$, $y_C = 0.943$, and $y_B = 0.018$ and the raffinate $x_A = 0.125$, $x_C = 0.025$, and $x_B = 0.850$. If 100 lb mass of mixture were originally added, (a) determine the lb mass of V and L using the inverse lever arm rule, (b) repeat using a material balance.

Solution

In Fig. 9.2-6, the distance ef is measured as 8.20 units and fg as 11.68. Then, by Eq. (9.3-8),

$$\frac{L}{\Sigma} = \frac{\overline{ef}}{\overline{fg}} = \frac{8.20}{11.68}$$

Hence, $L = 8.20(100)/11.68 = 70.3$ lb mass. $V = 29.7$ lb mass.
For part (b), substituting into Eq. (9.3-1),

$$V + L = 100$$

Substituting into Eq. (9.3-2),

$$V(0.039) + L(0.125) = 100(0.10)$$

Solving the two equations simultaneously, $L = 70.9$ and $V = 29.1$, which is a close check to part (a).

9.3B Single Equilibrium Stage Process

In a single equilibrium stage extraction two streams, L_0 and V_2, of known amounts and compositions enter a stage, mixing and equilibration occurs, and then the exit streams L_1 and V_1 leave in equilibrium with each other (Fig. 9.3-2a). Making a total balance as before,

$$L_0 + V_2 = \Sigma = L_1 + V_1 \qquad (9.3\text{-}12)$$

A balance on A and C gives

$$L_0 x_{A0} + V_2 y_{A2} = \Sigma z_A = L_1 x_{A1} + V_1 y_{A1} \qquad (9.3\text{-}13)$$

$$L_0 x_{C0} + V_2 y_{C2} = \Sigma z_C = L_1 x_{C1} + V_1 y_{C1} \qquad (9.3\text{-}14)$$

An equation for B is not needed, since $x_A + x_B + x_C = 1.0$. Using Eqs. (9.3-12) through (9.3-14), we can obtain values of Σ, z_A, z_C, and z_B.

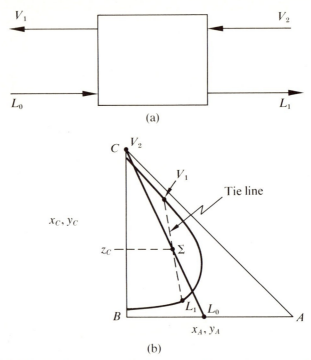

FIGURE 9.3-2 Single-stage equilibrium extraction: (a) process flow for one equilibrium stage, (b) phase-diagram plot for one equilibrium stage

The points representing L_0, V_2, and Σ can then be plotted as shown schematically in Fig. 9.3-2b. Then, by trial and error, a tie line is put through the Σ point, and this locates the compositions of L_1 and V_1. The amounts of L_1 and V_1 can be determined by the inverse lever arm rule or Eqs. (9.3-12) to (9.3-14).

Similar equations hold for an enthalpy-concentration diagram and a solvent-free extraction phase diagram.

9.4 MULTIPLE-STAGE PROCESSES

In some cases it is advantageous to divide the fresh solvent into several portions and use a series of equilibrium stage extractions rather than using all of the solvent for extraction in one stage. Suppose, for example, 100 lb mass L_0 of a 50 percent acetic acid solution in water is being extracted with 100 lb mass of isopropyl ether. The solvent could be used as follows. For a two-stage process 50.0 lb mass of fresh solvent would be used on the water layer in each stage. In this way a greater percentage removal of the acid is obtained.

9.4 Multiple-Stage Processes

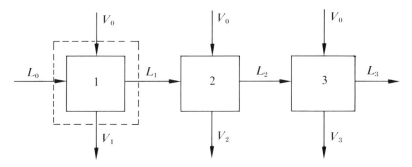

FIGURE 9.4-1 Three-stage multiple-extraction process

Figure 9.4-1 shows the flow diagram of a typical three-stage process with an equal volume V_0 of fresh solvent used in each stage. The entering solution L_0 is being extracted. Writing a material balance on stage 1,

$$V_0 + L_0 = \Sigma_1 = L_1 + V_1 \tag{9.4-1}$$

$$V_0 y_{A0} + L_0 x_{A0} = \Sigma_1 z_{A1} \tag{9.4-2}$$

$$V_0 y_{C0} + L_0 x_{C0} = \Sigma_1 z_{C1} \tag{9.4-3}$$

As before the points V_0, L_0, and z_1 are plotted on the phase diagram, as shown in Fig. 9.4-2. The tie line through point z_1 locates the two

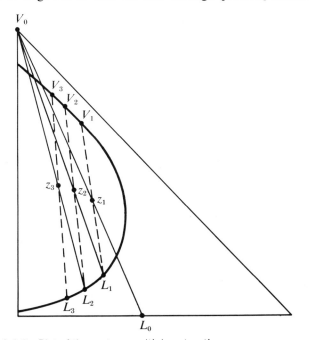

FIGURE 9.4-2 Plot of three-stage multiple-extraction process

streams L_1 and V_1 leaving in equilibrium with each other. The amounts of each can be determined by the inverse lever arm rule or by equations written similar to (9.4-1) to (9.4-3).

For the second stage the process is repeated and

$$V_0 + L_1 = \Sigma_2 = L_2 + V_2 \tag{9.4-4}$$

$$V_0 y_{A0} + L_1 x_{A1} = \Sigma_2 z_{A2} \tag{9.4-5}$$

$$V_0 y_{C0} + L_1 x_{C1} = \Sigma_2 z_{C2} \tag{9.4-6}$$

The points V_0, L_1, and z_2 are plotted and a tie line through z_2 is located to give L_2 and V_2. This is repeated to give L_3 and V_3. As the extraction proceeds, the composition of L shifts to the left.

The same theory and methods hold for multiple-stage extraction using a solvent-free equilibrium and phase diagram.

9.5 COUNTERCURRENT MULTISTAGE PROCESSES

9.5A Countercurrent Multistage Process and Overall Balances

In Section 9.4 we used multiple-stage extraction in order to remove more of the solute by the solvent. However, the solvent leaving each stage was relatively dilute. In order to conserve the solvent and get a more concentrated extract, countercurrent multistage contacting is generally used.

The process flow diagram is shown for a countercurrent process in Fig. 9.5-1. The equations will first be derived for extraction, then shown to hold for distillation. Making a total overall balance on all stages,

$$L_0 + V_{N+1} = L_N + V_1 = \Sigma \tag{9.5-1}$$

For a component balance on A, B, or C,

$$L_0 x_0 + V_{N+1} y_{N+1} = L_N x_N + V_1 y_1 = \Sigma z_\Sigma \tag{9.5-2}$$

Substituting the value of Σ from Eq. (9.5-1) to (9.5-2) and rearranging,

$$z_\Sigma = \frac{L_0 x_0 + V_{N+1} y_{N+1}}{L_0 + V_{N+1}} = \frac{L_N x_N + V_1 y_1}{L_N + V_1} \tag{9.5-3}$$

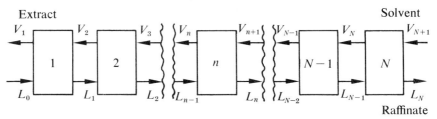

FIGURE 9.5-1 Countercurrent multistage process or cascade

9.5 Countercurrent Multistage Processes

Equation (9.5-3) is used to locate the Σ point on the phase diagram that ties together the two entering streams L_0 and V_{N+1} and the exit streams V_1 and L_N. Usually L_0 and V_{N+1} are known and the desired exit composition x_N is set. Then, if we plot the Σ point, L_N, Σ, and V_1 must lie on one line. Also L_N and V_1 must lie on the phase envelope and can be determined as shown in Fig. 9.5-2. The same methods can be used on a solvent-free phase diagram.

For distillation an overall enthalpy balance gives

$$L_0 h_0 + V_{N+1} H_{N+1} = L_N h_N + V_1 H_1 = \Sigma h_\Sigma \tag{9.5-4}$$

Substituting Eq. (9.5-1) into (9.5-4) and rearranging as before,

$$h_\Sigma = \frac{L_0 h_0 + V_{N+1} H_{N+1}}{L_0 + V_{N+1}} = \frac{L_N h_N + V_1 H_1}{L_N + V_1} \tag{9.5-5}$$

The use of this is shown in Fig. 9.5-3.

Example 9.5-1 Overall Material Balance in Countercurrent Extraction Tower

Pure isopropyl ether at the rate of $V_{N+1} = 300$ lb mass/hr is used to extract an aqueous solution of $L_0 = 100$ lb mass/hr containing 20 percent acetic acid by countercurrent extraction. The exit acetic acid concentration in the aqueous phase is 4 percent. Determine the compositions of the exit ether extract and the aqueous raffinate.

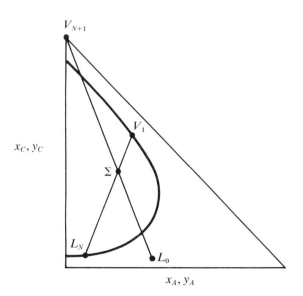

FIGURE 9.5-2 Use of sigma point for overall material balance in extraction

428 Mass Transport in Stage Processes

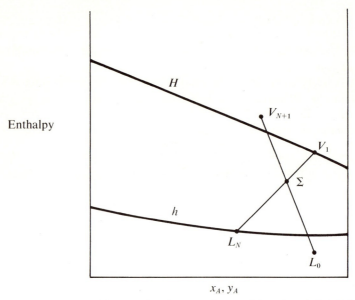

FIGURE 9.5-3 Use of sigma point for overall material and enthalpy balance in distillation

Solution

$V_{N+1} = 300$, $y_{AN+1} = 0$, $y_{CN+1} = 1.0$. $L_0 = 100$, $x_{A0} = 0.20$, $x_{B0} = 0.80$, $x_{C0} = 0$, $x_{AN} = 0.04$. In Fig. 9.5-4, L_0 and V_{N+1} are plotted. Also L_N can be plotted, since it lies on the phase boundary. To calculate the Σ point, Eq. (9.5-3) is used for component A:

$$z_{A\Sigma} = \frac{L_0 x_0 + V_{N+1} y_{N+1}}{L_0 + V_{N+1}} = \frac{100(0.20) + 300(0)}{100 + 300} = 0.050$$

For component C,

$$z_{C\Sigma} = \frac{100(0) + 300(1.00)}{100 + 300} = 0.75$$

Using the above coordinates of z_Σ, this point is plotted in Fig. 9.5-4. Drawing a line from L_N through Σ and extending it until it intersects the phase boundary, we locate V_1. This gives $y_{A1} = 0.052$ and $y_{C1} = 0.926$. For L_N a value of $x_{CN} = 0.017$ is obtained. By Eq. (9.5-1) and the inverse lever arm rule, $V_1 = 323$ and $L_N = 77$ lb mass/hr.

9.5B Stage-to-Stage Calculations in Countercurrent Processes

After an overall balance has been made for a countercurrent multistage process as in Eqs. (9.5-1) to (9.5-5), the next step is to go stage by stage

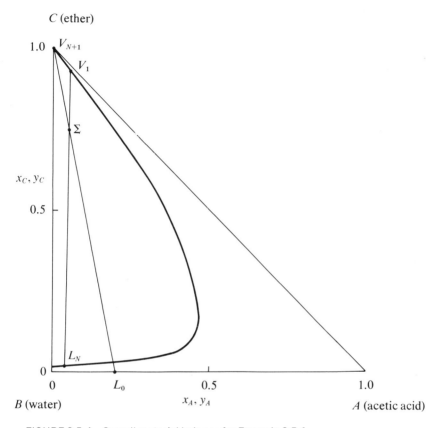

FIGURE 9.5-4 Overall material balance for Example 9.5-1

to determine the concentrations at each stage and the total number of stages N to reach L_N (Fig. 9.5-1).

Making a balance on stage 1,

$$L_0 + V_2 = L_1 + V_1 \tag{9.5-6}$$

For stage n,

$$L_{n-1} + V_{n+1} = L_n + V_n \tag{9.5-7}$$

Rearranging Eq. (9.5-6) and getting the difference in flows,

$$L_0 - V_1 = L_1 - V_2 = \Delta \tag{9.5-8}$$

The value of Δ in lb mass/hr is constant and

$$\Delta = L_0 - V_1 = L_n - V_{n+1} = L_N - V_{N+1} = \cdots \tag{9.5-9}$$

This also holds for a component balance on A, B, or C.

$$\Delta x_\Delta = L_0 x_0 - V_1 y_1 = L_n x_n - V_{n+1} y_{n+1} = L_N x_N - V_{N+1} y_{N+1} \tag{9.5-10}$$

For distillation the equation is

$$\Delta h_\Delta = L_0 h_0 - V_1 H_1 = L_n h_n - V_{n+1} H_{n+1} = L_N h_N - V_{N+1} H_{N+1} \quad (9.5\text{-}11)$$

Solving for x_Δ using Eqs. (9.5-9) and (9.5-10),

$$x_\Delta = \frac{L_0 x_0 - V_1 y_1}{L_0 - V_1} = \frac{L_n x_n - V_{n+1} y_{n+1}}{L_n - V_{n+1}} = \frac{L_N x_N - V_{N+1} y_{N+1}}{L_N - V_{N+1}} \quad (9.5\text{-}12)$$

Also,

$$h_\Delta = \frac{L_0 h_0 - V_1 H_1}{L_0 - V_1} = \frac{L_n h_n - V_{n+1} H_{n+1}}{L_n - V_{n+1}} = \frac{L_N h_N - V_{N+1} H_{N+1}}{L_N - V_{N+1}} \quad (9.5\text{-}13)$$

Now Eqs. (9.5-8) and (9.5-9) can be written as

$$L_0 = \Delta + V_1, \quad L_n = \Delta + V_{n+1}, \quad L_N = \Delta + V_{N+1} \quad (9.5\text{-}14)$$

This means L_0 is on a line between Δ and V_1, L_n is on a line between Δ and V_{n+1}, and so on. Hence Δ is a point common to all streams passing each other such as V_1 and L_0, V_{n+1} and L_n, V_{N+1} and L_N, and so on. The coordinates of the Δ point are given for $x_{\Delta C}$ and $x_{\Delta A}$ by an A or C balance in Eq. (9.5-12). Usually the end points V_{N+1}, L_N or V_1, L_0 are known so that x_Δ can be calculated. The graphical construction to locate Δ is shown in Fig. 9.5-5 for a typical extraction diagram. The intersection of the lines $L_0 V_1$ and $L_N V_{N+1}$ is the Δ point.

To step off the number of stages we start at L_0 and draw line $L_0 \Delta$, which locates V_1. Then a tie line through V_1 locates L_1, which is in equilibrium with V_1. Next, line $L_1 \Delta$ is drawn, which gives V_2. Then tie line $V_2 L_2$ is drawn. This procedure is repeated until the desired composition of L_N is reached. This gives the number of stages N needed to perform the extraction.

If a solvent rate of V is selected too low or is too small, a limiting case will be reached, with the line through Δ and the tie line being the same. Then the minimum solvent rate is reached where no separation is obtained.

A similar procedure for the Δ point is used in determining the number of stages on the enthalpy-concentration diagram in Fig. 9.5-3. The Δ point is on the intersection between the extensions of lines $L_0 V_1$ and $L_N V_{N+1}$. It should be pointed out that the Δ point can be located above or below the diagram, depending on the inlet conditions.

Treybal (T1), Perry (P1), McCabe and Smith (M1), and Brown (B1) give many examples and design procedures for multistage processes and processes with reflux using the methods presented in this chapter, and the reader is referred there for further study.

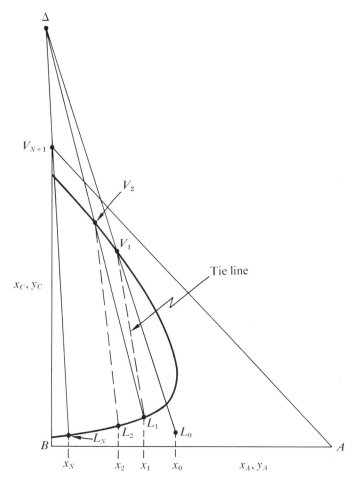

FIGURE 9.5-5 Location of the Δ point and number of theoretical stages needed for countercurrent extraction

9.5C Simplified Procedures for Countercurrent Stage Processes

1. *Introduction and Derivation of Basic Equation.* In absorption countercurrent processes the inert gas or "solvent" in the gas stream does not dissolve in the liquid, but only the solute A dissolves. Also, the liquid solvent usually does not vaporize in the gas phase. Hence, both solvents are "immiscible" in each other, and the material balance methods and phase diagrams given in this present chapter are not needed. In Section 8.2 simplified methods were given in order to determine the number of ideal stages or trays. Also, when the solvents were immiscible,

Section 8.2 gave a simplified method for multistage extraction. In these methods curved or straight operating and equilibrium lines were plotted and the trays or stages simply stepped off.

When the operating and equilibrium lines are linear over the tower, simplified analytical expressions can be derived to obtain the number of equilibrium stages or trays for a countercurrent process. Figure 9.5-1 is a sketch of a countercurrent stage process or cascade. Rewriting Eq. (9.5-10), which was a component balance,

$$L_n x_n - V_{n+1} y_{n+1} = L_N x_N - V_{N+1} y_{N+1} \tag{9.5-15}$$

Since the operating line is straight, $L_n = L_N =$ constant $= L$ and $V_{n+1} = V_{N+1} =$ constant $= V$. Then Eq. (9.5-15) becomes

$$L(x_n - x_N) = V(y_{n+1} - y_{N+1}) \tag{9.5-16}$$

Since y_{n+1} and x_{n+1} are in equilibrium, and the equilibrium line is straight, $y_{n+1} = mx_{n+1}$ and $y_{N+1} = mx_{N+1}$. Substituting mx_{n+1} for y_{n+1} and calling $A = L/(mV)$, Eq. (9.5-16) becomes

$$x_{n+1} - Ax_n = \frac{y_{N+1}}{m} - Ax_N \tag{9.5-17}$$

where A is the absorption factor and is a constant. All factors on the right-hand side of Eq. (9.5-17) are constant. Equation (9.5-17) is a linear first-order difference equation and will be solved by the calculus of finite difference methods (M2).

2. *Solution of Difference Equation by Calculus of Finite Differences.* In a multiple-stage process the transition from one stage to another is accompanied by a discontinuity in the value of x in Eq. (9.5-17), which is mass fraction and is the dependent variable, and stage number n, which is the independent variable. The calculus of finite differences is used when x does not vary continuously but has meaning for certain values of n of 0, 1, 2, and so on. The theory of these difference equations has many similarities to that of differential equations (M2). The method of solution will be illustrated by solving Eq. (9.5-17), which is non-homogeneous.

First, we use the following equation to obtain the solution to the homogeneous equation:

$$x_n = C_1 \beta^n, \qquad x_{n+1} = C_1 \beta^{n+1} \tag{9.5-18}$$

where C_1 is a constant. This is substituted into Eq. (9.5-17) where the right-hand side is set equal to zero for the homogeneous equation.

$$C_1 \beta^{n+1} - AC_1 \beta^n = 0 \tag{9.5-19}$$

9.5 Countercurrent Multistage Processes

Canceling out C_1 and β^n, and solving the characteristic equation,

$$\beta = A \tag{9.5-20}$$

Hence, the solution to the homogeneous equation is

$$x_n = C_1 A^n \tag{9.5-21}$$

For the particular solution, since the right-hand side of Eq. (9.5-17) is a constant, then by methods similar to differential equations,

$$x = C_2 \tag{9.5-22}$$

where C_2 is a constant. Substituting Eq. (9.5-22) into Eq. (9.5-17),

$$C_2 - AC_2 = \frac{1}{m} y_{N+1} - A x_N \tag{9.5-23}$$

Solving,

$$C_2 = \frac{\frac{1}{m} y_{N+1} - A x_N}{1 - A} \tag{9.5-24}$$

The complete solution is the sum of the homogeneous solution, Eq. (9.5-21), and Eq. (9.5-22):

$$x_n = C_1 A^n + C_2 = C_1 A^n + \frac{\frac{1}{m} y_{N+1} - A x_N}{1 - A} \tag{9.5-25}$$

To solve for the constant C_1, we set $n = 0$ and $x_n = x_0$ in Eq. (9.5-25).

$$x_0 = C_1 + \frac{\frac{1}{m} y_{N+1} - A x_N}{1 - A} \tag{9.5-26}$$

Solving for C_1,

$$C_1 = x_0 - \frac{\frac{1}{m} y_{N+1} - A x_N}{1 - A} \tag{9.5-27}$$

Finally, substituting Eq. (9.5-27) into (9.5-25),

$$x_n = \left(x_0 - \frac{\frac{1}{m} y_{N+1} - A x_N}{1 - A} \right) A^n + \frac{\frac{1}{m} y_{N+1} - A x_N}{1 - A} \tag{9.5-28}$$

Equation (9.5-28) is used to calculate x_n at any stage n.

Equation (9.5-28) can be rearranged as below after setting $n = N$ for transfer of solute from phase L to V (stripping).

$$\frac{x_0 - x_N}{x_0 - (y_{N+1}/m)} = \frac{(1/A)^{N+1} - (1/A)}{(1/A)^{N+1} - 1} \qquad (9.5\text{-}29)$$

$$N = \frac{\log\left[\dfrac{x_0 - (y_{N+1}/m)}{x_N - (y_{N+1}/m)}(1 - A) + A\right]}{\log(1/A)} \qquad (9.5\text{-}30)$$

When $A = 1$,

$$N = \frac{x_0 - x_N}{x_N - (y_{N+1}/m)} \qquad (9.5\text{-}31)$$

When the transfer of solute is from phase V to L as in absorption, the following can be derived (S1, T1):

$$\frac{y_{N+1} - y_1}{y_{N+1} - mx_0} = \frac{A^{N+1} - A}{A^{N+1} - 1} \qquad (9.5\text{-}32)$$

$$N = \frac{\log\left[\dfrac{y_{N+1} - mx_0}{y_1 - mx_0}\left(1 - \dfrac{1}{A}\right) + \dfrac{1}{A}\right]}{\log A} \qquad (9.5\text{-}33)$$

When $A = 1$,

$$N = \frac{y_{N+1} - y_1}{y_1 - mx_0} \qquad (9.5\text{-}34)$$

The series of equations above are often called the Kremser equations. They are very convenient to use. If A varies slightly, an average value can be used. These equations can be used with any consistent set of units, such as mass fraction, mole ratio, or mole fraction.

PROBLEMS

9-1 Boiling-Point Diagram and Equilibrium Diagram Using Raoult's Law

Use the vapor-pressure data of Table 9.2-1 for the system benzene-toluene.
(a) At 1.0 atm calculate the x-y data using Raoult's law.
(b) Plot the boiling-point diagram and x-y equilibrium diagram.
(c) If a mixture has a composition of 0.50 mole fraction and is at 80°C originally, is it at the boiling point? If not, at what temperature will it boil, and what will be the composition of the vapor first coming off?
(d) A mixture at 105°C has a composition of 0.20 mole fraction. What is its physical state?

9-2 Boiling-Point Diagram for Hexane-Octane

Use the data given below for the system hexane-octane.

T, °F	Vapor Pressure, mm Hg	
	Hexane	Octane
155.7	760	121
175	1025	173
200	1480	278
225	2130	434
250	3000	654
258.2	3420	760

(a) Calculate the x-y data at 1.0 atm using Raoult's law.
(b) Plot the boiling-point and equilibrium diagrams.

9-3 Enthalpy-Concentration Diagram

Refer to the enthalpy-concentration diagram for ethanol-water in Fig. 9.2-4 and perform the following. [Make a small sketch of each diagram and show approximate positions on this diagram. Use the text for actual calculations.]
(a) Find the compositions of the phases at 174.3°F.
(b) A liquid mixture contains $x_A = 0.40$ and is heated to the boiling point. What is its enthalpy at the boiling point? What are the composition and enthalpy of the first amount of vapor boiled off?
(c) A liquid mixture of $x_A = 0.10$ at the boiling point is completely vaporized. The vapor is kept in contact with the liquid. How much heat is added?
(d) A superheated vapor of $y_A = 0.60$ is cooled until a small drop of liquid is formed. What are the compositions and enthalpies of the two phases formed?

9-4 Extraction of Acetic Acid in a Single Stage Using the Inverse Lever Arm Rule

A mixture of 100 lb mass of acetic acid, 250 lb mass of water, and 150 lb mass of isopropyl ether at 20°C is mixed thoroughly in a single stage and separated. What are the compositions of the two layers? What is the amount of each phase? Find the answers by a graphical method using the inverse lever arm rule and by a material balance.

9-5 Solvent-Free Diagram for Extraction

Using the data for the system styrene-ethylbenzene-diethylene glycol in Table A.4-7 in the appendix, convert the fourth and sixth points to a solvent-free basis. Compare with the values in Example 9.2-4.

9-6 Heating a Water-Ethanol Solution

A mixture containing 0.40 mass fraction ethanol and 0.60 mass fraction water is partially vaporized in a closed container at 1 atm pressure. The resulting mixture is in equilibrium. The composition of the resulting liquid is $x_A = 0.200$

mass fraction. Using Fig. 9.2-4, graphically determine y_A in equilibrium and the fraction of vapor and liquid. What is the temperature of the final mixture? [See Table A.4-8 for boiling points.]

Ans.: $y_A = 0.65$

9-7 Single-Equilibrium-Stage Extraction

A liquid contains 0.50 mass fraction acetic acid and the rest water. This is to be extracted with an equal weight of pure isopropyl ether at 20°C in a single stage. Determine the amounts and compositions of the final extract and raffinate layers. What percentage of the acid is removed by the ether?

9-8 Single-Stage Extraction

Plot the equilibrium data for the system acetic acid-water-isopropyl ether given in Table A.4-6, using rectangular coordinates.

(a) A single-stage extraction is carried out in which 100 lb mass of a solution L_0 containing 35 percent acetic acid in water is contacted with 100 lb mass of isopropyl ether V_2. What are the amounts and compositions of the extract V_1 and raffinate layers L_1? Show all points on the graph in red. Solve for the amounts of both layers algebraically and by the inverse lever arm rule. What percentage of the acid is in the raffinate?

(b) A mixture L_0 of unknown composition and weighing 50.0 lb mass is contacted with 70.0 lb mass of a mixture V_2 containing 40 percent acid, 10 percent water, and 50 percent i-propyl ether in a single stage. The resulting raffinate layer L_1 weighs 80 lb mass and contains 30.0 percent acid, 66.0 percent water, and 4.0 percent i-propyl ether. Find the composition of the original mixture and the composition of the resulting extract layer V_1. Show points in pencil on the graph.

(c) Repeat part (a) with L_0 containing 50 percent acetic acid in water instead of 35 percent.

Ans.: (a) $L_1 = 89$ lb mass, $x_{A1} = 0.255$
(b) $x_{A0} = 0.036$, $x_{C0} = 0.014$

9-9 Multiple-Stage Extraction

Water is used to extract acetic acid from a solution of 35 percent acetic acid in isopropyl ether that weighs 100 lb mass. Plot the phase and tie line diagram using Table A.4-6 in the appendix.

(a) If 100 lb mass of water are used, calculate the percent recovery of acetic acid in a one-stage process. Show on the graph in pencil.

(b) Calculate the overall percent recovery if a multiple three-stage system is used and 33.3 lb mass of fresh water are used in each stage. Show this on the graph in red. Tabulate compositions of all streams and amounts in a table.

(c) Repeat part (a) for a solution containing 50 percent acetic acid in isopropyl ether.

(d) Repeat part (b) for ether containing 50 percent acetic acid.

Ans.: (a) 84 percent recovery; (b) 96 percent recovery

9-10 Countercurrent Extraction in Multiple Stages

Plot the phase and tie line data from Table A.4-6. Pure isopropyl ether (3000 lb mass/hr) is being used to extract the acetic acid from 1000 lb mass/hr of an aqueous solution containing 20 percent acid in water by countercurrent extraction. The concentration of the acetic acid is to be reduced to 4 percent in the exit raffinate. How many equilibrium stages are needed and what is the percent recovery of acetic acid? [*Note*: It may be necessary to plot the low range of the equilibrium diagram on an expanded scale for accuracy.]

Ans.: Six stages

9-11 Number of Stages in an Extraction Plate Tower

Example 8.2-4 gives data for extraction of nicotine from water by an oil where the two solvents are immiscible.
(a) Plot the operating and equilibrium lines and step off on the graph the number of theoretical stages.
(b) Use Eqs. (9.5-29) to (9.5-34) and calculate analytically the number of stages. Compare with part (a).

9-12 Calculus of Finite Differences

Derive Eq. (9.5-32) below for a stage process.

$$\frac{y_{N+1} - y_1}{y_{N+1} - mx_0} = \frac{A^{N+1} - A}{A^{N+1} - 1} \qquad (9.5\text{-}32)$$

[*Hint*: Start with Eq. (9.5-16) and convert the x values to y values before proceeding.]

9-13 Number of Stages Analytically

When the absorption factor A in Eq. (9.5-29) is 1.0, the equation is indeterminant and Eq. (9.5-31) must be used. Derive Eq. (9.5-31) from (9.5-29). [*Hint*: Use L'Hospital's rule.]

NOTATION

A absorption factor = $L/(mV)$
C number of components in phase rule
C_1 constant
C_2 constant
F number of degrees of freedom in phase rule
H enthalpy of vapor, Btu/lb mass or Btu/lb mole
H Henry's law constant, atm/mole frac or mm/mole frac
h enthalpy of liquid, Btu/lb mass or Btu/lb mole
L phase L, lb mass/hr or lb mole/hr; also lb mass or lb mole
m equilibrium constant = y_A/x_A
N stage number N

n stage number n
p_A partial pressure of A, mm or atm
P_A vapor pressure of pure A, mm or atm
P total pressure, mm or atm
P number of degrees of freedom in phase rule
S concentration ratio in V phase, lb mass C/lb mass A plus B or lb mole C/lb mole A plus B
s concentration ratio in L phase, lb mass C/lb mass A plus B or lb mole C/lb mole A plus B
T temperature, °C or °F
t time, sec or hr
V phase V, lb mass/hr or lb mole/hr; also lb mass or lb mole
X_A lb mass A/lb mass A plus B or lb mole A/lb mole A plus B in L phase
x_A mass fraction of A or mole fraction of A in L phase
Y_A lb mass A/lb mass A plus B or lb mole A/lb mole A plus B in V phase
y_A mass fraction of A or mole fraction of A in V phase
z_A mass fraction of A or mole fraction of A in Σ stream

Greek Letters

α angle
β angle
β constant
Δ difference of two streams, lb mass/hr or lb mole/hr; also lb mass or lb mole
Σ sum of two streams, lb mass/hr or lb mole/hr; also lb mass or lb mole

Subscripts

A component A
B component B
C component C
max   maximum
n stage number n
N stage number N
z point z
0   stage number 0
1   stage number 1
Δ Δ stream
Σ Σ stream

REFERENCES

(B1)   G. G. Brown *et al.*, *Unit Operations*. New York: John Wiley & Sons, Inc., 1950.

(M1)   W. L. McCabe and J. C. Smith, *Unit Operations of Chemical Engineering*, 2d ed. New York: McGraw-Hill, Inc., 1967.

(M2)   H. S. Mickley, T. K. Sherwood, and C. E. Reed, *Applied Mathematics in Chemical Engineering*, 2d ed. New York: McGraw-Hill, Inc., 1957.

(P1) J. H. Perry, *Chemical Engineers' Handbook*, 4th ed. New York: McGraw-Hill, Inc., 1963.
(S1) T. K. Sherwood and R. L. Pigford, *Absorption and Extraction*, 2d ed. New York: McGraw-Hill, Inc., 1952.
(T1) R. E. Treybal, *Mass Transfer Operations*, 2d ed. New York: McGraw-Hill, Inc., 1968.
(T2) R. E. Treybal, *Liquid Extraction*, 2d ed. New York: McGraw-Hill, Inc., 1963.

10 Analog Computer Methods

10.1 INTRODUCTION TO ANALOG COMPUTATION

This chapter offers the reader another computational tool to solve some of the equations in this text that are difficult or cumbersome to solve by other methods. The reader need not have an actual analog computer available to learn the basic analog programming methods. The programming can be done as shown in this chapter, and the wiring or "hardware" part can then be learned rather quickly afterward.

In solving engineering problems one first writes an equation or equations describing the physical system. Often the equation can be solved analytically. Then, by substitution of actual numerical values into the solution and by means of a slide rule, log tables, or even a digital computer for complicated or tedious calculations, a numerical solution is obtained. If the equations do not yield to analytic solutions, they frequently can be solved by numerical methods such as finite differences, which can be accomplished with a digital computer or by hand.

The analog computer is useful for solving single or simultaneous differential equations, linear or nonlinear, with constant or variable coefficients. The areas of usefulness of the analog and digital computers overlap. The analog computer represents the dependent variables by a voltage e and the independent variable by time. The analog computer is accurate to about three significant figures and the digital to about seven

10.2 BASIC ANALOG COMPUTER ELEMENTS OR FUNCTIONS

or eight. The results of an analog computer solution are generally presented as a continuous plot of the dependent variable as a voltage e versus time.

10.2A Introduction

The electronic analog computer consists mainly of high-gain amplifiers, fixed or variable resistors, capacitors, potentiometers, relays to make or break circuits, and a voltmeter. These components are prewired together in different combinations, called computer elements. The user of the computer "programs" the problem first by arranging the computer elements in proper sequence on paper. Then he wires together these elements by removable patch cords to solve different equations. Most computers are either 10-volt or 100-volt systems.

The high-gain amplifier is a basic part of several different computer elements. The reader need not understand the detailed circuits of the elements but only how to use them. The equation of the high-gain amplifier is

$$e_o = -A e_i \qquad (10.2\text{-}1)$$

where e_i is the input d-c voltage in volts, e_o the output in volts, and A a constant factor or gain of about 1×10^8, as shown in Fig. 10.2-1a. Note that the output voltage e_o always is reversed in sign. In a 10-volt computer, since e_o is ± 10 volts or less, e_i is almost zero.

10.2B Summer or Summation Element

When input resistances and a feedback resistance are used with the amplifier, as in Fig. 10.2-1b, the combination is called a summer. For deriving the equation of the summer, i_i in the amplifier is usually about 10^{-11} ampere, so it can be assumed $i_i = 0$. Then

$$i_1 + i_2 = i_f \qquad (10.2\text{-}2)$$

However, $i_1 = (e_1 - e_i)/R_1$, $i_2 = (e_2 - e_i)/R_2$, and $i_f = (e_i - e_o)/R_f$. Substituting these terms into Eq. (10.2-2) and also substituting $-e_o/A$ for e_i from Eq. (10.2-1),

$$\frac{e_1}{R_1} + \frac{e_o}{AR_1} + \frac{e_2}{R_2} + \frac{e_o}{AR_2} = \frac{-e_o}{AR_f} - \frac{e_o}{R_f} \qquad (10.2\text{-}3)$$

Discarding the terms containing $1/A$, which are negligible, the final equation for the summer is

$$e_o = -\left(\frac{R_f}{R_1} e_1 + \frac{R_f}{R_2} e_2\right) \qquad (10.2\text{-}4)$$

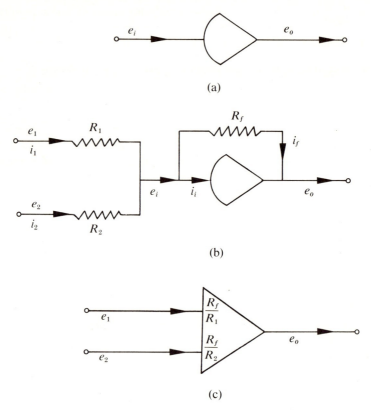

FIGURE 10.2-1 Use of high-gain amplifier: (a) programming symbol for high-gain amplifier, (b) summer element, (c) programming symbol for summer element

Hence, to sum the voltages e_1 and e_2 they are first multiplied by the "gains" R_f/R_1 and R_f/R_2, respectively, and the negative sum of the resultants taken. The gains are usually fixed values, such as 1 megohm/ 0.10 megohm or 10/1, 1/1, or 1/10. The typical programming symbol (Fig. 10.2-1c) has the actual gains written inside the amplifier symbol. The usual summer is permanently prewired with four or five inputs, which can all be summed at once. If a summer has one input and a gain of 1, it only changes sign and is called a sign changer or inverter.

Example 10.2-1 Summing Three Inputs

Three inputs of $e_1 = +5.5$ volts, $e_2 = -5.0$ volts, and $e_3 = +0.3$ volt are to be summed. The feedback resistor is 0.5 megohm and the input resistors are $R_1 = 0.5$ megohm, $R_2 = 2.5$ megohms, and $R_3 = 0.1$ megohm. Draw the programming symbol and calculate e_o.

10.2 Basic Analog Computer Elements or Functions

Solution

Rewriting Eq. (10.2-4) for three inputs,

$$e_o = -\left(\frac{R_f}{R_1}e_1 + \frac{R_f}{R_2}e_2 + \frac{R_f}{R_3}e_3\right) \quad (10.2\text{-}5)$$

Substituting the values for the resistors,

$$e_o = -\left(\frac{0.5}{0.5}e_1 + \frac{0.5}{2.5}e_2 + \frac{0.5}{0.1}e_3\right)$$

$$= -(1.0e_1 + 0.2e_2 + 5e_3)$$

The programming symbol is shown in Fig. 10.2-2 with the gains of 1.0, 0.2, and 5 shown. Solving for e_o,

$$e_o = -[1.0(5.5) + 0.2(-5.0) + 5(0.3)] = -6.0 \text{ volts}$$

10.2C Integrator Element

An integrator element can be obtained by putting in a capacitor C_f (in microfarads) in place of the feedback resistance of Fig. 10.2-1b. This is shown in Fig. 10.2-3a. To derive the equation for integration, since $i_i \cong 0$, $i_1 = i_f$. Then, since e_i is very small or approximately zero,

$$i_1 = i_f = \frac{e_1 - e_i}{R_1} = \frac{e_1 - 0}{R_1} \quad (10.2\text{-}6)$$

For the voltage drop across the capacitance C_f,

$$e_i - e_o = 0 - e_o = \frac{Q}{C_f} \quad (10.2\text{-}7)$$

where Q is the coulombs stored in the capacitance. Differentiating both sides of Eq. (10.2-7),

$$\frac{-de_o}{dt} = \frac{1}{C_f}\frac{dQ}{dt} \quad (10.2\text{-}8)$$

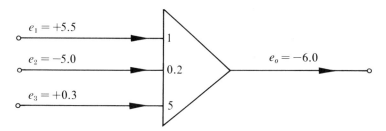

FIGURE 10.2-2 Programming symbol for summer of Example 10.2-1

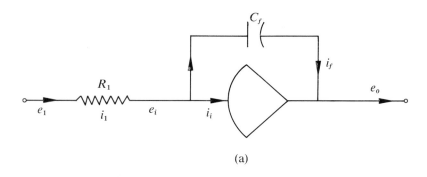

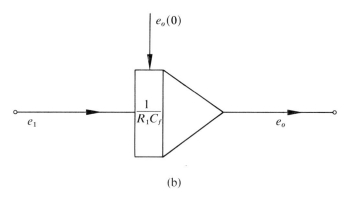

FIGURE 10.2-3 (a) Integrator element and (b) its programming symbol

However, i_f is equal to dQ/dt coulombs/sec or amperes, and i_f also is e_1/R_1 by Eq. (10.2-6). Substituting e_1/R_1 for dQ/dt and integrating, Eq. (10.2-8) becomes

$$\int_{e_o(0)}^{e_o} de_o = -\frac{1}{R_1 C_f} \int_0^t e_1 \, dt \tag{10.2-9}$$

$$e_o = -\frac{1}{R_1 C_f} \int_0^t e_1 \, dt + e_o(0) \tag{10.2-10}$$

Equation (10.2-10) shows that the input voltage e_1 is integrated with respect to time, multiplied by a gain $1/R_1 C_f$, and its sign changed. The programming symbol usually used for the integrator element is shown in Fig. 10.2-3b.

The following equation can be shown to be valid for an integrator network used to integrate several input voltages and sum them as shown in Fig. 10.2-4.

$$e_o = -\left(\frac{1}{R_1 C_f} \int_0^t e_1 \, dt + \frac{1}{R_2 C_f} \int_0^t e_2 \, dt\right) + e_o(0) \tag{10.2-11}$$

10.2 Basic Analog Computer Elements or Functions

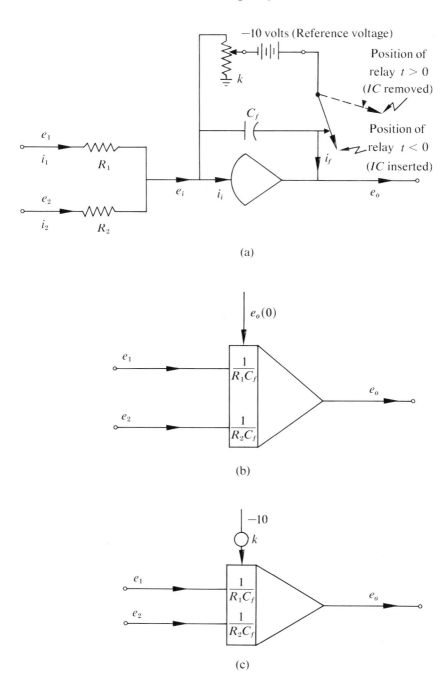

FIGURE 10.2-4 Integrator element with two inputs and initial condition: (a) integrator network used, (b) its programming symbol, (c) alternate programming symbol

Example 10.2-2 Integrator with Two Inputs

It is desired to integrate and sum $e_1 = -0.25$ volt and $e_2 = +0.15$ volt. The gains are $1/R_1C_f = 0.50$ and $1/R_2C_f = 2.0$. The initial condition is $e_o(0) = +2.0$ volts at $t = 0$. Calculate e_o at 10, 20, and 30 seconds.

Solution

Substituting into Eq. (10.2-11),

$$e_o = -\left(0.50 \int_0^t e_1\, dt + 2.0 \int_0^t e_2\, dt\right) + e_o(0)$$

$$= -[0.5e_1(t-0) + 2.0e_2(t-0)] + e_o(0)$$

For $t = 10$ sec,

$$e_o = -[0.5(-0.25)(10-0) + 2.0(+0.15)(10-0)] + 2.0 = +0.25 \text{ volt}$$

Similarly at 20 sec, $e_o = -1.50$, and at 30 sec, $e_o = -3.25$ volts. A plot of e_o versus t would be a straight line in this case.

Figure 10.2-4 gives the wiring diagram for the initial conditions IC for $e_o(0)$. Since $e_o(0)$ is the value of e_o in volts at $t = 0$, the capacitor is charged with an initial voltage of $-e_o(0)$ at $t < 0$ with the relay as indicated. To start the run at $t = 0$, the relay switch is opened, disconnecting the battery or voltage source, and the capacitor discharges during the run. To repeat the run, the relay is closed and the process repeated. The negative value of $e_o(0)$ is used since the integrator changes the sign of the input voltage.

10.2D Potentiometer Element

In the summer or integrator elements the input voltage e_1 can be multiplied by a fixed gain that is greater or less than 1.0. A potentiometer can be used to multiply an input voltage by a constant k that is less than 1 and can be varied from 0 to 1.0. The equation for a potentiometer is given below and shown in Fig. 10.2-5.

$$e_o = ke_1 \qquad (10.2\text{-}12)$$

The potentiometer is located before the input resistors to a summer or integrator.

10.2E Initial Conditions

Potentiometers are also used to set initial conditions, IC, when $t < 0$ on integrators. In Fig. 10.2-4 a potentiometer is used to give the desired voltage for the IC when $t < 0$. In a 10-volt computer a constant reference voltage of $+10$ or -10 volts is available.

10.3 Use of Computer in Solving Differential Equations

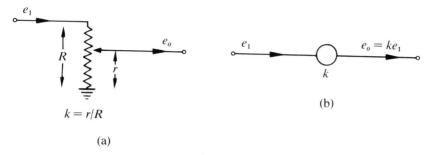

FIGURE 10.2-5 Potentiometer element: (a) wiring circuit, (b) programming symbol

As an illustration, suppose $e_o(0) = +8.6$ volts. Then the setting on the potentiometer is $k = 0.860$. The reference voltage is $0.860(-10.0)$ or -8.60 volts, since the integrator changes the sign of any input voltage. Two alternate methods of drawing the programming symbol for the *IC* are shown in Fig. 10.2-4b and c.

10.3 USE OF ANALOG COMPUTER IN SOLVING DIFFERENTIAL EQUATIONS

10.3A Solving Single Differential Equations

In solving a single differential equation the following general steps can be used. Given the equation

$$a\frac{dx}{dt} - bx + c = 0 \qquad (10.3\text{-}1)$$

where a, b, and c are constants and x the dependent variable:

1. Solve the equation for the highest derivative.

$$\frac{dx}{dt} = +\frac{b}{a}x - \frac{c}{a} \qquad (10.3\text{-}2)$$

2. Assuming this derivative exists, integrate it the number of times necessary to give x. If the highest derivative is d^2x/dt^2, integrate two times. For integration of dx/dt the programming symbol is drawn as in Fig. 10.3-1. The input voltage is $e_1 = dx/dt$. Substituting e_1 into Eq. (10.2-10), assuming a gain of 1.0, and omitting $e_o(0)$,

$$e_o = -\int e_1\, dt = -\int \frac{dx}{dt}\, dt = -\int dx = -x \qquad (10.3\text{-}3)$$

3. Using the $-x$ generated, generate the right side of Eq. (10.3-2) as follows and given in Fig. 10.3-1b. Convert $-x$ to $+x$ in an inverter. Then put it through a potentiometer to give bx/a. This then is one input to the integrator. For the constant $-c/a$, take -10 volts from a reference voltage and put it through a potentiometer setting of $k = \frac{1}{10}(c/a)$ so that $-10(k) = -10[\frac{1}{10}(c/a)] = -c/a$.

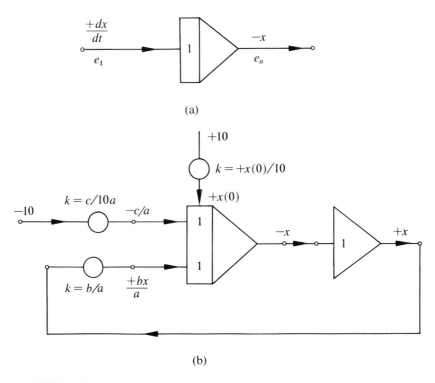

FIGURE 10.3-1 Programming symbols for solution of

$$\frac{dx}{dt} = +\frac{b}{a}x - \frac{c}{a}$$

(a) integration of dx/dt to obtain $-x$, (b) complete programming symbols for solution of differential equation

4. Using $+10$ volts from the reference source, put it through a potentiometer setting of $k = x(0)/10$ so that $x(0) = k10 = [x(0)/10]10 = x(0)$. The integrator changes its sign to $-x(0)$ which is the initial condition IC for $-x$.

Example 10.3-1 Solution of Differential Equation

Solve the following:

$$0.4\frac{d^2x}{dt^2} + 2\frac{dx}{dt} + x - 2.0 = 0$$

At $t = 0$, $x(0) = 3.30$ and $(dx/dt)_{t=0} = -0.55$.

Solution

Solving for the highest derivative,

$$\frac{d^2x}{dt^2} = -\frac{5\,dx}{dt} - 2.5x + 5.0$$

Next, integrate it twice as shown in Fig. 10.3-2a. In Fig. 10.3-2b generate the function d^2x/dt^2 by bringing to integrator number (1) $-0.5dx/dt$, $-0.25x$, and $+5.0$ as inputs. Note that $-0.25x$ is multiplied by a gain of 10 to give $-2.5x$ as the input. This also holds for the input $-0.5dx/dt$ to give $-5.0\ dx/dt$. The initial conditions to integrators (1) and (2) are shown.

10.3B Solving Simultaneous Differential Equations

To solve simultaneous differential equations we treat each equation separately and integrate it first. Then we combine the solutions by making

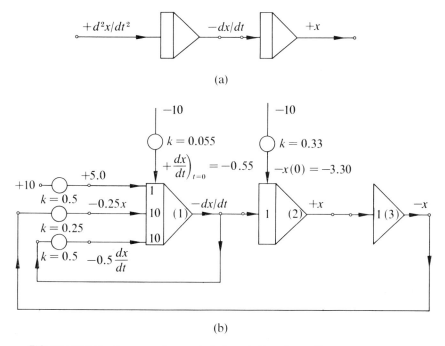

FIGURE 10.3-2 Programming symbols for solution of equation

$$\frac{d^2x}{dt^2} = -5\frac{dx}{dt} - 2.5x + 5.0$$

(a) integration to obtain x, (b) final solution of equation

suitable interconnections between them. The method is best illustrated by an example.

Example 10.3-2 Solution of Simultaneous Equations

Solve the following simultaneous equations:

$$\frac{d^2x}{dt^2} + a_1\frac{dy}{dt} + a_2 x = 0$$

$$\frac{d^2y}{dt^2} + a_3\frac{dx}{dt} - a_4 y = 0$$

Solution

First solve for the highest derivative in each equation.

$$\frac{d^2x}{dt^2} = -a_1\frac{dy}{dt} - a_2 x$$

$$\frac{d^2y}{dt^2} = -a_3\frac{dx}{dt} + a_4 y$$

Next, in Fig. 10.3-3 at the top of the diagram, d^2x/dt^2 is integrated twice in integrators (1) and (2) to give $+x$. Below it d^2y/dt^2 is integrated twice in integrators (4) and (5) to give $+y$. Then, to generate d^2x/dt^2, inputs of $-a_2 x$ and $-a_1 dy/dt$ are used to integrator (1). For generating d^2y/dt^2, inputs of $-a_3 dx/dt$ and $+a_4 y$ are used. Finally the IC values are added.

10.4 SCALING OF VARIABLES FOR COMPUTER

10.4A Introduction and Output Devices

All of the dependent variables are represented as voltages and the independent variable as time. Output devices to plot or indicate these variables are indicating or recording voltmeters of different types, which show x or e versus t.

Mechanical servo-driven devices are slow but accurate and plot the data with a pen. As a practical rule of thumb total times for complete recording of the solution should not be less than about 5 sec, nor should frequencies be over 2 or 3 cycles/sec, since the recorder may be too slow to accurately follow. Galvanometer high-speed recorders are extremely fast up to about 300 cycles/sec but have errors up to 5 percent of full scale. Cathode-type oscilloscopes are used to display the solution. The computer is wired for repetitive operation, which speeds up the solution by about a factor of 1000. The computer solves the problem, automatically

10.4 Scaling of Variables for Computer 451

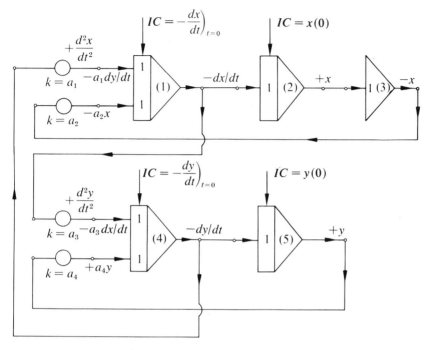

FIGURE 10.3-3 Solution of Example 10.3-2 for

$$\frac{d^2x}{dt^2} = -a_1\frac{dy}{dt} - a_2 x, \qquad \frac{d^2y}{dt^2} = -a_3\frac{dx}{dt} + a_4 y$$

stops it and reimposes the *IC*, and repeats the process many times per second.

In a 10-volt computer the magnitude of the variables x and y cannot exceed $+10$ or -10 volts. Hence, these variables x and y must be scaled to be in this range and represented as machine volts e. The real time t of the problem may vary from microseconds to years. It must be scaled to machine time τ, where τ is the actual number of seconds elapsed in the computer since the problem was started by removing the *IC*.

10.4B Time Scaling

Time scaling is necessary because the dynamic response of the recording equipment puts a lower limit on the time required to solve the problem, or because the actual problem runs too long to be practical. Actually the programmer has a wide range of choice in between these two extremes. The time-scale factor α is a constant defined as

$$\tau = \alpha t \qquad (10.4\text{-}1)$$

where τ is machine time in seconds and t real time in the actual problem or the independent variable. On differentiation,

$$d\tau = \alpha\, dt \tag{10.4-2}$$

Hence, if t appears in an actual equation, τ/α is substituted for t and $d\tau/\alpha$ for dt.

10.4C Amplitude Scaling

The dependent variable may have units of velocity, concentration, distance, and so on but for the computer it must be expressed as e in volts. The equations for amplitude scaling of x to convert it to volts e are

$$e = \beta x \tag{10.4-3}$$

$$de = \beta\, dx \tag{10.4-4}$$

where β is a constant that is selected to "scale" x, so the value of e is always between $+10$ and -10 volts on a 10-volt computer. If another dependent variable y is present,

$$e = \gamma y \tag{10.4-5}$$

where γ is a constant for scaling y.

To scale dx/dt, de/β is substituted for dx and $d\tau/\alpha$ for dt:

$$\frac{dx}{dt} = \frac{de/\beta}{d\tau/\alpha} = \frac{\alpha}{\beta}\frac{de}{d\tau} \tag{10.4-6}$$

For the second derivative,

$$\frac{d^2x}{dt^2} = \frac{d\left(\dfrac{dx}{dt}\right)}{dt} = \frac{d\left(\dfrac{\alpha}{\beta}\dfrac{de}{d\tau}\right)}{d\tau/\alpha} = \frac{\alpha^2}{\beta}\frac{d^2e}{d\tau^2} \tag{10.4-7}$$

Example 10.4-1 Scaling an Equation for Time and Amplitude

Given the differential equation

$$75\frac{dz}{dt} + 150z = 1.40t$$

where t is distance in feet and z is concentration as lb mole/ft^3. The maximum distance in the problem is 1.0 ft, and the maximum concentration to be obtained is 0.010 lb mole/ft^3. At $t = 0$, $z(0) = 0.005$ lb mole/ft^3. A mechanical recorder is to be used. Time- and amplitude-scale the equation.

Solution

For this type of recorder 5 sec is the fastest time normally used. Hence, select 10 sec for the run to be completed. From Eq. (10.4-1)

$$\tau = \alpha t \quad (10.4\text{-}1)$$

Substituting maximum values,

$$10 \text{ sec} = \alpha 1.0 \text{ ft}$$

Thus, $\alpha = 10.0$ sec/ft. For z, Eq. (10.4-3) is

$$e = \beta z \quad (10.4\text{-}3)$$

Or, substituting maximum values,

$$10 \text{ volts} = \beta 0.010 \text{ lb mole/ft}^3$$

Then, $\beta = 1000$ volts/(lb moles/ft^3). Substituting into the original equation values for dz/dt from Eq. (10.4-6), for z from Eq. (10.4-3), and for t from Eq. (10.4-1),

$$75\left(\frac{\alpha}{\beta}\frac{de}{d\tau}\right) + 150\left(\frac{e}{\beta}\right) = 1.40\left(\frac{\tau}{\alpha}\right)$$

$$\frac{de}{d\tau} = \frac{1.40}{75}\left(\frac{\beta}{\alpha^2}\right)\tau - \frac{150}{75\alpha}e$$

$$= \frac{1.40}{75}\frac{1000}{(10)^2}\tau - \frac{150}{75(10)}e$$

$$\frac{de}{d\tau} = 0.187\tau - 0.20e$$

For the initial conditions IC, at $t = 0$, $z(0) = 0.005$ lb mole/ft^3. Then $e(0) = \beta z(0) = 1000(0.005) = 5.0$ volts for the IC to the integrator on $de/d\tau$.

The complete programming symbols are given in Fig. 10.4-1. To generate the function of τ, which is $f(\tau) = 0.187\tau$, a constant voltage of -0.187 volt is integrated in integrator (1). This can be shown by first writing the equation for an integrator, Eq. (10.2-10), in terms of τ:

$$e_o = -\frac{1}{R_1 C_f}\int_0^\tau e_1 \, d\tau + e_o(0) \quad (10.4\text{-}8)$$

In this case $e_o(0) = 0$. The input voltage e_1 is a constant in this case. So, integrating,

$$e_o = -\frac{1}{R_1 C_f}\int_0^\tau e_1 \, d\tau = -\frac{1}{R_1 C_f}e_1(\tau - 0) \quad (10.4\text{-}9)$$

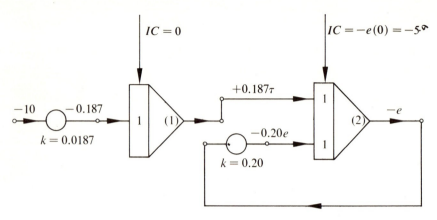

FIGURE 10.4-1   Solution of Example 10.4-1 for

$$\frac{de}{d\tau} = 0.187\tau - 0.20e$$

Setting the gain $1/R_1C_f = 1.0$ and $e_1 = -0.187$ volt, $e_o = +0.187\tau$, which is the output of integrator (1).

A few comments on practical applications are in order here. The potentiometer setting of 0.0187 in Fig. 10.4-1 is generally beyond the accuracy of three decimal places in a potentiometer. This can be avoided by using a setting of 0.187 and in integrator (1) using a gain of 0.10 for an input of 1.87τ. Also, it is not always easily evident on scaling that the voltage e will not exceed ± 10 volts during the run. If this occurs in a run, then the amplitude must be rescaled to reduce e. This means β in Eq. (10.4-3) must be decreased.

10.5 NONLINEAR AND OTHER FUNCTION ELEMENTS

Many nonlinear functions can be easily generated in the analog computer. Only a few of the more important elements will be described here; the details on others are available elsewhere (C1, J1, J2, M1, S1).

A function switch element allows a discontinuous or step function to be generated during a run at time t_1 by snapping a relay on or off, as in Fig. 10.5-1a. To multiply a variable x by another y or e_1 by e_2, a function multiplier is used, as in Fig. 10.5-1b.

$$e_o = -\frac{e_1 e_2}{10} = -\frac{xy}{10} \tag{10.5-1}$$

Note that the resultant is $-\frac{1}{10}$ times the product. The function multiplier can also be wired for division, as in Fig. 10.5-1c, and for square root and $\log_{10}$, as in Fig. 10.5-1d and e. A diode function generator in Fig.

10.5 Nonlinear and Other Function Elements

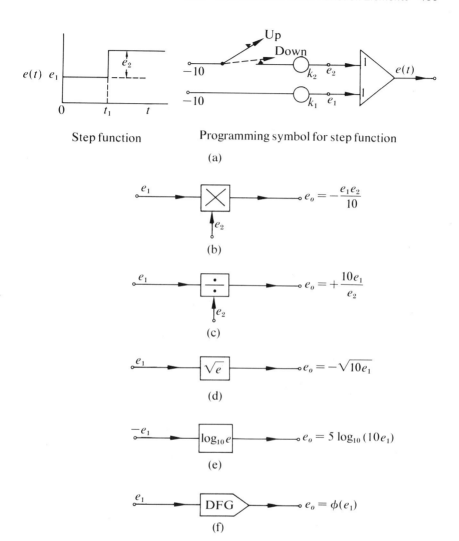

FIGURE 10.5-1 Programming symbols for other function and nonlinear elements: (a) step function, (b) multiplication, (c) division, (d) square root, (e) $\log_{10}$, (f) diode function generator

10.5-1f approximates an arbitrary function by a number of straight line segments.

Analytic functions such as $y = e^{-at}$, $y = A \sin at$, $y = e^{-x}$, $y = A(t+a)^n$, and so on can be easily generated; the reader is referred to Johnson (J1) for details.

In the previous chapters of this text a few more advanced examples are given of solving various mass transfer equations with the analog computer.

PROBLEMS

10-1 Summer Element

For summing four voltages of $e_1 = 2.0$ volts, $e_2 = 6.5$, $e_3 = -6.0$, $e_4 = 3.0$, input resistors are $R_1 = 1.0$ megohm, $R_2 = 0.5$, $R_3 = 0.25$, $R_4 = 2.0$, and the feedback resistor is 0.5 megohm.
(a) Draw the detailed diagram.
(b) Calculate the output voltage e_o.
(c) Draw the programming symbol.

Ans.: (b) $e_o = +3.75$ volts

10.2 Summers in Series

It is desired to sum the voltages $e_1 = 3.0$ volts and $e_2 = 2.5$ volts in a summer where the gains are 0.10 and 0.50, respectively. The feedback resistor is 0.10 megohm. However it is desired that the above sum be as a positive number.
(a) Draw the detailed diagram, showing the final output voltage as a positive number.
(b) Calculate the final output voltage.
(c) Draw the programming symbols.

Ans.: (b) Final output $= 1.55$ volts

10-3 Integrator Element

It is desired to integrate and sum several input voltages $e_1 = 0.20$ volt, $e_2 = -0.40$ volt, and $e_3 = 0.50$ volt, where the gains are respectively 1.0, 0.50, and 0.20 and the feedback capacitor is 1.0 microfarad. The initial condition at $t = 0$ is $e_o(0) = 1.5$ volts.
(a) Calculate e_o at times $t = 0, 5,$ and 10 sec. Plot e_o versus time.
(b) Calculate the values of $R_1, R_2,$ and R_3.
(c) Draw the detailed network used with initial condition.
(d) Draw the programming symbol.

Ans.: (a) At $t = 0$, $e_o = +1.5$; at $t = 5$, $e_o = +1.0$; at $t = 10$, $e_o = +0.5$

10-4 Summer and Potentiometer Elements

Two input voltages $e_1 = -3.1$ volts and $e_2 = 0.37$ volt are being summed in a summer with $R_f = 0.5$ megohm. It is desired to produce the sum $-(0.44e_1 + 3.2e_2)$.
(a) Draw the detailed network wiring diagram and calculate the output voltage, R_1 and R_2, and the potentiometer settings.
(b) Draw the programming symbols.

10-5 Solution of Differential Equation

Solve the following equation:

$$+0.75\frac{dy}{dt} - 1.1y + 0.83 = 0$$

At $t = 0$, $y(0) = 8.63$. Draw the detailed wiring diagram, including initial conditions and the complete programming symbols.

10-6   Solution of Second-Order Differential Equation
Solve the following differential equation:

$$-2.2\frac{d^2x}{dt^2} + 0.7\frac{dx}{dt} - x + 0.85 = 0$$

At $t = 0$, $x(0) = -1.70$ and $(dx/dt)_{t=0} = 0.93$. Draw the complete programming symbols.

10-7   Solution of Simultaneous Differential Equations
It is desired to solve the following two equations:

$$1.5\frac{d^2x}{dt^2} + 0.7\frac{dy}{dt} - x + 2.0 = 0$$

$$\frac{d^2y}{dt^2} - 1.5\frac{dx}{dt} - 3.0y = 0$$

Draw the complete programming symbols.

10-8   Scaling of Time
The following equation is given:

$$0.75\frac{dz}{dt} + 1.5z = 0.40t$$

It is desired to find the solution of the problem, where t is ft, up to a distance of 100 ft. A mechanical recorder is being used with a 10-in. chart. Using 1 in. as 1 sec, time-scale the equation and calculate the value of α.

Ans.: $\alpha = 0.10$ sec/ft; $0.075(dz/d\tau) + 1.5z = 4.0\tau$

10-9   Amplitude and Time scaling
The following equation is to be scaled completely for time and amplitude, where $\theta =$ hr and $w =$ ft.

$$\frac{d^2w}{d\theta^2} + 0.72\frac{dw}{d\theta} - 0.3w + 1.0\theta = 0$$

The maximum value of w is to be 20 ft, and the total time of the run is $\theta = 50$ hr. For the analog computer solution a mechanical plotter will be used in order to record the complete run in 10 sec. At $\theta = 0$, $w(0) = 0.10$ ft and $(dw/d\theta)_{\theta=0} = 1.20$.
   (a) Completely scale the equation, using the symbols α and β.
   (b) Solve for α and β and substitute numerical values into the equation and IC.
   (c) Draw the complete programming symbols.

NOTATION

- a constant
- A constant
- A constant gain
- b constant
- c constant
- C_f capacitance, microfarads
- e dependent variable on analog computer, volts
- $e_o(0)$ value of e_o at $\tau = 0$, volts
- i current, amperes
- IC initial conditions at $\tau = 0$
- k constant for potentiometer, less than 1.0
- Q coulombs
- R resistance, megohms
- t time, sec or hr
- x dependent variable
- y dependent variable
- z dependent variable

Greek Letters

- α time scale factor in $\tau = \alpha t$
- β amplitude scale factor in $e = \beta x$ or $e = \beta z$
- γ amplitude scale factor in $e = \gamma y$
- τ analog computer time, sec

Subscripts

- f feedback
- i input
- o output
- 1   position 1
- 2   position 2

REFERENCES

- (C1) D. R. Coughanowr and L. B. Koppel, *Process Systems Analysis and Control*. New York: McGraw-Hill, Inc., 1965.
- (J1) C. L. Johnson, *Analog Computer Techniques*. New York: McGraw-Hill, Inc., 1956.
- (J2) A. S. Jackson, *Analog Computation*. New York: McGraw-Hill, Inc., 1960.
- (M1) P. W. Murrill, *Automatic Control of Processes*. Scranton, Pa.: International Textbook Company, 1967.
- (S1) G. W. Smith and R. C. Wook, *Principles of Analog Computation*. New York: McGraw-Hill, Inc., 1959.

Appendix

APPENDIX A.1 FUNDAMENTAL CONSTANTS

A.1-1 Standard Constants

Physical Constants

The Gas-Law Constant R

Numerical Value	Units
1.987	g cal/(g mole)(°K)
1.987	Btu/(lb mole)(°R)
82.057	$(cm^3)(atm)/(g\ mole)(°K)$
0.08205	(liter)(atm)/(g mole)(°K)
10.731	$(ft^3)(lb_f)/(in.)^2(lb\ mole)(°R)$
0.7302	$(ft)^3(atm)/(lb\ mole)(°R)$

Avogadro constant, $N = 6.02380 \times 10^{23}$ atoms per gram-atom or molecules per gram-mole

Density and Volume

1 g mole of an ideal gas at 0°C, 760 mm Hg = 22.4140 liters
= 22,414.6 cc
1 lb mole of an ideal gas at 0°C, 760 mm Hg = 359.05 cu ft
Density of dry air at 0°C and 760 mm Hg = 1.2929 g mass per liter
= 0.080711 lb mass per cu ft
1 g mass per cc = 62.43 lb mass per cu ft
1 g mass per cc = 8.345 lb mass per U.S. gal

Length

1 in. = 2.540 cm
1 micron = 10^{-6} meter
1 angstrom = 10^{-10} meter

Standard Acceleration of Gravity

g = 980.665 cm/sec^2
  = 32.174 ft/sec^2

A.1-2 Transport Property Constants*

Conversion Factors for Quantities Having Dimensions[a] of g mass/cm-sec or lb force-sec/ft² or g mole/cm-sec

(Viscosity, density times diffusivity, concentration times diffusivity)

Given a Quantity in These Units ↓ / Multiply by Table Value to Convert to These Units →	g_m cm^{-1} sec^{-1} (poises)	kg_m m^{-1} sec^{-1}	lb_m ft^{-1} sec^{-1}	lb_f sec ft^{-2}	Centipoises	lb_m ft^{-1} hr^{-1}
g_m cm^{-1} sec^{-1}	1	10^{-1}	6.7197×10^{-2}	2.0886×10^{-3}	10^2	2.4191×10^2
kg_m m^{-1} sec^{-1}	10	1	6.7197×10^{-1}	2.0886×10^{-2}	10^3	2.4191×10^3
lb_m ft^{-1} sec^{-1}	1.4882×10^1	1.4882	1	3.1081×10^{-2}	1.4882×10^3	3.6000×10^3
lb_f sec ft^{-2}	4.7880×10^2	4.7880×10^1	32.1740	1	4.7880×10^4	1.1583×10^5
Centipoises	10^{-2}	10^{-3}	6.7197×10^{-4}	2.0886×10^{-5}	1	2.4191
lb_m ft^{-1} hr^{-1}	4.1338×10^{-3}	4.1338×10^{-4}	2.7778×10^{-4}	8.6336×10^{-6}	4.1338×10^{-1}	1

[a] When moles appear in the given and desired units, the conversion factor is the same as for the corresponding mass units.

Conversion Factors for Quantities Having Dimensions of cm²/sec or ft²/hr

(Momentum diffusivity, thermal diffusivity, molecular diffusivity)

Given a Quantity in These Units ↓ / Multiply by Table Value to Convert to These Units →	cm² sec^{-1}	m² sec^{-1}	ft² hr^{-1}	Centistokes
cm² sec^{-1}	1	10^{-4}	3.8750	10^2
m² sec^{-1}	10^4	1	3.8750×10^4	10^6
ft² hr^{-1}	2.5807×10^{-1}	2.5807×10^{-5}	1	2.5807×10^1
Centistokes	10^{-2}	10^{-6}	3.8750×10^{-2}	1

Conversion Factors for Quantities Having Dimensions[a] of g mass/sec-cm² or g mole/sec-cm² or lb force-sec/ft³

(Mass flux, molar flux, mass transfer coefficients k_x or k_y)

Given a Quantity in These Units ↓	Multiply by Table Value to Convert to These Units →	$g_m\,cm^{-2}\,sec^{-1}$	$kg_m\,m^{-2}\,sec^{-1}$	$lb_m\,ft^{-2}\,sec^{-1}$	$lb_f\,ft^{-3}\,sec$	$lb_m\,ft^{-2}\,hr^{-1}$
$g_m\,cm^{-2}\,sec^{-1}$		1	10^1	2.0482	6.3659×10^{-2}	7.3734×10^3
$kg_m\,m^{-2}\,sec^{-1}$		10^{-1}	1	2.0482×10^{-1}	6.3659×10^{-3}	7.3734×10^2
$lb_m\,ft^{-2}\,sec^{-1}$		4.8824×10^{-1}	4.8824	1	3.1081×10^{-2}	3600
$lb_f\,ft^{-3}\,sec$		1.5709×10^1	1.5709×10^2	32.1740	1	1.1583×10^5
$lb_m\,ft^{-2}\,hr^{-1}$		1.3562×10^{-4}	1.3562×10^{-3}	2.7778×10^{-4}	8.6336×10^{-6}	1

[a] When moles appear in the given and desired units, the conversion factor is the same as for the corresponding mass units.

*SOURCE: R. B. Bird, W. E. Stewart, and E. N. Lightfoot, *Transport Phenomena* (New York: John Wiley & Sons, Inc., 1960). With permission.

APPENDIX A.2 PHYSICAL PROPERTIES

Table A.2-1 Vapor Pressure of Liquid Water from 0 to 100°C

Temp., °C	Vapor Pressure mm Hg	Temp., °C	Vapor Pressure mm Hg
0	4.58	45	71.88
5	6.54	50	92.51
10	9.21	60	149.4
15	12.79	70	233.7
20	17.54	80	355.1
25	23.76	90	525.8
30	31.82	95	633.9
35	42.18	100	760.0
40	55.32		

Table A.2-2 Density and Volume of Water −10 to +250°C*

Temp., °C	Density	Volume	Temp., °C	Density	Volume	Temp., °C	Density	Volume
−10	0.99815	1.00186	15	0.99913	1.00087	40	0.99225	1.00782
−9	843	157	16	897	103	41	187	821
−8	869	131	17	880	120	42	147	861
−7	892	108	18	862	138	43	107	901
−6	912	088	19	843	157	44	066	943
−5	0.99930	1.00070	20	0.99823	1.00177	45	0.99025	1.00985
−4	945	055	21	802	198	46	0.98982	1.01028
−3	958	042	22	780	220	47	940	072
−2	970	031	23	757	244	48	896	116
−1	979	021	24	733	268	49	852	162
0	0.99987	1.00013	25	0.99708	1.00293	50	0.98807	1.01207
1	993	007	26	682	320	51	762	254
2	997	003	27	655	347	52	715	301
3	999	001	28	627	375	53	669	349
4	1.00000	1.00000	29	598	404	54	621	398
5	0.99999	1.00001	30	0.99568	1.00434	55	0.98573	1.01448
6	997	003	31	537	465	60	324	705
7	993	007	32	506	497	65	059	979
8	988	012	33	473	530	70	0.97781	1.02270
9	981	019	34	440	563	75	489	576
10	0.99973	1.00027	35	0.99406	1.00598	80	0.97183	1.02899
11	963	037	36	371	633	85	0.96865	1.03237
12	952	048	37	336	669	90	534	590
13	940	060	38	300	706	95	192	959
14	927	073	39	263	743	100	0.95838	1.04343

Table A.2-2 (continued)

Temp., °C	Density	Volume	Temp., °C	Density	Volume	Temp., °C	Density	Volume
110	0.9510	1.0515	160	0.9075	1.1019	210	0.850	1.177
120	0.9434	1.0601	170	0.8973	1.1145	220	0.837	1.195
130	0.9352	1.0693	180	0.8866	1.1279	230	0.823	1.215
140	0.9264	1.0794	190	0.8750	1.1429	240	0.809	1.236
150	0.9173	1.0902	200	0.8628	1.1590	250	0.794	1.259

*The mass of 1 cc at 4°C is taken as unity. Extracted from Table 290, "Smithsonian Physical Tables," 9th rev. ed., Washington, D.C., 1954.
SOURCE: J. H. Perry, *Chemical Engineers' Handbook*, 4th ed. (New York: McGraw-Hill, Inc., 1963). With permission.

Table A.2-3 Viscosities of Gases at 1 Atm (Coordinates for use with Fig. A.2-1)

No.	Gas	X	Y	No.	Gas	X	Y
1	Acetic acid	7.7	14.3	29	Freon-113	11.3	14.0
2	Acetone	8.9	13.0	30	Helium	10.9	20.5
3	Acetylene	9.8	14.9	31	Hexane	8.6	11.8
4	Air	11.0	20.0	32	Hydrogen	11.2	12.4
5	Ammonia	8.4	16.0	33	$3H_2 + 1N_2$	11.2	17.2
6	Argon	10.5	22.4	34	Hydrogen bromide	8.8	20.9
7	Benzene	8.5	13.2	35	Hydrogen chloride	8.8	18.7
8	Bromine	8.9	19.2	36	Hydrogen cyanide	9.8	14.9
9	Butene	9.2	13.7	37	Hydrogen iodide	9.0	21.3
10	Butylene	8.9	13.0	38	Hydrogen sulfide	8.6	18.0
11	Carbon dioxide	9.5	18.7	39	Iodine	9.0	18.4
12	Carbon disulfide	8.0	16.0	40	Mercury	5.3	22.9
13	Carbon monoxide	11.0	20.0	41	Methane	9.9	15.5
14	Chlorine	9.0	18.4	42	Methyl alcohol	8.5	15.6
15	Chloroform	8.9	15.7	43	Nitric oxide	10.9	20.5
16	Cyanogen	9.2	15.2	44	Nitrogen	10.6	20.0
17	Cyclohexane	9.2	12.0	45	Nitrosyl chloride	8.0	17.6
18	Ethane	9.1	14.5	46	Nitrous oxide	8.8	19.0
19	Ethyl acetate	8.5	13.2	47	Oxygen	11.0	21.3
20	Ethyl alcohol	9.2	14.2	48	Pentane	7.0	12.8
21	Ethyl chloride	8.5	15.6	49	Propane	9.7	12.9
22	Ethyl ether	8.9	13.0	50	Propyl alcohol	8.4	13.4
23	Ethylene	9.5	15.1	51	Propylene	9.0	13.8
24	Fluorine	7.3	23.8	52	Sulfur dioxide	9.6	17.0
25	Freon-11	10.6	15.1	53	Toluene	8.6	12.4
26	Freon-12	11.1	16.0	54	2,3,3-Trimethylbutane	9.5	10.5
27	Freon-21	10.8	15.3	55	Water	8.0	16.0
28	Freon-22	10.1	17.0	56	Xenon	9.3	23.0

SOURCE: J. H. Perry, *Chemical Engineers' Handbook*, 4th ed. (New York: McGraw-Hill, Inc., 1963). With Permission.

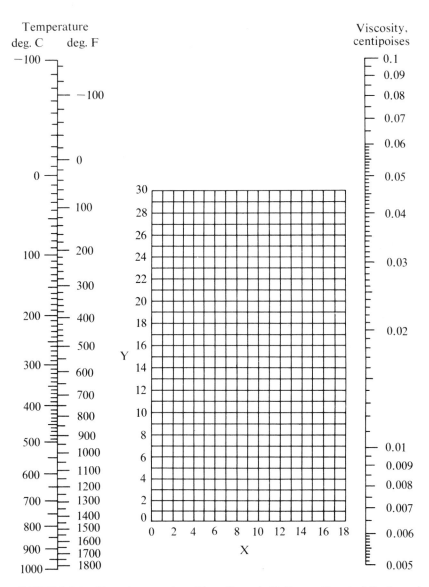

FIGURE A.2-1 Chart for gas viscosities. (From J. H. Perry, *Chemical Engineers' Handbook*. New York: McGraw-Hill, Inc., 1963. With permission.)

Appendix 467

Table A.2-4 Viscosities of Liquids (Coordinates for Use with Fig. A.2-2)

Liquid	X	Y	Liquid	X	Y
Acetaldehyde	15.2	14.8	Ethyl acetate	13.7	9.1
Acetic acid, 100%	12.1	14.2	Ethyl acrylate	12.7	10.4
Acetic acid, 70%	9.5	17.0	Ethyl alcohol, 100%	10.5	13.8
Acetic anhydride	12.7	12.8	Ethyl alcohol, 95%	9.8	14.3
Acetone, 100%	14.5	7.2	Ethyl alcohol, 40%	6.5	16.6
Acetone, 35%	7.9	15.0	Ethyl benzene	13.2	11.5
Acetonitrile	14.4	7.4	Ethyl bromide	14.5	8.1
Acrylic acid	12.3	13.9	2-Ethyl butyl acrylate	11.2	14.0
Allyl alcohol	10.2	14.3	Ethyl chloride	14.8	6.0
Allyl bromide	14.4	9.6	Ethyl ether	14.5	5.3
Allyl iodide	14.0	11.7	Ethyl formate	14.2	8.4
Ammonia, 100%	12.6	2.0	2-Ethyl hexyl acrylate	9.0	15.0
Ammonia, 26%	10.1	13.9	Ethyl iodide	14.7	10.3
Amyl acetate	11.8	12.5	Ethyl propionate	13.2	9.9
Amyl alcohol	7.5	18.4	Ethyl propyl ether	14.0	7.0
Aniline	8.1	18.7	Ethyl sulfide	13.8	8.9
Anisole	12.3	13.5	Ethylene bromide	11.9	15.7
Arsenic trichloride	13.9	14.5	Ethylene chloride	12.7	12.2
Benzene	12.5	10.9	Ethylene glycol	6.0	23.6
Brine, CaCl$_2$, 25%	6.6	15.9	Ethylidene chloride	14.1	8.7
Brine, NaCl, 25%	10.2	16.6	Fluorobenzene	13.7	10.4
Bromine	14.2	13.2	Formic acid	10.7	15.8
Bromotoluene	20.0	15.9	Freon-11	14.4	9.0
Butyl acetate	12.3	11.0	Freon-12	16.8	15.6
Butyl acrylate	11.5	12.6	Freon-21	15.7	7.5
Butyl alcohol	8.6	17.2	Freon-22	17.2	4.7
Butyric acid	12.1	15.3	Freon-113	12.5	11.4
Carbon dioxide	11.6	0.3	Glycerol, 100%	2.0	30.0
Carbon disulfide	16.1	7.5	Glycerol, 50%	6.9	19.6
Carbon tetrachloride	12.7	13.1	Heptane	14.1	8.4
Chlorobenzene	12.3	12.4	Hexane	14.7	7.0
Chloroform	14.4	10.2	Hydrochloric acid, 31.5%	13.0	6.6
Chlorosulfonic acid	11.2	18.1	Iodobenzene	12.8	15.9
Chlorotoluene, ortho	13.0	13.3	Isobutyl alcohol	7.1	18.0
Chlorotoluene, meta	13.3	12.5	Isobutyric acid	12.2	14.4
Chlorotoluene, para	13.3	12.5	Isopropyl alcohol	8.2	16.0
Cresol, meta	2.5	20.8	Isopropyl bromide	14.1	9.2
Cyclohexanol	2.9	24.3	Isopropyl chloride	13.9	7.1
Cyclohexane	9.8	12.9	Isopropyl iodide	13.7	11.2
Dibromomethane	12.7	15.8	Kerosene	10.2	16.9
Dichloroethane	13.2	12.2	Linseed oil, raw	7.5	27.2
Dichloromethane	14.6	8.9	Mercury	18.4	16.4
Diethyl ketone	13.5	9.2	Methanol, 100%	12.4	10.5
Diethyl oxalate	11.0	16.4	Methanol, 90%	12.3	11.8
Diethylene glycol	5.0	24.7	Methanol, 40%	7.8	15.5
Diphenyl	12.0	18.3	Methyl acetate	14.2	8.2
Dipropyl ether	13.2	8.6	Methyl acrylate	13.0	9.5
Dipropyl oxalate	10.3	17.7	Methyl i-butyrate	12.3	9.7

Table A.2-4 (continued)

Liquid	X	Y	Liquid	X	Y
Methyl n-butyrate	13.2	10.3	Propyl formate	13.1	9.7
Methyl chloride	15.0	3.8	Propyl iodide	14.1	11.6
Methyl ethyl ketone	13.9	8.6	Sodium	16.4	13.9
Methyl formate	14.2	7.5	Sodium hydroxide, 50%	3.2	25.8
Methyl iodide	14.3	9.3	Stannic chloride	13.5	12.8
Methyl propionate	13.5	9.0	Succinonitrile	10.1	20.8
Methyl propyl ketone	14.3	9.5	Sulfur dioxide	15.2	7.1
Methyl sulfide	15.3	6.4	Sulfuric acid, 110%	7.2	27.4
Napthalene	7.9	18.1	Sulfuric acid, 100%	8.0	25.1
Nitric acid, 95%	12.8	13.8	Sulfuric acid, 98%	7.0	24.8
Nitric acid, 60%	10.8	17.0	Sulfuric acid, 60%	10.2	21.3
Nitrobenzene	10.6	16.2	Sulfuryl chloride	15.2	12.4
Nitrogen dioxide	12.9	8.6	Tetrachloroethane	11.9	15.7
Nitrotoluene	11.0	17.0	Thiophene	13.2	11.0
Octane	13.7	10.0	Titanium tetrachloride	14.4	12.3
Octyl alcohol	6.6	21.1	Toluene	13.7	10.4
Pentachloroethane	10.9	17.3	Trichloroethylene	14.8	10.5
Pentane	14.9	5.2	Triethylene glycol	4.7	24.8
Phenol	6.9	20.8	Turpentine	11.5	14.9
Phosphorus tribromide	13.8	16.7	Vinyl acetate	14.0	8.8
Phosphorus trichloride	16.2	10.9	Vinyl toluene	13.4	12.0
Propionic acid	12.8	13.8	Water	10.2	13.0
Propyl acetate	13.1	10.3	Xylene, ortho	13.5	12.1
Propyl alcohol	9.1	16.5	Xylene, meta	13.9	10.6
Propyl bromide	14.5	9.6	Xylene, para	13.9	10.9
Propyl chloride	14.4	7.5			

SOURCE: J. H. Perry, *Chemical Engineers' Handbook*, 4th ed. (New York: McGraw-Hill, Inc., 1963). With permission.

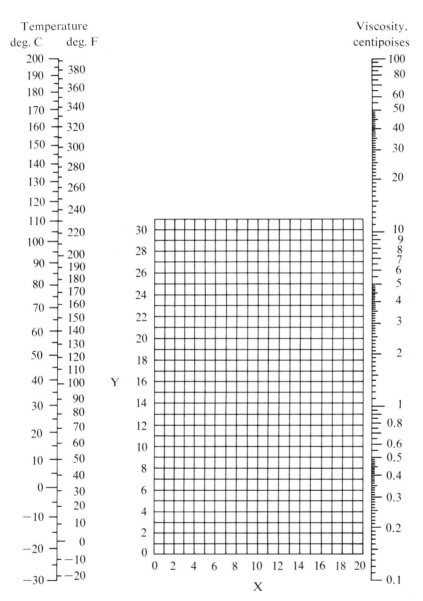

FIGURE A.2-2 Chart for liquid viscosities. (From J. H. Perry, *Chemical Engineers' Handbook*. New York: McGraw-Hill, Inc., 1963. With permission.)

Table A.2-5 Viscosity of Water

Temp., °C	Viscosity, centipoises	Temp., °C	Viscosity, centipoises	Temp., °C	Viscosity, centipoises
0	1.7921	33	0.7523	67	0.4233
1	1.7313	34	0.7371	68	0.4174
2	1.6728	35	0.7225	69	0.4117
3	1.6191	36	0.7085	70	0.4061
4	1.5674	37	0.6947	71	0.4006
5	1.5188	38	0.6814	72	0.3952
6	1.4728	39	0.6685	73	0.3900
7	1.4284	40	0.6560	74	0.3849
8	1.3860	41	0.6439	75	0.3799
9	1.3462	42	0.6321	76	0.3750
10	1.3077	43	0.6207	77	0.3702
11	1.2713	44	0.6097	78	0.3655
12	1.2363	45	0.5988	79	0.3610
13	1.2028	46	0.5883	80	0.3565
14	1.1709	47	0.5782	81	0.3521
15	1.1404	48	0.5683	82	0.3478
16	1.1111	49	0.5588	83	0.3436
17	1.0828	50	0.5494	84	0.3395
18	1.0559	51	0.5404	85	0.3355
19	1.0299	52	0.5315	86	0.3315
20	1.0050	53	0.5229	87	0.3276
20.20	*1.0000*	54	0.5146	88	0.3239
21	0.9810	55	0.5064	89	0.3202
22	0.9579	56	0.4985	90	0.3165
23	0.9358	57	0.4907	91	0.3130
24	0.9142	58	0.4832	92	0.3095
25	0.8937	59	0.4759	93	0.3060
26	0.8737	60	0.4688	94	0.3027
27	0.8545	61	0.4618	95	0.2994
28	0.8360	62	0.4550	96	0.2962
29	0.8180	63	0.4483	97	0.2930
30	0.8007	64	0.4418	98	0.2899
31	0.7840	65	0.4355	99	0.2868
32	0.7679	66	0.4293	100	0.2838

Calculated by the formula:

$$1/\mu = 2.1482[(t-8.435) + \sqrt{8078.4 + (t-8.435)^2}] - 120.$$

SOURCE: Bingham, *Fluidity and Plasticity*, p. 340 (New York: McGraw-Hill, Inc., 1922). With permission.

APPENDIX A.3 UNSTEADY-STATE CHARTS*

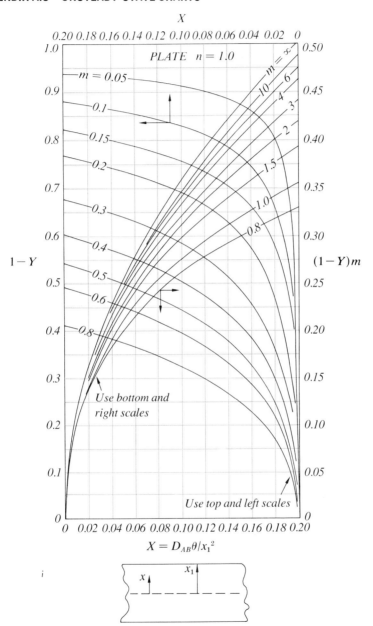

FIGURE A.3-1 Chart for determining concentration history at surface of flat plate ($n = 1.0$). [From H. P. Heisler, *Trans. ASME* **69**, 227 (1947). With permission.]

---

*SOURCE: The eight charts in this appendix are from M. P. Heisler, *Trans. ASME* **69**, 227 (1947). With permission.

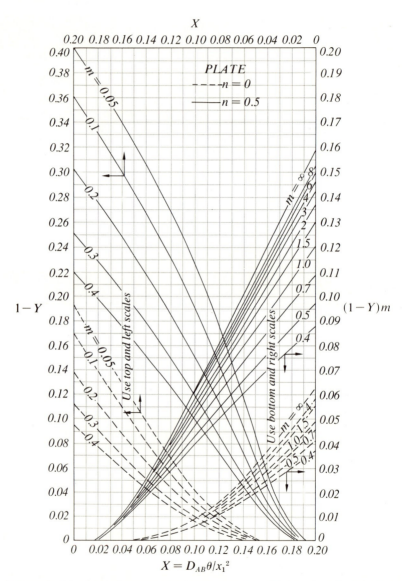

FIGURE A.3-2 Chart for determining concentration history at center ($n = 0$) and midplane ($n = 0.5$) of flat plate. [From H. P. Heisler, *Trans. ASME* **69**, 227 (1947). With permission.]

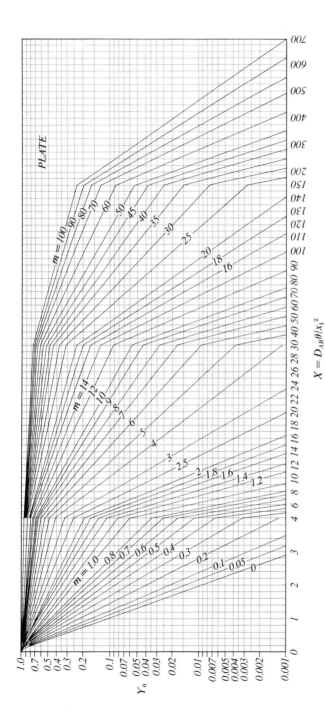

FIGURE A.3-3 Chart for determining concentration history at center of flat plate. [From H. P. Heisler, *Trans. ASME* **69**, 227 (1947). With permission.]

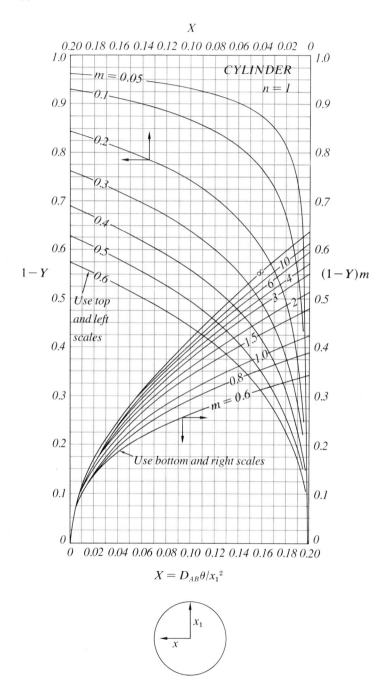

FIGURE A.3-4 Chart for determining concentration history at surface of infinitely long cylinder. [From H. P. Heisler, *Trans. ASME* **69**, 227 (1947). With permission.]

Appendix 475

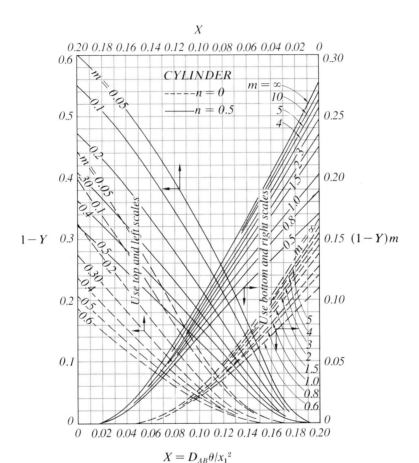

FIGURE A.3-5 Chart for determining concentration history at center ($n = 0$) and half-radius ($n = 0.5$) of infinitely long cylinder. [From H. P. Heisler, *Trans. ASME* **69**, 227 (1947). With permission.]

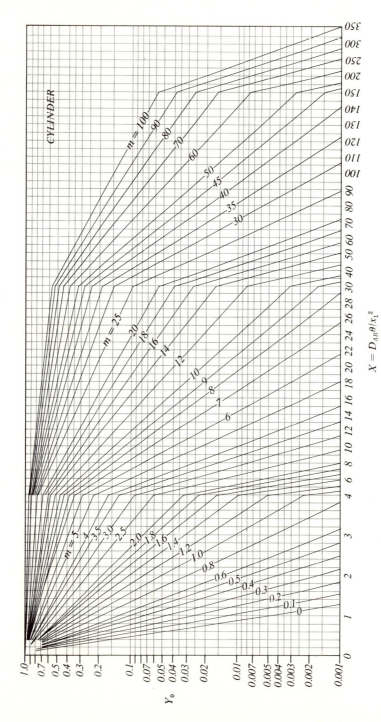

FIGURE A.3-6 Chart for determining concentration history at center of infinitely long cylinder. [From H. P. Heisler, *Trans. ASME* **69**, 227 (1947). With permission.]

Appendix 477

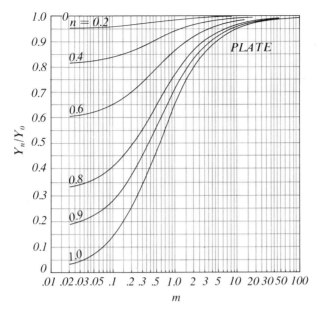

FIGURE A.3-7 Position correction factors for dimensionless concentration ratios for flat plate ($X > 0.2$). [From H. P. Heisler, *Trans. ASME* **69**, 227 (1947). With permission.]

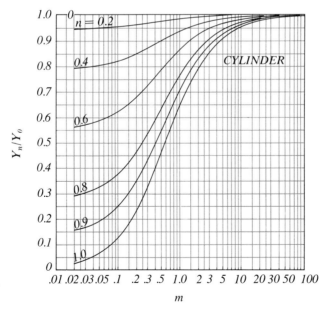

FIGURE A.3-8 Position correction factors for dimensionless concentration ratios for infinitely long cylinder ($X > 0.2$). [From H. P. Heisler, *Trans. ASME* **69**, 227 (1947). With permission.]

APPENDIX A.4 EQUILIBRIUM DATA

Table A.4-1 Henry's Law Constants for Gases in Water ($H \times 10^{-4}$)

T, °C	CO_2	CO	C_2H_6	C_2H_4	He	H_2	H_2S	CH_4	N_2	O_2
0	0.0728	3.52	1.26	0.552	12.9	5.79	2.68	2.24	5.29	2.55
10	0.104	4.42	1.89	0.768	12.6	6.36	3.67	2.97	6.68	3.27
20	0.142	5.36	2.63	1.02	12.5	6.83	4.83	3.76	8.04	4.01
30	0.186	6.20	3.42	1.27	12.4	7.29	6.09	4.49	9.24	4.75
40	0.233	6.96	4.23		12.1	7.51	7.45	5.20	10.4	5.35

$p_A = H x_A$
p_A = partial pressure of A in the gas, atm
x_A = mole fraction of A in the liquid
H = Henry's law constant, atm/mole frac
SOURCE: National Research Council, *International Critical Tables*, vol. III (New York: McGraw-Hill, Inc., 1929).

Table A.4-2 Equilibrium Data for Ammonia-Water System

x_A, Mole Frac. NH_3 in Liquid	p_A, mm Hg, Partial Pressure of NH_3 in Vapor		y_A, Mole Frac. NH_3 in Vapor, $P = 1$ Atm	
	20°C	30°C	20°C	30°C
0	0	0	0	0
0.0126		11.5		0.0151
0.0167		15.3		0.0201
0.0208	12	19.3	0.0158	0.0254
0.0258	15	24.4	0.0197	0.0321
0.0309	18.2	29.6	0.0239	0.0390
0.0405	24.9	40.1	0.0328	0.0527
0.0503	31.7	51.0	0.0416	0.0671
0.0737	50.0	79.7	0.0657	0.105
0.0960	69.6	110	0.0915	0.145
0.137	114	179	0.150	0.235
0.175	166	260	0.218	0.342
0.210	227	352	0.298	0.463
0.241	298	454	0.392	0.597
0.297	470	719	0.618	0.945

SOURCE: J. H. Perry, *Chemical Engineers' Handbook*, 4th ed. (New York: McGraw-Hill, Inc., 1963). With permission.

Table A.4-3 Equilibrium Data for SO$_2$-Water System

x_A, Mole Frac. SO$_2$ in Liquid	p_A, mm Hg, Partial Pressure of SO$_2$ in Vapor		y_A, Mole Frac. SO$_2$ in Vapor, $P = 1$ Atm	
	20°C	30°C	20°C	30°C
0	0	0	0	0
0.0000562	0.5	0.6	0.000658	0.000790
0.0001403	1.2	1.7	0.00158	0.00223
0.000280	3.2	4.7	0.00421	0.00619
0.000422	5.8	8.1	0.00763	0.01065
0.000564	8.5	11.8	0.01120	0.0155
0.000842	14.1	19.7	0.01855	0.0259
0.001403	26.0	36	0.0342	0.0473
0.001965	39.0	52	0.0513	0.0685
0.00279	59	79	0.0775	0.1040
0.00420	92	125	0.121	0.1645
0.00698	161	216	0.212	0.284
0.01385	336	452	0.443	0.594
0.0206	517	688	0.682	0.905
0.0273	698		0.917	

SOURCE: T. K. Sherwood, *Ind. Eng. Chem.* **17**, 745 (1925). With permission.

Table A.4-4 Equilibrium Data for Methanol-Water System

x_A, Mole Frac. Methanol in Liquid	p_A, mm Hg, Partial Pressure of Methanol in Vapor	
	39.9°C	59.4°C
0	0	0
0.05	25.0	50
0.10	46.0	102
0.15	66.5	151

SOURCE: National Research Council, *International Critical Tables*, vol. III (New York: McGraw-Hill, Inc., 1929).

Table A.4-5 Equilibrium Data for Acetone-Water System at 68°F

x_A, Mole Frac. Acetone in Liquid	p_A, mm Hg, Partial Pressure of Acetone in Vapor
0	0
0.0333	30.0
0.0720	62.8
0.117	85.4
0.171	103

SOURCE: T. K. Sherwood, *Absorption and Extraction*, 1st ed. (New York: McGraw-Hill, Inc., 1937). With permission.

Table A.4-6 Acetic Acid-Water-Isopropyl Ether System, Liquid-Liquid Equilibria at 20°C

Water Layer, Wt Percent			Isopropyl Ether Layer, Wt Percent		
Acetic Acid	Water	Isopropyl Ether	Acetic Acid	Water	Isopropyl Ether
0	98.8	1.2	0	0.6	99.4
0.69	98.1	1.2	0.18	0.5	99.3
1.41	97.1	1.5	0.37	0.7	98.9
2.89	95.5	1.6	0.79	0.8	98.4
6.42	91.7	1.9	1.93	1.0	97.1
13.30	84.4	2.3	4.82	1.9	93.3
25.50	71.1	3.4	11.40	3.9	84.7
36.70	58.9	4.4	21.60	6.9	71.5
44.30	45.1	10.6	31.10	10.8	58.1
46.40	37.1	16.5	36.20	15.1	48.7

SOURCE: *TAICHE* **36**, 601, 628 (1940). With permission.

Table A.4-7 Styrene-Ethyl Benzene-Diethylene Glycol System, Liquid-Liquid Equilibria; Equilibrium Compositions Weight Percent, at 25°C

Ethylbenzene Layer		Diethylene Glycol Layer	
Styrene	Diethylene Glycol	Styrene	Diethylene Glycol
0	0.67	0	89.60
8.63	0.81	1.64	88.51
18.67	0.93	3.49	87.20
28.51	1.00	5.48	85.80
37.98	1.09	7.45	84.48
45.84	1.20	9.49	83.20
57.09	1.40	12.54	81.40
76.60	1.80	18.62	77.65
97.50	2.50	27.07	72.93

SOURCE: M. G. Boobar et al., *Ind. Eng. Chem.* **43**, 2922 (1951). With permission.

Table A.4-8 Equilibrium Data for Ethanol-Water System at 1 Atm
(Reference state for enthalpy is pure liquid at 32°F)

Temp., °F	Vapor-Liquid Equilibria, Mass Fraction Ethanol		Temp., °F	Vapor-Liquid Equilibria, Mass Fraction Ethanol	
	x_A	y_A		x_A	y_A
212	0	0	177.8	0.600	0.794
208.5	0.020	0.192	176.2	0.700	0.822
203.4	0.050	0.377	174.3	0.800	0.858
197.2	0.100	0.527	173.0	0.900	0.912
189.2	0.200	0.656	172.8	0.940	0.942
184.5	0.300	0.713	172.7	0.960	0.959
181.7	0.400	0.746	172.8	0.980	0.978
179.6	0.500	0.771	173.0	1.00	1.00

Temp., °F	Mass Fraction	Enthalpy, Btu/Lb Mass of Mixture	
		Liquid	Vapor
212	0	180.1	1150
197.2	0.1	159.8	1082
184.5	0.3	135.0	943
179.6	0.5	122.9	804
176.2	0.7	111.1	664
173.0	0.9	96.6	526
173.0	1.0	89.0	457.5

SOURCE: Data from L. W. Cornell and R. E. Montonna, *Ind Eng. Chem.* **25**, 1331 (1933); and W. A. Noyes and R. R. Warfel, *J. Am. Chem. Soc.* **23**, 463 (1901), as given by G. G. Brown, *Unit Operations* (New York: John Wiley & Sons, Inc., 1950). With permission.

Author Index

References in parentheses after the page number indicate the author's work is cited but his name is not given in the text. The complete references are given at the end of each chapter.

Anderson, D. K., 126(A1)
Anderson, L. B., 349(F1)
Aris, R., 161
Armistead, F. C., 20, 21(N2)
Atkins, B. E., 108(A1)

Babb, A. L., 126(A1,J1)
Badgett, C. O., 359(C1)
Barrer, R. M., 144, 146
Bastick, R. E., 108(A1,B4)
Bedingfeld, C. H. Jr., 253(B1), 254(B1), 292
Bennett, C. O., 87(B2), 88(B2), 111, 261(B4), 262(B4), 272(B4), 304(B4)
Bennett, J. A. R., 299
Bidstrup, D. E., 124, 125, 126(B1), 139
Bingham, E. C., 470
Bird, R. B., 23, 25(H1), 29, 64, 65, 67, 73, 83(B1), 87(B1,B3), 88(B1), 92(B1), 96, 99(B1), 105, 108(B1,H2), 109(B1), 111, 132, 138, 254, 259(B2), 260, 261(B2), 267(B2), 274, 333, 463
Bonilla, C. F., 299(D1), 308(D1)
Boobar, M. G., 480
Brow, J. E., 22(S5)
Brown, G. G., 430, 481
Brown, H., 108(B5)
Burchard, J. K., 130, 138
Burris, L., 292, 299

Carmichael, L. T., 22(C1)
Carslaw, H. S., 239
Carswell, A. J., 22(C2)
Chakraborti, P. K., 22(C4)
Chang, Pin, 126(C3), 127(W1), 128
Chang, S. Y., 126(C2)
Chapman, S., 22(C3), 31, 32(C3)
Christian, W. J., 304(C2)
Claffey, J. B., 359(C1)
Clump, C. W., 349(F1)
Coe, J. R. Jr., 6(S4)
Conte, S. D., 52, 54(C5), 55(C5), 59, 286

Cornell, D., 395
Cornell, L. W., 481
Coughanowr, D. R., 454(C1)
Cowling, T. G., 22(C3), 31, 32(C3)
Craine, K., 128(D1)
Crank, J., 149, 185, 219, 225(C1), 239, 240, 241, 242
Cullinen, H. T., 130
Cunningham, R. S., 157, 159
Curtiss, C. F., 25(H1), 29, 108(H2), 138

Danckwerts, P. V., 134, 301, 333, 336, 341(D1), 342, 353
Darkin, L. S., 181, 182, 186(D2)
Davies, G. A., 128(D1)
De Groot, S. R., 138
Denbigh, K. G., 108, 138
Dobbins, W. E., 301(D3)
Doraiswamy, L. R., 127(R1), 128(R1), 129
Dougherty, E. L., 108(S2)
Drew, T. B., 253, 254(B1), 292
Drickamer, H. G., 108(S2,T1)
Dullien, F. A. L., 153, 156, 157
Dunn, W. E., 299, 308
Dusinberre, G. M., 176(D1)

Einstein, A., 127
Epstein, N., 353(E1)
Evnochides, S., 291

Fair, J. R., 395(C2)
Ferstenberg, C., 299(D1), 308(D1)
Foust, A. S., 349(F1)
Frazier, G., 22(W4)
Friedlander, S. K., 290
Fujita, H., 138
Fuller, E. N., 32, 33

Garner, F. H., 267, 291, 292(G2,G7)
Geankoplis, C. J., 124, 125, 126(B1,P1), 128, 130, 131, 139, 157, 158(R2),

483

484 Index

Geankoplis, C. J. (continued)
  159, 280, 287, 290, 292, 296, 299
Geertson, L. R., 22(S3)
Giddings, J. C., 22(S3), 32, 33
Gilliland, E. R., 22(G1), 288, 289
Godfrey, T. B., 6(S4)
Gordon, A. R., 124
Gosting, L., 138
Grafton, R. W., 291(G7), 292(G7)
Gray, P., 22(C4)
Grew, K. E., 108(G1)
Gross, B., 299(D1), 308(D1)
Gupta, A. S., 294(G4), 295, 296
Gurney, H. P., 204, 205, 206
Gurry, R. W., 181(D2), 182, 186(D2)

Hall, J. R., 126(A1)
Hallman, T. M., 94(S1)
Hammerton, D., 267
Hammond, B. R., 126(H1)
Harriott, P., 301(H3)
Hatta, S., 337
Hawkins, G. A., 171
Heath, H. R., 108(B4,H1)
Heisler, M. P., 203, 471-477
Henry, J. P., 159
Higbie, R., 254(H1), 301
Hirschfelder, J. O., 25(H1), 29, 108(H2), 138
Holloway, F. A. L., 397, 398
Holsen, J. N., 22(H2)
Hougen, O. A., 316(H1)
Hsu, H. W., 64, 65, 67, 73

Ibbs, T. L., 108(A1,B4,H1,G1)

Jackson, A. S., 454(J2)
Jacob, M., 171
Jaeger, J. C., 239
Johnson, C. L., 454(J1)
Johnson, P. A., 126(J1)
Jost, W., 163, 164, 176(J1), 242, 243

Kennedy, A. M., 341(D1), 342
Kent, E. R., 321(K1)
Kezios, S. P., 304(C2)
King, C. J., 126(V1)
Kirkaldy, J. S., 138
Kmak, W. S., 108(T1)
Knapp, W. G., 395(C2)
Koppel, L. B., 454(C1)

Lacey, W. N., 22(C1)
Lane, J. E., 138(K1)
Lange, N. A., 6(L2), 9(L2)
Lapidus, L., 52, 54(L3), 55(L3), 59(L3)
  170, 176(L1), 224(L1), 239(L1), 281
Le Bas, G., 30(L1), 32
Lee, C. Y., 32(W1)
Levenspiel, O., 162, 163, 353
Lewis, J. B., 299
Lewis, W. K., 319(L1), 333(L1)
Lightfoot, E. N., 23(B1), 83(B1), 87(B1),
  88(B1), 92(B1), 96, 99(B1), 105(B1),
  108(B1), 109(B1), 111(B1), 132(B3),
  254(B2), 259(B2), 260(B2), 261(B2),
  267(B2), 274(B2), 333(B1), 463
Linton, M., 292(L3)
Linton, W. H., Jr., 288, 289, 290, 293
Litt, M., 290
Lurie, J., 204, 205, 206

Marshello, J. M., 301(T3)
Mason, E. A., 22(M1), 31
Mason, G. R., 138(K1)
Maus, L., 349(F1)
McAdams, W. H., 111
McCabe, W. L., 1, 349(M1), 430
Mickley, H. S., 52, 55(M3), 167, 168(M1),
  176(M1), 224(M1), 225(M1), 281,
  432(M2)
Monchick, L., 22(M1), 31
Montonna, R. E., 481
Muller, W., 22(T4)
Murrill, P. W., 454(M1)
Myers, J. E., 87(B2), 88(B2), 111, 261(B4),
  262(B4), 272(B4), 304(B4)

Newman, A. B., 207
Ney, E. P., 20, 21(N2)
Noyes, W. A., 481

Othmer, D. F., 128

Park, G. S., 149
Perkins, L. R., 126(P1), 128, 130, 131
Perry, J. H., 1, 20, 23, 30(P2), 52, 54(P2),
  55(P2), 121, 266(P1), 274(P1), 316(P1),
  342, 349(P1), 350(P1), 395, 396(P1,P2),
  398(P1), 402, 408, 409, 430, 465, 466,
  468, 469
Perry, R. H., 341
Phillips, G. W., 359(C1)
Pigford, R. L., 261(S2), 262, 266(S2), 267,
  289(S2),290, 321(K1), 333, 336, 337,

Pigford, R. L. (continued)
    341(S1), 342, 366, 395(S1), 396(S1),
    397(S1), 398(S1), 402
Ponter, A. B., 128(D1)
Portalski, S., 266(T1)

Ragatz, R. A., 316(H1)
Reddy, K. A., 127(R1), 128(R1), 129
Reed, C. E., 52(M3), 55(M3), 167(M1),
    168(M1), 176(M1), 224(M1), 225(M1),
    281(M1), 432(M2)
Reid, R. C., 6(R1), 23, 26(R1), 27(R1),
    28, 31(R1), 33(R1), 298
Remick, R. S., 157, 158(R2)
Reynolds, O., 272
Rothfeld, L. B., 153, 156

Sage, B. H., 22(C1)
Saxton, R. L., 108(S2)
Schafer, K. L., 22(S6)
Schettler, P. D., 32, 33
Schneider, P. J., 202
Schwertz, F. A., 22(S5)
Scott, D. S., 153, 156
Scriven, L. E., 266
Seager, S. L., 22(S3)
Sherwood, T. K., 6(R1), 23, 26(R1),
    27(R1), 28, 31(R1), 33(R1), 52(M3),
    55(M3), 167(M1), 168(M1), 176(M1),
    224(M1), 225(M1), 261(S2), 262,
    266(S2), 267, 281(M1), 288, 289(S2),
    290, 293, 298, 333, 336, 337, 341(S1),
    342, 366, 395(S1), 396(S1), 397(S1,S3),
    398(S1, S3), 402, 432(M2), 479
Shewman, P. G., 165(S2), 185, 186(S2)
Shulman, H. L., 394(S2), 395
Siegel, R., 94(S1)
Skalamera, C. D., 359(C1)
Sleicher, C. A. Jr., 353(S4)
Smith, A. S., 20
Smith, G. W., 454(S1)
Smith, J. C., 1, 349(M1), 430
Smith, J. M., 159, 163
Sparrow, E. M., 94(S1)
Srivastava, B. N., 22(S7)
Srivastava, I. B., 22(S7)

Steele, L. R., 287, 290, 292
Steinberger, R. L., 290, 291, 292(S5)
Sternling, C. V., 266
Stewart, W. E., 23(B1), 83(B1), 87(B1),
    88(B1), 92(B1), 96, 99(B1), 105(B1),
    108(B1), 109(B1), 111(B1), 132(B3),
    254(S1,B2), 259(B2), 260(B2),
    261(B2), 267(B2), 274(B2), 333(B1),
    463
Stiel, L. I., 31
Stokes, R. H., 126(H1)
Strunk, M. R., 22(H2)
Stryland, J. E., 22(C2)
Suckling, R. D., 291, 292(G2)
Sutherland, L. L., 292(L3)
Swindells, J. F., 6(S4)

Thankar, M. S., 128
Thodos, G., 31, 291, 294(G4), 295, 296
Thomas, W. J., 266(T1)
Tichacek, L. J., 108(T1)
Toor, H. L., 61, 63, 130, 138, 301(T3)
Trautz, M., 22(T4)
Traylor, E. D., 292, 299
Treybal, R. E., 1, 30(T2), 120(T2),
    121(T1), 122, 123, 127(T1), 209,
    288(T4), 290, 291, 292(S5), 321(T1),
    349(T1), 350(T1), 366, 396, 398, 402,
    408, 430

Vivian, J. E., 126(V1), 374, 404

Wakao, N., 159
Warfel, R. R., 481
Watson, K, M., 316(H1)
Wenzel, L. A., 349(F1)
Westenberg, A. A., 22(W4)
Whitman, W. G., 319(L1), 333(L1)
Whitney, R. P., 374, 404
Wild, N. E., 108(H1)
Wilke, C. R., 32(W1), 38(W2), 60, 61(W2),
    126(C3), 127(W1), 128
Wilson, E. J., 280, 296
Wintergerst, V. E., 22(W3)
Wook, R. C., 454(S1)

Subject Index

Absorption of gases (*see* Gas absorption processes)
Activity coefficient, 130, 186, 317
Adsorption, 159
Analog computation, amplitude scaling, 47-48, 452-454
    initial conditions, 43, 48, 446-447
    integrator element, 443-446
    introduction to, 440-441
    nonlinear elements, 50, 454-455
    with numerical method, 238-239
    output devices, 450
    potentiometer element, 446
    for solving differential equations, 447-450, 452-454
    for solving transport problems, 7-8, 42-43, 47-50
    summer element, 441-443
    time scaling, 47-48, 451-453
Analogies, Chilton-Colburn J factor, 274-275, 301-304
    dimensionless groups, 269-270
    among equations of change, 101
    between momentum, heat, and mass transport, 1-2, 4, 18-19, 270-275
    Reynolds, 272-274
    Taylor and Prandtl, 273-274
    use of, 274-275
Association parameter, 128
Atomic volumes, 30, 33
Average (bulk) concentration, 251-253
Avogadro's number, 23, 460
Axial mixing or dispersion, 353
Azeotrope, 412-413

Berl saddles, 396-398
Boiling point, calculation of, 410-412
    diagram, 410-413
Boltzmann constant, 23, 26
Boltzmann equation, 25, 31
Boltzmann-Matano method, 185-186
Boundary conditions, 79-80
    See also Flux

Boundary layer theory, 254, 302-304, 331-333
Boundary-value differential equation, 53-59
Boundary-value problem, 168
Buckingham pi theorem, 111, 275-277
Buffer (transition) region, 251-252, 273
Bulk flow correction factor, 253-257, 320-325

Calculus of finite differences, 432-434
Capillaries (*see* Porous solids)
Centipoise, 6
Centrifugal force, 108
Chapman-Enskog theory, 25-34
Chemical diffusivity, 185-186
Chemical potential, 138, 165, 241-242
Chemical reaction (*see* Diffusion with chemical reaction)
Chilton-Colburn relations, 271, 274-275, 301-304
Closed and open systems, 156-157
Cocurrent processes, 367-369
Colburn-Drew coefficient, 253
Collision diameters, 25-28, 31
Collision integral, 25-29, 31
Concentration, 2-3, 37
Condensation, 105-107
Conduction (*see* Heat transport)
Conservation of mass equation, 78, 84-86, 99-102
Continuity equation, 84-86, 99-102
    See also Equation of continuity
Continuous two-phase transport processes (*see* Gas absorption processes; Cocurrent processes; Countercurrent processes; Liquid-liquid extraction processes)
Convective mass transport, 13-14, 35-40, 80, 202
    See also Mass transfer coefficients; Laminar flow mass transport
Convergence, 217, 225, 230-231
Conversion factors, 460-463

488 Index

Countercurrent processes, axial dispersion in, 353
  classification of, 348-349
  design methods, data needed, 350
    simplified procedure, 388-394
    using digital computation, 393-394
    using film coefficients, 369-380, 397-401
    using overall coefficients, 380-385, 399-401
    using transfer units, 385-392, 396-401
  material balances, 350-364
  multistage, 350, 365-367
  operating line slope, 352, 355, 360-364
  See also Gas absorption processes; Liquid-liquid extraction processes; Stage processes
Coupling, 136-138
Crank-Nicolson method, 224-229
Cubes, mass transfer to, 293-295
Cullinen equation, 130
Cylinder, catalyst, 161
  diffusion in, 144
  mass transfer to, 292-297
  surface/volume ratio, 293-294
  unsteady-state diffusion in, 202-203, 205, 208-209, 474-477
Cylindrical coordinates, 90, 94

Delta point, 429-430
Density, 37
Derivative following the motion, 81, 85-86
Dew point, 410
Differential operations, with scalars, 83-84
  with vectors, 83-84
Diffusion (mass), of A through stagnant B, 40-43, 103-104
  with chemical reaction (*see* Diffusion with chemical reaction)
  with convection, 13-14, 35-40
  with energy transport, 19, 104-107
  with equation of continuity, 99-104
  equimolar counterdiffusion, 14-18, 40, 102-103
  Fick's law, 10-11, 13-18, 97-99
  flux (*see* Flux)
  forced, 108
  interphase (*see* Interphase mass transport)
  introduction to, 4-5, 11-13
  Knudsen, 151-159
  in liquids (*see* Liquid diffusion)

Diffusion (mass) (continued)
  multicomponent (*see* Multicomponent transport)
  in pores (*see* Porous solids)
  pressure, 108
  in solids (*see* Solids)
  Stefan-Maxwell equations, 38-39, 59-61, 64
  thermal, 108, 136-138
  turbulent (*see* Turbulent mass transport)
  in wetted-wall tower, 262-267
  See also Steady-state mass transport; Unsteady-state mass transport
Diffusion barrier, 64, 157
Diffusion with chemical reaction, diffusion controlled, 45-46
  and formation of second phase, 164-165, 243
  heterogeneous, instantaneous reaction rate, 45-46, 64-68
    slow reaction rate, 46
  homogeneous, empirical reaction rate, 49, 55-59
    first-order reaction rate, 46-50, 79, 131-134
  in semiinfinite medium, 134-135
  in solids, 135, 163-165
  steady-state, in liquids, 131-134
  See also Mass transport with chemical reaction
Diffusion coefficient (*see* Diffusivity)
Diffusion in liquids (*see* Liquid diffusion)
Diffusion in metals, 148-149, 181, 185-186
Diffusion with phase change, 243
Diffusion-thermo effect, 136-138
Diffusivity (mass), conversion factor, 23, 462
  definition, 11, 15, 39, 120
  effect of composition, 31-32, 126-127, 130-131
  effective, for gas mixtures, 64-68, 259-261
    for porous solids, 149-150, 159
  estimation of, in gases, 23-34
    in liquids, 127-131
  experimental determination, for gases, 20-21
    for liquids, 124-126, 182-184
    for solids, 147, 180-186
  experimental values, for gases, 10, 22
    for liquid metals, 299
    for liquids, 10, 126
    for solids, 10, 146, 148-149

Diffusivity (mass) (continued)
  Knudsen, 151-152, 154
  molal average, 130
  pressure effect on, 25, 33
  temperature effect on, 32, 128-129
  turbulent (*see* Eddy transport)
  variable in three directions, 193-195
Diffusivity (momentum), 6-7, 18-19
Diffusivity (thermal), 9, 18-19, 127
Digital computation, for boundary-value differential equation, 53-59, 72
  comparison with analog computation, 440
  for integration, 282-286
  for tower design, 393-394
  for two-dimensional diffusion, 176-180
  for unsteady-state diffusion, explicit method, 222-224
    implicit method, 225-229
    slabs in series, 234-236
  See also Numerical methods
Dimensional analysis, Buckingham pi theorem, 111, 275-277
  in equations of change, 108-111
  in natural convection, 275-277
Dimensionless groups, analogous types, 202-203, 239, 269-271
  concentration, 197, 202-203
  friction factor, 158, 272-275
  Froude number, 110
  Grashof number, 270-271, 276-277
  J factors, 271, 274-275, 301-304
  Lewis number, 19
  Nusselt number, 269-271
  Peclet number, 269-271
  Prandtl number, 19, 110-111, 269-271
  Reynolds number, 110-111, 266, 268, 271
  Schmidt number, 19, 32, 110-111, 269-271
  Sherwood number, 269-271
  Stanton number, 269-271
  temperature, 203
Distillation, 93-94, 349, 357-359
  See also Gas absorption processes; Stage processes
Distribution coefficient, in liquids, 316-317
  in solids, 219-220, 234-235, 238-243
Divergence, 83
Drag, 275
Drying, 143, 193
Dufour effect, 136-138
Dyne, 6

Eddy transport, 250-254, 302
  See also Turbulent mass transport; Mass transfer coefficients
Effective diffusivity, for gases, 64-68, 259-261
  for porous solids, 149-150, 159-162
Effectiveness factor, 161-163
Energy flux (*see* Flux; Heat transport)
Enthalpy, 105-107
Enthalpy-composition diagram, 413-414, 427, 430
Equation of continuity (mixtures), analogy with equation of energy change, 101
  for constant concentration, 101
  for constant density, 101
  with energy change, 104-107
  general equation, 99-101
  use of, 102-104, 133-134
  for zero velocity, 102
  See also Diffusion
Equation of continuity (pure fluid), 84-86, 88-90, 108-110
Equation of energy change, analogy with equation of continuity, 101
  for constant density, 92, 101
  for constant pressure, 92
  general equation, 91-92
  for Newtonian fluids, 92
  for solids, 93
  use of, 93-95, 108-111
  for zero velocity, 92
  See also Heat transport; Unsteady-state heat transport
Equation of motion, for all fluids, 86-87
  for Newtonian fluids, 87-88
  use of, 88-90, 108-111
  See also Momentum transport
Equations of change, 78-79, 108-111
  See also Equation of continuity; Equation of energy change; Equation of motion
Equilibrium data (*see* Gas-liquid equilibrium; Liquid-liquid equilibrium; Vapor-liquid equilibrium)
Equimolar counterdiffusion, 14-18, 40, 102-104
Equivalent film, 237-238, 300
Error function, 214-215
Euler formulas, 201
Euler method, 52
Euler's equation, 111-112
Evaporation, of drops, 20
  of liquid in a tube, 20, 40-41

Evaporation (continued)
  in packed towers, 395
  from a pan, 42-43
Explicit method, 216-224, 229-238
Extraction, 316-317
  See also Liquid-liquid extraction processes
Eyring's theory, 127-128, 130

Falling film (*see* Wetted-wall tower)
Fick's law, forms of, 97-99
  steady-state, 10-11, 13-18
  unsteady-state, 102, 193-195
Film theory, for chemical reaction, 333-340, 342-343
  for mass transfer coefficients, 254, 299-301, 331-332
Flat plate, analogies for, 275
  mass transfer coefficient for, 287, 289-290, 296-298, 302-304
Fluidized bed, 296
Flux (energy), 5, 8-10, 91-92, 104-107
Flux (mass), convective plus diffusion, 13-14, 35-39, 99-101
  conversion factors, 463
  definition of, 2, 96-98
  diffusion, 10-11, 13-14
  ratio factor, 152-155
  ratios affected by, heat balance, 43-44, 80, 357-359
    solubility barrier, 40-41, 44, 79, 120-122
    stoichiometry, 43-46, 64-68, 79
  tabulation of, 37, 96, 463
  types of, 107-108
Force constants, 25-28, 31
Forced convection (*see* Laminar flow mass transport; Turbulent mass transport)
Forced diffusion, 108
Form drag, 275
Fourier series, 166-168, 196-202
Fourier's law, 8-10, 18, 92, 105
Fraction of change, 202-209
Friction factor, 158, 272-275
Froude number, 110
Fuller *et al.* method, 32-34
Fundamental constants, 460-463

Gas absorption processes, 348-349
  design methods, for dilute gases, 388-393
    using film coefficients, 369-380, 397-401
    using overall coefficients, 380-385,

Gas absorption processes (continued)
    399-401
    using transfer units, 385-392, 396-401
  distillation, 357-359
  with falling films, 258, 263-267, 286-289, 322-325
  heat effects, 401-402
  interface compositions, 372-378
  operating line, for A through stagnant B, 355-356
    effect of flux ratios, 353, 357-359
    for equimolar counterdiffusion, 356-357
    for general case, 350-353
    slope of, 352, 355, 360-364
  packing mass transfer coefficients, experimental methods, 394-395
    for gas film, 396-401
    for liquid film, 397-401
  for recovery of, ammonia, 364, 389-393, 397-400
    carbon dioxide, 400-401
    sulfur dioxide, 374-379, 382-385, 387
  stripping, of propane, 367
    theory, 349, 353, 363
  in tray and multistage towers, 350, 365-367, 432-434
  See also Countercurrent processes
Gas bubbles, 267-268
Gas law, 15, 460
Gas-liquid equilibrium, acetone-water, 479
  ammonia-water, 478
  discussion of, 314-317
  Henry's law, 315-316, 410, 478
  methanol-water, 479
  sulfur dioxide-water, 314, 479
Gauss method, 55
General transport equation, 2-4
Geometry correction factor, 294-296
Gibbs free energy, 137-138
Gilliland equation, 61-63
Graetz solution, 262
Graphical addition, 422-423
Graphical integration, 279-280, 378-379
Grashof number, 270-271, 276-277
Gravitational conversion factor, 6
Gravitational force, 87-90, 108
Gurney-Lurie charts, 204-206

Heat capacity, 9, 105-107
Heat conduction, 8-10, 91-95, 105-107
Heat generation, 93

Heat transfer coefficients, 291, 293-294
Heat transport, for a binary mixture, 104-107
   with continuity equation, 104-107
   flux, 8-10, 91, 104-107
   Fourier's law, 8, 18, 92, 105
   introduction to, 4-5, 91
   steady-state, 8-10, 93-95, 349-350
   See also Equation of energy change
Height of transfer unit, 385-393, 395-401
Heisler charts, 471-477
Henry's law, 315-316, 410, 478
Heterogeneous reaction (*see* Diffusion with chemical reaction; Porous solids)
High mass transfer rates, 254
Homogeneous chemical reaction (*see* Diffusion with chemical reaction; Solids)
Hsu and Bird method, 64-68
Humidification, 349

Ideal fluid, 111-112
Ideal solution, 316, 410
Implicit method, 224-229
Initial-value differential equation, 52-53
Initial-value problem, 168
Interactions between molecules, 25-32
Interfacial area, in packing, 348, 369-371, 394-395
   in stage processes, 408-409
Interfacial resistance, 237-238, 318-319
Internal energy, 91-92
Interphase mass transport, bulk flow correction factor, 320-325
   concentration profile, 318-319
   equilibrium between phases, 313, 318
   interface composition, 318-325, 372-378
   introduction to, 313
   use of film transfer coefficients, for diffusion of A through stagnant B, 320-325
      for equimolar counterdiffusion, 319-320, 322-324
      for general case, 321-322, 331-332
   use of overall transfer coefficients, 325-333
   use of phase rule, 314-315
   See also Gas absorption processes; Mass transport with chemical reaction; Two-resistance theory
Inverse lever-arm rule, 422-424, 426
Ionic diffusion, 108
Irreversible processes, 136-138
Isotopes, 23

J-factors, 271, 274-275, 301-304

Kinematic viscosity, 6
Kinetic energy, 91
Kinetic theory, for liquids, 119, 128
   for low density gases, 23-32, 38-39
Knudsen, diffusion, 151-159
   number, 152, 154
Kremser equations, 434

Laminar flow, 5-8, 88-90, 158
   See also Equation of motion
Laminar flow mass transport, analogy with heat transport, 261
   for cylinder, 292-293
   for flat plate, 289-290, 302-304
   for packed bed, 294-296
   for sphere, 290-291
   for tube, 261-262, 288-289
   for wetted-wall tower, 263-267
Laminar region, 251, 273
Laplace's equation, 166
Laplace transform, method, 210-215, 241-242, 265
   table, 211
Laplacian, 83
Leaching, 349
Le Bas molar volume, 29-32
Lennard-Jones function, 25
Leveque equation, 262, 288-289
Lewis number, 19
Liebmann method, 170-180
Liquid diffusion, description of, 12-13, 119-120
   effect of flux ratio, 120-124
   equation of change, 99-102
   molecular transport, of A through stagnant B, 120-122
   equimolar counterdiffusion, 124
   general equation, 119-120
   multicomponent, 130-131
   unsteady-state, 99-102, 196-201
   See also Diffusion with chemical reaction
Liquid-liquid equilibrium, acetic acid-water-isopropyl ether, 416-418, 480
   discussion, 315-317, 414-416
   nicotine-water-petroleum, 359-360
   phase diagrams, 414-419
   styrene-ethyl benzene-diethylene glycol, 419, 480
Liquid-liquid extraction processes, for acetic acid, 317
   for nicotine, 359-360
   operating lines, 350-357

492 Index

Liquid-liquid extraction processes (continued)
   See also Countercurrent processes; Stage processes
Liquid metal mass transport, mass transfer coefficients, 292, 298-299
   physical properties, 6, 299
Ln mean drive force, 41-42, 121-122, 277-279, 389-393

Mass diffusivity (see Diffusivity)
Mass flux (see Flux)
Mass transfer coefficients, conversion factors for, 257, 269-271, 463
   correlations, for cylinder, 292-296
      for flat plate, 287, 289-290, 296-298, 302-304
      for fluidized bed, 296
      geometry correction factor, 294-296
      for liquid metals, 292, 298-299
      for packed bed of spheres, 275, 293-298
      for packed tower, 396-398
      for sphere, 290-292, 296-298
      for tube, 261-262, 287-289
      for wetted-wall tower, 263-267, 288-289
   definitions of, 80, 202, 253-256
   experimental determination, 286-287
   film coefficients, for A diffusing through stagnant B, 255-258, 320-325
      for equimolar counterdiffusion, 255-257, 319-320, 322-324
      for general case, 253-255, 321-322, 331-332
      for multicomponent diffusion, 259-261
      tabulation of, 257, 271
   flux correction factor, 254-257, 320-325
   overall two-phase coefficients, 325-331
   See also Gas absorption processes; Interphase mass transport; Mass transport with chemical reaction; Turbulent mass transport
Mass transport (see Diffusion; Mass transfer coefficients)
Mass transport with chemical reaction, coefficient for, 335-338, 341-343
   introduction to two-phase reactions, 332-333
   slow first-order reaction, 333-336
   slow second-order reaction, 342-343
   very fast second-order reaction, film

Mass transport with chemical reaction (continued)
      theory, 336-340, 342
      penetration theory, 340-343
Matrix, 54-55, 57, 59, 176-180, 225, 228-229
Mean driving force (see Ln mean driving force)
Mean free path, 24-25, 151-154
Membranes, 124, 149
Microscopic reversibility, 136-137
Mixtures, concentrations, 37
   equations of change, 99-107
   fluxes, 37, 96
   velocities, 37
   See also Diffusion; Diffusion with chemical reaction; Mass transfer coefficients; Multicomponent transport
Molar volume, 29-32, 128
Molecular diameter, 24-25
Molecular diffusion (see Diffusion)
Molecular diffusivity (see Diffusivity)
Molecular velocity, 23-25
Momentum transport, with energy transport, 19
   equation of motion, 86-88
   flux, 5-8, 86-87
   molecular, 5-8
   for Newtonian fluids, 87-88
   Newton's law, 5-8, 18, 87-88
   in pipes, 88-90
   steady-state, 7-8, 88-90
   See also Equation of motion
Multicomponent transport, of component A in a stagnant mixture, 59-61
   of components A and B in stagnant C, 61-63, 67-68
   effective diffusivity, 64-68, 259-261
   equimolar, 63-64
   general case, 64-68
   in liquids, 130-131, 138
   Stefan-Maxwell equations, 38-39, 59-61, 64
   in transition region, 156-158
   turbulent, 259-261

Natural convection, 270, 276-277, 290-291
Navier-Stokes equation, 88
Newman's method, 207-209
Newton's law, 5-8, 18, 87-88
Notation, for analogous heat and mass transport, 202-203, 239

Notation (continued)
  for concentrations, 37
  for dimensionless numbers, 271
  for mass fluxes, 37, 96
  for mass transfer coefficients, 257
  for velocities, 37
  See also Notation at end of each chapter
Number of transfer units, 385-392
Numerical methods, for differential equations, 50-59
  for integration, 281-286
  similarity for heat and mass transport, 239
  for simultaneous linear equations, 54-59, 225-229
  for two-dimensional transport, 168-180
  for unsteady-state transport, in flat slab, 216-229
    for resistance between slabs, 237-238
    in slabs in series, 229-236
  See also Digital computation; Unsteady-state mass transport
Nusselt number, 269-271

Ohm's law, 2
Onsager's relations, 136-138
Orthogonal functions, 168, 196, 201
Osmotic diffusion, 157
Othmer-Thakar equation, 128-129

Packed bed, catalyst pellet, 159, 161-163
  mass transfer coefficients, 275, 293-298
  surface area, 293-294
  velocity, 268, 294
  void fraction, 268, 293-294
Packed towers (*see* Gas absorption processes)
Packing, Berl saddles, 396-398
  Raschig rings, 396-401
Parallel plates, 5-8, 10
Partial molal enthalpy, 105-107
Peclet number, 269-271
Penetration theory, 254, 266, 301, 331-332, 336, 340-343
Perkins and Geankoplis equation, 130-131
Permeability, 146-149
  See also Solids
Phase diagram (*see* Liquid-liquid equilibrium)
Phase rule, 314-315, 409-410, 414
Pipe (*see* Tube)
Pohlhausen method, 304
Poise, 6, 461

Poiseuille flow, 158-159
Polar molecule, 31
Pores (*see* Porous solids; Solids)
Porous solids, diffusion dependent on structure, 149-163
  diffusion of gases in pores, flux ratios, 156-157
    Knudsen, 151-159
    molecular, 151-156
    multicomponent, 156-158
    transition, 151-158
  diffusion independent of structure, 143-149
  diffusion of liquids, 149-150
  forced flow of gases, 158-159
  reaction in pores, effectiveness factor, 161-163, 187
  overall reaction rate, 163
  void fraction, 149-150
  See also Solids
Potential energy function, 25-26, 31
Poundal, 6
Prandtl number, 19, 110-111, 269-271
Pressure diffusion, 108
Pressure drop, 89-90, 158-159
Pressure force, 87, 90

Quadratic law, 164, 184
Quasi steady-state, 20, 124, 164

Radiotracers, 186
Random diffusion, 4-6, 11-13, 19
Raoult's law, 316, 410-412
Reaction (*see* Diffusion with chemical reaction)
Reaction zone, 337, 339-341
Reciprocal relations, 136-138
Reddy and Doraiswamy equations, 129-130
Relaxation method, 168-180
Residuals, 171-180
Resistance, definition, 2
  interfacial, 237-238, 242-243, 318-319
  in two-phase system, 326-331, 400-401
Reverse diffusion, 64
Reynolds, analogy, 272-274
  number, 110-111, 266, 268, 271
Rubber, diffusion in, 143-148

Scalar, 81-84
Schmidt method, 217
Schmidt number, 19, 32, 110-111, 269-271
Self-diffusion, 127, 186
Semiinfinite solid, 202-203, 212-215

494 Index

Separation of variables, 166-168, 195-202
Shear stress, 5, 87
Sherwood number, 269-271
Sigma point, 421-428
Simpson's method, 281-286
Simultaneous energy and mass transport, 104-107
Simultaneous linear equations, 54-59, 225-229
Skin friction, 275
Solids, diffusion dependent on structure, 149-159
　diffusion in metals, 10-11, 181, 185-186
　diffusion with reaction, 135, 163-165
　dissolution to a fluid, 113-114, 277-280, 282-286
　permeability, classification of, 148-149
　　equation for, 147-149
　　relation to diffusivity, 147-148
　　for rubber, 144-148
　　tabulation of, 146
　steady-state diffusion in two dimensions, 166-180
　structure independent diffusion, 143-149
　See also Porous solids; Unsteady-state mass transport
Solubility barrier, 40-41, 44, 79, 120-122
Solubility of gases in liquids (see Gas-liquid equilibrium)
Solubility of gases in solids, 144-149
Solvent extraction (see Liquid-liquid extraction processes)
Soret effect, 108, 136-138
Sphere, catalyst pellet, 161-163
　drag, 127, 275
　mass transfer from, 290-292, 296-298
　unsteady-state diffusion in, 202-203, 206
　See also Packed bed
Stability, 54, 217, 224-225
Stage processes, countercurrent multistage, analytical solution, 432-434
　overall balance, 426-428
　stage-to-stage calculation, 428-430
　enthalpy-concentration diagram, 413-414, 427, 430
　introduction to, 408-409
　multiple stage, 424-426
　single stage, 421-424
　theoretical stage, 366, 409
　tray tower, 350, 432-434
Stagnant film, 105, 254, 299-301, 333, 340
Stagnant fluid, 35

Stanton number, 269-271
Steady-state mass transport, definition, 2-3, 79
　in gases, 11, 16-18, 39-45, 102-104
　in liquids, 120-124
　in solids, 144-148, 180-181
　in two dimensions, 166-180
　See also Diffusion; Solids; Countercurrent processes
Stefan-Maxwell equations, 38-39, 59-61, 64
Stockmayer potential, 31
Stokes-Einstein equation, 127
Streamline flow (see Laminar flow)
Stress tensor, 87
Stripping (see Gas absorption processes)
Subroutine, 55-56, 228, 285-286
Substantial derivative, 81, 85-86
Surface flow, 148, 159
Surface renewal theory, 301
Symbols (see Notation)

Tarnishing of silver, 163-164
Taylor and Prandtl analogy, 273-274
Tensor, 87
Theoretical tray, 366
Thermal conductivity, 8-10
Thermal diffusion, 108, 136-138
Thermodynamics of irreversible processes, chemical potential, 138, 165, 241-242, 317
　entropy balance, 137
　Gibbs free energy, 137-138
　linear laws, 136-138
　microscopic reversibility, 136-137
　Onsager's relations, 136-138
Tie line, 414-416, 424
Time derivatives, 80-81
Tortuosity, 149-150, 158
Transfer coefficients (see Mass transfer coefficients)
Transient diffusion (see Unsteady-state mass transport)
Transition region diffusion, 151-158
Transport phenomena, 1-4
Transport property conversion factors, 462-463
Transport properties (see Diffusivity; Viscosity; Thermal conductivity)
Tray tower, 350, 365-367, 432-434
Tridiagonal matrix, 59, 229
Truncation error, 170
Tube, energy and momentum transport, 93-94

Tube (continued)
  laminar flow, 88-90, 158
  mass transfer coefficients, 261-262, 287-289
  turbulent mass transport, 250-252, 289
  velocity, 88-90
Turbulent heat transport, for cylinder, 293
  for flat plate, 275, 290
  for packed bed, 294
  for sphere, 291
  for tube, 275
Turbulent mass transport, description of, 12-13, 37, 250-254, 299-301
  theories of, boundary layer theory, 254, 291, 302-304, 331-333
    eddy diffusivity theory, 251-253, 302
    film theory, 254, 299-301, 331-332, 333-340, 342-343
    J factor model, 274-275, 301-304
    penetration theory, 254, 266, 301, 331-332, 336, 340-343
  See also Mass transfer coefficients
Turbulent momentum transport, 250-252, 270-275
Two-resistance theory, gas-phase resistance controlled, 327
  general case, 325-328
  liquid-phase resistance controlled, 327

Unsteady-state heat transport, charts, 202-206, 471-477
  equation, 92-95, 202, 239
Unsteady-state mass transport, analytical methods, Laplace transform, 210-215, 239-241
  separation of variables, 195-202
  basic equations, 99-102, 193-195
  charts for, 202-206, 209, 219, 471-477
  with chemical reaction, 134-135, 163-165, 333, 340-343
  in composite media, slabs with interface resistance, 242-243
  slabs in series, 239-242
  definition, 79
  for determination of diffusivity, 182-186
  numerical methods, for cylinder, 246
    explicit for slab, 216-224
    implicit for slab, 224-229
    for slabs with interface resistance, 237-238, 242-243

Unsteady-state mass transport (continued)
  for slabs in series, 229-236
  with phase change, 243
  in slabs, 201-209
  in two or more directions, 207-209
  See also Solids; Equation of continu for mixtures
Unsteady-state momentum transport, 8
Unsteady-state transport equation, 2-4

Vapor-liquid equilibrium, benzene-tolue 410-412
  enthalpy-concentration plot, 413-41, 427, 430
  ethanol-water, 414-415, 481
  Henry's law, 315-316, 410
  Hexane-octane, 435
  Raoult's law, 316, 410-412
Vapor pressure, 410-412, 464
Vaporization from a solid, 256
Varying cross-sectional area, 17-18, 144
Vector, 81-84
Velocity, average, 112, 268
  boundary layer, 302-304
  discussion of, 5-6, 23, 95-97, 268
  individual species, 13, 37
  interstitial, 268
  mass average, 37, 95-96
  maximum, 90, 112
  molar average, 13, 23, 35-37, 95-97
  molecular, 23-25
  parabolic profile, 90, 113-114, 262
  superficial, 268, 294
  tabulation of, 37
Viscosity, conversion factors for, 6
  definition of, 5
  prediction of, 127
  table of, 6, 465-470
Viscous dissipation term, 92, 110
Void fraction, 149-150, 268, 293-294
von Kármán integral method, 304

Water, density, 464
  vapor pressure, 464
  viscosity, 470
Wetted-wall tower, 258, 263-267, 286-28 322-325
Wilke-Chang correlation, 128-129